Theory of Nonclassical States of Light

Theory of Nonclassical States of Light

Edited by

V. V. Dodonov

and

V. I. Man'ko

CRC Press is an imprint of the
Taylor & Francis Group, an **informa** business

CRC Press
Taylor & Francis Group
6000 Broken Sound Parkway NW, Suite 300
Boca Raton, FL 33487-2742

First issued in paperback 2019

ISBN-13: 978-0-415-28413-4 (hbk)
ISBN-13: 978-0-367-39535-3 (pbk)

British Library Cataloguing in Publication Data
A catalogue record for this book is available from the British Library

Library of Congress cataloging in Publication Data
A catalog record for this book has been requested

Visit the Taylor & Francis Web site at
http://www.taylorandfrancis.com

and the CRC Press Web site at
http://www.crcpress.com

Contents

Preface

The term "nonclassical states" appeared in the language of physics in the early 1960s. Roughly speaking, it is used to distinguish quantum states produced in the "usual" sources of light, like lasers or lamps (which give states close to the coherent and thermal ones, respectively), from other conceivable states, which require a more sophisticated apparatus for their production. Actually, such states had been considered sporadically since the first years of quantum mechanics. However, only after the remarkable experiments carried out at the end of 1980s, did they attract the attention of many theoreticians and experimentalists working in different fields of quantum optics. As a result, the total number of related publications up to the end of the 20th century is estimated at several thousand. Almost every modern textbook or monograph on quantum optics now includes sections devoted at least to the best known examples of nonclassical states, such as the *squeezed* and *"Schrödinger's cat"* states. There exist several review articles, special issues of different journals, and proceedings of conferences devoted to the squeezed and some other nonclassical states of light and their applications. Nonetheless, these reviews do not exhaust all aspects of the problem.

Our intention is to give the modern view of the subject, together with its history. Initially, we intended to describe in detail all known kinds of "nonclassical states" and their interrelations. However, we quickly recognized that it is impossible to put all the available material on nonclassical states in a mere 400 pages. For this reason, we have decided to confine ourselves to the detailed description of the most important types of nonclassical states, such as the *squeezed* and *even/odd* (*"Schrödinger's cat"*) states, which are considered in the majority of publications on quantum optics, giving some fresh insight into the subject and discussing the available generalizations (for example, the "nonunitary approach" to squeezed states or the "classical-like" formulation of quantum mechanics). The third class of states combines the *binomial states* and their generalizations, which are representatives of "intermediate states" in quantum mechanics and quantum optics. In addition we consider in detail such topics as the evolution of nonclassical states in linear and nonlinear (Kerr) media, including the results related to the problems of relaxation and decoherence, and the interaction between the field and atomic degrees of freedom in the framework of the Jaynes–Cummings model and its generalizations. However, details of several important topics have been left aside. They include, e.g., methods of generation and detection of nonclassical states (known nowadays under the names *quantum state engineering* and *quantum state tomography*), specific features of nonclassical states in various traps and in Bose–Einstein condensates, the problem of *phase* in quantum mechanics (although many relevant results based on the most popular approaches are given in Chapters 4–6), entanglement, and some others. Instead, we have confined ourselves to giving references to the sources, where the related subjects are discussed in detail.

This flaw is compensated partially in the introductory chapter, which contains a brief description of the majority (we hope) of known families of "nonclassical" states (including "relativistic", "super", "para", "deformed", etc.) and their applications discussed during the 20th century, not only in quantum optics but also in other areas, from condensed matter to cosmology. Many of these states now look like toys or artificial constructions without any physical sense. One should remember, however, that this is the typical fate of "nonclassical states". For example, the squeezed states were studied (under different names) in various publications as far back as the 1920s and 1950s, before they acquired "respectable" status in the 80s. The same thing happened with the even/odd states: although introduced in 1974, they became the subject of intensive studies only about 15 years later. It is difficult to guess which states will be "popular" in 10 or 20 years' time.

This book would not have appeared unless we had the encouragement and support of many people and institutes. In this connection, we would like to thank, in the first place, our colleagues from the institutes of our permanent affiliations: Lebedev Physics Institute of Russian Academy of Sciences and Moscow Institute of Physics and Technology. Personally, we wish to thank Drs. V.A. Isakov and M.A. Man'ko. Also, we thank our colleagues from the Physics Departments of the Federal University of São Carlos (Brazil), the University of Naples (Italy) and the University of Lisbon (Portugal), because the book was prepared during the period when we worked as visiting professors at these institutes, using all their facilities. We are extremely grateful to Prof. S.S. Mizrahi for his fruitful collaboration and invaluable help. Checking several thousand references, one of us enjoyed the excellent facilities and rich magazine fund of the library of the Institute of Physics of São Carlos of the University of São Paulo (IFSC/USP). The support of the Brazilian agencies CNPq and FAPESP was also very important.

Thinking about this book several years ago, we supposed that one of its authors would be Dr. Andrei Vinogradov from Lebedev Institute, who made a significant and recognized contribution to the theory of nonclassical states. Unfortunately, his premature death in 1997 interfered with our plans. We would like to dedicate this book to his memory.

V.V.Dodonov, V.I.Man'ko São Carlos – Moscow, August 2002

List of authors

Sergei M. Chumakov
Departamento de Física, Universidad de Guadalajara,
Revolución 1500, 44420, Guadalajara, Jalisco, México

Victor V. Dodonov
Physics Department, Faculty of Aeromechanics and Flying Technics,
Moscow Institute of Physics and Technology
Lebedev Physics Institute of Russian Academy of Sciences, Moscow
Departamento de Física, Universidade Federal de São Carlos,
Via Washington Luiz, km 235, 13565-905 São Carlos, SP, Brasil

Andrei B. Klimov
Departamento de Física, Universidad de Guadalajara,
Revolución 1500, 44420, Guadalajara, Jalisco, México

Maciej Kozierowski
Institute of Physics, Adam Mickiewicz University,
Umultowska 85, 61-614 Poznań, Poland

Vladimir I. Man'ko
Lebedev Physics Institute of Russian Academy of Sciences,
Leninsky Prospect 53, 117924 Moscow, Russia

Ryszard Tanaś
Nonlinear Optics Division, Institute of Physics,
Adam Mickiewicz University, Umultowska 85, 61-614 Poznań, Poland

Antonio Vidiella Barranco
Depto. de Eletrónica Quântica, Instituto de Física Gleb Wataghin,
Universidade Estadual de Campinas, Brasil

Alfred Wünsche
Institut für Physik, Humboldt-Universität Berlin,
Invalidenstr. 110, 10115 Berlin, Germany

Chapter 1

"Nonclassical" states in quantum physics: brief historical review

V.V. Dodonov and **V.I. Man'ko**

1 Introduction

In modern monographs on quantum optics [1] the "nonclassical" states of the electromagnetic field are defined as those for which the so called Glauber–Sudarshan P-function (used by Glauber [2] to represent only thermal states and introduced by Sudarshan [3] for arbitrary density matrices) "is less well behaved than a probability density", e.g., takes on negative values and becomes more singular than a delta function. Actually, this concept has its origin in the paper by Glauber [4], where one can find the following sentence: *"If the singularities of $P(\alpha)$ are of types stronger than those of delta function, e.g., derivatives of delta function, the field represented will have no classical analog."* In this sense, the thermal (chaotic) fields are classical, since their P-functions are usual positive probability distributions. The states of such fields, being quantum mixtures, are described in terms of the density matrices. An example of the "well-behaved" *pure* quantum state is the Klauder–Glauber–Sudarshan coherent state [2–5]

$$|\alpha\rangle = \exp\left(-|\alpha|^2/2\right) \sum_{n=0}^{\infty} \frac{\alpha^n}{\sqrt{n!}} |n\rangle, \tag{1}$$

where the Fock state $|n\rangle$ [6] is the eigenstate of the number operator $\hat{a}^\dagger \hat{a}$, and $\hat{a}^\dagger, \hat{a}$ are the bosonic creation and annihilation operators satisfying the canonical commutation relation

$$[\hat{a}, \hat{a}^\dagger] = 1. \tag{2}$$

The coherent state is the eigenstate of the annihilation operator $\hat{a}$:

$$\hat{a}|\alpha\rangle = \alpha|\alpha\rangle \tag{3}$$

(where α may be an arbitrary complex number), and it can be obtained from the vacuum state $|0\rangle$ by means of the unitary *displacement operator*,

$$|\alpha\rangle = \hat{D}(\alpha)|0\rangle, \qquad \hat{D}(\alpha) = \exp\left(\alpha\hat{a}^\dagger - \alpha^*\hat{a}\right). \tag{4}$$

Introducing the (dimensionless) *quadrature components* as

$$\hat{x} = \left(\hat{a} + \hat{a}^{\dagger}\right)/\sqrt{2}, \qquad \hat{p} = \left(\hat{a} - \hat{a}^{\dagger}\right)/(i\sqrt{2}), \tag{5}$$

one can write the "coordinate" representation of the coherent state (1)

$$\langle x|\alpha\rangle = \pi^{-1/4}\exp\left(-\tfrac{1}{2}x^2 + \sqrt{2}x\alpha - \tfrac{1}{2}\alpha^2 - \tfrac{1}{2}|\alpha|^2\right). \tag{6}$$

The dynamics of the state (6) is also very simple in the case of harmonic oscillator with a constant unit frequency: one should simply replace the coefficient α by $\alpha\exp(-it)$. Such "nonspreading" wave packets were constructed for the first time in 1926 by Schrödinger [7]. They possess several remarkable properties. In particular, the *variances* of the "coordinate" and "momentum" operators, $\sigma_x = \langle\hat{x}^2\rangle - \langle\hat{x}\rangle^2$, $\sigma_p = \langle\hat{p}^2\rangle - \langle\hat{p}\rangle^2$, have equal values, $\sigma_x = \sigma_p = 1/2$, so their product assumes the minimal value permitted by the Heisenberg uncertainty relation, $\sigma_x\sigma_p = 1/4$. The average values $\langle\hat{x}\rangle$ and $\langle\hat{p}\rangle$ follow the classical trajectory. Gaussian packets describing the motion of a free particle and a charged particle in uniform constant electric and magnetic fields were considered by Darwin [8].

Prototypes of state (1) can be found in the papers by Fock (e.g., [9]). The eigenstates of the non-Hermitian annihilation operator $\hat{a}$ in forms (1) and (6) were considered explicitly in 1951 by Iwata [10]. The displacement operator (4) was actually used by Feynman and Glauber in 1951 [11, 12] in their studies of quantum transitions caused by the classical currents. This operator was also intensively exploited in 1951–1953 by Friedrichs, who called the states analogous to $\hat{D}(\alpha)|0\rangle$ "modified vacuum states" and wrote the expansions similar to (1) in a series of papers combined in the book [13]. Schwinger [14] used (1) as some auxiliary states, permitting the simplification of calculations (rather detailed discussion of such states can be found also in Schwinger's book of 1970 [15], whose main part was written, according to the author's preface, as far back as 1955). Several other authors [5, 16–21] also studied properties of such states from different points of view (mainly in the framework of the quantum field theory) during the decade preceding 1963.

However, only after publications by Glauber, Sudarshan and Klauder [2–4,22–24], did the "coherent states" become widely known and intensively used by physicists, because they provided an effective and adequite tool for studying different physical problems (in particular, the problem of coherence of electromagnetic fields), based on two main properties of these states: the (over)completeness (a continuous decomposition of the unit operator, called "resolution of unity") [2–5]

$$\pi^{-1}\int|\alpha\rangle\langle\alpha|\,d\,\mathrm{Re}\alpha\,d\,\mathrm{Im}\alpha = \hat{I} \tag{7}$$

and the existence of the "diagonal" representation of arbitrary statistical operator $\hat{\rho}$ over the coherent state basis [3, 4, 23, 24]

$$\hat{\rho} = \int P(\alpha)|\alpha\rangle\langle\alpha|\,d\,\mathrm{Re}\alpha\,d\,\mathrm{Im}\alpha, \qquad \int P(\alpha)\,d\,\mathrm{Re}\alpha\,d\,\mathrm{Im}\alpha = 1. \tag{8}$$

The real function of the complex argument $P(\alpha)$ in (8) is frequently called [1] "Glauber–Sudarshan P-distribution".

The name *coherent states* appeared for the first time in Glauber's papers [2] and [4]. After that, the first papers published in journals with the words "coherent states" in their titles were [24–27]. A detailed bibliography of the first years of the "coherent states era" can be found in [28, 29]. In this connection it seems appropriate to mention two papers which apparently did not receive much attention in that time. The first one [30] was published by Anderson in 1958. In that study he used the name "coherent excited states" for the states obtained by the action of the operator (in the original notation) $\rho_Q = \sum_n \exp(i\mathbf{Q} \cdot \mathbf{r}_n)$ on the BCS ground state in the theory of superconductivity. Here Q is the momentum of the excited pair of electrons. Rewriting $\mathbf{Q}$ and $\mathbf{r}_n$ in terms of the operators $\hat{a}$ and $\hat{a}^\dagger$ according to (5), one can see a formal coincidence of operators $\hat{D}(\alpha)$ and each term in the sum defining ρ_Q. The second paper was published by Senitzky [31] about six weeks before Glauber's article [2]. In that paper, the state (1) was explicitly considered, and it was written that "for a given energy expectation value the coherence is a maximum when the oscillator is in the state (1)" (although the definition of "coherence" was different from that given later in [2, 4]).

Coherent states have been always considered as "the most classical" ones (among the *pure* quantum states, of course): see, e.g., [32]. In particular, they are "classical" in the sense of the definition given above, since the P-distribution for the state $|\beta\rangle$ is, obviously, the delta function, $P_\beta(\alpha) = \pi\delta(\alpha-\beta)$. On the other hand, these states lie "on the border" of the set of "classical" states, because the delta function is the most singular distribution admissible in the classical theory. Consequently, it is enough to make slight modifications in each of the definitions (1), (3), (4), and (6) to arrive at various families of states which will be "nonclassical". For this reason, many classes of states which are nowadays labelled as "nonclassical" appeared in the literature as some kind of *"generalized coherent states"*. At least five different kinds of generalizations exist.

(I). One can generalize equations (1) or (6), choosing different sets of coefficients $\{c_n\}$ in the expansion

$$|\psi\rangle = \sum_n c_n |n\rangle \tag{9}$$

or writing different explicit wave functions in other (continuous) representations (coordinate, momentum, Wigner–Weyl, etc.). Of course, some additional principles or restrictions should be imposed, e.g., the requirement of "temporal stability", which means that initial states characterized by some label α are transformed with the course of time to the states of the same nature (analytical form), but labeled with a time-dependent parameter $\alpha(t)$.

(II). One can replace the exponential form of the operator $\hat{D}(\alpha)$ in equation (4) by some other operator function, using also some other set of operators $\hat{A}_k$ instead of operator $\hat{a}$ (e.g., making some kinds of "deformations" of the fundamental commutation relations like (2)), and taking the "initial" state $|\varphi\rangle$ different from the vacuum state. This line leads to the *"displacement-operator coherent states"* (DOCS) of the form (λ is some continuous parameter)

$$|\psi_\lambda\rangle = f(\hat{A}_k, \lambda)|\varphi\rangle. \tag{10}$$

(III). One can look for the "continuous" families of eigenstates of the new operators $\hat{A}_k$, trying to find solutions to the equation

$$\hat{A}_k|\psi_\mu\rangle = \mu|\psi_\mu\rangle. \tag{11}$$

Such states can be called "*annihilation-operator coherent states*" (AOCS).

(IV). One can try to "minimize" the generalized Heisenberg uncertainty relation for Hermitian operators $\hat{A}$ and $\hat{B}$ different from $\hat{x}$ and $\hat{p}$,

$$\Delta A \Delta B \geq \tfrac{1}{2}\,|\langle[\hat{A},\hat{B}]\rangle|, \tag{12}$$

looking, for instance, for the "*minimum-uncertainty coherent states*" (MUCS), for which (12) becomes the equality. The abbreviations DOCS, AOCS and MUCS were proposed in the paper by Nieto and Simmons [33].

(V). One can look for the "wave packets" which are localized near the classical trajectory or give the same evolution of the mean values as predicted by the equations of classical mechanics (for nonquadratic Hamiltonians).

Actually, all these approaches have been used for several decades of studies of "generalized coherent states". The goal of this chapter is to give a review of concrete families of states found in the framework of each method.

The "nonclassicality" of the Fock states and their finite superpositions was mentioned in [34]. A simple criterion of "nonclassicality" can be established if one considers a generic Gaussian wave packet with *unequal* variances of two quadratures, whose P-function reads [35] (in the special case of the statistically uncorrelated quadrature components)

$$P_G(\alpha) = \mathcal{N}\exp\left[-\frac{(\mathrm{Re}\alpha - a)^2}{\sigma_x - 1/2} - \frac{(\mathrm{Im}\alpha - b)^2}{\sigma_p - 1/2}\right] \tag{13}$$

(a and b give the position of the centre of the distribution in the α-plane; the symbol $\mathcal{N}$ hereafter is used for the normalization factor). Since the function (13) exists as a normalizable distribution only for $\sigma_x \geq 1/2$ and $\sigma_p \geq 1/2$, the states possessing one of the quadrature component variances less than $1/2$ are *nonclassical*. This statement holds for any (not only Gaussian) state. Indeed, one can easily express the quadrature variance in terms of the P-function:

$$\sigma_x = \frac{1}{2}\int P(\alpha)\left[\left(\alpha+\alpha^* - \langle\hat{a}+\hat{a}^\dagger\rangle\right)^2 + 1\right] d\,\mathrm{Re}\alpha\, d\,\mathrm{Im}\alpha.$$

If $\sigma_x < 1/2$, then function $P(\alpha)$ must assume negative values, thus it cannot be interpreted as a classical probability [36]. Later the states with $\sigma_x < 1/2$ or $\sigma_p < 1/2$ were named *squeezed states* [37–39].

Another example of a "nonclassical" state is a superposition of two different coherent states [40]. As a matter of fact, *all pure states*, excepting the coherent states, are "nonclassical", both from the point of view of their physical contents [41] and the formal definition in terms of the P-function given above [42]. However, speaking of "nonclassical" states, people usually have in mind not an arbitrary pure state, but members of some families of quantum states possessing more or less useful or distinctive properties. The aim of this chapter is to provide a brief review of the majority of known families, not only in quantum optics, but in other areas of physics, too, trying to follow the historical order of their appearance.

According to the *Web of Science* electronic database of journal articles, the combination of words "nonclassical states" appeared for the first time in the titles of the papers by Helstrom [43], Hillery [44], and Mandel [45]. The first three papers containing the combination of words "nonclassical effects" were published by Loudon [46], Zubairy [47], and Lugiato and Strini [48]. "Nonclassical light" was the subject of the first papers by Schubert [49], Yuen [50], Janszky *et al.* [51], Huttner and Ben-Aryeh [52], and Gea-Banacloche [53]. The first conference under the title "Squeezed and non-classical light" seems to be held in 1988 [54, 55]. To supply the complete list of publications in this area seems impossible due to their immense number.

According to the search in *Web of Science* performed in March 2002, the total number of articles containing in their titles, abstracts, or keywords the combination "coherent state(s)" was equal to 3964, including 303 articles published in 1999, 352 in 2000 and 341 in 2001. The total number of articles containing the combination "squeezed state(s)" was 1430 (of these, 92 papers were published in 1999, 104 in 2000 and 98 in 2001). Having added the word "squeezing", we obtained a total of 4100. Although not all of these publications are related to quantum mechanics or quantum optics, the general picture is clear.

The words "nonclassical state(s)" can be met in titles, abstracts, or keywords of 167 papers (19 in 1999, 18 in 2000 and 21 in 2001), and "nonclassical effects" (in optics) in 92 papers (7 in 1999, 6 in 2000 and 8 in 2001). "Nonclassical light" was considered in 97 papers (11 in 1999, 8 in 2000 and 9 in 2001) and "sub-Poissonian light" in 89 papers (13 in 1999, 7 in 2000 and 4 in 2001). 331 papers (38 in 1999, 42 in 2000 and 48 in 2001) were devoted to "cat states" (less than 10% of them were related to animals), 391 to "Schrödinger cat(s)" (50 in 1999, 48 in 2000 and 55 in 2001), 167 to "generalized coherent states" (15 in 1999, 11 in 2000 and 11 in 2001), 149 to "two-photon coherent states" (8 in 1999, 10 in 2000 and 6 in 2001), and so on.

We concentrate on papers introducing *new types of quantum states* and studying their properties, omitting many publications devoted to *applications* in optics (although we give references related to the use of such states in other areas of physics) and to the methods of generation and experimental schemes. The corresponding references can be found in the known special issues of journals, conference proceedings, reviews and monographs cited in Chapter 2 (see also [56–58]). For example, we pay attention to various families of *phase states* proposed in the literature, but we avoid the numerous references to *phase properties* of nonclassical states and to the problem of phase in quantum mechanics as a whole, since a comprehensive discussion of this problem can be found in many good sources [59–64], and the detailed list of publications up to 1996 was given in the tutorial review [65]. The same policy was used with respect to various superpositions of different quantum states: we tried not to duplicate an extensive reference list given in the review [66]. We tried also to avoid the repetition of references given in the subsequent chapters. Besides, we have included in our review mainly papers written "at the physical level". References to numerous publications devoted to "mathematical" aspects of the theory of (generalized) coherent states can be found in [67,68].

It seems reasonable to divide the history of "nonclassical states" into three periods. The first began in 1927 with Kennard's paper [69] and ended approximately in 1976/77. This border is not only connected to the magic number 50; in that period it was recognized that nonclassical (in particular, *squeezed*) states could play important role in various applications. Also, the terms "squeezed states", "nonclassical states" and "nonclassical effects" (in optics) appeared in 1979-1980.

The border between the second and the third periods can be drawn approximately in 1990/91. Up to that time, several fundamental experiments on generation of squeezed states were performed and the first conferences on "squeezed" and "nonclassical" states were held. Moreover, some "phase transition" in the number of related publications happened in 1991. If in 1990 there were published 57 articles on "coherent states", 40 on "squeezed states", and 2 on "nonclassical states" (according to titles or abstracts), the numbers of publications in 1991 were, respectively, 234, 138 and 9. During the 1990s, the total number of publications related to "nonclassical" states was at the level of 400–500 per year, with the only gap (of about 25%) in 1995.

We have tried to pay more attention to studies made in the first period, because many publications have been undeservedly forgotten since that time. As to the works that appeared in the last quarter of the 20th century, we tried to find the earliest publications in each area and to cite the most recent papers, where other references could be found. No doubt we missed (undeliberately) many interesting publications. Nonetheless, our list seems to be much more representative than "standard citation packages" used by many authors.

2 First 50 years

Actually, people dealt with "the most nonclassical" states (namely, energy eigenstates) from the very beginning of quantum mechanics. However, over the first 30–40 years, quantum states were considered by the majority of researchers (including the founders of quantum mechanics) not as objects possessing some "intrinsic value", but only as a tool for calculating the "physical" quantities, like energy spectra, mean values, transition probabilities, cross-sections, etc., The explanation is very simple: in that period practically nobody thought seriously about the possibility of experimental generation of the given quantum state, and this subject seemed rather scholastic. Only after the invention of the laser did the different types of quantum states begin to acquire some value in themselves, although initially they were also considered mainly as the new useful basis for performing calculations (permitting discrete sums over the Fock states to be replaced by some integrals in the complex plane). On the contrary, modern quantum optics, born simultaneously with the laser, used the language of coherent states from the beginning. Perhaps this circumstance explains why other types of states were considered by the "optical community" as something "nonclassical".

2.1 Childhood of squeezed states

Generic Gaussian wave packets

Historically, the first example of the nonclassical (squeezed) states was given as far back as 1927 (one year after Schrödinger's paper [7]) by Kennard [69], who considered, in particular, the evolution in time of the *generic Gaussian wave packet*

$$\psi(x) = \exp\left(-ax^2 + bx + c\right) \tag{14}$$

of the harmonic oscillator. In this case, the quadrature variances may be arbitrary (they are determined by the real and imaginary parts of the parameter a: see Chapter 3), but they satisfy the Heisenberg inequality $\sigma_x\sigma_p \geq 1/4$.

Two decades later, Gaussian packets in a parabolic potential were studied by Coulson and Rushbrooke [70]. In 1953, Husimi [71] found the following family of solutions to the time-dependent Schrödinger equation for the harmonic oscillator (we assume here $\omega = m = \hbar = 1$):

$$\begin{aligned}\psi(x,t) &= [2\pi\sinh(\beta+it)]^{-1/2}\exp\Big\{-\tfrac{1}{4}\Big[\tanh\left[\tfrac{1}{2}(\beta+it)\right](x+x')^2 \\ &\quad + \coth\left[\tfrac{1}{2}(\beta+it)\right](x-x')^2\Big]\Big\},\end{aligned} \tag{15}$$

where x' and $\beta > 0$ are arbitrary real parameters. He showed that the quadrature variances oscillate with twice the oscillator frequency between the extreme values $\frac{1}{2}\tanh\beta$ and $\frac{1}{2}\coth\beta$. A detailed study of the Gaussian states of the harmonic oscillator was performed by Takahasi [72].

The first appearance of the squeezing operator

An important contribution to the theory of squeezed states was made by Infeld and Plebański in studies [73–75], whose results were summarized in a short article [76]. Plebański introduced the following family of states:

$$|\tilde{\psi}\rangle = \exp\left[i\left(\eta\hat{x} - \xi\hat{p}\right)\right]\exp\left[\frac{i}{2}\log a\left(\hat{x}\hat{p} + \hat{p}\hat{x}\right)\right]|\psi\rangle, \tag{16}$$

where ξ, η, and $a > 0$ are real parameters, and $|\psi\rangle$ is an *arbitrary* initial state. Evidently, the first exponential in the right-hand side of (16) is nothing but the *displacement operator* (4) written in terms of the Hermitian quadrature operators. Its properties were studied in the first article [73]. The second exponential is the special case of the *squeezing operator* (see Eq. (22) below). For the initial *vacuum* state, $|\psi_0\rangle = |0\rangle$, the state $|\tilde{\psi}_0\rangle$ (16) is exactly the squeezed state in modern terminology (cf. the formulae in Chapter 2), whereas choosing other initial states one can obtain various *generalized squeezed states*. In particular, the choice $|\psi_n\rangle = |n\rangle$ results in the family of *squeezed number states*, which were considered in [75, 76]. In the case $a = 1$ (considered in [73]) we arrive at the states known nowadays under the name *displaced number states*. Plebański gave the explicit expressions describing the time evolution of the state (16) for the harmonic oscillator with a constant frequency and proved the completeness of the set of "displaced" number states. Infeld and Plebański [74] performed a detailed study of the properties of the unitary operator $\exp(i\hat{T})$, where $\hat{T}$ is a generic inhomogeneous quadratic form of the canonical operators $\hat{x}$ and $\hat{p}$ with constant c-number coefficients, giving some classification and analyzing various special cases (some special cases of this operator were discussed briefly by Bargmann [18]). However, even before the paper [74], Friedrichs [13] considered the states

$$|\psi\rangle = \mathcal{N}\exp\left(-{}^1\!/_2\hat{\mathbf{A}}^\dagger G\hat{\mathbf{A}}^\dagger\right)|vac\rangle, \tag{17}$$

where $\hat{\mathbf{A}} = (\hat{a}_1, \hat{a}_2, \ldots)$ is the (infinite-dimensional) vector, whose components are the (Bose or Fermi) "annihilation" operators. From the modern point of view, these states are nothing but a special case of the (vacuum) *multimode squeezed states* (see section 3). Unfortunately, the publications cited appeared practically unknown or forgotten over many years.

From "characteristic" to "two-photon" coherent states

In 1966, Miller and Mishkin [77] deformed the defining equation (3), introducing the "characteristic states" as the eigenstates of the operator

$$\hat{b} = u\hat{a} + v\hat{a}^\dagger \tag{18}$$

(for real u and v). In order to preserve the commutation relation $[\hat{b}, \hat{b}^\dagger] = 1$ one should impose the constraint $|u|^2 - |v|^2 = 1$.

Similar states were considered by Lu [78], who called them "new coherent states", and by I. Bialynicki-Birula [79]. The general structure of the wave functions of these states in the coordinate representation is

$$\langle x|\beta\rangle = (\pi w_-^2)^{-1/4}\exp\left[-\frac{w_+x^2}{2w_-} + \frac{\sqrt{2}\beta x}{w_-} - \frac{w_-^*\beta^2}{2w_-} - \frac{|\beta|^2}{2}\right], \tag{19}$$

where $w_\pm = u \pm v$ and $\hat{b}|\beta\rangle = \beta|\beta\rangle$.

Wave function (19) also arises if one looks for states in which the uncertainty product $\sigma_x\sigma_p$ assumes the minimal possible value ${}^1\!/_4$. This approach was used in [32, 80–82]. The variances in the state (19) are given by

$$\sigma_x = \tfrac{1}{2}|u - v|^2, \qquad \sigma_p = \tfrac{1}{2}|u + v|^2 \tag{20}$$

so that for *real* parameters u and v one has $\sigma_x\sigma_p = {}^1\!/_4$ due to the constraint $|u|^2 - |v|^2 = 1$. In this case one can express the "transformed" operator $\hat{b}$ in (18) in terms of the quadrature operators (5) as

$$\hat{b} = \left(\lambda^{-1}\hat{x} + \lambda\hat{p}\right)/\sqrt{2}, \qquad \lambda^{-1} = u + v, \qquad \lambda = u - v. \tag{21}$$

Stoler [81] showed that the "minimum uncertainty states" (MUS) can be obtained from the oscillator ground state by means of the unitary operator

$$\hat{S}(z) = \exp\left[\tfrac{1}{2}\left(z\hat{a}^2 - z^*\hat{a}^{\dagger 2}\right)\right], \tag{22}$$

which was later named the "squeezed operator" [37–39]. In the second paper of [81] the operator $\hat{S}(z)$ was written for real z in terms of the quadrature operators as $\exp\left[ir\left(\hat{x}\hat{p}+\hat{p}\hat{x}\right)\right]$, which is exactly the form given by Plebański [76]. The conditions under which the MUS preserve their forms were studied in [81,83–85].

The first detailed review of the states defined by the relations (18)–(19) was given by Yuen [86], who proposed the name "two-photon states" and obtained, in particular, the photon distribution function

$$p_n \equiv |\langle n|\psi\rangle|^2 = \frac{|v/(2u)|^n}{n!|u|} \exp\left[\operatorname{Re}\left(\beta^2 v^*/u\right) - |\beta|^2\right] \left|H_n\left(\frac{\beta}{\sqrt{2uv}}\right)\right|^2 \tag{23}$$

for a generic state (19) (here $H_n(z)$ is the Hermite polynomial). In some sense, the paper [86] concluded the "prehistoric" period of studies on squeezed and "nonclassical" states, which were considered up to that time, as a rule, only as objects of mathematical curiosity.

Coherent states of nonstationary oscillators and Gaussian packets

The operators like (18) and the state (19) arise naturally in the process of dynamical evolution governed by the Hamiltonian

$$\hat{H} = \omega\hat{a}^\dagger\hat{a} + \left(\kappa\hat{a}^{\dagger 2}e^{-2i\omega t-i\varphi} + \text{h.c.}\right), \tag{24}$$

which describes the *degenerate parametric amplifier*. This problem was studied in detail by Takahasi in 1965 [72]. Similar *two-mode* states were considered implicitly in 1967 by Mollow and Glauber [35], who developed the quantum theory of the *nondegenerate parametric amplifier*, described by the interaction Hamiltonian between two modes of the form $\hat{H}_{int} = \hat{a}^\dagger\hat{b}^\dagger e^{-i\omega t} + \text{h.c.}$. For other references see Chapter 3.

In the case of an oscillator with arbitrary time-dependent frequency, the operator $\exp\left[\left(\hat{a}^{\dagger 2}+\hat{a}^2\right)s\right]$ naturally appeared in [87], where the evolution of initially coherent states was considered. The generalizations of the coherent state (3) as the eigenstates of the *linear time-dependent integral of motion operator* were introduced by Malkin, Man'ko and Trifonov [88–90] (the most general form in the case of one degree of freedom was described in detail in [91]). This integral of motion has the structure (18), $u(t)$ and $v(t)$ being certain solutions of the *classical* equations of motion (see Chapter 3). The eigenstates have the form (19). They are coherent with respect to the time-dependent operator $\hat{b}(t)$, but they are *squeezed* with respect to the quadrature components of the "initial" operator $\hat{a}$.

Coherent states of a charged particle in a constant homogeneous magnetic field were introduced by Malkin and Man'ko in [92] (see also [93, 94]). In the early years of quantum mechanics, the nonspreading wave packets in this case were considered by Darwin [8]. Generalizations of the nonstationary oscillator coherent states to systems with several degrees of freedom, such as a charged particle in nonstationary homogeneous magnetic and electric fields plus a harmonic potential, were studied in detail in [88, 89, 95–97]. Multidimensional time-dependent coherent states for arbitrary quadratic Hamiltonians were introduced in [98]. Their wave functions were expressed in the form of generic Gaussian exponentials of N variables. From the modern point of view, all those states ("breathing modes" [99]) can be considered as squeezed states, since the variances of different canonically conjugated variables can assume values which are less than the ground state variances. Similar one-dimensional and multidimensional quantum Gaussian states were studied in connection with the problems of the theories of photodetection, measurements, and information transfer in [100–102].

Approximate quasiclassical solutions to the Schrödinger equation with arbitrary potentials, in the form of the Gaussian packets whose centres move along the classical trajectories, have been extensively studied in many papers by Heller and his coauthors, beginning with [103]. The first coherent states for the *relativistic* particles obeying the Klein-Gordon or Dirac equations were introduced in [92, 104–107].

2.2 Non-Gaussian oscillator states

First finite superpositions of coherent states

Titulaer and Glauber [108] introduced the "generalized coherent states", multiplying each term of the expansion (1) by arbitrary phase factors:

$$|\alpha\rangle_g = \exp\left(-|\alpha|^2/2\right) \sum_{n=0}^{\infty} \frac{\alpha^n}{\sqrt{n!}} \exp\left(i\vartheta_n\right) |n\rangle. \tag{25}$$

These are the most general states satisfying Glauber's criterion of "coherence" [4]. Their properties and special cases corresponding to the concrete dependences of the phases ϑ_n on n were studied in [109, 110]. In particular, Z. Bialynicka-Birula [109] showed that in the *periodic case*, $\vartheta_{n+N} = \vartheta_n$ (with arbitrary values $\vartheta_0, \vartheta_1, \ldots, \vartheta_{N-1}$), the state (25) is the superposition of N Glauber's coherent states, whose labels are uniformly distributed along the circle $|\alpha| = const$:

$$|\phi\rangle = \sum_{k=1}^{N} c_k |\alpha_0 \exp(i\phi_k)\rangle, \qquad \phi_k = 2\pi k/N. \tag{26}$$

The amplitudes c_k are determined from the system of N equations

$$\sum_{k=1}^{N} c_k \exp(im\phi_k) = \exp(i\vartheta_m), \qquad m = 0, 1, \ldots, N-1, \tag{27}$$

and they are different in the generic case. Stoler [110] has noticed that any sum (26) is an eigenstate of the operator $\hat{a}^N$ with the eigenvalue α_0^N, therefore it can be represented as a superposition of N orthogonal states, each one being a certain combination of N coherent states $|\alpha_0 \exp(i\phi_k)\rangle$.

Mathews and Eswaran [111] gave the name "semi-coherent states" to the states of a harmonic oscillator, possessing time-independent values of the variances σ_x and σ_p, different from the vacuum (or coherent states) values (so that $\sigma_x\sigma_p > 1/4$).

A trivial example is the Fock state. A nontrivial example, briefly discussed in [111], is the superposition of two coherent states of the form

$$|\alpha;\beta\rangle = \frac{|\alpha\rangle - |\beta\rangle\langle\beta|\alpha\rangle}{\left(1-|\langle\beta|\alpha\rangle|^2\right)^{1/2}}, \qquad \sigma_x = \sigma_p = \tfrac{1}{2} + \frac{|\langle\beta|\alpha\rangle|^2|\alpha-\beta|^2}{1-|\langle\beta|\alpha\rangle|^2}. \tag{28}$$

The mean value of the annihilation operator in the state (28) equals $\langle\hat{a}\rangle = \alpha$; it does not depend on parameter β.

Even and odd thermal and coherent states

An example of "nonclassical" *mixed* states was given by Cahill and Glauber [40], who discussed in detail the *displaced thermal states* (such states, which are obviously "classical", were also studied in [35, 112]) and compared them with the following quantum mixture:

$$\hat{\rho}_{ev-th} = \frac{2}{2+\overline{n}}\sum_{n=0}^{\infty}\left(\frac{\overline{n}}{2+\overline{n}}\right)^{n}|2n\rangle\langle 2n|. \tag{29}$$

The *even and odd coherent states*

$$|\alpha\rangle_{\pm} = \mathcal{N}_{\pm}\left(|\alpha\rangle \pm |-\alpha\rangle\right), \qquad \mathcal{N}_{\pm} = \left(2\left[1\pm\exp(-2|\alpha|^2)\right]\right)^{-1/2} \tag{30}$$

were introduced by Dodonov, Malkin and Man'ko in [113]. They are not reduced to the superpositions given in (26), since coefficients $c_1 = \pm c_2$ can never satisfy the equations (27) for real phases ϑ_0, ϑ_1. Besides, the photon statistics of the states (30) is quite different from the Poissonian statistics inherent to all states of the form (25). Even/odd states possess many remarkable properties. Being eigenstates of the operator $\hat{a}^2$ (cf. Eq. (11)), the states $|\alpha\rangle_{\pm}$ are the simplest special cases of the *multiphoton states* (see section 3). They can be obtained from the vacuum by the action of *nonexponential* displacement operators $\hat{D}_+(\alpha) = \cosh\left(\alpha\hat{a}^\dagger - \alpha^*\hat{a}\right)$ and $\hat{D}_-(\alpha) = \sinh\left(\alpha\hat{a}^\dagger - \alpha^*\hat{a}\right)$ (cf. Eq. (10)). From the modern point of view, the special cases of the time-dependent wave packet solutions of the Schrödinger equation for the (singular) oscillator with a time-dependent frequency found in [113] are nothing but the *odd squeezed states*. The states $|\alpha\rangle_{\pm}$ with $|\alpha| \gg 1$ are frequently considered as the simplest examples of *macroscopic quantum superpositions* [66], which can be interpreted as well defined models of the famous *"Schrödinger cat"* [114]. The states (30) and their generalizations are discussed in Chapter 4.

Displaced and squeezed number states

These states are obtained by applying the displacement operator $\hat{D}(\alpha)$ (4) or the squeezing operator $\hat{S}(z)$ (22) to the states different from the vacuum oscillator state $|0\rangle$. As a matter of fact, the first examples of *displaced* states were given by Plebański [73, 76], who studied the properties of the state $|n,\alpha\rangle = \hat{D}(\alpha)|n\rangle$, and by Senitzky [115]. These results were rediscovered by Boiteux and Levelut in [116], where the name *semicoherent state* was used (however, the states of [116] are different from the Mathews–Eswaran states [111] bearing the same name). The general construction $\hat{D}(\alpha)|\psi_0\rangle$ for an arbitrary *fiducial state* $|\psi_0\rangle$ was considered by Klauder in [5].

The special case of the states $|n, z\rangle = \hat{S}(z)|n\rangle$ (known now under the name *squeezed number states*) for real z was considered in [75, 76]. These states were also discussed briefly by Yuen [86]. The change of the form of the related Gauss–Hermite wave functions in the process of evolution was considered in [117].

2.3 Coherent phase states

The state of the classical oscillator can be described either in terms of its quadrature components x and p, or in terms of the amplitude and phase, so that $x + ip = A\exp(i\varphi)$. Moreover, in the classical case one can introduce the action and angle variables, which have the same Poisson brackets as the cordinate and momentum: $\{p, x\} = \{I, \varphi\} = 1$. However, in the quantum case we meet serious mathematical difficulties in trying to define the phase operator in such a way that the commutation relation $[\hat{n}, \hat{\varphi}] = i$ is fulfilled, if the photon number operator is defined as $\hat{n} = \hat{a}^\dagger\hat{a}$. These difficulties originate in the fact that the spectrum of operator $\hat{n}$ is *bounded* from below.

The first solution to the problem was given by Susskind and Glogower [118], who introduced the *exponential phase operator*

$$\hat{E}_- \equiv \hat{C} + i\hat{S} \equiv \sum_{n=1}^{\infty} |n-1\rangle\langle n| = \left(\hat{a}\hat{a}^\dagger\right)^{-1/2}\hat{a}, \tag{31}$$

which can be considered as a quantum analogue of the classical phase factor $e^{i\varphi}$ [32, 59]. Operator $\hat{E}_-$ and its Hermitian "cosine" and "sine" components satisfy the "classical" commutation relations

$$[\hat{C}, \hat{n}] = i\hat{S}, \qquad [\hat{S}, \hat{n}] = -i\hat{C}, \qquad [\hat{E}_-, \hat{n}] = \hat{E}_-. \tag{32}$$

However, operator $\hat{E}_-$ is *nonunitary*, since the commutator with its Hermitially conjugated partner $\hat{E}_+ \equiv \hat{E}_-^\dagger$ is not equal to zero:

$$[\hat{E}_-, \hat{E}_+] = 1 - \hat{C}^2 - \hat{S}^2 = |0\rangle\langle 0|. \tag{33}$$

The annihilation operator $\hat{a}$ has no inverse operator in the full meaning of this term. Nonetheless, it possesses the *right inverse operator*

$$\hat{a}^{-1} = \sum_{n=0}^{\infty} \frac{|n+1\rangle\langle n|}{\sqrt{n+1}} = \hat{a}^\dagger\left(\hat{a}\hat{a}^\dagger\right)^{-1} = \hat{E}_+\left(\hat{a}\hat{a}^\dagger\right)^{-1/2}, \tag{34}$$

which satisfies, among many others, the following relations:

$$\hat{a}\hat{a}^{-1} = 1, \qquad [\hat{a}, \hat{a}^{-1}] = |0\rangle\langle 0|, \qquad [\hat{a}^\dagger, \hat{a}^{-1}] = \hat{a}^{-2}.$$

This operator was discussed briefly by Dirac [119], who noticed that Fock had considered it long before. However, it found applications in quantum optics only in the last decade of the 20th century (see the paragraph on *photon-added states* in section 3). Lerner [120] noticed that the commutation relations (32) do not determine the operators $\hat{C}, \hat{S}$ uniquely. Earlier, the same observation was made by Wigner [121] with respect to the triple $\{\hat{n}, \hat{a}, \hat{a}^\dagger\}$ (see the next section). In the general case, besides the "polar decomposition" (31), which is equivalent to the relations

$$\hat{E}_-|n\rangle = (1-\delta_{n0})|n-1\rangle, \qquad \hat{E}_+|n\rangle = |n+1\rangle, \tag{35}$$

one can define operator $\hat{U} = \hat{C} + i\hat{S}$ via the relation $\hat{U}|n\rangle = f(n)|n-1\rangle$, where function $f(n)$ may be arbitrary enough, being restricted by the requirement $f(0) = 0$ and certain other constraints which ensure that the spectra of the "cosine" and "sine" operators belong to the interval $(-1, 1)$. The properties of Lerner's construction were studied in [122].

The states with the definite phase

$$|\varphi\rangle = \sum_{n=0}^{\infty} \exp(in\varphi)|n\rangle \tag{36}$$

were considered in [32, 118]. However, they are not normalizable (like the coordinate or momentum eigenstates). The normalizable *coherent phase states* were introduced in [123, 124] as the eigenstates of the operator $\hat{E}_-$:

$$|\varepsilon\rangle = \sqrt{1-\varepsilon\varepsilon^*}\sum_{n=0}^{\infty}\varepsilon^n|n\rangle, \qquad \hat{E}_-|\varepsilon\rangle = \varepsilon|\varepsilon\rangle, \qquad |\varepsilon| < 1. \tag{37}$$

The *pure* quantum state (37) has the same probability distribution $|\langle n|\varepsilon\rangle|^2$ as the mixed *thermal state* described by the statistical operator

$$\hat{\rho}_{th} = \frac{1}{1+\overline{n}}\sum_{n=0}^{\infty}\left(\frac{\overline{n}}{1+\overline{n}}\right)^n |n\rangle\langle n|, \tag{38}$$

if one identifies the mean photon number $\overline{n}$ with $|\varepsilon|^2/\left(1-|\varepsilon|^2\right)$ [125].

A two-parameter set of states

$$|z;\kappa\rangle = \mathcal{N}\sum_{n=0}^{\infty}\left[\frac{\Gamma(n+\kappa+1)}{n!\Gamma(\kappa+1)}\right]^{1/2}\left(\frac{z}{\sqrt{\kappa+1}}\right)^n|n\rangle, \qquad \kappa \neq -1, \tag{39}$$

was introduced in [126] and studied in detail from different points of view in [113, 125, 127]. These states are eigenstates of the operator

$$\hat{A}_\kappa = \hat{E}_-\left[\frac{(\kappa+1)\hat{n}}{\kappa+\hat{n}}\right]^{1/2} \equiv \left[\frac{(\kappa+1)}{\kappa+1+\hat{n}}\right]^{1/2}\hat{a}. \tag{40}$$

If $\kappa = 0$, $\hat{A}_0 = \hat{E}_-$, and the state $|z;0\rangle$ coincides with (37). If $\kappa \to \infty$, then $\hat{A}_\infty = \hat{a}$, and the state (39) goes to the *coherent state* $|z\rangle$ (1). In two decades, the state (39) appeared again under the name *negative binomial state* (see the next section). In the special case $\kappa = -p-1$, $p = 1, 2, \ldots$, the state (39) goes to the superposition of the first $p+1$ Fock states

$$|z;p\rangle = \mathcal{N} \sum_{n=0}^{p} \left[\frac{p!}{n!(p-n)!} \right]^{1/2} \left(\frac{z}{\sqrt{p}} \right)^n |n\rangle, \tag{41}$$

which is nothing but the *binomial state* introduced 12 years later.

2.4 Coherent states for arbitrary algebras

Angular momentum (quantum top) and spin (Bloch) coherent states

The first family of coherent states for the angular momentum operators was constructed in [128], actually as some special two-dimensional oscillator coherent states, using the Schwinger representation of the angular momentum operators in terms of the auxiliary bosonic annihilation operators $\hat{a}_+$ and $\hat{a}_-$:

$$\hat{J}_+ = \hat{a}_+^\dagger \hat{a}_-, \qquad \hat{J}_- = \hat{a}_-^\dagger \hat{a}_+, \qquad \hat{J}_3 = \tfrac{1}{2}\left(\hat{a}_+^\dagger \hat{a}_+ - \hat{a}_-^\dagger \hat{a}_-\right). \tag{42}$$

Such "oscillator-like" angular momentum coherent states were studied in [129–133]. A possibility of constructing "continuous" families of states using different modifications of the unitary displacement (4) and squeezing (22) operators was recognized at the beginning of the 1970s. The first explicit example is frequently related to the *coherent spin states* [134] (also called *atomic coherent states* [135] and *Bloch coherent states* [136, 137])

$$|\vartheta, \varphi\rangle = \exp(\zeta \hat{J}_+ - \zeta^* \hat{J}_-)|j, -j\rangle, \qquad \zeta = (\vartheta/2)\exp(-i\varphi), \tag{43}$$

where $\hat{J}_\pm$ are the standard raising and lowering spin (angular momentum) operators, $|j, m\rangle$ are the standard eigenstates of the operators $\hat{\mathbf{J}}^2$ and $\hat{J}_z$, and ϑ, φ are the angles in the spherical coordinates (a similar construction was used in [138], for later works see, e.g., [139]). However, the special case of these states for spin–$1/2$ was considered much earlier (in 1960) by Klauder [5], who introduced the simplest variant of *fermion coherent states* in the form

$$|\beta\rangle = \sqrt{1 - |\beta|^2}\,|0\rangle + \beta\,|1\rangle, \tag{44}$$

where β could be an arbitrary complex number satisfying the inequality $|\beta| \le 1$. In the same 1960, Thouless [140] briefly discussed the importance of the states

$$|\Phi\rangle = \exp\left[\sum_{i,j=1}^{N} C_{ij}\hat{a}_i\hat{a}_j + \sum_{i=1}^{N} \sum_{m=N+1}^{\infty} \left(C_{mi} - C_{im}\right) \hat{a}_m^\dagger \hat{a}_i + \sum_{m=N+1}^{\infty} \sum_{n=N+1}^{\infty} \hat{a}_m^\dagger \hat{a}_n^\dagger \right] |\Phi_0\rangle \tag{45}$$

(with fermionic operators $\hat{a}_m^\dagger$ and $\hat{a}_i$) for the theory of superconductivity. More general multimode constructions, including the use of Grassman's anticommuting variables, were introduced in [15, 141]. Also, Klauder studied generic states of the form (43) in [142]. Rowe [99] called the states of the form (45) with *bosonic* operators $\hat{a}_m^\dagger$, $\hat{a}_i$ and $|\Phi_0\rangle = |vac\rangle$ "coherent states of the symplectic group".

Group coherent states

The operators $\hat{J}_\pm, \hat{J}_z$ are the generators of the algebra $su(2)$. A general construction looks like

$$|\xi_1, \xi_2, \ldots, \xi_n\rangle = \exp\left(\xi_1 \hat{A}_1 + \xi_2 \hat{A}_2 + \ldots + \xi_n \hat{A}_n\right)|\psi\rangle, \tag{46}$$

where $\{\xi_k\}$ is a continuous set of complex or real parameters, $\hat{A}_k$ are the generators of some algebra, and $|\psi\rangle$ is some "basic" (*"fiducial"*) state. This scheme was proposed by Klauder as far back as in [142] (under the name "continuous representation theory"), and later it was rediscovered in [143]. A great number of different families of "generalized coherent states" can be obtained, choosing different algebras and basic states. Perhaps, the first examples are the "coherent soft-photon states" [144] and "affine coherent states" [145]. More familiar examples are related to the $su(1,1)$ algebra [143]

$$|\zeta; k\rangle \sim \exp(\zeta \hat{K}_+)|0\rangle \sim \sum_{n=0}^{\infty} \left[\frac{\Gamma(n+2k)}{n!\Gamma(2k)}\right]^{1/2} \zeta^n |n\rangle, \tag{47}$$

where k is the so-called Bargmann index labelling the concrete representation of the algebra, and $\hat{K}_+$ is the corresponding rising operator (see Chapter 2 for details). Evidently, the states (39) and (47) are the same, although their interpretation may be different. Solomon [146] considered the states of the form $\exp(i\vartheta \hat{J}_2)|n\rangle$, where $\hat{J}_2$ was one of generators of the $su(1,1)$ algebra, chosen in the bosonic realization as (cf. equation (42)) $\hat{J}_2 = \frac{1}{2}i(\hat{a}_+^\dagger \hat{a}_-^\dagger - \hat{a}_+\hat{a}_-)$. Some particular realizations of the state (47) connected with the problem of quantum "singular oscillator" (described by the Hamiltonian $H = p^2 + x^2 + gx^{-2}$) were considered in [113, 127]. The "generalized phase state" (39) and its special case (41) can also be considered as coherent states for the groups $O(2,1)$ or $O(3)$, with the "displacement operator" of the form [125]

$$\hat{D}_\kappa(z) = \exp\left(z\sqrt{\hat{n}(\hat{n}+\kappa)}\hat{E}_+ - z^*\hat{E}_-\sqrt{\hat{n}(\hat{n}+\kappa)}\right).$$

A comparison of the coherent states for the Heisenberg–Weyl and $su(2)$ algebras was made in [147] (see also [148]). Coherent states for the group $SU(n)$ were studied in [136, 149].

There exists an abuse of mathematical terminology in the theory of "coherent states". This name was introduced by Glauber [4] in connection with the clear physical phenomenon of coherence of electromagnetic radiation. Later on, however, the same name was used frequently in the physical literature for pure mathematical objects, well known in group representation theory, in which abstract mathematical results were obtained using specific mathematical tools and terminology (such as "orbits", etc.). In this context, the physical notion of coherence is completely lost (as, for instance, in the case of "vector coherent states" [150] frequently used in the representation theory). For this reason, we would prefer to use names reflecting the physical properties of states, such as "squeezed states", etc., trying to minimize the abuse of mathematical terminology, though it is present in the current physical literature (cf. discussions on different meanings of the word "coherent" in the introductory chapter of [29] and in [31]).

The name *Barut–Girardello coherent states* (BGCS) is used in the modern literature for states which are eigenstates of some non-Hermitian *lowering operators*. The first example was given in [151], where the eigenstates of the lowering operator $\hat{K}_-$ of the $su(1,1)$ algebra were introduced in the form

$$|z;k\rangle = \mathcal{N}\sum_{n=0}^{\infty}\frac{z^n|n\rangle}{\sqrt{n!\Gamma(n+2k)}}. \tag{48}$$

The corresponding wave packets in the coordinate representation, related to the problem of the nonstationary "singular oscillator", were expressed in terms of the Bessel functions in [113]. Later, BGCS were studied, e.g., in [152].

The first two-dimensional analogues of the Barut–Girardello states, namely, eigenstates of the product of commuting boson annihilation operators $\hat{a}\hat{b}$,

$$\hat{a}\hat{b}|\xi,q\rangle = \xi|\xi,q\rangle, \qquad \hat{Q}|\xi,q\rangle = q|\xi,q\rangle, \qquad \hat{Q} = \hat{a}^\dagger\hat{a} - \hat{b}^\dagger\hat{b}, \tag{49}$$

were introduced by Horn and Silver [153] in connection with the problem of pions production (see also [154,155]), where the internal isospin degree of freedom must be taken into account. In the simplified variant these states have the form (actually, the authors of [153–155] considered the infinite-dimensional case, being interested in the quantum field theory applications)

$$|\xi,q\rangle = \mathcal{N}\sum_{n=0}^{\infty}\frac{\xi^n|n+q,n\rangle}{[n!(n+q)!]^{1/2}}. \tag{50}$$

Operator $\hat{Q}$ can be interpreted as the "charge operator"; for this reason the state (50) appeared in [156] under the name "charged boson coherent state." Joint eigenstates of the operators $\hat{a}_+\hat{a}_+$ and $\hat{a}_+\hat{a}_-$, which generate the angular momentum operators according to (42), were studied in [157]. The properties of states (50) were discussed in [158]. Their infinite-dimensional generalizations, named "quasi-coherent states", were considered in [159–161]. In the modern quantum-optical literature the states like (50) are usually known as "pair coherent states" (see section 3). They are also discussed in Chapter 2.

2.5 "Intelligent" and "minimum uncertainty" states

For the operators $\hat{A}$ and $\hat{B}$ different from $\hat{x}$ and $\hat{p}$, the right-hand side of the uncertainty relation (12) depends on the quantum state. The problem of finding the states for which (12) becomes the *equality* was discussed by Jackiw [80], who showed that it is reduced essentially to solving the equation

$$(\hat{A} - \langle\hat{A}\rangle)|\psi\rangle = \lambda(\hat{B} - \langle\hat{B}\rangle)|\psi\rangle. \tag{51}$$

In particular, one of several "phase–number" uncertainty relations, namely

$$(\Delta\hat{n})^2(\Delta\hat{C})^2/\langle\hat{S}\rangle^2 \geq 1/4 \tag{52}$$

(operators $\hat{C}$ and $\hat{S}$ were defined in (31) and (32)), is minimized on the states [80]

$$|\psi\rangle = \mathcal{N}\sum_{n=0}^{\infty}(-i)^n I_{n-\lambda}(\gamma)|n\rangle, \tag{53}$$

where $I_\nu(\gamma)$ is the modified Bessel function, and λ, γ are some parameters.

Eswaran [122] considered the $\hat{n}$–$\hat{C}$ pair of operators and solved the equation $(\hat{n} + i\gamma\hat{C})|\psi\rangle = \lambda|\psi\rangle$. The ratio $(\Delta C)^2(\Delta S)^2/|\langle[\hat{C},\hat{S}]\rangle|^2$ was minimized by Mathews and Eswaran [162], who solved the equation (v and λ are parameters)

$$\left[(1+v)\hat{E}_- + (1-v)\hat{E}_+\right]|\psi\rangle = \lambda|\psi\rangle. \tag{54}$$

The case of the spin (angular momentum) operators was considered in [163], where the name *intelligent states* was introduced. The relations between coherent spin states, intelligent spin states, and minimum-uncertainty spin states were discussed in [164–169]. Nowadays the "intelligent" states are understood as the states for which the Heisenberg uncertainty relation (12) becomes the equality, whereas the "minimum uncertainty states" are those for which the "uncertainty product" $\Delta A \Delta B$ attains the minimal possible value (for arbitrary operators $\hat{A}, \hat{B}$ such states may not exist [80]).

3 "Ugly duckling" turns into a "beautiful swan"

Until the mid-1970s, "nonclassical" states were mainly considered either as useful auxiliary constructions, which could help in calculating various characteristics of quantum systems, or as "mathematical toys" having nothing in common with the physical reality. For these reasons, they did not attract much attention from the majority of the quantum optics community.

3.1 Maturity of squeezed states

Possible applications of squeezed states and first experiments

The situation began to change at the end of 1970s, when it was recognized that the "two-photon" or "squeezed" states can be useful for solving various fundamental physical and technological problems. In that period, many theoretical and experimental studies were devoted to the phenomena of *antibunching* or *sub-Poissonian photon statistics*, which are unequivocal features of the quantum nature of light (the first publications were [170–176]. The relations between antibunching and squeezing were discussed in [171], and the first experiment was reported in 1977 [177] (for the detailed story see, e.g., [46, 178–180]). The words "antibunched light" appeared in the titles of articles [181–184], and the term "sub-Poissonian light" was used for the first time in the titles of studies [182, 185, 186].

In 1978, Yuen and Shapiro [187] proposed to use the two-photon states in order to improve optical communications by reducing the quantum fluctuations in one (signal) quadrature component of the field at the expense of the amplified fluctuations in another (unobservable) component. The states with the reduced quantum noise in one of the quadrature components appeared to be necessary [37, 38] in order to overcome the so-called "standard quantum limit" in the problem of measurement of weak forces, important for the detectors of the gravitational waves [188–190] (this is a special case of the so-called "reproducible measurements," the idea of which originated in the paper by Landau and Peirls [191], and which were named *quantum nondemolition measurements* in [192]: see, e.g., [193–196]).

The words "squeezing", "squeezed states", and "squeezed operator" were introduced by Hollenhorst and Caves [37, 38] in connection with the problem of reducing quantum noise, but they became especially popular and generally accepted after Walls' paper [39]. Independently, exactly the same term "squeezing" was proposed by Brosa and Gross [197] in connection with the problem of nuclear collisions (they considered a model of an oscillator with time-dependent *mass* and fixed elastic constant, whereas in the simplest quantum optical oscillator models squeezed states appear usually in the case of time-dependent *frequency* and fixed mass). Note also that "non-classical behaviour" of the harmonic oscillator in the "dilated (or compressed)" state $\hat{S}(z)|0\rangle$ was mentioned by Rowe [99] in 1978.

Many different schemes of generating squeezed states were proposed, such as four-wave mixing [198], resonance fluorescence [199], the use of the free-electron laser or the Josephson junction [200], cavities [201], harmonic generation [202, 203], etc.. The words "squeezed states" were used for the first time in the titles of papers [200, 204–206]. In 1985–1986 the results of the first successful experiments in the generation and detection of squeezed states were reported in [207] (backward four-wave mixing), [208] (forward four-wave mixing), [209] (parametric down conversion). For the details of the theory and experiment see, e.g., [210–213]. The first proposals for different schemes of generating "the most nonclassical" n-photon (Fock) states appeared in [214–217].

Correlated, multimode, thermal and displaced states

The states (19) with *complex* parameters u, v do not minimize the Heisenberg uncertainty product. But they are minimum uncertainty states for the *Schrödiner–Robertson uncertainty relation* [218, 219]

$$\sigma_x\sigma_p - \sigma_{px}^2 \geq \hbar^2/4, \qquad \sigma_{px} = \tfrac{1}{2}\langle \hat{p}\hat{x} + \hat{x}\hat{p}\rangle - \langle\hat{p}\rangle\langle\hat{x}\rangle, \tag{55}$$

which can be written also in the form

$$\sigma_p\sigma_x \geq \hbar^2/\left[4\left(1-r^2\right)\right], \qquad r = \sigma_{px}/\sqrt{\sigma_p\sigma_x}\,, \tag{56}$$

where r is the *correlation coefficient* between the coordinate and momentum. The left-hand side of (56) takes on the minimal possible value $\hbar^2/4$ for *any pure* Gaussian state (this fact was known to Kennard [69]), which can be written as

$$\psi(x) = \mathcal{N}\exp\left[-\frac{x^2}{4\sigma_{xx}}\left(1 - \frac{ir}{\sqrt{1-r^2}}\right) + bx\right]. \tag{57}$$

In order to emphasise the role of the correlation coefficient, the state (57) was named the *correlated coherent state* [220]. It is the unitary equivalent of the squeezed states with complex squeezing parameters, which were considered, e.g., in [221–224] (see also Chapter 2). A detailed review of the properties of correlated states (including multidimensional systems, e.g., a charge in a magnetic field) was given in [225]. For arbitrary Hermitian operators, inequality (55) should be replaced by [218, 219]

$$\sigma_A\sigma_B - \sigma_{AB}^2 \geq \tfrac{1}{4}\,|\langle[\hat{A},\hat{B}]\rangle|. \tag{58}$$

For detailed reviews of generalized uncertainty relations, see [226–228].

Various *two-mode squeezed states* of the form

$$|\mathrm{TMSS}\rangle \sim \exp\left(z\hat{a}\hat{b} - z^*\hat{a}^\dagger\hat{b}^\dagger + \cdots\right)|\varphi\rangle \tag{59}$$

were studied (sometimes implicitly or under different names, such as "cranked oscillator states") in [229–235]. In particular, the "two-mode squeezed state" introduced in [236] (see also [237], where similar states were studied in the frameworks of the "thermofield" representation) has the "diagonal" expansion over the Fock states

$$|\lambda\rangle = \sqrt{1-|\lambda|^2}\sum_{n=0}^{\infty}\lambda^n|n,n\rangle, \qquad |\lambda| < 1. \tag{60}$$

Multimode generalizations of the (Gaussian) minimum uncertainty states were discussed in [238, 239]. *Multimode squeezed states* were considered in [240, 241], whereas *multiatom squeezed states* were introduced in [242]. The properties of generic (pure and mixed) Gaussian states were studied in [243–253]. Their special cases, sometimes called *mixed squeezed states* or *squeezed thermal states*, were discussed in [254–258].

Implicitly, the squeezed states of a charged particle moving in a homogeneous (stationary and nonstationary) magnetic field were considered in [88, 89, 95, 96]. In the explicit form they were introduced and studied in [225] and independently in [259, 260]. Coherent states in magnetic field were the subject of studies [261, 262]. Approximate solutions in the form of quasiclassical Gaussian packets moving in a one-dimensional anharmonic potential were considered in [263], where they were named "pseudo-coherent states". In the case of an arbitrary inhomogeneous electromagnetic field or an arbitrary multidimensional anharmonic potential, such packets were studied in [264] and [265] (where they were named *trajectory-coherent states*).

The *displaced number states* were considered in connection with the time-energy uncertainty relation in [266]. They can be expressed as [267]

$$|n,\alpha\rangle = \hat{D}(\alpha)|n\rangle = \mathcal{N}\left(\hat{a}^\dagger - \alpha^*\right)^n|\alpha\rangle. \tag{61}$$

"Maximally symmetric" coherent states were considered in [268].

General concepts of squeezing

The first definition of squeezing for arbitrary Hermitian operators $\hat{A}, \hat{B}$ was given by Walls and Zoller [199]. Taking into account the uncertainty relation (12), they said that fluctuations of the observable A are "reduced" if

$$(\Delta A)^2 < \tfrac{1}{2}\,|\langle[\hat{A},\hat{B}]\rangle|. \tag{62}$$

This definition was extended to the case of several variables (whose operators are generators of some algebra) in [269] and specified in [270] (see also [271]).

Hong and Mandel [272] introduced the concept of higher-order squeezing. The state $|\psi\rangle$ is squeezed to the $2n$th order in some quadrature component, say $\hat{x}$, if the mean value $\langle\psi|(\Delta\hat{x})^{2n}|\psi\rangle$ is less than the mean value of $(\Delta\hat{x})^{2n}$ in the coherent state. If $\hat{x}$ is defined as in (5), then the condition of squeezing reads $\langle(\Delta\hat{x})^{2n}\rangle < 2^{-n}(2n-1)!!$. In particular, for $n = 2$ we have the requirement $\langle(\Delta\hat{x})^4\rangle < 3/4$. Hong and Mandel showed that the usual squeezed states are squeezed to any even order $2n$. The methods of producing such squeezing were proposed in [273]. Other definitions of higher-order squeezing are usually based on the Walls-Zoller approach. Hillery [274] defined second-order squeezing taking in (62) $\hat{A} = \left(\hat{a}^2 + \hat{a}^{\dagger 2}\right)/2$ and $\hat{B} = \left(\hat{a}^2 - \hat{a}^{\dagger 2}\right)/(2i)$. A generalization to kth-order squeezing, based on the operators $\hat{A} = \left(\hat{a}^k + \hat{a}^{\dagger k}\right)/2$ and $\hat{B} = \left(\hat{a}^k - \hat{a}^{\dagger k}\right)/(2i)$, was made in [275]. The uncertainty relations for higher-order moments were considered in [226].

The concepts of *sum squeezing* and *difference squeezing* were introduced by Hillery [276], who considered two-mode systems and used sums and differences of various bilinear combinations constructed from the creation and annihilation operators. Physical bounds on squeezing due to the finite energy of real systems were discussed in [277]. "Semicoherent states" related to the group $ISp(2, R)$ were considered in [278].

One of the first explicit treatments of the squeezed states as the $SU(1,1)$ *coherent states* was given in [269], where the realization of the $su(1,1)$ algebra in terms of the operators

$$\hat{K}_+ = \hat{a}^{\dagger 2}/2, \qquad \hat{K}_- = \hat{a}^2/2, \qquad \hat{K}_3 = \left(\hat{a}^\dagger\hat{a} + \hat{a}\hat{a}^\dagger\right)/4, \tag{63}$$

was considered. The state was named $SU(1,1)$ squeezed, if

$$(\Delta K_1)^2 < |\langle\hat{K}_3\rangle|/2 \quad \text{or} \quad (\Delta K_2)^2 < |\langle\hat{K}_3\rangle|/2, \qquad \hat{K}_\pm = \hat{K}_1 \pm i\hat{K}_2. \tag{64}$$

In [274] such states were named *amplitude-squared squeezed states*. The relations between the squeezed states and the Bogolyubov transformations were considered in [279]. An algebraic approach to the squeezed states was developed in [280] in connection with the theory of linear optical devices, such as Mach–Zender or Fabry–Perot interferometers and four-wave mixers (see also [281] for the detailed theory of transformations on the beam splitters). Nonclassical properties of the states (50) (renamed as *pair coherent states*) were studied in [282].

The quantum phase can be relatively well defined for the coherent states with $\langle n\rangle = |\alpha|^2 \gg 1$. In this case the product of the number and phase uncertainties is $\Delta n\Delta\varphi \sim 1$ [32]. Moreover, due to the Poissonian photon number statistics in the coherent states, $\Delta n = \sqrt{\langle n\rangle}$. The concept of *amplitude squeezing* was introduced in [283]. It means states possessing the property $\Delta n < \sqrt{\langle n\rangle}$. If the product $\Delta n\Delta\varphi$ remains of the order of 1, then we have the *number-phase minimum uncertainty state* [284]. The methods of generating the minimum uncertainty states for the operators $\hat{N}, \hat{C}, \hat{S}$ (31) were discussed in [285].

Oscillations of the distribution functions

While the photon distribution function $p_n = \langle n|\hat{\rho}|n\rangle$ is rather smooth for the "classical" thermal and coherent states (being given by the Planck and Poisson distributions, respectively), it reveals strong oscillations for many "nonclassical" states. This is obvious for the even coherent states or the squeezed vacuum states, for which $p_n \equiv 0$ if n is an odd number, or for the odd coherent states, if n is an even number. The graphical analysis of the "two-photon" distribution (23), made in [286, 287], showed that for certain relations between the squeezing and displacement parameters, the photon distribution function exhibits strong irregular oscillations, whereas for other values of the parameters it remains rather regular. Ten years later, these oscillations were rediscovered in [288–292], and since that time they have attracted the attention of many researchers, mainly due to their interpretation [288, 289, 292, 293] as the manifestation of the *interference in phase space* (a method of calculating quasiclassical distributions, based on the areas of overlapping in the phase plane, was used earlier in [294]). It is worth mentioning that as far back as 1970, Walls and Barakat [295] discovered strong oscillations of the photon distribution functions, calculating eigenstates of the *trilinear* Hamiltonian $\hat{H}_{WB} = \sum_{k=1}^{3} \omega_k \hat{a}_k^\dagger \hat{a}_k + \kappa(\hat{a}_1 \hat{a}_2^\dagger \hat{a}_3^\dagger + \text{h.c.})$ The parametric amplifier time-dependent Hamiltonian (24), which "produces" squeezed states, can be considered as the semiclassical approximation to $\hat{H}_{WB}$. For recent studies on trilinear Hamiltonians and references to other publications, see [296].

3.2 Non-Gaussian states

Coherent states for anharmonic potentials

There exist different constructions of "coherent states" for a particle moving in an arbitrary potential. *Minimum uncertainty states* whose time evolution is as close as possible to the trajectory of a *classical* particle have been studied by Nieto and Simmons [33, 297]. In [298, 299] "coherent" states were defined as eigenstates of operators like $\hat{A} = f(x) + i\sigma(x)d/dx$. However, such packets do not preserve their forms in the process of evolution.

At least three anharmonic potentials are of special interest in quantum mechanics. The closest to the harmonic oscillator is the "singular oscillator" potential $x^2 + gx^{-2}$ (called also as "isotonic," "pseudoharmonical," "centrifugal" oscillator, or "oscillator with centripetal barrier"). Different coherent states for this potential have been constructed in [33, 113, 127, 300].

Minimum uncertainty states for the Morse potential $U_0\left(1 - e^{-ax}\right)^2$ were constructed in [301]; the cases of the Pöschl–Teller and Rosen–Morse potentials, $U_0 \tan^2(ax)$ and $U_0 \tanh^2(ax)$, were considered in [33]. Coherent states based on the algebras $su(1,1)$ or $so(2,1)$ have been proposed in [302–304].

Coherent states and packets in the hydrogen atom

Introducing the nonspreading Gaussian wave packets for the harmonic oscillator, Schrödinger wrote in the same paper [7]:

"We can definitely foresee that, in a similar way, wave groups can be constructed which move round highly quantized Kepler ellipses and are the representation by wave mechanics of the hydrogen electron. But the technical difficulties in the calculation are greater than in the especially simple case which we have treated here."

Indeed, this problem turned out to be much more complicated than the oscillator one. The *spreading* of initial packets resulting in their destruction was evaluated (within the frameworks of the quasiclassical approach) by Darwin [8] in 1927. One of the first attempts to construct the quasiclassical packets moving along the Kepler orbits was made by Brown [305] in 1973 and by Eswaran and Mathews [306] in 1974. A similar approach was used in [307]. The symmetry explaining degeneracy of hydrogen energy levels was found by Fock [308] (see also Bargmann [309]): it is $O(4)$ for the discrete spectrum and the Lorentz or $O(3,1)$ group for the continuous one. The symmetry combining all the discrete levels into one irreducible representation (the dynamical group $O(4,2)$) was found in [310] (see also [311,312]). Different group-related coherent states connected to these dynamical symmetries were discussed in [313–319]. The Kustaanheimo–Stiefel transformation [320], which reduces the 3-dimensional Coulomb problem to the 4-dimensional constrained harmonic oscillator [321,322], was applied to obtain various coherent states of the hydrogen atom in [323–325].

Relativistic oscillator coherent states

One of the first papers on the relativistic equations with internal degrees of freedom was published by Ginzburg and Tamm [326]. The model of "covariant relativistic oscillator" obeying the modified Dirac equation

$$\left(\gamma_\mu \partial/\partial x_\mu + a\left[\xi_\mu \xi_\mu - \partial^2/\partial\xi_\mu \partial\xi_\mu\right] + m_0\right)\psi(x,\xi) = 0, \tag{65}$$

where x is the 4-vector of the "centre of mass", and 4-vector ξ is responsible for the "internal" degrees of freedom of the "extended" particle, was studied by many authors [327–332]. Coherent states for this model, related to the representations of the $SU(1,1)$ group and to the "singular oscillator" coherent states of [113], have been constructed in [333]. Coherent states of relativistic particles obeying the standard Dirac or Klein–Gordon equations were discussed in [334]. "Generalized coherent states" related to the Poincaré group were considered in [335] (see [67] for other references).

Supersymmetric states

The concept of supersymmetry was introduced by Gol'fand and Likhtman [336]. The nonrelativistic supersymmetric quantum mechanics was proposed by Witten [337] and studied in [338–340]. Its super-simplified model can be described in terms of the Hamiltonian which is a sum of the free oscillator and spin parts, so that the lowering operator can be conceived as a matrix

$$\hat{H} = \hat{a}^\dagger \hat{a} - \tfrac{1}{2}\sigma_3, \qquad \hat{A} = \left\| \begin{matrix} \hat{a} & 1 \\ 0 & \hat{a} \end{matrix} \right\|$$

(σ_3 is the Pauli matrix). Then one can try to construct various families of states applying operators like $\exp(\alpha\hat{A}^\dagger)$ to the ground (or another) state, looking for eigenstates of $\hat{A}$ or some functions of this operator, and so on. The first scheme was applied by Bars and Günaydin [341], who constructed group *supercoherent states*. The same (displacement operator) approach was used in [342, 343]. The second way was chosen by Aragone and Zypman [344], who constructed the eigenstates of some "supersymmetric" non-Hermitian operator.

Sub-Poissonian and subcoherent states

The name *subpoissonian states* was proposed in [345, 346] for the states showing sub-Poissonian fluctuations (antibunching) of the photon number, viz., for the states satisfying the relation

$$\left[\langle\hat{a}^\dagger\hat{a}\left(\hat{a}^\dagger\hat{a}-1\right)\rangle - \langle\hat{a}^\dagger\hat{a}\rangle^2\right]/\langle\hat{a}^\dagger\hat{a}\rangle^2 = -A, \qquad A>0, \tag{66}$$

where the positive constant A measures the relative deviation from the Poissonian statistics. Two different generating functions of the factorial moments,

$$Q_1(x) = \sum_{\nu=0}^{\infty}(-x)^\nu\langle\hat{a}^{\dagger\nu}\hat{a}^\nu\rangle/\nu!\,, \tag{67}$$

satisfying (66) for the given values of A and $\overline{n}\equiv\langle\hat{a}^\dagger\hat{a}\rangle$, were found in [345, 346]:

$$Q_1^{(c)}(x) = \exp\left[A^{-1}\ln(1-\overline{n}Ax)\right], \tag{68}$$

$$Q_1^{(s)}(x) = (A-1)^{-1}\ln\left[1+(1-A)\overline{n}x)\right]+1. \tag{69}$$

These distributions exist if $0\le\overline{n}A\le 1$. The corresponding factorial moments are as follows,

$$\langle\hat{a}^{\dagger\nu}\hat{a}^\nu\rangle_{(c)} = (\overline{n}A)^\nu\frac{\Gamma\left(1+A^{-1}\right)}{\Gamma\left(1-\nu+A^{-1}\right)}, \tag{70}$$

$$\langle\hat{a}^{\dagger\nu}\hat{a}^\nu\rangle_{(s)} = (\nu-1)!\,\overline{n}^\nu(1-A)^{\nu-1}. \tag{71}$$

In the case (c), the limit $A\to 0$ yields the Poissonian distribution (i.e., the coherent states, if one considers pure quantum states with appropriately chosen phase factors). On the contrary, in the case (s), the limit $A\to 0$ results in the state with the same second-order moments as in the coherent state,

$$\langle\hat{a}^{\dagger 2}\hat{a}^2\rangle = \langle\hat{a}^\dagger\hat{a}\rangle^2 = \overline{n}^2, \tag{72}$$

but with different higher-order moments: $\langle\hat{a}^{\dagger\nu}\hat{a}^\nu\rangle_{(s)0} = (\nu-1)!\,\overline{n}^\nu$. Such limit states, satisfying relation (72) but different from (1), were named in [345, 346] *sub-coherent states*.

After Mandel's work [347], the sub-Poissonian properties of quantum states are characterized usually by the *Mandel parameter*

$$\mathcal{Q} = \overline{n^2}/\bar{n} - \bar{n} - 1 \equiv -\overline{n}A \tag{73}$$

(a similar parameter $\mathcal{F} = \overline{n^2}/\bar{n} - \bar{n} \equiv \mathcal{Q}-1$ was introduced by Fano [348]).

Binomial states

These states are *finite* combinations of the first $M+1$ Fock states:

$$|p, M; \theta_n\rangle = \sum_{n=0}^{M} e^{i\theta_n} \left[\frac{M!}{n!\,(M-n)!} p^n (1-p)^{M-n}\right]^{1/2} |n\rangle, \quad 0 \le p < 1. \quad (74)$$

In this form they appeared for the first time in [349, 350], although they were studied much earlier in the paper by Aharonov *et al.* [125]: see equation (41). Binomial states go to the coherent states in the limit $p \to 0$ and $M \to \infty$, provided $\overline{n} = pM = const$, so they are representatives of a wider class of intermediate, or interpolating, states. The special case of $M = 1$ (i.e., combinations of the ground and the first excited states) was named in [349] *Bernoulli states*. Formally, these states coincide with Klauder's *fermion coherent states* (44), although their physical meanings are quite different. The binomial states belong to the family of sub-Poissonian states with Mandel's parameter $\mathcal{Q} = -p$. Actually, using this property one can derive the distribution (74).

Other "intermediate" states are (they are superpositions of an *infinite* number of Fock states) *logarithmic states* [351]

$$|\psi\rangle_{log} = c|0\rangle + \mathcal{N} \sum_{n=1}^{\infty} \frac{z^n}{\sqrt{n}} |n\rangle, \qquad \mathcal{Q} = \frac{(1-|\mathcal{N}|^2)|z|^2}{1-|z|^2}, \quad \overline{n} = \frac{|\mathcal{N}|^2|z|^2}{1-|z|^2}, \quad (75)$$

and *negative binomial states* [352, 353]

$$|\xi, \mu; \theta_n\rangle = \sum_{n=0}^{\infty} e^{i\theta_n} \left[(1-\xi)^{\mu} \frac{\Gamma(\mu+n)}{\Gamma(\mu) n!} \xi^n\right]^{1/2} |n\rangle, \quad \mu > 0, \quad 0 \le \xi < 1. \quad (76)$$

For $\theta_n = n\theta_0$ the state (76) coincides with the generalized phase state (39) introduced in [125, 126]. The negative binomial states are super-Poissonian for all values of their parameters: $\mathcal{Q} = \xi/(1-\xi)$, $\overline{n} = \mu\mathcal{Q}$, whereas the logarithmic states may be both sub- and super-Poissonian. The binomial states and their generalizations are discussed in detail in Chapter 5.

Kerr states and other "macroscopic superpositions"

The main suppliers of nonclassical states are the media with *nonlinear* optical characteristics. One of the simplest examples is the so-called *Kerr nonlinearity*, which can be modelled in the single-mode case by means of the Hamiltonian $\hat{H}_{Kerr} = \omega\hat{n} + \chi\hat{n}^2$ (where $\hat{n} = \hat{a}^\dagger\hat{a}$). The sub-Poissonian statistics in the equilibrium state $\hat{\rho} = \exp(-\beta\hat{H}_{Kerr})$ was shown in [354]. In 1984, Tanaš [355] demonstrated a possibility of obtaining squeezing using this kind of nonlinearity. In 1986, considering the time evolution of the initial coherent state under the action of the "Kerr Hamiltonian", Kitagawa and Yamamoto [284] showed that the states whose initial shapes in the complex phase plane α were circles (these shapes are determined by the equation $Q(\alpha) = const$, where the *Q-function* is defined as $Q(\alpha) = \langle\alpha|\hat{\rho}|\alpha\rangle$), are transformed to some "crescent states", with essentially reduced fluctuations of the number of photons: $\Delta n \sim \langle n \rangle^{1/6}$ (whereas in the case of the squeezed states one has $\Delta n \geq \langle n \rangle^{1/3}$).

The behaviour of the Q-function was also studied by Milburn [356], who (besides confirming the squeezing effect) discovered that under certain conditions, the initial single Gaussian function is split into several well-separated Gaussian peaks. Yurke and Stoler [357] gave the analytical treatment to this problem, generalizing the nonlinear term in the Hamiltonian as $\hat{n}^k$ (k being an integer). The initial coherent state is transformed under the action of the Kerr Hamiltonian to the Titulaer–Glauber state (25):

$$|\alpha; t\rangle = \exp\left(-|\alpha|^2/2\right) \sum_{n=0}^{\infty} \frac{\alpha_t^n}{\sqrt{n!}} \exp\left(-i\chi n^k t\right) |n\rangle, \tag{77}$$

where $\alpha_t = \alpha \exp(-i\omega t)$. Yurke and Stoler noticed that at the special value of time $t_* = \pi/2\chi$, the state (77) becomes the superposition of two or four coherent states, depending on the parity of the exponent k:

$$|\alpha; t_*\rangle = \begin{cases} \left[e^{-i\pi/4}|\alpha_t\rangle + e^{i\pi/4}|-\alpha_t\rangle\right]/\sqrt{2}, & k \text{ is even} \\ \left[|\alpha_t\rangle - |i\alpha_t\rangle + |-\alpha_t\rangle + |-i\alpha_t\rangle\right]/2, & k \text{ is odd} \end{cases} \tag{78}$$

Note that the first superposition (for k even) is different from the even/odd states (30). The even and odd states arise in the case of the two-mode nonlinear interaction $\hat{H} = \omega(\hat{a}^\dagger\hat{a} + \hat{b}^\dagger\hat{b}) + \chi(\hat{a}^\dagger\hat{b} + \hat{b}^\dagger\hat{a})^2$ considered by Mecozzi and Tombesi [358, 359]. In this case, the initial state $|0\rangle_a|\beta\rangle_b$ is transformed at the moment $t_* = \pi/4\chi$ to the superposition

$$|\text{out}, t_*\rangle = \frac{1}{2}\left(e^{-i\pi/4}\left[|\beta_t\rangle_b - |-\beta_t\rangle_b\right]|0\rangle_a + |0\rangle_b\left[|-i\beta_t\rangle_a + |i\beta_t\rangle_a\right]\right).$$

In course of time, such "macroscopic superpositions" of quantum states attracted great attention, being considered as simple models of the "Schrödinger cat states". They are discussed in Chapters 4 and 6.

Other superpositions of quantum states of the electromagnetic field, which can be created in a cavity due to the interaction with a beam of two-level atoms passing one after another, were studied in [360]. They have the form (9) with a finite number of terms, the coefficients c_n being determined by the recurrence relations $c_n = \gamma \tan(\beta\sqrt{n})c_{n-1}$ or $c_n = \gamma \cot(\beta\sqrt{n})c_{n-1}$, therefore they were named "tangent" and "cotangent" states.

3.3 Appearance of "deformations" and related states

Para-coherent states

In 1951 Wigner [121] pointed out that to obtain an equidistant spectrum of the harmonic oscillator, one could use, instead of the canonical bosonic commutation relation (2), weaker conditions

$$[\hat{a}_\epsilon, \hat{H}] = \hat{a}_\epsilon, \qquad [\hat{a}_\epsilon^\dagger, \hat{H}] = -\hat{a}_\epsilon^\dagger, \qquad \hat{H} = \left(\hat{a}_\epsilon^\dagger\hat{a}_\epsilon + \hat{a}_\epsilon\hat{a}_\epsilon^\dagger\right)/2. \tag{79}$$

Then the spectrum of the oscillator becomes $E_n = n + \epsilon$ (in the dimensionless units), with $n = 0, 1, \ldots$ and an arbitrary possible lowest energy ϵ (for the usual oscillator $\epsilon = 1/2$). Wigner's observation is closely related to the problem of *parastatistics* (intermediate between the Bose and Fermi ones) [361], which was studied by many authors in the 1950s and 60s; see [362, 363] and references therein. In 1978, Sharma, Mehta and Sudarshan [364, 365] introduced *para-Bose coherent states* as the eigenstates of the operator $\hat{a}_\epsilon$ satisfying the relations (79):

$$|\alpha\rangle_\epsilon = \mathcal{N} \sum_{n=0}^{\infty} \{2^n \Gamma([[n/2]] + 1)\Gamma([[(n+1)/2]] + \epsilon)\}^{-1/2} \alpha^n |n\rangle_\epsilon, \tag{80}$$

where $|n\rangle_\epsilon \sim \hat{a}_\epsilon^{\dagger n}|0\rangle_\epsilon$, $|0\rangle_\epsilon$ is the ground state, and $[[u]]$ means the greatest integer less than or equal to u. Introducing the Hermitian "quadrature" operators in the same manner as in (5) (but with a different physical meaning), one can check that $\Delta x_\epsilon \Delta p_\epsilon = |\langle[\hat{x}_\epsilon, \hat{p}_\epsilon]\rangle|/2$ in the state (80), i.e., this state is "intelligent" for the operators $\hat{x}_\epsilon, \hat{p}_\epsilon$.

Another kind of "para-Bose operators" was considered in [366] on the basis of a nonlinear transformation of the canonical bosonic operators

$$\hat{A} = F^*(\hat{n}+1)\hat{a}, \qquad \hat{A}^\dagger = \hat{a}^\dagger F(\hat{n}+1), \qquad \hat{n} = \hat{a}^\dagger \hat{a}. \tag{81}$$

The transformed operators satisfy the relations

$$[\hat{A}, \hat{n}] = \hat{A}, \qquad [\hat{A}^\dagger, \hat{n}] = -\hat{A}^\dagger, \qquad \hat{n} = \hat{a}^\dagger \hat{a}. \tag{82}$$

Defining the "para-coherent state" $|\lambda\rangle$ as an eigenstate of the operator $\hat{A}$, one obtains

$$|\lambda\rangle = \mathcal{N}\left(|0\rangle + \sum_{n=1}^{\infty} \frac{\lambda^n |n\rangle}{\sqrt{n!} F^*(1)\cdots F^*(n)}\right), \qquad \hat{A}|\lambda\rangle = \lambda|\lambda\rangle. \tag{83}$$

Evidently, the choice $F(n) = \sqrt{n + 2k - 1}$ reduces the state (83) to the Barut–Girardello state (48). Nowadays the states (83) are known mainly under the name *nonlinear coherent states*: see section 4.2.

q-coherent states

One of the most popular directions in mathematical physics of the last decades of the 20th century was related to various *deformations* of the canonical commutation relations (2) (and others). Perhaps the first study was performed in 1951 by Iwata [367], who found eigenstates of the operator $\hat{a}_q^\dagger \hat{a}_q$, assuming that $\hat{a}_q$ and $\hat{a}_q^\dagger$ satisfy the relation (he used letter p instead of q)

$$\hat{a}_q \hat{a}_q^\dagger - q\,\hat{a}_q^\dagger \hat{a}_q = 1, \qquad q = const. \tag{84}$$

A quarter of a century later, the same relation (84) and its generalizations to the case of several dimensions were considered in [368–372]. A realization of the commutation relation (84) in terms of the usual bosonic operators $\hat{a}, \hat{a}^\dagger$ by means of the *nonlinear* transformation was found in [372]:

$$\hat{a}_q = F(\hat{n})\hat{a}, \qquad \hat{n} = \hat{a}^\dagger \hat{a}. \tag{85}$$

For *real* functions $F(n)$ equation (84) results in the following recurrence relation and its solution:

$$(n+1)[F(n)]^2 - qn[F(n-1)]^2 = 1, \quad F(n) = \{[n+1]_q/(n+1)\}^{1/2}, \tag{86}$$

where symbol $[n]_q$ means

$$[n]_q = (q^n - 1)/(q-1) \equiv 1 + q + \cdots + q^{n-1}. \tag{87}$$

The operators given by (85) obey the relations (82). Using the transformation (85) one can obtain the realizations of more general relations than (84):

$$\hat{A}\hat{A}^\dagger = 1 + \sum_{k=1}^{K} q_k \hat{A}^{\dagger k} \hat{A}^k. \tag{88}$$

The corresponding recurrence relations for $F(n)$ were given in [372].

Coherent states of the *pseudo-oscillator*, defined by the "inverse" commutation relation $[\hat{a}, \hat{a}^\dagger] = -1$, were studied in [373–375]. They satisfy relations (1) and (3), but in the right-hand side of (1) one should write $-\alpha$ instead of α.

The *q-coherent state* was introduced in [369, 372] as

$$|\alpha\rangle_q = \exp_q\left(-|\alpha|^2/2\right) \exp_q(\alpha \hat{a}_q^\dagger)|0\rangle_q, \qquad \exp_q(x) \equiv \sum_{n=0}^{\infty} \frac{x^n}{[n]_q!}, \tag{89}$$

where $[n]_q! \equiv [n]_q[n-1]_q \cdots [1]_q$.

Assuming other definitions of the symbol $[n]_q$, but the same equations (85) and (86), one can construct other "deformations" of the canonical commutation relations. For example, making the choice

$$[x]_q \equiv \left(q^x - q^{-x}\right) / \left(q - q^{-1}\right) \tag{90}$$

one arrives at the relation

$$\hat{a}_q \hat{a}_q^\dagger - q\, \hat{a}_q^\dagger \hat{a}_q = q^{-\hat{n}} \tag{91}$$

considered by Biedenharn in 1989 in his famous paper [376], which gave rise (together with several other publications [377–379]) to the "boom" in the field of "q-deformed" states in quantum optics.

Multi-photon coherent and squeezed states

The usual coherent states are generated from the vacuum by the displacement operator of the form $\hat{U}_1 = \exp(z\hat{a}^\dagger - z^*\hat{a})$, whereas the squeezed states are generated from the vacuum by the squeeze operator $\hat{U}_2 = \exp(z\hat{a}^{\dagger 2} - z^*\hat{a}^2)$. It seems natural to suppose that one could define a much more general class of states, acting on the vacuum by the operator [380]

$$\hat{U}_k = \exp\left(z\hat{a}^{\dagger k} - z^*\hat{a}^k\right). \tag{92}$$

However, such a simple definition leads to certain troubles [381,382], for instance, the vacuum expectation value $\langle 0|\hat{U}_k|0\rangle$ has zero radius of convergence as a power series with respect to z, for any $k > 2$ [381]. Although this phenomenon was considered as a mathematical artifact in [383], many people preferred to modify the operators $\hat{a}^{\dagger k}$ and $\hat{a}^k$ in the argument of the exponential in such a way that no questions on the convergence would arise.

One of the possible ways was studied in the series of papers [384–388]. Instead of a simple power $\hat{a}^{\dagger k}$ in $\hat{U}_k$, it was proposed to use the *k-photon generalized boson operator* (introduced in [363])

$$\hat{A}^\dagger_{(k)} = \left([[\hat{n}/k]]\frac{(\hat{n}-k)!}{\hat{n}!}\right)^{1/2}(\hat{a}^\dagger)^k, \qquad \hat{n} = \hat{a}^\dagger\hat{a}, \qquad k = 1, 2, \cdots, \tag{93}$$

which satisfies the relations (note that $\hat{A}_{(1)} = \hat{a}$, but $\hat{A}_{(k)} \neq \hat{a}^k$ for $k \geq 2$)

$$\left[\hat{A}_{(k)}\,, \hat{A}^\dagger_{(k)}\right] = 1, \qquad \left[\hat{n}\,, \hat{A}_{(k)}\right] = -k\hat{A}_{(k)}. \tag{94}$$

Bužek and Jex [389] used Hermitian combinations of the operators $\hat{A}_{(k)}$ and $\hat{A}^\dagger_{(k)}$ in order to define the kth-order squeezing in the frameworks of the Walls-Zoller scheme (62). The multiphoton squeezing operator was defined as (see also [390])

$$\hat{S}_{(k)} = \exp\left[z\hat{A}^\dagger_{(k)} - z^*\hat{A}_{(k)}\right], \tag{95}$$

which means that we have, in fact, an infinite series of the products of the operators $(\hat{a}^\dagger)^l\hat{a}^{l+k}$ and their hermitially conjugated partners in the argument of the exponential function. The generic multiphoton squeezed state can be written as $|\alpha, z\rangle_{(k)} = \hat{D}(\alpha)\hat{S}_{(k)}(z)|\psi\rangle$ (in the cited papers the vacuum initial state $|\psi\rangle$ was considered). An alternative definition may be (in the simplest case of $\alpha = \psi = 0$)

$$|z; k\rangle = \exp\left(-|z|^2/2\right)\sum_{n=0}^{\infty}\frac{z^n}{n!}\hat{A}^{\dagger n}_{(k)}|0\rangle, \tag{96}$$

which is the superposition of the states with 0, k, $2k$, etc., photons:

$$|z; k\rangle = \exp\left(-|z|^2/2\right)\sum_{n=0}^{\infty}\frac{z^n}{\sqrt{n!}}|kn\rangle. \tag{97}$$

Among the other states considered in [384–388], we mention here the k-photon analogues of the binomial and negative binomial states, which were treated as the multiphoton squeezed $SU(2)$ and $SU(1,1)$ states, respectively

$$|z; k, \sigma\rangle_{SU(2)} = \left(1 + |z|^2\right)^{-\sigma}\sum_{n=0}^{\infty}\left[\frac{(2\sigma)!}{n!(2\sigma-n)!}\right]^{1/2} z^n|kn\rangle, \tag{98}$$

$$|z;k,\sigma\rangle_{SU(1,1)} = \left(1-|z|^2\right)^\sigma \sum_{n=0}^{\infty}\left[\frac{(2\sigma+n-1)!}{n!(2\sigma-1)!}\right]^{1/2} z^n|kn\rangle. \qquad (99)$$

Another construction was considered by Sukumar [391], who has introduced the states ($\alpha = (r/m)e^{i\varphi}$, $\beta = \tanh r$, $m = 1,2,\ldots$)

$$\exp\left(\alpha\hat{T}_m^\dagger - \alpha^*\hat{T}_m\right)|0\rangle = \sqrt{1-\beta^2}\sum_{n=0}^{\infty}\left(\beta e^{i\varphi}\right)^n |mn\rangle, \qquad (100)$$

where $\hat{T}_m = \hat{E}_-^m\hat{n} \equiv (\hat{n}+m)\hat{E}_-^m$, operator $\hat{E}_-$ being defined in (31). Operators $\hat{T}_\pm$ act on the Fock states as follows:

$$\hat{T}_m|n\rangle = \begin{cases} n|n-m\rangle, & n \ge m \\ 0, & n < m \end{cases}, \qquad \hat{T}_m^\dagger|n\rangle = (n+m)|n+m\rangle. \qquad (101)$$

For $m = 1$ the state (100) coincides with the *phase coherent state* (37).

Tombesi and Mecozzi [392] showed that the unitary operator

$$\tilde{U}_4(t) = \exp\left[i\chi\left(\hat{a}^2 - \hat{a}^{\dagger 2}\right)^2 t\right]$$

has no singularities, resulting in normal and higher-order squeezing for all but vacuum initial coherent states (for the "standard" operators U_k (92) with $k > 2$ this was proved in [380]). However, all the moments of the quadrature components diverge at some instants of time. The properties of operators U_3, U_4, and $\tilde{U}_4$ were studied also in [393] and [394] ("three-photon squeezing").

4 Flourishing of "nonclassical states"

Squeezed states

After the papers by Schleich and Wheeler [288, 289, 292], oscillations of the photon distribution function p_n are frequently considered as a signature of nonclassicality. In the case of correlated coherent states, such oscillations were studied in [395, 396]. For the squeezed states, this subject was considered in [397–401]. The photocount distributions and oscillations in the two-mode nonclassical states were studied in [402–408]. The influence of thermal noise was studied in [258, 290, 409–414]. The cumulants and factorial moments of the squeezed state photon distribution function were considered in [398, 409]. They exhibit strong oscillations even in the case of slightly squeezed states [415]. For more recent publications see [416, 417].

The specific properties of the coherent, correlated and squeezed states of a charged particle moving in a homogeneous magnetic field were considered in [418–425] (as usual, they were "rediscovered" several times). In particular, it was shown in [422] that these states can be interpreted as "geometrical" squeezed states, in the sense that in the presence of magnetic field, the canonical momenta are related to the coordinates of the centre of the orbit and the radius-vector of the relative motion, i.e., all variances assume the "geometrical" meaning in the configuration (not phase) space.

As was noticed in [220], inequality (56) permits us to suppose that certain quantum effects, such as, for instance, the tunnelling rate through a potential barrier, could be enhanced in the correlated states due to the appearance of the "renormalized Planck constant" $\hbar_* = \hbar/\sqrt{1-r^2}$. This subject was analyzed in [396,426]. In [427] the name "correlated squeezed state" was used as a synonim of two-mode squeezed states (59) (with the vacuum fiducial vector $|\varphi\rangle$). The dynamics of Gaussian wavepackets possessing nonzero correlation coefficients was studied in [428]. Correlated states of the free particle were considered (under the name "contractive states") in [429–431]. Gaussian correlated packets were applied to many physical problems: from neutron interferometry [432] to cosmology [433]. For the most recent publications on correlated states see [434,435].

Parity-dependent squeezed states $[\hat{S}(z_1)\hat{\Pi}_+ + \hat{S}(z_2)\hat{\Pi}_-]|\alpha\rangle$, where $\hat{\Pi}_\pm$ are the projectors to the even/odd subspaces of the Hilbert space of Fock states, were studied in [436]. Higher order squeezing was studied, e.g., in [437–443]. Various characteristics of squeezed states (phase properties, photon statistics, etc.) and their modifications, such as displaced and squeezed number states (sometimes called "two-photon Fock states" [246]), thermal squeezed states, "sheared states", "squashed states", etc., were studied in [444–469]. The label "successive squeezed states" was used in [470] with respect to the states arising upon multiple application of the squeezing operator to different initial states.

Two-mode and multimode (squeezed) Gaussian states (pure and mixed) were studied in [471–483] (see also Chapter 3). Recently, the properties of multidimensional Gaussian "continuous variable systems" attracted great attention in connection with the problems of quantum teleportation, entanglement, quantum communications and quantum computations [484–494]. Sum and difference squeezing, including multimode generalizations, were considered in [495–498]. Gaussian packets centered on the classical trajectories as approximate solutions to the Schrödinger equation for arbitrary potentials (*"trajectory-coherent states"*) were considered in [499–503]. In the special case of x^4-anharmonicity such packets were considered in [504] under the name "quasi-coherent states" (cf. [263]), which have nothing in common with Skagerstam's [160, 161] "quasi-coherent states". Applications of multimode (entangled) squeezed states for the problems of *quantum imaging* were considered in [505–509]. For reviews of theoretical and experimental achievements in different areas related to squeezed states and their applications see, e.g., [510–513]. *Squeezed fermion states* were introduced in [514] and studied in [515,516].

New and old "intelligent" and minimum uncertainty states

Noise minimum states, which give the minimal value of the photon number operator fluctuation σ_n for the fixed values of the lower-order moments, e.g., $\langle\hat{a}\rangle$, $\langle\hat{a}^2\rangle$, $\langle\hat{n}\rangle$, were considered in [517]. This study was continued in [518], where eigenstates of operator $\hat{a}^\dagger\hat{a} - \xi^*\hat{a} - \xi\hat{a}^\dagger$ were found in the form

$$|\xi; M\rangle = (\hat{a}^\dagger + \xi^*)^M |\xi\rangle_{coh} \tag{102}$$

(these states differ from (61), due to the opposite sign of the term ξ^*). Due to the form of the quasiprobability distribution in the phase space, similar to that discovered in [284], these states were also called *crescent states*. The eigenstates of the most general linear combination of the operators $\hat{a}$, $\hat{a}^\dagger$, $\hat{a}^2$, $\hat{a}^{\dagger 2}$, and $\hat{a}^\dagger\hat{a}$ were studied by Brif in [519], where the general concept of *algebra eigenstates* was introduced. These states were defined as eigenstates of the linear combination $\xi_1\hat{A}_1 + \xi_2\hat{A}_2 + \ldots + \xi_n\hat{A}_n$, where $\hat{A}_k$ are generators of some algebra and ξ_k are complex coefficients. In the case of *two-photon* algebra these states are expressed in terms of the confluent hypergeometric function or the Bessel function, and they contain as special cases many other families of nonclassical states [519–521]. The same general construction (and its special cases related to the $su(1,1)$ algebra) was studied by Trifonov [481, 522, 523] under the names "algebraic coherent states" or "algebra-related" coherent states (see also [524]. "Algebraic" states differ from Klauder's [142] "group" coherent states (46): they can be considered sooner as generalizations of the "Barut–Girardello" states. For other studies on different kinds of "minimum uncertainty states" and "intelligent states" see the papers by Trifonov [227, 271, 525, 526] and other authors and [527–539]. The intelligent states for arbitrary potentials were considered in [540, 541].

Phase states

The phase coherent state (37) was rediscovered in the early 1990s in [542] as a pure analogue of the thermal state. It was also noticed that this state yields a strong squeezing effect. Vourdas [543] introduced the $SU(2)$ and $SU(1,1)$ phase states as the eigenstates of the operators similar to the Sussking–Glogower operator $\hat{E}_-$ (31). For example, the $SU(1,1)$ phase states were defined as the eigenstates of the operator $\hat{\mathcal{E}}_- = \hat{K}_-(\hat{K}_+\hat{K}_-)^{-1/2}$, where operators $\hat{K}_\pm$ are the generators of the algebra $su(1,1)$. The $SU(2)$ phase states were defined in a similar way.

By analogy with the usual squeezed states, which are eigenstates of the linear combination of the operators $\hat{a}$, $\hat{a}^\dagger$ (18), the *phase squeezed states* (PSS) were constructed in [544] (where the state (37) was rediscovered) as the eigenstates of the operator $\hat{B} = \mu\hat{E}_- + \nu\hat{E}_+$ (see also [545]). The coefficients of the decomposition of PSS over the Fock basis are given by $c_n = const\left(z_+^{n+1} - z_-^{n+1}\right)$, where $z_\pm = \left(\beta \pm \sqrt{\beta^2 - 4\mu\nu}\right)/(2\mu)$ and β is the complex eigenvalue of $\hat{B}$. It is worth noting that PSS was actually introduced as far back as in [162]: see Eq. (54). Various "Phase-optimized states" minimizing products of the number variance by different measures of the phase uncertainty were considered in [546–549].

The continuous representation of arbitrary quantum states by means of the phase coherent states (37) (an analogue of the Klauder–Glauber–Sudarshan coherent state representation) was considered in [550] (where it was called the *analytic representation in the unit disk*), [551] (under the name *harmonious state representation*), and [552, 553]. Eigenstates of operator $\hat{Z}(\sigma) = \hat{E}_-(\hat{n} + \sigma)$,

$$|z;\sigma\rangle = \sum_{n=0}^{\infty} \frac{z^n|n\rangle}{\Gamma(n+\sigma+1)}, \quad \hat{Z}(\sigma)|z;\sigma\rangle = z|z;\sigma\rangle, \tag{103}$$

were named *philophase states* in [554]. If $\sigma = 0$, then (103) goes to the special case of the Barut–Girardello state (48) with $k = 1/2$ (see also [555]).

The name "pseudothermal state" was given to the state (37) in [556], where it was shown that this state arises naturally as an exact solution to certain nonlinear modifications of the Schrödinger equation. For other studies related to the phase states see [557–567].

Group coherent and squeezed states

The algebraic approaches to studying the squeezing phenomenon have been used in [568–570] (see Chapter 2). An example of two-mode $SU(1,1)$ coherent states was given in [571]. $SU(1,1)$ interferometers were considered in [491, 534, 573]. Barut–Girardello coherent states were studied in [572], and the "subcoherent P-representation" based on these states was introduced in [574]. A method of constructing squeezed states for general systems (different from the harmonic oscillator) was described in [575], where the eigenstates of the operator $\mu\hat{a}^2 + \nu\hat{a}^{\dagger 2}$ were considered as an example (see also [576, 577]). The generalization of the "charged boson" coherent states (50) in the form of the eigenstates of the operator $\mu\hat{a}\hat{b} + \nu\hat{a}^{\dagger}\hat{b}^{\dagger}$, satisfying the constraint $(\hat{a}^{\dagger}\hat{a} - \hat{b}^{\dagger}\hat{b})|\psi\rangle = 0$, was studied in [578] under the name *pair excitation-deexcitation coherent states*. Nonclassical properties of the *even and odd charge coherent states* and *k-component charge coherent states* were studied in [579, 580].

A detailed discussion of the *spin-coherent states* was given in [581]. Fermion coherent states were studied in [582] (sometimes they were called "Thouless coherent states" [583]). *Spin squeezed states* were discussed in [271, 584, 585]. Coherent states for the groups $E(n)$, $SU(N)$ and $SU(m,n)$ were studied in [586–589]. *Multilevel atomic coherent states* were introduced in [590]. The *squeezed Kerr states* were considered in [591]. Squeezing and displacement of the $SU(1,1)$ intelligent states were performed in [519, 592]. Analytic representations based on the $SU(1,1)$ group coherent states and Barut–Girardello states were compared in [593]. $SU(3)$-coherent states were considered in [594, 595]. For different applications of *affine coherent states* see, e.g., [596, 597].

Different kinds of coherent states for the *Coulomb potential* were studied in [598–608]. *Nonspreading and squeezed Rydberg packets* were considered in [609–616]. The "Bessel" and "hypergeometric" coherent states for this problem were introduced in [617]. Different families of coherent states for the *Morse potential* were given in [618–624]. The case of the Pöschl–Teller potential was studied in [540, 601, 625, 626]. The *singular* oscillator was considered (frequently under other names, such as, e.g., "*pseudoharmonical*," "*isotonic*," or "Calogero–Sutherland" oscillator) in [523, 592, 601, 627–631]. Coherent states for the *reflectionless potentials* were constructed in [632]. Coherent states defined *on a circle* and *on a sphere* were studied in [633–637]. For other publications on group and algebraic states and their applications see [638–642].

Deformations, deformations, deformations

Different q-analogues of the coherent states based on the equations (9)–(11) were introduced in [379]. Two families of coherent states for the *difference analogue* of the harmonic oscillator were studied in [643], and one of them coincided with the q-coherent states. Squeezing properties of *q-coherent states* and different types of *q-squeezed states* were studied in [557, 644–651]. For example, in [644] the q-squeezed state was defined as a solution to equation $\left(\hat{a}_q - \alpha\hat{a}_q^\dagger\right)|\alpha\rangle_{q-sq} = 0$ [cf. (18) and (19)]. It has the following structure:

$$|\alpha\rangle_{q-sq} = \mathcal{N}\sum_{n=0}^{\infty}\alpha^n\left(\frac{[2n-1]_q!!}{[2n]_q!!}\right)^{1/2}|2n\rangle. \tag{104}$$

The "quasi-coherent" state $\exp_q(z\hat{a}_q^\dagger)\exp_q(z^*\hat{a}_q^\dagger)|0\rangle$ was considered in [652]. Coherent states of the two-parameter quantum algebra $su_{pq}(2)$ were introduced in [653], on the basis of the definition

$$[x]_{pq} = \left(q^x - p^{-x}\right)/\left(q - p^{-1}\right).$$

Analogous construction, characterized by the deformations of the form

$$\hat{a}\hat{a}^\dagger = q_1^2\hat{a}^\dagger\hat{a} + q_2^{2\hat{n}} = q_2^2\hat{a}^\dagger\hat{a} + q_1^{2\hat{n}}, \qquad [n] = \left(q_1^{2n} - q_2^{2n}\right)/\left(q_1^2 - q_2^2\right),$$

has been studied in [654] under the name "*Fibonacci oscillator*". The relations between the quantum phase and the q-deformation in the form (91) with $q = \exp(2\pi i/N)$ were established in [655]. *Even and odd q-coherent states* were studied in [656]. The *q-binomial states* were considered in [657]. Multi-mode q-coherent states were studied in [658]. Various families of "q-states", including q-analogues of the "standard sets" of nonclassical states, have been studied in [659–678].

The sets of coherent and squeezed states connected with the "λ-deformations" of the form

$$\hat{a}_\lambda^\dagger = \hat{a}^\dagger + \lambda\hat{I}, \qquad [\hat{a}, \hat{a}_\lambda^\dagger] = \hat{I}, \qquad \hat{H}_\lambda = \hat{H}_0 + \lambda\hat{a} \tag{105}$$

(the Hamiltonian $\hat{H}_\lambda$ is non-Hermitian) were considered in [679–681]. Different "C_λ-deformations" and corresponding coherent states were studied in [682, 683]. They are based on the deformations of the commutation relations of the type

$$[\hat{a}, \hat{a}^\dagger] = \hat{I} + \lambda^{-1}\sum_{\mu,\nu=0}^{\lambda-1}\alpha_\mu\exp\left[2\pi i\nu(\hat{N} - \mu\hat{I})/\lambda\right], \quad \sum\alpha_\mu = 0. \tag{106}$$

Various "*para-states*" were considered in [684–688]. In particular, the generalized "para-commutator relations"

$$[\hat{n}, \hat{a}^\dagger] = \hat{a}^\dagger, \qquad [\hat{n}, \hat{a}] = -\hat{a}, \qquad \hat{a}^\dagger\hat{a} = \phi(\hat{n}), \qquad \hat{a}\hat{a}^\dagger = \phi(\hat{n}+1)$$

were resolved in [686] by means of the nonlinear transformation to the usual bosonic operators $\hat{b}, \hat{b}^\dagger$:

$$\hat{a} = \sqrt{\frac{\phi(\hat{n}+1)}{\hat{n}+1}}\hat{b}, \qquad \hat{A} = \frac{\hat{n}+1}{\phi(\hat{n}+1)}\hat{a}, \qquad [\hat{A}, \hat{a}^\dagger] = 1, \tag{107}$$

and coherent states were defined as $|\alpha\rangle \sim \exp(\alpha\hat{A}^\dagger)|0\rangle$. Interrelations between "para-" and "q-coherent" states were elucidated in [648,686].

Different coherent states for the Hamiltonians obtained from the harmonic oscillator Hamiltonian through various deformations of the potential (by the *Darboux transformation*, for example) were studied in [689–698]. *"Supercoherent"* and *"supersqueezed"* states were studied in [539,699–705]. "PT-symmetric" coherent states were constructed in [706]. The concept of supersymmetry was used for constructing coherent states in the double well and linear ($|x|$) potentials in [707]. The physical meaning of "deformed" oscillators as special kinds of nonlinear oscillators was revealed in [708,709].

4.1 New quantum superpositions

Searching for the states giving maximal squeezing, various superpositions of states were considered. In particular, the superpositions of the Fock states $|0\rangle$ and $|1\rangle$ (*Bernoulli states*) were studied in [349,710,711]; similar superpositions (plus state $|2\rangle$) were studied in [712,713]. The superposition of two squeezed vacuum states was considered in [714]. The superpositions of the even/odd states with the vacuum state were considered in [715,716]. The superpositions of coherent states with equal amplitudes but opposite phases, $|\alpha e^{i\varphi}\rangle$ and $|\alpha e^{-i\varphi}\rangle$ (α real), were considered in [717–719], whereas the superpositions of states with equal phases but different amplitudes, $|\alpha+\delta\alpha\rangle$ and $|\alpha-\delta\alpha\rangle$ (the ratio $\delta\alpha/\alpha$ is real), were introduced in [720] under the name "phase-cat states". "Real" and "imaginary" cat states $|\alpha\rangle \pm |\alpha^*\rangle$ were studied in [721]. Specific superpositions of the vacuum and coherent states $|\alpha\rangle - \exp\left(-|\alpha|^2/2\right)|0\rangle$, for which the total probability of finding the system in the ground state is zero, were considered in [722] as examples of the "most nonclassical" states. *Entangled coherent states* were introduced in [723,724] (see Chapter 4 for other references). Various superpositions of coherent, squeezed, Fock, Kerr, $su(1,1)$, deformed, and other states, as well as methods of their generation, were considered in [725–744].

Representations of nonclassical states (including quadrature squeezed and amplitude squeezed) via linear integrals over some curve C (closed or open, infinite or finite) in the complex plane of parameters,

$$|g\rangle = \int_C g(z)|z\rangle dz, \tag{108}$$

were studied in [715,716,745–748]. Originally, $|z\rangle$ was the coherent state, but it is clear that one may choose other families of states, as well.

"Multiphoton", "circular" and generalized even/odd states

Squeezing in the *fractional photon states* was discussed in [749]. *Multiphoton states*, defined as eigenstates of operator $\hat{a}^k$, were studied in [750–753]. The special case of $k = 2$ is closely related to the *even/odd* states (30) considered, e.g., in [754–760]. The actions of the squeezing and displacement operators on the superpositions of the form $|\alpha, \tau, \varphi\rangle = \mathcal{N}\left(|\alpha\rangle + \tau e^{i\varphi}|-\alpha\rangle\right)$ were studied in [519]. The special case $\tau = 1$ of such superpositions (which contain even, odd, and Yurke–Stoler states for $\varphi = 0, \pi$, and $\pi/2$, respectively) was considered in [761]).

The case of $k = 3$ was considered in [762]. Another type of the three-photon states, named "star states", which arise during the evolution described by the third-order squeezing operator $\hat{U}_3$ (92), was discussed in [763] (cf. [393, 394]). The quantized interaction Hamiltonian $\hat{H}_{int} = i\chi[\hat{b}\hat{a}^{\dagger 3} - \hat{b}^{\dagger}\hat{a}^3]$ was studied in [764]. The generation of such states was considered in [765, 766].

Four-photon states ($k = 4$) [753, 767] can be represented as superpositions of two even/odd coherent states (30) $|\alpha\rangle_{\pm}$ and $|i\alpha\rangle_{\pm}$ whose labels are rotated by 90° in the parameter complex plane (see also [713]). For this reason, the state $|\alpha\rangle_{+} + |i\alpha\rangle_{+}$ was called in [768] the *orthogonal-even state*. A transformation of such states into a pair of two practically disentangled even or odd states due to the interaction with an engineered reservoir was studied in [769].

General schemes of constructing multiphoton coherent and squeezed states of arbitrary orders were given in [770–774]. There exist k orthogonal multiphoton states of the form

$$|\alpha; k, j\rangle = \mathcal{N} \sum_{n=0}^{\infty} \frac{\alpha^{nk+j}}{\sqrt{(nk+j)!}} |nk+j\rangle, \qquad j = 0, 1, \ldots, k-1, \tag{109}$$

which can also be expressed as the superpositions of k coherent states uniformly distributed along the circle,

$$|\alpha; k, j\rangle = \mathcal{N} \sum_{n=0}^{k-1} \exp(-2\pi i n j/k)\, |\alpha \exp(2\pi i n/k)\rangle. \tag{110}$$

Such orthogonal "circular" states were studied in [775, 776]. The difference between the Bialynicki-Birula states (26) and the states (110) is in the "weights" of each coherent state in the superposition. In (26) these weights are different, to ensure the *Poissonian* photon statistics, whereas in the "circular" case all the coefficients have the same absolute value, resulting in non-Poissonian statistics.

"Crystallized" Schrödinger cat states were introduced in [778]. They are superpositions of coherent states $|\alpha_r\rangle$, whose labels are obtained from the given complex number α by the action of elements g_r of some finite group G on the complex plane ($r = 1, 2, \ldots, N$). The relative amplitude of each member of the superposition is given either by the matrix element of an irreducible representaion of the group involved or by the character of this representation $\chi^{(l_k)}(g_r)$:

$$|\psi; l_k\rangle_{crys} \sim \sum_{r=1}^{N} \chi^{(l_k)}(g_r)|\alpha_r\rangle, \qquad \alpha \xrightarrow{g_r} \alpha_r \tag{111}$$

(index l_k labels irreducible representations). If the group G consists of two elements, identity and inversion, then (111) goes to the even or odd state (30). If G is the cyclic group $\mathcal{C}_n$, then (111) goes to the "circular" state (110). A detailed study of various special cases related to different irreducible representations of $\mathcal{C}_n$ for $n \leq 5$ can be found in [778]. Another special case studied in detail in [778] was related to the group of six elements $G = \mathcal{C}_{3v} \sim \mathcal{D}_3$. Then one has various superpositions of the states (110) with $k = 3$ and different j, and similar states with α replaced by $-\alpha^*$. The construction (111) can be extended to continuous groups by replacing sums over group elements with integrals over group parameters (using corresponding invariant measures). In this way one can arrive, in particular, at the "quasi-coherent" states considered in [159–161].

Eigenstates of the operator $(\mu\hat{a} + \nu\hat{a}^\dagger)^k$ were considered in [777] (with the emphasis on the case $k = 2$), whereas eigenstates of the operator $\mu\hat{a}^k + \nu\hat{a}^{\dagger k}$ were discussed in [779]. Eigenstates of products of annihilation operators of different modes have been found in [780] in the form of finite superpositions of products of coherent states. For example, in the two-mode case one has (M is an integer)

$$|\eta\rangle = \mathcal{N} \sum_{n=0}^{Mkl-1} c_j |\alpha \exp\left[(2\pi i n/k)\left(1 + M^{-1}\right)\right]\rangle |\beta \exp[-2\pi i n/(lM)]\rangle,$$

with $\hat{a}^k\hat{b}^l|\eta\rangle = \alpha^k\beta^l|\eta\rangle$. The case of three modes ("trio coherent states") was considered in [781].

The photon distribution functions of even/odd or multiphoton (circular) states have infinite number of regularly distributed zeros ("holes" in the photon-number distribution). The methods of constructing various superpositions of *two* coherent states with equal intensities, resulting in given periodic sets of zeros ("holes") in the photon distributions were studied in [782, 783]. It appears that such states are closely connected with the Euler arithmetic function, therefore they were named "Euler coherent states" in [783]. For the most recent studies on the "multiphoton" or "circular" states and their generalizations see [682, 784–792].

4.2 Zoo of nonclassical states

Nonclassical polarization states

The specific families of two- and multi-mode quantum states connected with the polarization degrees of freedom of the electromagnetic field (*biphotons, unpolarized light, polarization squeezing*) have been introduced by Karassiov [793–795]. For other studies see, e.g., [796–805].

Temporally stable coherent states for arbitrary Hamiltonians

Klauder [806–808] has proposed a general construction of coherent states

$$|z;\gamma\rangle = \sum_n \frac{z^n}{\sqrt{\rho_n}} \exp(-ie_n\gamma)|n\rangle, \qquad \hat{H}|n\rangle = e_n|n\rangle, \tag{112}$$

where positive coefficients $\{\rho_n\}$ satisfy certain conditions, while the discrete energy spectrum $\{e_n\}$ may be quite arbitrary. This construction was applied to the hydrogen atom in [809], where the name "temporally stable coherent states" (TSCS) actually appeared. Another special case of the states (112) is the *Mittag–Leffler coherent state* [810]

$$|z;\alpha,\beta\rangle = \mathcal{N}\sum_{n=0}^{\infty}\frac{z^n}{\sqrt{\Gamma(\alpha n+\beta)}}|n\rangle. \tag{113}$$

The Gaussian exponential form of the coefficients ρ_n was used to construct localized wave packets for the Coulomb problem [811], the planar rotor and the particle in a box [812, 813]. For the most recent studies of TSCS (also called "Gazeau–Klauder coherent states") see, e.g., [540, 601, 605, 624, 625, 814–816].

New sets of "relativistic", "string" and "gauge" coherent states

Different families of coherent states for several new models of the *relativistic oscillator*, different from the Yukawa–Markov type (65), were studied during the 1990s. Mir-Kasimov [817] constructed *intelligent* states (in terms of the Macdonald function $K_\nu(z)$) for the coordinate and momentum operators obeying the "deformed relativistic uncertainty relation"

$$[\hat{x},\hat{p}] = i\hbar\cosh[(i\hbar/2mc)d/dx].$$

(The "usual" uncertainty relations for relativistic particles were analyzed in [818].) Coherent states for another model, described by the equation

$$i\partial\psi/\partial t = \left[\alpha_k\left(\hat{p}_k - im\omega\hat{x}_k\beta\right) + m\beta\right]\psi \tag{114}$$

and named the "Dirac oscillator" in [819] (although similar equations were considered earlier in [820–822]), have been studied in [823, 824]. Coherent states introduced by Aldaya and Guerrero [825] were based on the modified "relativistic" commutation relations

$$[\hat{E},\hat{x}] = -i\hbar\hat{p}/m, \quad [\hat{E},\hat{p}] = im\omega^2\hbar\hat{x}, \quad [\hat{x},\hat{p}] = i\hbar(1+\hat{E}/mc^2). \tag{115}$$

They were studied in [826]. Another family of "relativistic coherent states" for spin-0 particles was constructed in [827, 828]. They have close connections with *nonlinear coherent states* discussed on page 41.

"String coherent states" were considered in [829, 830]. Actually, they are certain specific realizations of the $su(2)$ or $su(1,1)$ coherent states, both of the Klauder type [829] and the Barut–Girardello type [830]. "Gauge" coherent states were discussed in [831, 832].

Photon-added, subtracted and depleted states

An interesting class of nonclassical states consists of the *photon-added* states

$$|\psi,m\rangle_{add} = \mathcal{N}_m\hat{a}^{\dagger m}|\psi\rangle, \tag{116}$$

where $|\psi\rangle$ may be an arbitrary quantum state, m is a positive integer – the number of added quanta (photons or phonons), and $\mathcal{N}_m$ is a normalization constant (which depends on the basic state $|\psi\rangle$). Agarwal and Tara [833] introduced these states for the first time as the *photon-added coherent states* (PACS) $|\alpha, m\rangle$, identifying $|\psi\rangle$ with Glauber's coherent state $|\alpha\rangle$ (1). These states differ from the displaced number state $|n, \alpha\rangle$ (61). Taking the initial state $|\psi\rangle$ in the form of a squeezed state, one obtains *photon-added squeezed states* [834–836]. The *even/odd photon-added states* were studied in [837, 838]. One can easily generalize the definition (116) to the case of mixed quantum states described in terms of the statistical operator $\hat{\rho}$: the statistical operator of the mixed photon-added state has the form (up to the normalization factor) $\hat{\rho}_m = \hat{a}^{\dagger m}\hat{\rho}\hat{a}^m$. A concrete example is the *photon-added thermal state* [839,841]. A distinguishing feature of all photon-added states is that the probability of detecting n quanta (photons) in these states equals exactly zero for $n < m$. These states arise in a natural way in the processes of the field–atom interaction in a cavity [833]. Replacing the creation operator $\hat{a}^\dagger$ in the definition (116) by the annihilation operator $\hat{a}$ one arrives at the *photon-subtracted state* $|\psi, m\rangle_{subt} = \mathcal{N}_m \hat{a}^m |\psi\rangle$. The methods of creating photon-added/subtracted states via conditional measurements on a beam splitter were discussed in [842, 843].

Photon-added states are closely related to the *boson inverse operator* (34), whose properties were studied in [844–846]. It was shown in [847] that PACS $|\alpha, m\rangle_{add}$ is an eigenstate of the operator $\hat{a} - m\hat{a}^{\dagger -1}$ with eigenvalue α. The *photon-depleted coherent states* $|\alpha, m\rangle_{depl} = \mathcal{N}_m \hat{a}^{\dagger -m}|\alpha\rangle$ were introduced in [847] together with the modified photon-added states $|\alpha, m\rangle'_{add} = \mathcal{N}_m \hat{a}^{-m}|\alpha\rangle$. The difference between the four families of states is seen distinctly in their decompositions over the Fock basis (for the initial coherent state):

$$|\alpha, m\rangle_{add} \sim \sum_{n=0}^{\infty} \alpha^n \frac{\sqrt{(m+n)!}}{n!}|m+n\rangle, \quad |\alpha, m\rangle_{subt} \sim \sum_{n=0}^{\infty} \frac{\alpha^{m+n}}{\sqrt{n!}}|n\rangle,$$

$$|\alpha, m\rangle'_{add} \sim \sum_{n=0}^{\infty} \frac{\alpha^{m+n}}{\sqrt{(m+n)!}}|m+n\rangle, \quad |\alpha, m\rangle_{depl} \sim \sum_{n=0}^{\infty} \frac{\alpha^n \sqrt{n!}}{(m+n)!}|n\rangle.$$

The states $\hat{E}_+^m|\alpha\rangle$, which could be named "shifted coherent states" (because their quantum number distribution is transformed as $p_n^{(m)} = p_{n-m}^{(0)}$ for $n \geq m$), were considered in [848]. For the most recent studies on photon-added states see [849–855].

Nonclassical states and special functions

Different special functions have been used for constructing various families of nonclassical states. The *Bessel coherent state*

$$|\lambda\rangle = \mathcal{N} \sum_{n=0}^{\infty} \frac{\lambda^n}{n!}|n\rangle = \mathcal{N} \sum_{n=0}^{\infty} \frac{(\lambda \hat{A}^\dagger)^n}{(n!)^2}|0\rangle = \mathcal{N} I_0\left(2\lambda \hat{A}^\dagger\right)|0\rangle \tag{117}$$

(which is in fact a special case of the Barut–Girardello state (48) with $k = 1/2$, but quite different from the hydrogen atom "Bessel states" of [617]) was introduced in [366]. The states whose coefficients of the expansion over the Fock basis are expressed in terms of Bessel functions were considered for the first time by Jackiw [80]: see Eq. (53). Similar states of the form $|\psi\rangle = \mathcal{N}\sum_n J_{\mu+n}(\gamma)e^{in\phi}|n\rangle$, introduced in [547] under the name "phase-optimized states", were renamed "Lommel states" in [566, 567]. The exact eigenstates of the Hamiltonian $\omega\hat{n} + \lambda\hat{C}$ (where the operator $\hat{C}$ was defined in Eq. (31)) were expressed in terms of the Lommel polynomials in [856]. The "phase-optimized" states of the form $|\psi\rangle \approx \mathcal{N}\sum_n \mathrm{Ai}\left(a[n+b]+b_0\right)|n\rangle$, where $\mathrm{Ai}(z)$ is the *Airy* function, were considered in [546, 549].

The *intelligent states* for the generators $\hat{K}_{1,2}$ of the $su(1,1)$ algebra in the form $\hat{S}(z)H_m(\xi\hat{a}^\dagger)|0\rangle$ (obviously, they are finite superpositions of the *squeezed number states*) were named in [857] the *Hermite polynomial states.* Such states were studied in [858]. The *minimum uncertainty state* for *sum squeezing* in the form $\hat{S}(\xi)H_{pq}(\mu\hat{a}_1^\dagger, \mu\hat{a}_2^\dagger)|00\rangle$ was found in [859]. Here H_{pq} is a special case of the family of *two-dimensional Hermite polynomials* (these polynomials are discussed in Chapter 3). The *Laguerre polynomial state* $L_M(-y\hat{R}^\dagger)|0\rangle$, where $\hat{R}^\dagger = \hat{a}^\dagger\sqrt{\hat{N}+1}$, was introduced in [860]. A more general definition $L_M(\xi\hat{J}_+)|k,0\rangle$ (where $\hat{J}_+$ is one of the generators of the $su(1,1)$ algebra, and $|k,n\rangle$ is the discrete basis state labelled by the representation index k) has been given in [592]. The properties of these states were studied in [861].

The states (9) with coefficients c_n given by the *Chebyshev polynomials* of the first and second kinds were discussed in [567,862]. *Jacobi polynomial states* were introduced in [843]. The *Gegenbauer polynomials* appear as the coefficients of the decomposition of the eigenstates of the *cosine phase operator* (31) over the Fock basis [122]. Several modifications of the binomial distributions have been used to define *Pólya states* [863], *hypergeometric states* [864] (quite different from the "hypergeometric states" of [617]), *negative hypergeometric states* [865, 866], *Tricomi states* [867], and so on. For example, the states introduced in [868]

$$|N,\alpha,\beta\rangle \sim \sum_{n=0}^{N}\left[\frac{(\alpha+1)_n(\beta+1)_{N-n}}{n!(N-n)!}\right]^{1/2}|n\rangle$$

are related to the *Hahn polynomials*. They contain as special or limit cases the usual binomial and negative binomial states. Nonclassical states related to the *Meixner polynomials* and their special cases (e.g., Charlier and Kravchuk polynomials) were considered in [869]. Penson and Solomon [870] have introduced the state $|q,z\rangle = \varepsilon(q,z\hat{a}^\dagger)|0\rangle$, where function $\varepsilon(q,z)$ is a generalization of the exponential function given by the relations

$$\mathrm{d}\varepsilon(q,z)/\mathrm{d}z = \varepsilon(q,qz), \quad \varepsilon(q,z) = \sum_{n=0}^{\infty} q^{n(n-1)/2}z^n/n!. \tag{118}$$

The Euler arithmetic function was used in [783].

Nonclassical mixed states

Mixed quantum states with *negative binomial distributions* of the diagonal elements of the density matrix in the Fock basis appeared in the theory of photodetectors and amplifiers in [871,872]. Mixed analogues of different families of coherent and squeezed states were studied in [873–875]. *Grey-body states* were considered in [876, 877]. *Maximum-entropy states*, which are mixed analogues of multiphoton states, were discussed in [878], while analogues of the even and odd states were considered in [879]. The name *shadowed states* was given in [880] to the mixed states whose statistical operators have the form $\hat{\rho}_{sh} = \sum p_n|2n\rangle\langle 2n|$ (i.e., generalizing the even thermal state (29)). Examples of such states were considered in [881]. *Shifted thermal states*, which can be written as $\hat{\rho}_{th}^{(shift)} = \hat{E}_+^m \hat{\rho}_{th} \hat{E}_-^m$ (where the operators $\hat{E}_\pm$ and $\hat{\rho}_{th}$ are given by (35) and (38)), have been considered in [882]. Their mixtures with the usual thermal states ("partially shifted thermal states") were also studied in [882], whereas methods of generating such states in a micromaser were discussed in [883].

"Truncated" and "intermediate" states

After the papers by Pegg and Barnett [884, 885], where the finite-dimensional "truncated" Hilbert space was used to define the phase operator, various "truncated" versions of nonclassical states have been studied. Properties of the state $\sum_{n=0}^{M}(1+n)^{-1}|n\rangle$ were discussed in [544, 886, 887]. *Quasi-photon phase states* $\sum_{n=0}^{s} \exp(in\vartheta)|n\rangle_g$, where $|n\rangle_g$ is the squeezed number state, were considered in [888]. The "finite-dimensional" and "discrete" coherent states were considered in [889–892]. The *generalized geometric state* $|y; M\rangle = \sum_{n=0}^{M} y^{n/2}|n\rangle$ was introduced in [893], and its *even* variant in [894]. In the limit $M \to \infty$ this state goes to the *phase coherent states* (37) (called *geometric states* in [893, 894], due to the geometric progression form of the coefficients in the Fock basis). A "truncated" q-oscillator was considered in [895]. For other examples see [896–904].

Another class of "truncated" states, obtained by removing the vacuum contribution from the superposition over the Fock basis and renormalizing the remaining coefficients as $c_n \to (1-|c_0|^2)^{-1/2}c_n$, $n \geq 1$, was considered in [722].

Reciprocal binomial states

$$|\Phi; M\rangle = \mathcal{N}\sum_{n=0}^{M} e^{in\phi}\sqrt{n!(M-n)!}\,|n\rangle \tag{119}$$

were introduced in [905], and schemes of their generation were discussed in [906, 907]. *Intermediate number squeezed states* $\hat{S}(z)|\eta, M\rangle$ (where $|\eta, M\rangle$ is the binomial state) were introduced in [908] and generalized in [909]. *Intermediate number coherent states* $\sqrt{\xi}|n\rangle + \sqrt{1-\xi}e^{i\varphi}|\alpha\rangle$ were considered in [910]. *Multinomial states* were introduced in [911]. Barnett [912] has introduced the modification of the *negative binomial states* of the form

$$|\eta, -(M+1)\rangle = \sum_{n=M}^{\infty}\left[\frac{n!}{(n-M)!M!}\eta^{M+1}(1-\eta)^{n-M}\right]^{1/2}|n\rangle, \tag{120}$$

which was studied in [913]. A generalization of the binomial state proposed in [914] consists in replacing factor $p^n(1-p)^{M-n}$ in the definition (74) by the product

$p(p+nq)^{n-1}[1-p+(M-n)q]^{M-n}$. The *binomial coherent states*

$$|\lambda; M\rangle \sim (\hat{a}^\dagger+\lambda)^M|0\rangle \sim \sum_{n=0}^{M} \frac{\lambda^{M-n}|n\rangle}{(M-n)!\sqrt{n!}} \sim \hat{D}(-\lambda)\hat{a}^{\dagger M}\hat{D}(\lambda)|0\rangle \qquad (121)$$

(here λ is real, $\hat{D}(\lambda)$ is the displacement operator) were studied in [915,916]. The state (121) is the eigenstate of operator $\hat{a}^\dagger\hat{a}+\lambda\hat{a}$ with the eigenvalue M. It can be also called the "displaced photon-added coherent state" [915]. The superposition states $|\lambda; M\rangle + \exp(i\varphi)|-\lambda; M\rangle$ were studied in [917]. Replacing operator $\hat{a}^\dagger$ in the right-hand side of (121) by the spin raising or lowering operators $\hat{S}_\pm$ one arrives at the *binomial spin-coherent state* [916]. For other studies on "new" and "old" classes of "intermediate" states, like *binomial, negative binomial* states and their generalizations, see [524,550,918–927].

Nonlinear coherent states

The *nonlinear coherent states* (NLCS) were defined in [928,929] as the right-hand eigenstates of the product of the boson annihilation operator $\hat{a}$ and some function $f(\hat{n})$ of the number operator $\hat{n}$,

$$f(\hat{n})\hat{a}|\alpha, f\rangle = \alpha|\alpha, f\rangle. \qquad (122)$$

Actually, such states have been known for many years under other names. The first example is the *phase state* (37) or its generalization (39) (known nowadays as the *negative binomial state* or the $SU(1,1)$ group coherent state), for which $f(n) = [(1+\kappa)/(1+\kappa+n)]^{1/2}$. The decomposition of the NLCS over the Fock basis has the form (83), consequently, NLCS coincide with "para-coherent states" of Ref. [366]. Many nonclassical states turn out eigenstates of some "nonlinear" generalizations of the annihilation operator. It was mentioned in section 2 that the "Barut–Girardello states" belong to the family (83). Another example is the *photon-added state* (116), which corresponds to the function $f(n) = 1-m/(1+n)$ [930]. The physical meaning of NLCS was elucidated in [928,929], where it was shown that such states may appear as stationary states of the centre-of-mass motion of a trapped ion [928], or may be related to some nonlinear processes (such as a hypothetical "frequency blue shift" in high intensity photon beams [929]). Even and odd nonlinear coherent states, introduced in [931], were studied in [932–935], and applications to the ion in the Paul trap were considered in [936]. Further generalizations, namely, NLCS on the circle, were given in [937] (also with applications to the trapped ions). Nonclassical properties of NLCS and their generalizations have been studied in [938–947]. *Nonlinear entangled states* were introduced in [948] as eigenstates of the operators of the form $f(\hat{a}^\dagger\hat{a} - \hat{b}^\dagger\hat{b})(\hat{a} - \hat{b}^\dagger)$.

Quantum state engineering and quantum tomography

The problem of creating Fock states and their arbitrary superpositions in a cavity (in particular, via the interaction between the field and atoms passing through the cavity), named in [949] *quantum state engineering*, was studied in [949–954].

Generation of squeezed and "cat" states was considered in [771, 955, 956] (see Chapter 4 for more references). The use of *beam splitters* to create various types of nonclassical states was considered in [843, 957–961], and the role of "nonclassicality" of input states for creating *entangled states* in beam splitters was discussed in [962]. The methods of generating superpositions of the two first Fock states (44) by means of "optical state truncation" were discussed in [963], and their generalizations to the case of the first N Fock states were considered in [964]. The "quantum scissors" for truncating quantum states were considered in [965]. A possibility of creating nonclassical states in a cavity with *moving mirrors* (which can mimic a Kerr-like medium) was studied in [966–968].

Descriptions of proposed and performed experiments on generating nonclassical states can be found, e.g., in [58, 66, 969–977] (nonclassical states of the electromagnetic field inside a cavity) and [58, 973, 974, 978–984] (nonclassical motional states of trapped ions); see also Chapter 5 of this book. Various aspects of the problem of *detecting* nonclassical quantum states and their "recognition" or reconstruction were treated in [58, 717, 973, 976, 985–996]. The conditional generation of special states via continuous measurements was discussed in [997–1000].

Miscellanea: nonclassical states in various areas of physics

The squeezed and "cat" states in *high energy physics* were considered in [1001–1012]. Squeezed states of quantum bosonic strings were considered in [1013]. Applications of the squeezed and other nonclassical states to *cosmological problems* were studied in [1014–1024]. Nonclassical states in neutron physics were discussed in [1025, 1026]. A possibility of observing macroscopic superpositions in superconducting circuits was discussed in [1027]). Squeezed and "cat" states in the *Josephson junctions* were considered in [1028–1033]. Squeezed states of *phonons* and other bosonic excitations (*polaritons*, *excitons*, etc.) in condensed matter were studied in [1034–1050]. Spin-squeezed and spin-coherent states were used in [1051–1054]. For applications of squeezed atomic states in *spectroscopy* see, e.g., [1055]. Nonclassical states in *molecules* were studied in [616, 1056–1062]. Nonclassical states of the *Bose-Einstein condensate* were considered in [982, 1063–1067].

Several other names of specific kinds of quantum states are frequently met in the current literature. *Dark states* [1068] are certain superpositions of the atomic eigenstates, whose typical common feature is the existence of some sharp "dips" in the spectra of absorption, fluorescence, etc., due to the destructive quantum interference of transition amplitudes between the different energy levels involved. These states are connected with the *nonlinear coherent states* [928]. For reviews and references see, e.g., [1069–1072]. *Greenberger-Horne-Zeilingler states* (or *GHZ-states*) are certain states of three or more correlated spin–$1/2$ particles; they are popular in the studies related to the EPR-paradox [1073], Bell inequalities [1074], quantum teleportation, and so on. They were introduced in [1075] and described in detail in [1076] (for methods of generation and other references see, e.g., [1077]). Relations between *Bell states* and (generalized) coherent states were discussed in [1078, 1079]. Wave functions $\psi(x,y) \sim (x-iy)^m \exp[-(x^2+y^2)/\sigma^2]$

correspond to *vortex states* [1080–1082] with nonzero angular momentum.

The name "sub-Poissonian state" was used by many authors [450, 574, 725, 729, 731, 998, 1040, 1083] in connection with the states having sub-Poissonian statistics. However, in contradistinction to papers [345,346], no specific new states were derived in that studies (remember that all "binomial" states (74) are sub-Poissonian). Examples of "super-chaotic" or "super-random" states, possessing positive Mandel's parameter exceeding the mean number of quanta, $\mathcal{Q} > \langle n \rangle$ (in other terms, the second-order correlation parameter $g^{(2)}(0)$ must be greater than its value for the thermal states $g^{(2)}_{therm}(0) = 2$), were considered, e.g., in studies [173, 1084]. Note that all negative binomial states (76) with parameter $\mu < 1$ belong to this class of states.

4.3 Measures of squeezing and nonclassicality

Invariant ("principal") squeezing

The instantaneous values of variances σ_x, σ_p and σ_{xp} cannot serve as true measures of squeezing in all cases, since they depend on time in the course of the free evolution of the oscillator. For example (in dimensionless units),

$$\sigma_x(t) = \sigma_x(0)\cos^2(t) + \sigma_p(0)\sin^2(t) + \sigma_{xp}(0)\sin(2t),$$

and it can happen that both variances σ_x and σ_p are large, but nonetheless the state is highly squeezed due to the large nonzero covariance σ_{xp}. It is reasonable to introduce some invariant characteristics which do not depend on time in the course of free evolution (or on phase angle in the definition of the field quadrature as $\hat{E}(\varphi) = \left[\hat{a}\exp(-i\varphi) + \hat{a}^\dagger\exp(i\varphi)\right]/\sqrt{2}$ [1085]). They are related to the invariants of the total variance matrix or to the lengths of the principal axes of the ellipse of equal quasiprobabilities, which gives a graphical image of the squeezed state in the phase space (this explains the name "principal squeezing" used in [1086]). Following [422], we introduce the minimal σ_- and maximal σ_+ values of the variances σ_x or σ_p, which can be found by looking for extremal values of $\sigma_x(t)$ as a function of time:

$$\sigma_\pm = \frac{1}{2}\left(T \pm \sqrt{T^2 - 4d}\right), \quad T = \sigma_x + \sigma_p, \quad d = \sigma_x\sigma_p - \sigma_{xp}^2. \qquad (123)$$

$T \equiv 1 + 2\left(\langle\hat{a}^\dagger\hat{a}\rangle - |\langle\hat{a}\rangle|^2\right)$ is the double *energy of quantum fluctuations* (which is conserved in the process of free evolution), whereas parameter d, coinciding with the left-hand side of the *Schrödinger–Robertson uncertainty relation* (55), determines (for Gaussian states) the quantum "purity": $\mathrm{Tr}\hat{\rho}^2_{Gauss} = 1/\sqrt{4d}$ (it is the so-called *universal quantum invariant*: see Chapter 3). Equivalent expressions were found in [1085–1087]. Evidently, $T \le 1 + 2\bar{n}$, where $\bar{n}$ is the mean photon number. Then one can easily derive from (123) the inequalities

$$\sigma_- \ge \bar{n} + \tfrac{1}{2} - \sqrt{(\bar{n} + \tfrac{1}{2})^2 - d} > \frac{d}{1 + 2\bar{n}}. \qquad (124)$$

For pure states ($d = 1/4$) these inequalities were found in [1088]. For quantum mixtures, squeezing ($\sigma_- < 1/2$) is possible provided $\bar{n} \geq d^2 - 1/4$. The invariant squeezing parameters for *two-mode* (pair coherent) states were studied in [1089].

Mandel's parameter and its generalizations

One of the simplest quantitative "measures of nonclassicality" is the Mandel parameter (73) [347]. If $Q < 0$, then the state is undoubtedly "nonclassical", having the sub-Poissonian photon statictics. Several well-known sets of nonclassical states are sub-Poissonian for all values of their parameters. The first example is the set of binomial states (74), for which $Q = -p$ for any M. Another family consists of the *photon-added coherent states* $|\alpha, m\rangle \sim \hat{a}^{\dagger m}|\alpha\rangle$ [833]. It was shown in [574] that the Barut–Girardello coherent states (48) also have $Q < 0$ for arbitrary values of parameters z and k. These states were named in [574] "subcoherent states". However, they are quite different from the "subcoherent states" of studies [345, 346], defined in accordance with equation (72).

There are many "nonclassical" states, for which $Q \geq 0$. For example, for all Glauber–Titulaer generalized coherent states (25), $Q = 0$, and for all even coherent states (30), $Q > 0$ (while $Q < 0$ for the odd coherent states). Squeezed states may possess Poissonian, sub- and super-Poissonian statistics, depending on relative phases of their Fock state components [396, 713, 1090]. It is not difficult to constract obviously nonclassical states with $Q = 0$ and without squeezing [839, 840]. Therefore many people tried to find other more or less simple and convenient "measures of nonclassicality". Agarwal and Tara [839] have noticed that for the "classical" state (possessing non-singular positive P-distribution), the matrix

$$\mathcal{M}^{(n)} = \left\| \begin{array}{ccccc} 1 & m_1 & m_2 & \cdots & m_{n-1} \\ m_1 & m_2 & m_3 & \cdots & m_n \\ \vdots & \vdots & \vdots & \vdots & \vdots \\ m_{n-1} & m_n & m_{n+1} & \cdots & m_{2n-2} \end{array} \right\|, \qquad m_n \equiv \langle \hat{a}^{\dagger n} \hat{a}^n \rangle$$

must be positively definite. Thus one immediately obtains many parameters, like $\det \mathcal{M}^{(n)}$ and the principal minors of this matrix, which must be positive in the "classical" case. If some of these parameters are negative, the state is "nonclassical". The special case of such "generalized Mandel parameters",

$$Q^{(k)} = \mathcal{N}^{(k+1)}/\langle n \rangle - \mathcal{N}^{(k)}, \qquad \mathcal{N}^{(k)} \equiv \sum_{n=k}^{\infty} p_n n!/(n-k)!\,,$$

was discussed in [1091] ($\mathcal{N}^{(k)}$ is the kth factorial moment). Klyshko [1092] proposed using the moments $a_k(s)$ of a modified distribution $P(n)e^{-sn}$ with $s \geq 0$. One of his criteria is as follows: the state is nonclassical if $a_0(s) > 1$, $a_k(s) < 0$, and $a_{k-1}(s)a_{k+1}(s) < [a_k(s)]^2$ for some $s \geq 0$, $k = 1, 2, \ldots$. This approach was used in [882, 1093]. The criterion $\mathcal{N}^{(l+k)}\mathcal{N}^{(|j-k|)} < \mathcal{N}^{(l)}\mathcal{N}^{(j)}$, quantifying higher order nonclassical effects, was considered in [1094] (although its special case $k = 1$ was studied in [1095]).

Other measures of nonclassicality and entanglement

Some authors tried to give quantitative measures of "nonclassicality," taking into account the dimensions of the regions in the phase space where different quantum quasiprobabilities assume negative values. Lee [1096] considered the Glauber–Sudarshan function $P(\alpha)$ and introduced the concept of the *nonclassical depth* τ_m, defined as the minimal value of the parameter τ for which the function

$$R(z,\tau) = (\pi\tau)^{-1} \int d^2w \exp(-|z-w|^2/\tau)P(w) \tag{125}$$

is positive and has no singularities (worse than the delta-function). For coherent states, $\tau_m = 0$, while $\tau_m = 1$ for the Fock states. Other states are characterized by the values $0 < \tau_m < 1$. For instance, $\tau_m = \frac{1}{2}\left(1-e^{-2r}\right)$ for the vacuum squeezed state $\hat{S}(z)|0\rangle$ with $|z| = r$ [1096], and $\tau_m = p$ for the binomial state (74) [1097]. For other examples see [722, 1098]. Similar measures of nonclassicality, based on the regions of negativity of the s-ordered (Cahill–Glauber) quasiprobabilities, are discussed in Chapter 2. The concept of "weak" and "strong" nonclassicality was introduced in [1099, 1100]. The state is "weakly nonclassical", if the Glauber–Sudarshan function $P(\alpha)$ is not a "good" distribution (e.g., nonpositive or more singular than the delta-function), but the *phase averaged* function $\mathcal{P}(I) = \int_0^{2\pi} P(\sqrt{I}e^{i\vartheta})d\vartheta/(2\pi)$ is still a "good" one. If both $P(\alpha)$ and $\mathcal{P}(I)$ are "bad" (from the classical point of view), then the state is "strongly nonclassical". It was shown in [1099] that the Gaussian states may be either "classical" (if they are highly mixed and close to the thermal states) or "strongly nonclassical" (if they exhibit squeezing), without the intermediate regime. On the contrary, the Titulaer–Glauber generalized coherent states (25) are "weakly nonclassical", since their function $\mathcal{P}(I)$ is the Poissonian distribution. A quasiprobability approach to the problem of nonclassicality of *spin states* was used in [1101].

Hillery [1102, 1103] proposed to use the concept of *distance* in the Hilbert space between the given quantum state and the family of "classical" states (coherent or thermal) to give quantitative measures of the "degree of nonclassicality". The "polarized" distances were introduced in [1104]. For other references and the recent results concerning different approaches to the problem of quantifying the degree of nonclassicality see Chapter 2 and [1105–1109].

A specific class of *quantum mixtures* exhibiting "nonclassical" correlations between components (although satisfying the Bell inequalities) was considered by Werner in 1989 [1110]. The simplest example in the case of two spin-$\frac{1}{2}$ particles is given by the statistical operator $\hat{\rho}_W = \frac{1}{2}(\hat{\rho}_{singlet} + \frac{1}{4}\hat{I})$. Nowadays such states and their generalizations are frequently refered to as "Werner states" [1111]. They can be met in studies of the problem of separability of entangled quantum mixtures: see, e.g., [1112–1114].

Measures of *entanglement* (this term was introduced by Schrödinger in [114, 1115]) based on different kinds of *entropies* (von Neumann's, Shannon's, Wehrl's, etc.) were considered in [1116–1126]. The Hilbert-Schmidt distance between the given state and different "disentangled" states was used in [1120, 1127–1130]. Several authors [1131–1133] considered the "linear entropy of entanglement" and similar quantities related to the "quantum purity" $\text{Tr}\hat{\rho}^2$. Different kinds of "negativities" were used in [1134]. "Concurrences" were considered in [1135, 1136]. "Covariance entanglement measures" were introduced in [1137].

Bibliography

[1] Mandel, L. and Wolf, E., *Optical Coherence and Quantum Optics*, Cambridge Univ. Press., Cambridge, 1995.

[2] Glauber, R.J., Photon correlations. *Phys. Rev. Lett.* (1963) **10** 84–86.

[3] Sudarshan, E.C.G., Equivalence of semiclassical and quantum mechanical descriptions of statistical light beams. *Phys. Rev. Lett.* (1963) **10** 277–279.

[4] Glauber, R.J., Coherent and incoherent states of the radiation field. *Phys. Rev.* (1963) **131** 2766–2788.

[5] Klauder, J.R., The action option and a Feynman quantization of spinor fields in terms of ordinary *c*-numbers. *Ann. Phys.* (NY) (1960) **11** 123–168.

[6] Fock, V., Verallgemeinerung und Lösung der Diracschen statistischen Gleichung. *Z. Phys.* (1928) **49** 339–357.

[7] Schrödinger, E., Der stetige Übergang von der Mikro- zur Makromechanik. *Naturwissenschaften* (1926) **14** 664–666 [The continuous transition from micro- to macro-mechanics. In: Schrödinger, E., *Collected Papers on Wave Mechanics* (second edition), pp. 41–44. Chelsea Publ. Corp., New York, 1978].

[8] Darwin, C.G., Free motion in the wave mechanics. *Proc. Roy. Soc.* A (1927) **117** 258–293.

[9] Fock, V., Zur Quantenelektrodynamik. *Phys. Z. Sowjetunion* (1934) **6** 425–469.

[10] Iwata, G., Non-hermitian operators and eigenfunction expansions. *Prog. Theor. Phys.* (1951) **6** 216–226.

[11] Feynman, R.P., An operator calculus having applications in quantum electrodynamics. *Phys. Rev.* (1951) **84** 108–128.

[12] Glauber, R.J., Some notes on multiple-boson processes. *Phys. Rev.* (1951) **84** 395–400.

[13] Friedrichs, K.O., *Mathematical Aspects of the Quantum Theory of Fields*, Interscience, New York, 1953.

[14] Schwinger, J., The theory of quantized fields. III. *Phys. Rev.* (1953) **91** 728–740.

[15] Schwinger, J., *Quantum Kinematics and Dynamics*. W.A. Benjamin, New York, 1970.

[16] Novozhilov, Y.V. and Tulub, A.V., Functional method in quantum field theory. *Usp. Fiz. Nauk* (1957) **61** 53–102 (in Russian) [in English: *The Method of Functionals in the Quantum Theory of Fields*. Gordon & Breach, London, 1961].

[17] Rashevskii, P.K., On mathematical foundations of quantum electrodynamics. *Uspekhi Mat. Nauk* (1958) **13** no. 3(81) pp. 3–110 (in Russian).

[18] Bargmann, V., On a Hilbert space of analytic functions and an associated integral transform. Part I. *Commun. Pure Appl. Math.* (1961) **14** 187–214.

[19] Bargmann, V., Remarks on a Hilbert space of analytic functions. *Proc. Nat. Acad. Sci. USA* (1962) **48** 199–204.

[20] Schweber, S.S., On Feynman quantization. *J. Math. Phys.* (1962) **3** 831–842.

[21] Henley, E.M. and Thirring, W., *Elementary Quantum Field Theory*, p. 15. McGraw-Hill, New York, 1962.

[22] Glauber, R.J., The quantum theory of optical coherence. *Phys. Rev.* **130** (1963) 2529–2539.

[23] Mehta, C.L. and Sudarshan, E.C.G., Relation between quantum and semiclassical description of optical coherence. *Phys. Rev.* **138** (1965) B274–B280.

[24] Klauder, J.R., McKenna J., and Currie, D.G., On "diagonal" coherent-state representations for quantum-mechanical density matrices. *J. Math. Phys.* (1965) **6** 734–739.

[25] Carruthers, P. and Nieto, M., Coherent states and number-phase uncertainty relation. *Phys. Rev. Lett.* (1965) **14** 387-389.

[26] Carruthers, P. and Nieto, M., Coherent states and the forced quantum oscillator. *Amer. J. Phys.* (1965) **33** 537-544.

[27] Cahill, K.E., Coherent-state representations for the photon density operator. *Phys. Rev.* (1965) **138** B1566–B1576.

[28] Klauder, J.R. and Sudarshan, E.C.G., *Fundamentals of Quantum Optics*. W.A. Benjamin, New York, 1968.

[29] *Coherent States Applications in Physics and Mathematical Physics* (J.R. Klauder and B.-S. Skagerstam, eds.). World Scientific, Singapore, 1985.

[30] Anderson, P.W., Coherent excited states in the theory of superconductivity: Gauge invariance and the Meissner effect. *Phys. Rev.* (1958) **110** 827–835.

[31] Senitzky, I.R., Incoherence, quantum fluctuations, and noise. *Phys. Rev.* (1962) **128** 2864–2870.

[32] Carruthers, P. and Nieto, M., The phase-angle variables in quantum mechanics. *Rev. Mod. Phys.* (1968) **40** 411–440.

[33] Nieto, M.M. and Simmons, L.M., Jr., Coherent states for general potentials. I. Formalism. II. Confining one-dimensional examples. III. Nonconfining one-dimensional examples. *Phys. Rev.* D (1979) **20** 1321-1331,1332–1341,1342–1350.

[34] Titulaer, U.M. and Glauber, R.J., Correlation functions for coherent fields. *Phys. Rev.* (1965) **140** B676–B682.

[35] Mollow, B.R. and Glauber, R.J., Quantum theory of parametric amplification. I. *Phys. Rev.* (1967) **160** 1076–1096.

[36] Mollow, B.R., Quantum statistics of coupled oscillator systems. *Phys. Rev.* (1967) **162** 1256–1273.

[37] Hollenhorst, J.N., Quantum limits on resonant-mass gravitational-radiation detectors. *Phys. Rev.* D (1979) **19** 1669–1679.

[38] Caves, C.M., Quantum-mechanical noise in an interferometer. *Phys. Rev.* D (1981) **23** 1693–1708.

[39] Walls, D.F., Squeezed states of light. *Nature* (1983) **306** 141–146.

[40] Cahill, K.E. and Glauber, R.J., Density operators and quasiprobability distributions. *Phys. Rev.* (1969) **177** 1882-1902.

[41] Aharonov, Y., Falkoff, D., Lerner, E., and Pendleton, H., A quantum characterization of classical radiation. *Ann. Phys.* (NY) (1966) **39** 498–512.

[42] Hillery, M., Classical pure states are coherent states. *Phys. Lett.* A (1985) **111** 409–411.

[43] Helstrom, C.W., Nonclassical states in optical communication to a remote receiver. *IEEE Trans. Inform. Theory* (1980) **26** 378–382.

[44] Hillery, M., Conservation laws and nonclassical states in nonlinear optical systems. *Phys. Rev.* A (1985) **31** 338–342.

[45] Mandel, L., Nonclassical states of the electromagnetic field. *Phys. Scripta* (1986) **T12** 34–42.

[46] Loudon, R., Non-classical effects in the statistical properties of light. *Rep. Prog. Phys.* (1980) **43** 913–949.

[47] Zubairy, M.S., Nonclassical effects in a two-photon laser. *Phys. Lett.* A (1982) **87** 162–164.

[48] Lugiato, L.A. and Strini, G., On nonclassical effects in two-photon optical bistability and two-photon laser. *Opt. Commun.* (1982) **41** 374–378.

[49] Schubert, M., The attributes of nonclassical light and their mutual relationship. *Ann. Phys.* (Leipzig) (1987) **44** 53–60.

[50] Yuen, H.P., Nonclassical light. In: *Photons and Quantum Fluctuations* (E.R. Pike and H. Walther, eds.), pp. 1-9. Adam Hilger, Bristol, 1988.

[51] Janszky, J., Sibilia, C., Bertolotti, M., and Yushin, Y., Switch effect of nonclassical light. *J. Phys.* (Paris) (1988) **49**C 337–339.

[52] Huttner, B. and Ben-Aryeh, Y., Photodetection of non-classical light. *Opt. Commun.* (1988) **69** 93–97.

[53] Gea-Banacloche, G., Two-photon absorption of nonclassical light. *Phys. Rev. Lett.* (1989) **62** 1603–1606.

[54] Bullough, R.K., Squeezed and non-classical light. *Nature* (1988) **333** 601–602.

[55] *Squeezed and Nonclassical Light* (P. Tombesi and E.R. Pike, eds.). Plenum, New York, 1989.

[56] *Interaction of Electromagnetic Field with Condensed Matter* (N.N. Bogolubov, A.S. Shumovsky, and V.I. Yukalov, eds.). World Scientific, Singapore, 1990.

[57] *Quantum Noise Reduction in Optical Systems — Experiments* (E. Giacobino and C. Fabre, eds.), special issue of *Appl. Phys.* B (1992) **55** N 3.

[58] Vogel, W., Welsch, D.-G., and Wallentowitz, S., *Quantum Optics. An Introduction.* Wiley–VCH, Berlin, 2001.

[59] Paul, H., Phase of a microscopic electromagnetic field and its measurement. *Fortschr. Phys.* (1974) **22** 657–689.

[60] Bergou, J. and Englert, B.-G., Operators of the phase. Fundamentals. *Ann. Phys.* (NY) (1991) **209** 479–505.

[61] *Quantum Phase and Phase Dependent Measurements* (W.P. Schleich and S.M. Barnett, eds.), special issue of *Phys. Scripta* (1993) **T48** 1–142.

[62] Lukš, A. and Peřinová, V., Presumable solutions of quantum phase problem and their flaws. *Quant. Opt.* (1994) **6** 125–167.

[63] Lynch, R., The quantum phase problem: a critical review. *Phys. Rep.* (1995) **256** 367–437.

[64] Tanaś, R., Miranowicz, A., and Gantsog, T., Quantum phase properties of nonlinear optical phenomena. In: *Progress in Optics, vol. XXXV* (E. Wolf, ed.), pp. 355–446. North Holland, Amsterdam, 1996.

[65] Pegg, D.T. and Barnett, S.M., Tutorial review: Quantum optical phase. *J. Mod. Opt.* (1997) **44** 225–264.

[66] Bužek, V. and Knight, P.L., Quantum interference, superposition states of light, and nonclassical effects. In: *Progress in Optics, vol. XXXIV* (E. Wolf, ed.), pp. 1–158. North Holland, Amsterdam, 1995.

[67] Ali, S.T., Antoine, J.-P., Gazeau, J.-P., and Mueller, U.A., Coherent states and their generalizations: A mathematical overview. *Rev. Math. Phys.* (1995) **7** 1013–1104.

[68] Ali, S.T., Antoine, J.-P., and Gazeau, J.-P., *Coherent States, Wavelets and Their Generalizations.* Springer, Berlin, 1999.

[69] Kennard, E.H., Zur Quantenmechanik einfacher Bewegungstypen. *Z. Phys.* (1927) **44** 326–352.

[70] Coulson, C.A. and Rushbrooke, G.S., On the motion of a Gaussian wave-packet in a parabolic potential field. *Proc. Camb. Phil. Soc.* (1946) **42** 286–291.

[71] Husimi, K., Miscellanea in elementary quantum mechanics. I. *Prog. Theor. Phys.*

(1953) **9** 238–244.

[72] Takahasi, H., Information theory of quantum-mechanical channels. In: *Advances in Communication Systems. Theory and Applications, vol. 1* (A.V. Balakrishnan, ed.), pp. 227–310. Academic, New York, 1965.

[73] Plebański, J., Classical properties of oscillator wave packets. *Bulletin de l' Académie Polonaise des Sciences* (1954) **2** 213–217.

[74] Infeld, L. and Plebański, J., On a certain class of unitary transformations. *Acta Phys. Polon.* (1955) **14** 41–75.

[75] Plebański, J., On certain wave packets. *Acta Phys. Polon.* (1955) **14** 275–293.

[76] Plebański, J., Wave functions of a harmonic oscillator. *Phys. Rev.* (1956) **101** 1825–1826.

[77] Miller, M.M. and Mishkin, E.A., Characteristic states of the electromagnetic radiation field. *Phys. Rev.* (1966) **152** 1110–1114.

[78] Lu, E.Y.C., New coherent states of the electromagnetic field. *Lett. Nuovo Cim.* (1971) **2** 1241–1244.

[79] Bialynicki-Birula, I., Solutions of the equations of motion in classical and quantum theories. *Ann. Phys.* (NY) (1971) **67** 252–273.

[80] Jackiw, R., Minimum uncertainty product, number-phase uncertainty product, and coherent states. *J. Math. Phys.* (1968) **9** 339–346.

[81] Stoler, D., Equivalence classes of minimum uncertainty packets. I-II. *Phys. Rev.* D (1970) **1** 3217–3219; (1971) **4** 1925–1926.

[82] Bertrand, P.P., Moy, K., and Mishkin, E.A., Minimum uncertainty states $|\gamma\rangle$ and the states $|n\rangle_\gamma$ of the electromagnetic field. *Phys. Rev.* D (1971) **4** 1909–1912.

[83] Trifonov, D.A., Coherent states and uncertainty relations. *Phys. Lett.* A (1974) **48** 165–166.

[84] Stoler, D., Most-general minimality-preserving Hamiltonian. *Phys. Rev.* D (1975) **11** 3033–3034.

[85] Canivell, V. and Seglar, P., Minimum-uncertainty states and pseudoclassical dynamics. *Phys. Rev.* D (1977) **15** 1050–1054; (1978) **18** 1082–1094.

[86] Yuen, H.P., Two-photon coherent states of the radiation field. *Phys. Rev.* A (1976) **13** 2226–2243.

[87] Solimeno, S., Di Porto, P., and Crosignani, B., Quantum harmonic oscillator with time-dependent frequency. *J. Math. Phys.* (1969) **10** 1922–1928.

[88] Malkin, I.A., Man'ko, V.I., and Trifonov, D.A., Invariants and evolution of coherent states for charged particle in time-dependent magnetic field. *Phys. Lett.* A (1969) **30** 414–416.

[89] Malkin, I.A., Man'ko, V.I., and Trifonov, D.A., Coherent states and transition probabilities in a time-dependent electromagnetic field. *Phys. Rev.* D (1970) **2** 1371–1385.

[90] Malkin, I.A. and Man'ko, V.I., Coherent states and excitation of N-dimensional nonstationary forced oscillator. *Phys. Lett.* A (1970) **32** 243–244.

[91] Dodonov, V.V. and Man'ko, V.I., Coherent states and the resonance of a quantum damped oscillator. *Phys. Rev.* A (1979) **20** 550–560.

[92] Malkin, I.A. and Man'ko, V.I., Coherent states of a charged particle in a magnetic field. *Zh. Eksp. Teor. Fiz.* (1968) **55** 1014–1025 [*Sov. Phys. – JETP* (1969) **28** 527–532].

[93] Feldman, A. and Kahn, A.H., Landau diamagnetism from the coherent states of an electron in a uniform magnetic field. *Phys. Rev.* B (1970) **1** 4584–4589.

[94] Tam, W.G., Coherent states and the invariance group of a charged particle in a uniform magnetic field. *Physica* (1971) **54** 557–572.

[95] Holz, A., N-dimensional anisotropic oscillator in a uniform time-dependent electromagnetic field. *Lett. Nuovo Cim.* (1970) **4** 1319–1323.

[96] Dodonov, V.V., Malkin, I.A., and Man'ko, V.I., Coherent states of a charged particle in a time-dependent uniform electromagnetic field of a plane current. *Physica* (1972) **59** 241–256.

[97] Kim, H.Y. and Weiner, J.H., Gaussian-wave packet dynamics in uniform magnetic and quadratic potential fields. *Phys. Rev.* B (1973) **7** 1353–1362.

[98] Malkin, I.A., Man'ko, V.I., and Trifonov, D.A., Linear adiabatic invariants and coherent states. *J. Math. Phys.* (1973) **14** 576–582.

[99] Rowe, D.J., The two-photon laser beam as a breathing mode of the electromagnetic field. *Canad. J. Phys.* (1978) **56** 442–446.

[100] Picinbono, B. and Rousseau, M., Generalizations of Gaussian optical fields. *Phys. Rev.* A (1970) **1** 635–643.

[101] Holevo, A.S., Some statistical problems for quantum Gaussian states. *IEEE Trans. Inform. Theory* (1975) **21** 533–543.

[102] Helstrom, C.W., *Quantum Detection and Estimation Theory*. Academic, New York, 1976.

[103] Heller, E.J., Time-dependent approach to semiclassical dynamics. *J. Chem. Phys.* (1975) **62** 1544–1555.

[104] Bagrov, V.G., Buchbinder, I.L., and Gitman, D.M., Coherent states of a relativistic particle in an external electromagnetic field. *J. Phys.* A (1976) **9** 1955–1965.

[105] Dodonov, V.V., Malkin, I.A., and Man'ko, V.I., Coherent states and Green functions of relativistic quadratic systems. *Physica* A (1976) **82** 113–133.

[106] Bagrov, V.G., Bukhbinder, I.L., Gitman, D.M., and Lavrov, P.M., Coherent states of an electron in a quantized electromagnetic wave. *Teor. Mat. Fiz.* (1977) **33** 419–426 [*Theor. & Math. Phys.* (1977) **33** 1111–1116].

[107] Kaiser, G., Phase-space approach to relativistic quantum mechanics. 1. Coherent-state representation for massive scalar particles. *J. Math. Phys.* (1977) **18** 952–959.

[108] Titulaer, U.M. and Glauber, R.J., Density operators for coherent fields. *Phys. Rev.* (1966) **145** 1041–1050.

[109] Bialynicka-Birula, Z., Properties of the generalized coherent state. *Phys. Rev.* (1968) **173** 1207–1209.

[110] Stoler, D., Generalized coherent states. *Phys. Rev.* D (1971) **4** 2309–2312.

[111] Mathews, P.M. and Eswaran, K., Semi-coherent states of quantum harmonic oscillator. *Nuovo Cim.* B (1973) **17** 332–335.

[112] Glassgold, A.E. and Holliday, D., Quantum statistical dynamics of laser amplifiers. *Phys. Rev.* (1965) **139** A1717–A1734.

[113] Dodonov, V.V., Malkin, I.A., and Man'ko, V.I., Even and odd coherent states and excitations of a singular oscillator. *Physica* (1974) **72** 597–615.

[114] Schrödinger, E., Die gegenwärtige Situation in der Quantenmechanik. *Naturwissenschaften* (1935) **23** 807–812, 823–828, 844–849 [The present situation in quantum mechanics. In: *Quantum Theory and Measurement* (J.A. Wheeler and W.H. Zurek, eds.), pp. 152–167. Princeton Univ. Press, Princeton, 1983].

[115] Senitzky, I.R., Harmonic oscillator wave functions. *Phys. Rev.* (1954) **95** 1115–1116.

[116] Boiteux, M. and Levelut, A., Semicoherent states. *J. Phys.* A (1973) **6** 589–596.

[117] Marhic, M.E., Oscillating Hermite–Gaussian wave functions of the harmonic oscillator. *Lett. Nuovo Cim.* (1978) **22** 376–378.

[118] Susskind, L. and Glogower, J., Quantum mechanical phase and time operator. *Physics* (1964) **1** 49–62.

[119] Dirac, P.A.M., *Lectures on Quantum Field Theory*, p. 18. Academic, New York, 1966.

[120] Lerner, E.C., Harmonic-oscillator phase operators. *Nuovo Cim.* B (1968) **56** 183–186.

[121] Wigner, E.P., Do the equations of motion determine the quantum mechanical commutation relations. *Phys. Rev.* (1951) **77** 711–712.

[122] Eswaran, K., On generalized phase operators for the quantum harmonic oscillator. *Nuovo Cim.* B (1970) **70** 1–11.

[123] Lerner, E.C., Huang, H.W., and Walters, G.E., Some mathematical properties of oscillator phase operator. *J. Math. Phys.* (1970) **11** 1679–1684.

[124] Ifantis, E.K., States, minimizing the uncertainty product of the oscillator phase operator. *J. Math. Phys.* (1972) **13** 568–575.

[125] Aharonov, Y., Lerner, E.C., Huang, H.W., and Knight, J.M., Oscillator phase states, thermal equilibrium and group representations. *J. Math. Phys.* (1973) **14** 746–756.

[126] Aharonov, Y., Huang, H.W., Knight, J.M., and Lerner, E.C., Generalized destruction operators and a new method in group theory. *Lett. Nuovo Cim.* (1971) **2** 1317–1320.

[127] Onofri, E. and Pauri, M., Analyticity and quantization. *Lett. Nuovo Cim.* (1972) **3** 35–42.

[128] Bonifacio, R., Kim, D.M., and Scully, M.O., Description of many-atom system in terms of coherent boson states. *Phys. Rev.* (1969) **187** 441–445.

[129] Atkins, P.W. and Dobson, J.C., Angular momentum coherent states. *Proc. Roy. Soc. London* A (1971) **321** 321–340.

[130] Mikhailov, V.V., Transformations of angular momentum coherent states under coordinate rotations. *Phys. Lett.* A (1971) **34** 343–344.

[131] Makhviladze, T.M. and Shelepin, L.A., Coherent states of rotating bodies. *Sov. J. Nucl. Phys.* (1972) **15** 601–602.

[132] Doktorov, E.V., Malkin, I.A., and Man'ko, V.I., Coherent states and asymptotic behavior of symmetric top wave-functions. *Int. J. Quant. Chem.* (1975) **9** 951–968.

[133] Gulshani, P., Generalized Schwinger boson realization and the oscillator-like coherent states of the rotation groups and the asymmetric top. *Canad. J. Phys.* (1979) **57** 998–1021.

[134] Radcliffe, J.M., Some properties of coherent spin states. *J. Phys.* A (1971) **4** 313–323.

[135] Arecchi, F.T., Courtens, E., Gilmore, R., and Thomas, H., Atomic coherent states in quantum optics. *Phys. Rev.* A (1972) **6** 2211–2237.

[136] Gilmore, R., Geometry of symmetrized states. *Ann. Phys.* (NY) (1972) **74** 391–463.

[137] Lieb, E.H., Classical limit on quantum spin systems. *Commun. Math. Phys.* (1973) **31** 327–340.

[138] Thompson, B.V., A variational estimate of the lowest eigenvalue of the multiatom, multimode Hamiltonian of quantum optics. *J. Phys.* A (1972) **5** 1453–1460.

[139] Janssen, D., Coherent states of quantum-mechanical top. *Yad. Fiz.* (1977) **25** 897–907 [*Sov. J. Nucl. Phys.* (1978) **25** 479–484].

[140] Thouless, D.J., Stability conditions and nuclear rotations in the Hartree–Fock theory. *Nucl. Phys.* (1960) **21** 225–232.

[141] Ohnuki, Y. and Kashiwa, T., Coherent states of Fermi operators and path integral. *Prog. Theor. Phys.* (1978) **60** 548–564.

[142] Klauder, J.R., Continuous representation theory. II. Generalized relation between quantum and classical dynamics. *J. Math. Phys.* (1963) **4** 1058–1073.
[143] Perelomov, A.M., Coherent states for arbitrary Lie group. *Commun. Math. Phys.* (1972) **26** 222-236.
[144] Kibble, T.W.B., Coherent soft-photon states and infrared divergencies. I. Classical currents. *J. Math. Phys.* (1968) **9** 315–324.
[145] Aslaksen, E.W. and Klauder, J.R., Continuous representation theory using the affine group. *J. Math. Phys.* (1969) **10** 2267-2275.
[146] Solomon, A.I., Group theory of superfluidity. *J. Math. Phys.* (1971) **12** 390–394.
[147] Hioe, F.T., Coherent states and Lie algebras. *J. Math. Phys.* (1974) **15** 1174–1177.
[148] Ducloy, M., Application du formalisme des états cohérentes de moment angulaire a quelques problèmes de physique atomique. *J. Phys.* (Paris) (1975) **36** 927-941.
[149] Karasev, V.P. and Shelepin, L.A., Theory of generalized coherent states of the group SU_n. *Teor. Mat. Fiz.* (1980) **45** 54–63 [*Theor. & Math. Phys.* (1980) **45** 879–886].
[150] Rowe, D.J., Rosensteel, G., and Gilmore, R., Vector coherent state representation theory. *J. Math. Phys.* (1985) **26** 2787-2791.
[151] Barut, A.O. and Girardello, L., New 'coherent' states associated with noncompact groups. *Commun. Math. Phys.* (1971) **21** 41-55.
[152] Hongoh, M., Coherent states associated with the continuous spectrum of noncompact groups. *J. Math. Phys.* (1977) **18** 2081–2084.
[153] Horn, D. and Silver, R., Coherent production of pions. *Ann. Phys.* (NY) (1971) **66** 509–541.
[154] Botke, J.C., Scalapino, D.J., and Sugar, R.L., Coherent states and particle production. *Phys. Rev.* D (1974) **9** 813–823.
[155] Botke, J.C., Scalapino, D.J., and Sugar, R.L., Coherent states and particle production. II. Isotopic spin. *Phys. Rev.* D (1974) **10** 1604–1612.
[156] Bhaumik, D., Bhaumik, K., and Dutta-Roy, B., Charged bosons and the coherent state. *J. Phys.* A (1976) **9** 1507-1512.
[157] Bhaumik, D., Nag, T., and Dutta-Roy, B., Coherent states for angular momentum. *J. Phys.* A (1975) **8** 1868–1874.
[158] Skagerstam, B.-S.K., Some remarks on coherent states and conserved charges. *Phys. Lett.* A (1978) **69** 76–78.
[159] Skagerstam, B.-S., Coherent-state representation of a charged relativistic boson field. *Phys. Rev.* D (1979) **19** 2471-2476.
[160] Eriksson, K.-E., Mukunda, N., and Skagerstam, B.-S., Coherent-state representation of a non-Abelian charged quantum field. *Phys. Rev.* D (1981) **24** 2615–2625.
[161] Skagerstam, B.-S., Quasi-coherent states for unitary groups. *J. Phys.* A (1985) **18** 1-13.
[162] Mathews, P.M. and Eswaran, K., Simultaneous uncertainties of the cosine and sine operators. *Nuovo Cim.* B (1974) **19** 99–104.
[163] Aragone, C., Guerri, G., Salamó, S., and Tani, J.L., Intelligent spin states. *J. Phys.* A (1974) **7** L149–L151.
[164] Aragone, C., Chalbaud, E., and Salamó, S., On intelligent spin states. *J. Math. Phys.* (1976) **17** 1963–1971.
[165] Delbourgo, R., Minimal uncertainty states for the rotation and allied groups. *J. Phys.* A (1977) **10** 1837-1846.
[166] Bacry, H., Eigenstates of complex linear combinations of $J_1 J_2 J_3$ for any representation of $SU(2)$. *J. Math. Phys.* (1978) **19** 1192–1195.
[167] Bacry, H., Physical significance of minimum uncertainty states of an angular mo-

mentum system. *Phys. Rev.* A (1978) **18** 617–619.

[168] Rashid, M.A., The intelligent states. I. Group-theoretic study and the computation of matrix elements. *J. Math. Phys.* (1978) **19** 1391-1396.

[169] Van den Berghe, G. and De Meyer, H., On the existence of intelligent states associated with the non-compact group $SU(1,1)$. *J. Phys.* A (1978) **11** 1569–1578.

[170] Chandra, N. and Prakash, H., Anticorrelation in two-photon attenuated laser beam. *Phys. Rev.* A (1970) **1** 1696–1698.

[171] Stoler, D., Photon antibunching and possible ways to observe it. *Phys. Rev. Lett.* (1974) **33** 1397-1400.

[172] Agarwal, G.S., Comments on bunching and antibunching effects in thermal beams. *Z. Phys.* B (1975) **22** 207-209.

[173] McNeil, K.J. and Walls, D.F., Possibility of observing enhanced photon bunching from two photon emission. *Phys. Lett.* A (1975) **51** 233–234.

[174] Simaan, H.D. and Loudon, R., Quantum statistics of single-beam two-photon absorption. *J. Phys.* A (1975) **8** 539–554.

[175] Every, I.M., The production of photon antibunching by two-photon absorption. *J. Phys.* A (1975) **8** L69–L72.

[176] Kimble, H.J. and Mandel, L., Theory of resonance fluorescence. *Phys. Rev.* A (1976) **13** 2123–2144.

[177] Kimble, H.J., Dagenais, M., and Mandel, L., Photon antibunching in resonance fluorescence. *Phys. Rev. Lett.* (1977) **39** 691-695.

[178] Swain, S., Theory of atomic processes in strong resonant electromagnetic fields. *Adv. At. Mol. Phys.* (1980) **16** 159–200.

[179] Smirnov, D.F., and Troshin, A.S., New phenomena in quantum optics: photon antibunching, sub-Poisson photon statistics, and squeezed states. *Uspekhi Fiz. Nauk* (1987) **153** 233–271 *Sov. Phys. – Uspekhi* (1987) **30** 851-874.

[180] Teich, M.C. and Saleh, B.E.A., Photon bunching and antibunching. In: *Progress in Optics, vol. XXVI* (E. Wolf, ed.), pp. 1-104. North Holland, Amsterdam, 1988.

[181] Knight, P.L. and Pegg, D.T., Double photon transitions induced by antibunched light. *J. Phys.* B (1982) **15** 3211-3222.

[182] Teich, M.C., Saleh, B.E.A., and Peřina, J., Role of primary excitation statistics in the generation of antibunched and sub-Poisson light. *J. Opt. Soc. Am.* B (1984) **1** 366–389.

[183] Srinivasan, S.K., Generation of antibunched light by an inhibited Poisson stream. *Opt. Acta* (1986) **33** 835–842.

[184] Stoler, D. and Yurke, B., Generating antibunched light from the output of a nondegenerate frequency converter. *Phys. Rev.* A (1986) **34** 3143–3147.

[185] Saleh, B.E.A. and Teich, M.C., Sub-Poisson light generation by selective deletion from cascaded atomic emissions. *Opt. Commun.* (1985) **52** 429–432.

[186] Golubev, Y.M. and Gorbachev, V.N., Formation of sub-Poisson light statistics in parametric absorption in a resonant medium. *Opt. Spektrosk.* (1986) **60** 785–787 [*Opt. Spectrosc.* (1986) **60** 483–484].

[187] Yuen, H.P. and Shapiro, J.H., Optical communication with two-photon coherent states - part I: Quantum-state propagation and quantum-noise reduction. *IEEE Trans. Inf. Theory* (1978) **IT-24** 657–668.

[188] Caves, C.M., Thorne, K.S., Drever, R.W.P., Sandberg, V.D., and Zimmermann, M. On the measurement of a weak classical force coupled to a quantum mechanical oscillator. I. Issue of principle. *Rev. Mod. Phys.* (1980) **52** 341–392.

[189] Dodonov, V.V., Man'ko, V.I., and Rudenko, V.N., Nondemolition measurements in gravity wave experiments. *Zhurn. Exper. Teor. Fiz.* (1980) **78** 881–896 [*Sov. Phys.*

– *JETP* (1980) **51** 443–450].

[190] Grishchuk, L.P. and Sazhin, M.V., Squeezed quantum states of a harmonic oscillator in the problem of gravitational-wave detector. *Zhurn. Eksp. Teor. Fiz.* (1983) **84** 1937–1950 [*Sov. Phys. – JETP* (1984) **57** 1128–1135].

[191] Landau, L. and Peirls, R., Erweiterung des Unbestimmtheitsprinzips für die relativistische Quantentheorie. *Z. Phys.* (1931) **69** 56–69 [Extension of the uncertainty principle to relativistic quantum theory. In: *Collected papers of L.D. Landau* (D. Ter Haar, ed.), pp. 40–52. Gordon & Breach, New York, 1965].

[192] Braginsky, V.B., Vorontsov, Y.I., and Thorne, K.S., Quantum nondemolition measurements. *Science* (1980) **209** 547-557.

[193] Braginskii, V. B., Resolution in macroscopic measurements: progress and prospects. *Uspekhi Fiz. Nauk* (1988) **156** 93–115 [*Sov. Phys. - Uspekhi* (1988) **31** 836–849].

[194] Belavkin, V.P. and Staszewski, P., Nondemolition observation of a free quantum particle. *Phys. Rev.* A (1992) **45** 1347–1356.

[195] Vorontsov, Y.I., Standard quantum limits of measurement errors and methods of overcoming them. *Uspekhi Fiz. Nauk* (1994) **164** 89–104 [*Physics – Uspekhi* (1994) **37** 81–96].

[196] Bocko, M.F. and Onofrio, R., On the measurement of a weak classical force coupled to a harmonic oscillator: experimental progress. *Rev. Mod. Phys.* (1996) **68** 755–799.

[197] Brosa, U. and Gross, D.H.E., Squeezing of quantal fluctuations. *Z. Phys.* A (1980) **294** 217-220.

[198] Yuen, H.P. and Shapiro, J.H., Generation and detection of two-photon coherent states in degenerate four-wave mixing. *Opt. Lett.* (1979) **4** 334–336.

[199] Walls, D.F. and Zoller, P., Reduced quantum fluctuations in resonance fluorescence. *Phys. Rev. Lett.* (1981) **47** 709–711.

[200] Becker, W., Scully, M.O., and Zubairy, M.S., Generation of squeezed coherent states via a free-electron laser. *Phys. Rev. Lett.* (1982) **48** 475–477.

[201] Yurke, B., Use of cavities in squeezed-state generation. *Phys. Rev.* A (1984) **29** 408–410.

[202] Mandel, L., Squeezing and photon antibunching in harmonic generation. *Opt. Commun.* (1982) **42** 437–439.

[203] Kozierowski, M. and Kielich, S., Squeezed states in harmonic generation of a laser beam. *Phys. Lett.* A (1983) **94** 213–216.

[204] Milburn, G. and Walls, D.F., Production of squeezed states in a degenerate parametric amplifier. *Opt. Commun.* (1981) **39** 401–404.

[205] Meystre, P. and Zubairy, M.S., Squeezed states in the Jaynes-Cummings model. *Phys. Lett.* A (1982) **89** 390–392.

[206] Mandel, L., Squeezed states and sub-Poissonian photon statistics. *Phys. Rev. Lett.* (1982) **49** 136–138.

[207] Slusher, R.E., Hollberg, L.W., Yurke, B., Mertz, J.C., and Valley, J.F. Observation of squeezed states generated by four-wave mixing in an optical cavity. *Phys. Rev. Lett.* (1985) **55** 2409–2412.

[208] Shelby, R.M., Levenson, M.D., Perlmutter, S.H., DeVoe, R.G., and Walls, D.F. Broad-band parametric deamplification of quantum noise in an optical fiber. *Phys. Rev. Lett.* (1986) **57** 691–694.

[209] Wu, L.A., Kimble, H.J., Hall, J.L., and Wu, H., Generation of squeezed states by parametric down conversion. *Phys. Rev. Lett.* (1986) **57** 2520–2523.

[210] Reid, M.D. and Walls, D.F., Generation of squeezed states via degenerate four-

wave mixing. *Phys. Rev.* A (1985) **31** 1622–1635.

[211] Yamamoto, Y. and Haus, H.A., Preparation, measurement and information capacity of optical quantum states. *Rev. Mod. Phys.* (1986) **58** 1001–1020.

[212] *Squeezed Light* (R. Loudon and P.L. Knight, eds.), special issue of *J. Mod. Opt.* (1987) **34** N 6/7.

[213] *Squeezed States of the Electromagnetic Field* (H.J. Kimble and D.F. Walls, eds.), feature issue of *J. Opt. Soc. Am* B (1987) **4** N 10.

[214] Yuen, H.P., Generation, detection, and application of high-intensity photon-number-eigenstate fields. *Phys. Rev. Lett.* (1986) **56** 2176–2179.

[215] Filipowicz, P., Javanainen, J., and Meystre, P., Quantum and semiclassical steady states of a kicked cavity. *J. Opt. Soc. Am* B (1986) **3** 906–910.

[216] Krause, J., Scully, M.O., and Walther, H., State reduction and $|n\rangle$-state preparation in a high-Q micromaser. *Phys. Rev.* A (1987) **36** 4546–4550.

[217] Holmes, C.A., Milburn, G.J., and Walls, D.F., Photon-number-state preparation in nondegenerate parametric amplification. *Phys. Rev.* A (1989) **39** 2493–2501.

[218] Schrödinger, E., Zum Heisenbergschen Unschärfeprinzip. *Ber. Kgl. Akad. Wiss. Berlin* (1930) **24** 296–303.

[219] Robertson, H.P., A general formulation of the uncertainty principle and its classical interpretation. *Phys. Rev.* (1930) **35** 667.

[220] Dodonov, V.V., Kurmyshev, E.V., and Man'ko, V.I., Generalized uncertainty relation and correlated coherent states. *Phys. Lett.* A (1980) **79** 150–152.

[221] Jannussis, A.D., Brodimas, G.N., and Papaloucas, L.C., New creation and annihilation operators in the Gauss plane and quantum friction. *Phys. Lett.* A (1979) **71** 301–303.

[222] Fujiwara, I. and Miyoshi, K., Pulsating states for quantal harmonic oscillator. *Prog. Theor. Phys.* (1980) **64** 715–718.

[223] Rajagopal, A.K. and Marshall, J.T., New coherent states with applications to time-dependent systems. *Phys. Rev.* D (1982) **26** 2977-2980.

[224] Remaud, B., Dorso, C., and Hernandez, E.S., Coherent state propagation in open systems. *Physica* A (1982) **112** 193–213.

[225] Dodonov, V.V., Kurmyshev, E.V., and Man'ko, V.I., Correlated coherent states. In: *Classical and Quantum Effects in Electrodynamics. Proc. Lebedev Phys. Inst., vol. 176* (A.A. Komar, ed.), pp. 128–150. Nauka, Moscow, 1986 [translated by Nova Science, Commack, 1988, pp. 169–199].

[226] Dodonov, V.V. and Man'ko, V.I., Generalization of uncertainty relation in quantum mechanics. In: *Invariants and the Evolution of Nonstationary Quantum Systems. Proc. Lebedev Phys. Inst., vol. 183* (M.A. Markov, ed.), pp. 5–70. Nauka, Moscow, 1987 [translated by Nova Science, Commack, 1989, pp. 3–101].

[227] Trifonov, D.A., Generalized uncertainty relations and coherent and squeezed states. *J. Opt. Soc. Am.* A (2000) **17** 2486–2495.

[228] Dodonov, V.V., Purity- and entropy-bounded uncertainty relations for mixed quantum states. *J. Opt.* B (2002) **4** S98–S108.

[229] Yuen, H.P. and Shapiro, J.H., Optical communication with two-photon coherent states – part III: Quantum measurements realizable with photoemissive detectors. *IEEE Trans. Inf. Theory* (1980) **IT-26** 78–92.

[230] Gulshani, P. and Volkov, A.B., Heisenberg-symplectic angular-momentum coherent states in two dimensions. *J. Phys.* A (1980) **13** 3195–3204.

[231] Gulshani, P. and Volkov, A.B., The cranked oscillator coherent states. *J. Phys.* G (1980) **6** 1335–1346.

[232] Caves, C.M., Quantum limits on noise in linear amplifiers. *Phys. Rev.* D (1982) **26**

1817–1839.

[233] Schumaker, B.L., Quantum-mechanical pure states with Gaussian wave-functions. *Phys. Rep.* (1986) **135** 317–408.

[234] Jannussis, A. and Bartzis, V., General properties of the squeezed states. *Nuovo Cim.* B (1987) **100** 633–650.

[235] Bishop, R.F. and Vourdas, A., A new coherent paired state with possible applications to fluctuation-dissipation phenomena. *J. Phys.* A (1987) **20** 3727–3741.

[236] Schumaker, B.L. and Caves, C.M., New formalism for two-photon quantum optics. 2. Mathematical foundation and compact notation. *Phys. Rev.* A (1985) **31** 3093-3111.

[237] Barnett, S.M. and Knight, P.L., Thermofield analysis of squeezing and statistical mixtures in quantum optics. *J. Opt. Soc. Am.* B (1985) **2** 467–479.

[238] Marburger, J.H. and Power, E.A., Minimum-uncertainty states of systems with many degrees of freedom. *Found. Phys.* (1980) **10** 865–874.

[239] Milburn, G.J., Multimode minimum uncertainty squeezed states. *J. Phys.* A (1984) **17** 737–745.

[240] Mølmer, K. and Slowfkowski, W., A new Hilbert-space approach to the multimode squeezing of light. *J. Phys.* A (1988) **21** 2565-2571.

[241] Huang, J. and Kumar, P., Photon-counting statistics of multimode squeezed light. *Phys. Rev.* A (1989) **40** 1670–1673.

[242] Barnett, S.M. and Dupertuis, M.-A., Multiatom squeezed states: a new class of collective atomic states. *J. Opt. Soc. Am.* B (1987) **4** 505–511.

[243] Bastiaans, M.J., Wigner distribution function and its application to first-order optics. *J. Opt. Soc. Am.* (1979) **69** 1710–1716.

[244] Dodonov, V.V., Man'ko, V.I., and Semjonov, V.V., The density matrix of the canonically transformed multidimensional Hamiltonian in the Fock basis. *Nuovo Cim.* B (1984) **83** 145–161.

[245] Mizrahi, S.S., Quantum mechanics in the Gaussian wave packet phase space representation. *Physica* A (1984) **127** 241–264.

[246] Peřinová, V., Křepelka, J., Peřina, J., Lukš, A., and Szlachetka, P., Entropy of optical fields. *Opt. Acta* (1986) **33** 15–32.

[247] Dodonov, V.V. and Man'ko, V.I., Density matrices and Wigner functions of quasiclassical quantum systems. In: *Group Theory, Gravitation and Elementary Particle Physics. Proc. Lebedev Phys. Inst., vol. 167* (A.A. Komar, ed.), pp. 7–79. Nauka, Moscow, 1986 [translated by Nova Science, Commack, 1987, pp. 7–101].

[248] Agarwal, G.S., Wigner-function description of quantum noise in interferometers. *J. Mod. Opt.* (1987) **34** 909–921.

[249] Janszky, J. and Yushin, Y., Many-photon processes with the participation of squeezed light. *Phys. Rev.* A (1987) **36** 1288–1292.

[250] Simon, R., Sudarshan, E.C.G., and Mukunda, N., Gaussian-Wigner distributions in quantum mechanics and optics. *Phys. Rev.* A (1987) **36** 3868–3880.

[251] Turner, R.E. and Snider, R.F., A comparison of local and global single Gaussian approximations to time dynamics: One-dimensional systems. *J. Chem. Phys.* (1987) **87** 910–920.

[252] Mizrahi, S.S., Quantum mechanics in the Gaussian wave packet phase space representation. 3. From phase space probability functions to wave functions. *Physica* A (1988) **150** 541–554.

[253] Mizrahi, S.S. and Galetti, D., On the equivalence between the wave packet phase space representation (WPPSR) and the phase space generated by the squeezed states. *Physica* A (1988) **153** 567–572.

[254] Bishop, R.F. and Vourdas, A., Coherent mixed states and a generalized P representation. *J. Phys.* A (1987) **20** 3743–3769.

[255] Fearn, H. and Collett, M.J., Representations of squeezed states with thermal noise. *J. Mod. Opt.* (1988) **35** 553–564.

[256] Aliaga, J. and Proto, A.N., Relevant operators and non-zero temperature squeezed states. *Phys. Lett.* A (1989) **142** 63–67.

[257] Kireev, A., Mann, A., Revzen, M., and Umezawa, H., Thermal squeezed states in thermo field dynamics and quantum and thermal fluctuations. *Phys. Lett.* A (1989) **142** 215–221.

[258] Kim, M.S., de Oliveira, F.A.M., and Knight, P.L., Properties of squeezed number states and squeezed thermal states. *Phys. Rev.* A (1989) **40** 2494–2503.

[259] Bechler, A., Generation of squeezed states in a homogeneous magnetic field. *Phys. Lett.* A (1988) **130** 481–482.

[260] Jannussis, A., Vlahas, E., Skaltsas, D., Kliros, G., and Bartzis, V., Squeezed states in the presence of a time-dependent magnetic field. *Nuovo Cim.* B (1989) **104** 53–66.

[261] Varró, S., Coherent states of an electron in a homogeneous constant magnetic field and the zero magnetic field limit. *J. Phys.* A (1984) **17** 1631–1638.

[262] Varró, S. and Ehlotzky, F., Generalized coherent states for electrons in external fields and application to potential scattering. *Phys. Rev.* A (1987) **36** 497–510.

[263] Namiot, V.A. and Finkelshtein, V.Y., Method of pseudo-coherent states in nonlinear quantum systems. *Zh. Eksp. Teor. Fiz.* (1979) **77** 884–898 [*Sov. Phys. – JETP* (1979) **50** 446–453].

[264] Hagedorn, G.A., Semiclassical quantum mechanics. I. The $\hbar \to 0$ limit for coherent states. *Commun. Math. Phys.* (1980) **71** 77–93.

[265] Bagrov, V.G., Belov, V.V., and Ternov, I.M., Quasiclassical trajectory-coherent states of a non-relativistic particle in an arbitrary electromagnetic field. *Teor. Mat. Fiz.* (1982) **50** 390–396 [*Theor. & Math. Phys.* (1982) **50** 256–261].

[266] Roy, S.M. and Singh, V., Generalized coherent states and the uncertainty principle. *Phys. Rev.* D (1982) **25** 3413–3416.

[267] Satyanarayana, M.V., Generalized coherent states and generalized squeezed coherent states. *Phys. Rev.* D (1985) **32** 400–404.

[268] Beckers, J. and Debergh, N., On generalized coherent states with maximal symmetry for the harmonic oscillator. *J. Math. Phys.* (1989) **30** 1732–1738.

[269] Wódkiewicz, K. and Eberly, J.H., Coherent states, squeezed fluctuations, and the $SU(2)$ and $SU(1,1)$ groups in quantum-optics applications. *J. Opt. Soc. Am.* B (1985) **2** 458–466.

[270] Barnett, S.M., General criterion for squeezing. *Opt. Commun.* (1987) **61** 432–436.

[271] Trifonov, D.A., Generalized intelligent states and squeezing. *J. Math. Phys.* (1994) **35** 2297–2308.

[272] Hong, C.K. and Mandel, L., Generation of higher-order squeezing of quantum electromagnetic field. *Phys. Rev.* A (1985) **32** 974–982.

[273] Kozierowski, M., Higher-order squeezing in kth-harmonic generation. *Phys. Rev.* A (1986) **34** 3474–3477.

[274] Hillery, M., Amplitude-squared squeezing of the electromagnetic field. *Phys. Rev.* A (1987) **36** 3796–3802.

[275] Zhang, Z.M., Xu, L., Cai, J.L., and Li, F.L., A new kind of higher-order squeezing of radiation field. *Phys. Lett.* A (1990) **150** 27–30.

[276] Hillery, M., Sum and difference squeezing of the electromagnetic field. *Phys. Rev.* A (1989) **40** 3147–3155.

[277] Barnett, S.M. and Gilson, C.R., Quantum electrodynamic bound on squeezing. *Europhys. Lett.* (1990) **12** 325–328.

[278] Kramer, P. and Saraceno, M., Semicoherent states and the group $ISp(2, R)$. *Physica* A (1982) **114** 448–453.

[279] Bishop, R.F. and Vourdas, A., Generalised coherent states and Bogoliubov transformations. *J. Phys.* A (1986) **19** 2525–2536.

[280] Yurke, B., McCall, S.L., and Klauder, J.R., $SU(2)$ and $SU(1,1)$ interferometers. *Phys. Rev.* A (1986) **33** 4033–4054.

[281] Campos, R.A., Saleh, B.E.A., and Teich, M.C., Quantum-mechanical lossless beam splitter: $SU(2)$ symmetry and photon statistics. *Phys. Rev.* A (1989) **40** 1371–1384.

[282] Agarwal, G.S., Nonclassical statistics of fields in pair coherent states. *J. Opt. Soc. Am.* B (1988) **5** 1940–1947.

[283] Yamamoto, Y., Imoto, N., and Machida, S., Amplitude squezing in a semiconductor laser using quantum nondemolition measurement and negative feedback. *Phys. Rev.* A (1986) **33** 3243–3261.

[284] Kitagawa, M. and Yamamoto, Y., Number-phase minimum uncertainty state with reduced number uncertainty in a Kerr nonlinear interferometer. *Phys. Rev.* A (1986) **34** 3974–3988.

[285] Yamamoto, Y., Machida, S., Imoto, N., Kitagawa, M., and Bjork, G., Generation of number-phase minimum-uncertainty states and number states. *J. Opt. Soc. Am.* B (1987) **4** 1645–1661.

[286] Mišta, L., Peřinová, V., Peřina, J., and Braunerová, Z., Quantum statistical properties of degenerate parametric amplification process. *Acta Phys. Polon.* A (1977) **51** 739–751.

[287] Mišta, L. and Peřina, J., Quantum statistics of parametric amplification. *Czechosl. J. Phys.* B (1978) **28** 392–404.

[288] Schleich, W. and Wheeler, J.A., Oscillations in photon distribution of squeezed states and interference in phase space. *Nature* (1987) **326** 574–577.

[289] Schleich, W. and Wheeler, J.A., Oscillations in photon distribution of squeezed states. *J. Opt. Soc. Am.* B (1987) **4** 1715–1722.

[290] Vourdas, A. and Weiner, R.M., Photon-counting distribution in squeezed states. *Phys. Rev.* A (1987) **36** 5866–5869.

[291] Agarwal, G.S. and Adam, G. Photon-number distributions for quantum fields generated in nonlinear optical processes. *Phys. Rev.* A (1988) **38** 750–753.

[292] Schleich, W., Walls, D.F., and Wheeler, J.A., Area of overlap and interference in phase space versus Wigner pseudoprobabilities. *Phys. Rev.* A (1988) **38** 1177–86.

[293] Schleich, W., Walther, H., and Wheeler, J.A., Area in phase space as determiner of transition probability: Bohr-Sommerfeld bands, Wigner ripples, and Fresnel zones. *Found. Phys.* (1988) **18** 953–968.

[294] Dodonov, V.V., Man'ko, V.I., and Rudenko, V.N., Quantum properties of high-Q macroscopic resonators. *Sov. J. Quant. Electron.* (1980) **10** 1232–1238.

[295] Walls, D.F. and Barakat, R., Quantum-mechanical amplification and frequency conversion with a trilinear Hamiltonian. *Phys. Rev.* A (1970) **1** 446–453.

[296] Brif, C., Coherent states for quantum systems with a trilinear boson Hamiltonian. *Phys. Rev.* A (1996) **54** 5253–5261.

[297] Nieto, M.M. and Simmons, L.M., Jr., Coherent states for general potentials. *Phys. Rev. Lett.* (1978) **41** 207–210.

[298] Jannussis, A., Filippakis, P., and Papaloucas, L.C., Commutation relations and coherent states. *Lett. Nuovo Cim.* (1980) **29** 481–484.

[299] Jannussis, A., Papatheou, V., Patargias, N., and Papaloucas, L.C., Coherent states for general potentials. *Lett. Nuovo Cim.* (1981) **31** 385–389.

[300] Gutschik, V.P., Nieto, M.M. and Simmons, L.M., Jr., Coherent states for the "isotonic oscillator". *Phys. Lett.* A (1980) **76** 15–18.

[301] Nieto, M.M. and Simmons, L.M., Jr., Eigenstates, coherent states, and uncertainty products for the Morse oscillator. *Phys. Rev.* A (1979) **19** 438–444.

[302] Levine, R.D., Representation of one-dimensional motion in a Morse potential by a quadratic Hamiltonian. *Chem. Phys. Lett.* (1983) **95** 87–90.

[303] Gerry, C.C., Coherent states and a path integral for the Morse oscillator. *Phys. Rev.* A (1986) **33** 2207–2211.

[304] Kais, S. and Levine, R.D., Coherent states for the Morse oscillator. *Phys. Rev.* A (1990) **41** 2301–2305.

[305] Brown, L.S., Classical limit of the hydrogen atom. *Am. J. Phys.* (1973) **41** 525–530.

[306] Eswaran, K. and Mathews, P.M., Evolution of wave packets for particles in central potentials. *J. Phys.* A (1974) **7** 1547–1556.

[307] Gaeta, Z.D. and Stroud, C.R., Jr., Classical and quantum-mechanical dynamics of a quasiclassical state of the hydrogen atom. *Phys. Rev.* A (1990) **42** 6308–6313.

[308] Fock, V., Zur Theorie des Wasserstoffatoms. *Z. Phys.* (1935) **98** 145–154.

[309] Bargmann, V., Zur Theorie des Wasserstoffatoms. Bemerkungen zur gleichnamigen Arbeit von V.Fock. *Z. Phys.* (1936) **99** 576–582.

[310] Malkin, I.A. and Man'ko, V.I., Symmetry of the hydrogen atom *Pis'ma v ZhETF* (1965) **2** 230–234 [*Sov. Phys. – JETP Lett.* (1965) **2** 146–148].

[311] Sudarshan, E.C.G., Mukunda, N., and O'Raifeartaigh, L., Group theory of the Kepler problem. *Phys. Lett.* A (1965) **19** 322–326.

[312] Barut, A.O. and Kleinert, H., Transition probabilities of the hydrogen atom from noncompact dynamical groups. *Phys. Rev* (1967) **156** 1541–1545.

[313] Mostowski, J., On the classical limit of the Kepler problem. *Lett. Math. Phys.* (1977) **2** 1–5.

[314] Gerry, C.C., On coherent states for the hydrogen atom. *J. Phys.* A (1984) **17** L737–L740.

[315] Gerry, C.C. and Kiefer, J., Radial coherent states for the Coulomb problem. *Phys. Rev.* A (1988) **37** 665–671.

[316] Gay, J.-C., Delande, D., and Bommier, A., Atomic quantum states with maximum localization on classical elliptical orbits. *Phys. Rev.* A (1989) **39** 6587–6590.

[317] Nauenberg, M., Quantum wave packets on Kepler elliptic orbits. *Phys. Rev.* A (1989) **40** 1133–1136.

[318] de Prunelé, E., $SO(4,2)$ coherent states and hydrogenic atoms. *Phys. Rev.* A (1990) **42** 2542–2549.

[319] McAnally, D.S. and Bracken, A.J., Quasi-classical states of the Coulomb system and and $SO(4,2)$. *J. Phys.* A (1990) **23** 2027-2047.

[320] Kustaanheimo, P. and Stiefel, E., Perturbation theory of Kepler motion based on spinor regularization. *J. Reine. Angew. Math.* (1965) **218** 204–219

[321] Moshinsky, M., Seligman, T.H., and Wolf, K.B., Canonical transformations and radial oscillator and Coulomb problems. *J. Math. Phys.* (1972) **13** 901–907.

[322] Moshinsky, M. and Seligman, T.H., Canonical transformations relating the oscillator and Coulomb problems and their relevance for collective motions. *J. Math. Phys.* (1981) **22** 1526–1535.

[323] Gerry, C.C., Coherent states and the Kepler-Coulomb problem. *Phys. Rev.* A (1986) **33** 6–11.

[324] Bhaumik, D., Dutta-Roy, B., and Ghosh, G., Classical limit of the hydrogen atom. *J. Phys.* A (1986) **19** 1355–1364.

[325] Nandi, S. and Shastry, C.S., Classical limit of the two-dimensional and the three-dimensional hydrogen atom. *J. Phys.* A (1989) **22** 1005–1016.

[326] Ginzburg, V.L. and Tamm, I.E., On the theory of spin. *Zhurn. Eksp. Teor. Fiz.* (1947) **17** 227–237 (in Russian).

[327] Yukawa, H., Structure and mass spectrum of elementary particles. 2. Oscillator model. *Phys. Rev.* (1953) **91** 416–417.

[328] Markov, M.A., On dynamically deformable form factors in the theory of elementary particles. *Nuovo Cim.* (Suppl.) (1956) **3** 760–772.

[329] Ginzburg, V.L. and Man'ko, V.I., Relativistic oscillator models of elementary particles. *Nucl. Phys.* (1965) **74** 577–588.

[330] Feynman, R.P., Kislinger, M., and Ravndal, F., Current matrix elements from a relativistic quark model. *Phys. Rev.* D (1971) **3** 2706–2732.

[331] Kim, Y.S. and Noz, M.E., Covariant harmonic oscillators and quark model. *Phys. Rev.* D (1973) **8** 3521–3527.

[332] Kim, Y.S. and Wigner, E.P., Covariant phase-space representation for harmonic oscillator. *Phys. Rev.* A (1988) **38** 1159–1167.

[333] Atakishiev, N.M., Mir-Kasimov, R.M., and Nagiev, S.M., Quasipotential models of a relativistic oscillator. *Teor. Mat. Fiz.* (1980) **44** 47–62 [*Theor. & Math. Phys.* (1981) **44** 592–602].

[334] Ternov, I.M. and Bagrov, V.G., On coherent states of relativistic particles. *Ann. Phys.* (Leipzig) (1983) **40** 2–9.

[335] Ali, S.T., Stochastic localization, quantum mechanics on phase space and quantum space-time. *Riv. Nuovo Cim.* (1985) **8** 1–128.

[336] Gol'fand, Y.A. and Likhtman, E.P., Extension of the algebra of Poincare group generators and violation of P invariance. *Pis'ma v ZhETF* (1971) **13** 452–455 [*JETP Lett.* (1971) **13** 323–326].

[337] Witten, E., Dynamical breaking of supersymmetry. *Nucl. Phys.* B (1981) **188** 513–554.

[338] Salomonson, P. and van Holten, J.W., Fermionic coordinates and supersymmetry in quantum mechanics. *Nucl. Phys.* B (1982) **196** 509–531.

[339] Cooper, F. and Freedman, B., Aspects of supersymmetric quantum mechanics. *Ann. Phys.* (NY) (1983) **146** 262–288.

[340] De Crombrugghe, M. and Rittenberg, V., Supersymmetric quantum mechanics. *Ann. Phys.* (NY) (1983) **151** 99–126.

[341] Bars, I. and Günaydin, M., Unitary representations of non-compact supergroups. *Commun. Math. Phys.* (1983) **91** 31–51.

[342] Balantekin, A.B., Schmitt, H.A., and Barrett, B.R., Coherent states for the harmonic oscillator representations of the orthosymplectic supergroup $Osp(1/2N, R)$. *J. Math. Phys.* (1988) **29** 1634–1639.

[343] Orszag, M. and Salamó, S., Squeezing and minimum uncertainty states in the supersymmetric harmonic oscillator. *J. Phys.* A (1988) **21** L1059–L1064.

[344] Aragone, C. and Zypman, F., Supercoherent states. *J. Phys.* A (1986) **19** 2267–2279.

[345] Baltes, H.P., Quattropani, A., and Schwendimann, P., Subpoissonian and subcoherent radiation fields. *Helv. Phys. Acta* (1978) **51** 534–535.

[346] Baltes, H.P., Quattropani, A., and Schwendimann, P., Construction of sub-Poissonian radiation fields. *J. Phys.* A (1979) **12** L35–L37.

[347] Mandel, L., Sub-Poissonian photon statistics in resonance fluorescence. *Opt. Lett.*

(1979) **4** 205–207.

[348] Fano, U., Ionization yield of radiations. II. The fluctuations of the number of ions. *Phys. Rev.* (1947) **72** 26–29.

[349] Stoler, D., Saleh, B.E.A., and Teich, M.C., Binomial states of the quantized radiation field. *Opt. Acta* (1985) **32** 345–355.

[350] Lee, C.T., Photon antibunching in a free-electron laser. *Phys. Rev.* A (1985) **31** 1213–1215.

[351] Simon, R. and Satyanarayana, M.V., Logarithmic states of the radiation field. *J. Mod. Opt.* (1988) **35** 719–725.

[352] Joshi, A. and Lawande, S.V., The effects of negative binomial field distribution on Rabi oscillations. *Opt. Commun.* (1989) **70** 21–24.

[353] Matsuo, K., Glauber–Sudarshan P-representation of negative binomial states. *Phys. Rev.* A (1990) **41** 519–522.

[354] Quattropani, A., Schwendimann, P., and Baltes, H.P., Sub-Poissonian statistics of an anharmonic oscillator in thermal equilibrium. *Opt. Acta* (1980) **27** 135–138.

[355] Tanaš, R., Squeezed states of an anharmonic oscillator. In: *Coherence and Quantum Optics V* (Proceedings of the conference held in Rochester in 1983) (L. Mandel and E. Wolf, eds.), pp. 645–648. Plenum, New York, 1984.

[356] Milburn, G.J., Quantum and classical Liouville dynamics of the anharmonic oscillator. *Phys. Rev.* A (1986) **33** 674–685.

[357] Yurke, B. and Stoler, D., Generating quantum mechanical superpositions of macroscopically distinguishable states via amplitude dispersion. *Phys. Rev. Lett.* (1986) **57** 13–16.

[358] Mecozzi, A. and Tombesi, P., Distinguishable quantum states generated via nonlinear birefringence. *Phys. Rev. Lett.* (1987) **58** 1055–1058.

[359] Tombesi, P. and Mecozzi, A., Generation of macroscopically distinguishable quantum states and detection by the squeezed-vacuum technique. *J. Opt. Soc. Am.* B (1987) **4** 1700–1709.

[360] Slosser, J.J., Meystre, P., and Braunstein, S.L., Harmonic oscillator driven by a quantum current. *Phys. Rev. Lett.* (1989) **63** 934–937.

[361] Green, H.S., A generalized method of field quantization. *Phys. Rev.* (1953) **90** 270–273.

[362] Greenberg, O.W. and Messiah, A.M.L., Selection rules for parafields and the absence of para particles in nature. *Phys. Rev.* (1965) **138B** 1155–1167.

[363] Brandt, R.A. and Greenberg, O.W., Generalized Bose operators in the Fock space of a single Bose operator. *J. Math. Phys.* (1969) **10** 1168–1176.

[364] Sharma, J.K., Mehta, C.L., and Sudarshan, E.C.G., Para-Bose coherent states. *J. Math. Phys.* (1978) **19** 2089–2093.

[365] Sharma, J.K., Mehta, C.L., Mukunda, N., and Sudarshan, E.C.G., Representation and properties of para-Bose oscillator operators. II. Coherent states and the minimum uncertainty states. *J. Math. Phys.* (1981) **22** 78–90.

[366] Brodimas, G., Jannussis, A., Sourlas, D., Zisis, V., and Poulopoulos, P., Para-Bose operators. *Lett. Nuovo Cim.* (1981) **31** 177–182.

[367] Iwata, G., Transformation functions in the complex domain. *Prog. Theor. Phys.* (1951) **6** 524–528.

[368] Arik, M., Coon, D.D., and Lam, Y.M., Operator algebra of dual resonance models. *J. Math. Phys.* (1975) **16** 1776–1779.

[369] Arik, M. and Coon, D.D., Hilbert spaces of analytic functions and generalized coherent states. *J. Math. Phys.* (1976) **17** 524–527.

[370] Kuryshkin, V.V., On some generalization of creation and annihilation operators in

quantum theory (μ-quantization). Deposit no. 3936–76, Moscow, VINITI, 1976 (in Russian).

[371] Kuryshkin, V., Opérateurs quantiques généralisés de création et d'annihilation. *Ann. Fond. Louis de Broglie* (1980) **5** 111–125.

[372] Jannussis, A., Brodimas, G., Sourlas, D., and Zisis, V., Remarks on the q-quantization. *Lett. Nuovo Cim.* (1981) **30** 123–127.

[373] Gundzik, M.G., A continuous representation of an indefinite metric space. *J. Math. Phys.* (1966) **7** 641–651.

[374] Jaiswal, A.K. and Mehta, C.L., Phase space representation of pseudo oscillator. *Phys. Lett.* A (1969) **29** 245–246.

[375] Agarwal, G.S., Generalized phase-space distributions associated with a pseudo-oscillator. *Nuovo Cim.* B (1970) **65** 266–279.

[376] Biedenharn, L.C., The quantum group $SU_q(2)$ and a q-analog of the boson operator. *J. Phys.* A (1989) **22** L873–L878.

[377] Macfarlane, A.J., On q-analogs of the quantum harmonic oscillator and the quantum group $SU(2)_q$. *J. Phys.* A (1989) **22** 4581–4588.

[378] Kulish, P.P. and Damaskinsky, E.V., On the q-oscillator and the quantum algebra $su_q(1,1)$. *J. Phys.* A (1990) **23** L415–L419.

[379] Chaichan, M., Ellinas, D., and Kulish, P., Quantum algebra as the dynamical symmetry of the deformed Jaynes–Cummings model. *Phys. Rev. Lett.* (1990) **65** 980–983.

[380] Hillery, M., Zubairy, M.S., and Wódkiewicz, K., Squeezing in higher order nonlinear optical processes. *Phys. Lett.* A (1984) **103** 259–261.

[381] Fisher, R.A., Nieto, M.M., and Sandberg, V.D., Impossibility of naively generalizing squeezed coherent states. *Phys. Rev.* D (1984) **29** 1107–1110.

[382] Witschel, W., On Baker–Campbell–Hausdorff operator disentangling by similarity transformations. *Phys. Lett.* A (1985) **111** 383–388.

[383] Braunstein, S.L. and McLachlan, R.I., Generalized squeezing. *Phys. Rev.* A (1987) **35** 1659–1667.

[384] D'Ariano, G., Rasetti, M., and Vadacchino, M., New type of two-photon squeezed coherent states. *Phys. Rev.* D (1985) **32** 1034–1037.

[385] Katriel, J., Solomon, A.I., D'Ariano, G., and Rasetti, M., Multiboson Holstein–Primakoff squeezed states for $SU(2)$ and $SU(1,1)$. *Phys. Rev.* D (1986) **34** 2332–2338.

[386] D'Ariano, G. and Rasetti, M., Non-Gaussian multiphoton squeezed states *Phys. Rev.* D (1987) **35** 1239–1247.

[387] Katriel, J., Rasetti, M., and Solomon, A.I., Squeezed and coherent states of fractional photons. *Phys. Rev.* D (1987) **35** 1248–1254.

[388] D'Ariano, G., Morosi, S., Rasetti, M., Katriel, J., and Solomon, A.I., Squeezing versus photon-number fluctuations. *Phys. Rev.* D (1987) **36** 2399–2407.

[389] Bužek, V. and Jex, I., Amplitude kth-power squeezing of k-photon coherent states. *Phys. Rev.* A (1990) **41** 4079–4082.

[390] Luis, A. and Sánchez-Soto, L.L., Non-classical states of light and canonical transformations. *J. Phys.* A (1991) **24** 2083–2092.

[391] Sukumar, C.V., Revival Hamiltonians, phase operators and non-Gaussian squeezed states. *J. Mod. Opt.* (1989) **36** 1591–1605.

[392] Tombesi, P. and Mecozzi, A., Four-photon squeezed states: An exactly solvable model. *Phys. Rev.* A (1988) **37** 4778–4784.

[393] Braunstein, S.L. and Caves, C.M., Phase and homodyne statistics of generalized squeezed states. *Phys. Rev.* A (1990) **42** 4115–4119.

[394] Elyutin, P.V. and Klyshko, D.N., Three-photon squeezing: exploding solutions and possible experiments. *Phys. Lett.* A (1990) **149** 241–247.

[395] Dodonov, V.V., Klimov, A.B., and Man'ko, V.I., Photon number oscillation in correlated light. *Phys. Lett.* A (1989) **134** 211–216.

[396] Dodonov, V.V., Klimov, A.B., and Man'ko, V.I., Physical effects in correlated quantum states. In: *Squeezed and Correlated States of Quantum Systems. Proc. Lebedev Phys. Inst., vol. 200* (M.A. Markov, ed.), pp. 56–105. Nauka, Moscow, 1991 [translated by Nova Science, Commack, 1993, as vol. 205, pp. 61–107].

[397] Agarwal, G.S. and Adam, G., Photon distributions for nonclassical fields with coherent components. *Phys. Rev.* A (1989) **39** 6259–6266.

[398] Chaturvedi, S. and Srinivasan, V., Photon-number distributions for fields with Gaussian Wigner functions. *Phys. Rev.* A (1989) **40** 6095–6098.

[399] Peřina, J. and Bajer, J., Origin of oscillations in photon distribution of squeezed states. *Phys. Rev.* A (1990) **41** 516–518.

[400] Zahler, M. and Ben Aryeh, Y., Photon number distribution of detuned squeezed states. *Phys. Rev.* A (1991) **43** 6368–6378.

[401] Dutta, B., Mukunda, N., Simon, R., and Subramaniam, A., Squeezed states, photon-number distributions, and $U(1)$ invariance. *J. Opt. Soc. Am.* B (1993) **10** 253–264.

[402] Caves, C.M., Zhu, C., Milburn, G.J., and Schleich, W., Photon statistics of two-mode squeezed states and interference in four-dimensional phase space. *Phys. Rev.* A (1991) **43** 3854–3861.

[403] Dowling, J.P., Schleich, W.P., and Wheeler, J.A., Interference in phase space. *Ann. Phys.* (Leipzig) (1991) **48** 423–478.

[404] Artoni, M., Ortiz, U.P., and Birman, J.L., Photocount distribution of two-mode squeezed states. *Phys. Rev.* A (1991) **43** 3954–3965.

[405] Schrade, G., Akulin, V.M., Man'ko, V.I., and Schleich, W., Photon distribution for two-mode squeezed vacuum. *Phys. Rev.* A (1993) **48** 2398–2406.

[406] Selvadoray, M., Kumar, M.S., and Simon, R., Photon distribution in two-mode squeezed coherent states with complex displacement and squeeze parameters. *Phys. Rev.* A (1994) **49** 4957–4967.

[407] Arvind, B. and Mukunda, N., Non-classical photon statistics for two-mode optical fields. *J. Phys.* A (1996) **29** 5855–5872.

[408] Man'ko, O.V. and Schrade, G., Photon statistics of two-mode squeezed light with Gaussian Wigner function. *Phys. Scripta* (1998) **58** 228–234.

[409] Marian, P. and Marian, T.A., Squeezed states with thermal noise. I. Photon-number statistics. *Phys. Rev.* A (1993) **47** 4474–4486.

[410] Dodonov, V.V., Man'ko, O.V., Man'ko, V.I., and Rosa, L., Thermal noise and oscillations of photon distribution for squeezed and correlated light. *Phys. Lett.* A (1994) **185** 231–237.

[411] Dodonov, V.V. and Man'ko, V.I., New relations for two-dimensional Hermite polynomials. *J. Math. Phys.* (1994) **35** 4277–4294.

[412] Dodonov, V.V., Asymptotic formulae for two-variable Hermite polynomials. *J. Phys.* A (1994) **27** 6191–6203.

[413] Musslimani, Z.H., Braunstein, S.L., Mann, A., and Revzen, M., Destruction of photocount oscillations by thermal noise. *Phys. Rev.* A (1995) **51** 4967–4974.

[414] Marian, P. and Marian, T.A., Photon number and counting statistics for a field with Gaussian characteristic function. *Ann. Phys.* (NY) (1996) **245** 98–112.

[415] Dodonov, V.V., Dremin, I.M., Polynkin, P.G., and Man'ko, V.I., Strong oscillations of cumulants of photon distribution function in slightly squeezed states. *Phys. Lett.*

A (1994) **193** 209–217.

[416] Marchiolli, M.A., Mizrahi, S.S., and Dodonov, V.V., Marginal and correlation distribution functions in the squeezed-states representation. *J. Phys.* A (1999) **32** 8705–8720.

[417] Doebner, H.D., Man'ko, V.I., and Scherer, W., Photon distribution and quadrature correlations in nonlinear quantum mechanics. *Phys. Lett.* A (2000) **268** 17–24.

[418] Abdalla, M.S., Statistical properties of a charged oscillator in the presence of a constant magnetic field. *Phys. Rev.* A (1991) **44** 2040–2047.

[419] Kovarskiy, V.A., Coherent and squeezed states of Landau oscillators in a solid. Emission of nonclassical light. *Sov. Phys. - Solid State* (1992) **34** 1900–1902.

[420] Baseia, B., Mizrahi, S., and Moussa, M.H., Generation of squeezing for a charged oscillator and a charged particle in a time dependent electromagnetic field. *Phys. Rev.* A (1992) **46** 5885–5889.

[421] Aragone, C., New squeezed Landau states. *Phys. Lett.* A (1993) **175** 377–381.

[422] Dodonov, V.V., Man'ko, V.I., and Polynkin, P.G., Geometrical squeezed states of a charged particle in a time-dependent magnetic field. *Phys. Lett.* A (1994) **188** 232–238.

[423] Hadjioannou, F.T. and Sarlis, N.V., Coherent states for the two-dimensional magnetic-electric Euclidean group MEE(2). *Phys. Rev.* B (1997) **56** 9406–9413.

[424] Ozana, M., and Shelankov, A.L., Squeezed states of a particle in magnetic field. *Solid State Physics* (1998) **40** 1276–1282.

[425] Delgado, F.C. and Mielnik, B., Magnetic control of squeezing effects. *J. Phys.* A (1998) **31** 309–320.

[426] Dodonov, V.V., Klimov, A.B., and Man'ko, V.I., Low energy wave packet tunneling from a parabolic potential well through a high potential barrier. *Phys. Lett.* A (1996) **220** 41–48.

[427] Lo, C.F. and Sollie, R., Correlated-squeezed-state approach for phonon coupling in a tunneling system. *Phys. Rev.* B (1991) **44** 5013–5015.

[428] Heller, E., Wavepacket dynamics and quantum chaology. In: *Chaos and Quantum Physics* (M.-J. Giannoni, A. Voros and J. Zinn-Justin, eds.), pp. 547–663. Elsevier, Amsterdam, 1991.

[429] Yuen, H.P., Contractive states and the standard quantum limit for monitoring free-mass positions. *Phys. Rev. Lett.* (1983) **51** 719–722.

[430] Storey, P., Sleator, T., Collett, M., and Walls, D., Contractive states of a free atom. *Phys. Rev.* A (1994) **49** 2322–2328.

[431] Walls, D., Quantum measurements in atom optics. *Austral. J. Phys.* (1996) **49** 715–743.

[432] Lerner, P.B., Rauch, H., and Suda M., Wigner-function calculations for the coherent superposition of matter wave. *Phys. Rev.* A (1995) **51** 3889–3895.

[433] Barvinsky, A.O. and Kamenshchik, A.Y., Preferred basis in quantum theory and the problem of classicalization of the quantum universe. *Phys. Rev.* D (1995) **52** 743–757.

[434] Campos, R.A., Quantum correlation coefficient for angular momentum and spin. *Phys. Lett.* A (1999) **256** 141–146.

[435] Campos, R.A., Quantum correlation coefficient for position and momentum. *J. Mod. Opt.* (1999) **46** 1277–1294.

[436] Brif, C., Mann, A., and Vourdas, A., Parity-dependent squeezing of light. *J. Phys.* A (1996) **29** 2053–2067.

[437] Marian, P., Higher order squeezing properties and correlation functions for squeezed number states. *Phys. Rev.* A (1991) **44** 3325–3330.

[438] Marian, P., Higher order squeezing and photon statistics for squeezed thermal states. *Phys. Rev.* A (1992) **45** 2044–2051.

[439] Hillery, M., Phase-space representation of amplitude-square squeezing. *Phys. Rev.* A (1992) **45** 4944–4950.

[440] Du, S.D. and Gong, C.D., Higher-order squeezing for the quantized light field: Kth-power amplitude squeezing. *Phys. Rev.* A (1993) **48** 2198–2212.

[441] Wang, J.S., Liu, T.K., and Zhan, M.S., Higher order squeezed states of anharmonic oscillators. *Int. J. Theor. Phys.* (2000) **39** 2583–2593.

[442] An, N.B., Multi-directional higher-order amplitude squeezing. *Phys. Lett.* A (2001) **284** 72–80.

[443] Rath, B., An interesting new revelation on simultaneous higher order squeezing in an electro-magnetic field. *Prog. Theor. Phys.* (2001) **105** 697–705.

[444] Fan, H.-Y., Squeezed states: operators for two types of one- and two-mode squeezing transformations. *Phys. Rev.* A (1990) **41** 1526–1532.

[445] Hamilton, J.J., Higher-n squeezed states and factorization of the squeezing operator. *Phys. Rev.* A (1990) **42** 1–5.

[446] Klyshko, D.N., Two-photon (squeezed) light: classical and quantum effects. *Phys. Lett.* A (1990) **146** 93–101.

[447] Král, P., Displaced and squeezed Fock states. *J. Mod. Opt.* (1990) **37** 889–917.

[448] Bykov, V.P., Basic properties of squeezed light. *Uspekhi Fiz. Nauk* (1991) **161** 145–173 [*Sov. Phys. – Uspekhi* (1991) **34** 910–924].

[449] Oz-Vogt, J., Mann, A., and Revzen, M., Thermal coherent states and thermal squeezed states. *J. Mod. Opt.* (1991) **38** 2339–2347.

[450] Bogolubov, N.P., Damaskinsky, E.V., Shumovsky, A.S., and Yarunin, V.S., Quantum statistics of radiation in thermostat. *Czechosl. J. Phys.* (1991) **41** 1031–1036.

[451] Shumovskii, A.S., Bogolyubov canonical transformation and collective states of bosonic fields. *Teor. Mat. Fiz.* (1991) **89** 438–445 [*Theor. & Math. Phys.* (1991) **89** 1323–1329].

[452] Kim, Y.S. and Yeh, L., $E(2)$-symmetric two-mode sheared states. *J. Math. Phys.* (1992) **33** 1237–1246.

[453] Reynaud, S., Heidmann, A., Giacobino, E., and Fabre, C., Quantum fluctuations in optical systems. In: *Progress in Optics, vol. XXX* (E. Wolf, ed.), pp. 1–85. North Holland, Amsterdam, 1992.

[454] Mizrahi, S.S. and Daboul, J., Squeezed states, generalized Hermite polynomials and pseudo-diffusion equation. *Physica* A (1992) **189** 635–650.

[455] Tuyls, P. and Verbeure, A., Spectrum squeezing operator. *Nuovo Cim.* B (1993) **108** 103–107.

[456] Mann, A. and Revzen, M., Gaussian density matrices: quantum analogs of classical states. *Fortschr. Phys.* (1993) **41** 431–446.

[457] Chizhov, A.V. and Murzakhmetov, B.K., Photon statistics and phase properties of two-mode squeezed number states. *Phys. Lett.* A (1993) **176** 33–40.

[458] Freyberger, M. and Schleich, W., Phase uncertainties of a squeezed state. *Phys. Rev.* A (1994) **49** 5056–5066.

[459] Arnoldus, H.F. and George, T.F., Squeezing in resonance fluorescence and Schrödinger's uncertainty relation. *Physica* A (1995) **222** 330–346.

[460] Hillery, M., Freyberger, M., and Schleich, W., Phase distributions and large-amplitude states. *Phys. Rev.* A (1995) **51** 1792–1803.

[461] Bužek, V. and Hillery, M., Operational phase distributions via displaced squeezed states. *J. Mod. Opt.* (1996) **43** 1633–1651.

[462] Weigert, S., Spatial squeezing of the vacuum and the Casimir effect. *Phys. Lett.* A

(1996) **214** 215–220.

[463] Møller, K.B., Jørgensen, T.G., and Dahl, J.P., Displaced squeezed number states: Position space representation, inner product, and some applications. *Phys. Rev.* A (1996) **54** 5378–5385.

[464] Bialynicki-Birula, I., Nonstandard introduction to squeezing of the electromagnetic field. *Acta Phys. Pol.* B (1998) **29** 3569–3590.

[465] Lu, W.-F., Thermalized displaced and squeezed number states in the coordinate representation. *J. Phys.* A (1999) **32** 5037–5051.

[466] Wiseman, H.M., Squashed states of light: theory and applications to quantum spectroscopy. *J. Opt.* B (1999) **1** 459–463.

[467] Mancini, S., Vitali, D., and Tombesi, P., Motional squashed states. *J. Opt.* B (2000) **2** 190–195.

[468] Nieto, M.M. and Truax, D.R., Coherent states sometimes look like squeezed states and vice versa: the Paul trap. *New J. Phys.* (2000) **2** 1–9.

[469] Fan, H.Y. and Wünsche, A., Eigenstates of boson creation operator. *Eur. Phys. J.* D (2001) **15** 405–412.

[470] Popescu, V.A., Second-order correlation function for successive squeezed states. *Mod. Phys. Lett.* B (1999) **13** 1063–1073.

[471] Slowíkowski, W. and Mølmer, K., Squeezing of free Bose fields *J. Math. Phys.* (1990) **31** 2327-2333.

[472] Ma,X. and Rhodes, W., Multimode squeeze operators and squeezed states. *Phys. Rev.* A (1990) **41** 4625–4631.

[473] Tucci, R.R., Two-mode Gaussian density matrices and squeezing of photons. *Int. J. Mod. Phys.* B (1992) **6** 1657–1709.

[474] Yeoman, G. and Barnett, S.M., Two-mode squeezed gaussons. *J. Mod. Opt.* (1993) **40** 1497–1530.

[475] Huang, H. and Agarwal, G.S., General linear transformations and entangled states. *Phys. Rev.* A (1994) **49** 52–60.

[476] Simon, R., Mukunda, N., and Dutta, B., Quantum-noise matrix for multimode systems: $U(n)$-invariance, squeezing, and normal forms. *Phys. Rev.* A (1994) **49** 1567–1583.

[477] Caves, C.M. and Drummond, P.D., Quantum limits on bosonic communication rates. *Rev. Mod. Phys.* (1994) **66** 481–537.

[478] Hall M J W 1994 Gaussian noise and quantum optical communication *Phys. Rev.* A **50** 3295-3303

[479] Arvind, Dutta, B., Mukunda, N., and Simon, R., Two-mode quantum systems: Invariant classification of squeezing transformations and squeezed states. *Phys. Rev.* A (1995) **52** 1609–1620.

[480] Honegger, R. and Rieckers, A., Squeezing operations in Fock space and beyond. *Physica* A (1997) **242** 423–438.

[481] Trifonov, D.A., On the squeezed states for n observables. *Phys. Scripta* (1998) **58** 246–255.

[482] Fiurasek, J. and Peřina, J., Phase properties of two-mode Gaussian light fields with application to Raman scattering. *J. Mod. Opt.* (2000) **47** 1399–1417.

[483] Fan, H.Y. and Yu, G., Three-mode squeezed vacuum state in Fock space as an entangled state. *Phys. Rev.* A (2002) **65** 033829.

[484] Slater, P.B., Essentially all Gaussian two-party quantum states are a priori nonclassical but classically correlated. *J. Opt.* B (2000) **2** L19–L24.

[485] Lindblad, G., Cloning the quantum oscillator. *J. Phys.* A (2000) **33** 5059–5076.

[486] Duan, L.M., Giedke, G., Cirac, J.I., and Zoller, P., Inseparability criterion for

continuous variable systems. *Phys. Rev. Lett.* (2000) **84** 2722–2725.

[487] Simon, R., Peres–Horodecki separability criterion for continuous variable systems. *Phys. Rev. Lett.* (2000) **84** 2726–2729.

[488] Werner, R.F. and Wolf, M.M., Bound entangled gaussian states. *Phys. Rev. Lett.* (2001) **86** 3658–3661.

[489] Holevo, A.S. and Werner, R.F., Evaluating capacities of bosonic Gaussian channels. *Phys. Rev.* A (2001) **63** 032312.

[490] Wang, X.B., Keiji, M., and Akihisa, T., Detecting the inseparability and distillability of continuous variable states in Fock space. *Phys. Rev. Lett.* (2001) **87** 137903.

[491] Marian, P., Marian, T.A., and Scutaru, H., Inseparability of mixed two-mode Gaussian states generated with a $SU(1,1)$ interferometer. *J. Phys.* A (2001) **34** 6969–6980.

[492] Giedke, G., Kraus, B., Duan, L.-M., Zoller, P., Cirac, J.I., and Lewenstein, M., Separability and distillability of bipartite Gaussian states — the complete story. *Fortschr. Phys.* (2001) **49** 973-980.

[493] Giedke, G., Kraus, B., Lewenstein, M., and Cirac, J.I., Separability properties of three-mode Gaussian states. *Phys. Rev.* A (2001) **64** 052303.

[494] Scheel, S. and Welsch, D.-G., Entanglement generation and degradation by passive optical devices. *Phys. Rev.* A (2001) **64** 063811.

[495] Chizhov, A.V., Haus, J.W., and Yeong, K.C., Higher-order squeezing in a boson-coupled two-mode system. *Phys. Rev.* A (1995) **52** 1698–1703.

[496] Chizhov, A.V., Haus, J.W., and Yeong, K.C., Higher-order squeezing in a boson-coupled three-mode system. *J. Opt. Soc. Am.* B (1997) **14** 1541–1549.

[497] An, N.B. and Tinh, V., General multimode sum-squeezing. *Phys. Lett.* A (1999) **261** 34–39.

[498] An, N.B. and Tinh, V., General multimode difference-squeezing. *Phys. Lett.* A (2000) **270** 27–40.

[499] Dodonov, V.V., Man'ko, V.I., and Ossipov, D.L., Quantum evolution of the localized states. *Physica* A (1990) **168** 1055–1072.

[500] Combescure, M., The squeezed state approach of the semiclassical limit on the time-dependent Schrödinger equation. *J. Math. Phys.* (1992) **33** 3870–3880.

[501] Bagrov, V.G., Belov, V.V., and Trifonov, A.Y., Semiclassical trajectory-coherent approximation in quantum mechanics. I. High-order corrections to multidimensional time-dependent equations of Schrödinger type. *Ann. Phys.* (NY) (1996) **246** 231–290.

[502] Hagedorn, G.A., Raising and lowering operators for semiclassical wave packets. *Ann. Phys.* (NY) (1998) **269** 77–104.

[503] Belov, V.V., Trifonov, A.Y., and Shapovalov, A.V., Semiclassical trajectory-coherent approximations of Hartree-type equations. *Theor. & Math. Phys.* (2002) **130** 391–418.

[504] Sauerzapf, A. and Wagner, M., Quantal properties of self-localized solitons in anharmonic oscillatory systems. *J. Luminescence* (1998) **76** 599–603.

[505] Kolobov, M.I. and Sokolov, I.V., Squeezed states of light and noise-free optical images. *Phys. Lett.* A (1989) **140** 101–104.

[506] Gatti, A., Brambilla, E., Lugiato, L.A., and Kolobov, M.I., Quantum entangled images. *Phys. Rev. Lett.* (1999) **83** 1763–1766.

[507] Kolobov, M.I. and Fabre, C., Quantum limits on optical resolution. *Phys. Rev. Lett.* (2000) **85** 3789–3792.

[508] Szwaj, C., Oppo, G.-L., Gatti, A., and Lugiato, L.A., Quantum images in non degenerate optical parametric oscillators. *Eur. Phys. J.* D (2000) **10** 433–448.

[509] Lugiato, L.A., Gatti, A., and Brambilla, E., Quantum imaging. *J. Opt.* B (2002) **4** S176–S183.

[510] Milonni, P.W. and Singh, S., Some recent developments in the fundamental theory of light. *Adv. At. Mol. Opt. Phys.* (1990) **28** 75–142.

[511] Turchette, Q.A., Georgiades, N.P., Hood, C.J., Kimble, H.J., and Parkins, A.S., Squeezed excitation in cavity QED: Experiment and theory. *Phys. Rev.* A (1998) **58** 4056–4077.

[512] Dalton, B.J., Ficek, Z., and Swain, S., Atoms in squeezed light fields. *J. Mod. Opt.* (1999) **46** 379–474.

[513] Kolobov, M.I., The spatial behavior of nonclassical light. *Rev. Mod. Phys.* (1999) **71** 1539–1589.

[514] Svozil, K., Squeezed fermion states. *Phys. Rev. Lett.* (1990) **65** 3341–3343.

[515] Liu, W.S. and Li, X.P., BCS states as squeezed fermion-pair states. *Eur. Phys. J.* D (1998) **2** 1–4.

[516] Fan, H.Y. and Sun, Z.H., New formulation for squeezed fermion-pair states as a counterpart of Grassmannian evolution. *Eur. Phys. J.* D (2000) **12** 11–13.

[517] Hradil, Z., Noise minimum states and the squeezing and antibunching of light. *Phys. Rev.* A (1990) **41** 400–407.

[518] Hradil, Z., Extremal properties of near-photon-number eigenstate fields. *Phys. Rev.* A (1991) **44** 792–795.

[519] Brif, C., Two-photon algebra eigenstates. A unified approach to squeezing. *Ann. Phys.* (NY) (1996) **251** 180–207.

[520] Brif, C., SU(2) and SU(1,1) algebra eigenstates: a unified analytic approach to coherent and intelligent states. *Int. J. Theor. Phys.* (1997) **36** 1651–1682.

[521] Brif, C. and Mann, A., Generation of single-mode SU(1,1) intelligent states and an analytic approach to their quantum statistical properties. *Quant. Semiclass. Opt.* (1997) **9** 899–920.

[522] Trifonov, D.A., Algebraic coherent states and squeezing. *Los Alamos Preprint* quant-ph/9609001.

[523] Trifonov, D.A., Exact solutions for the general nonstationary oscillator with a singular perturbation. *J. Phys.* A (1999) **32** 3649–3661.

[524] Sivakumar, S., Interpolating coherent states for Heisenberg–Weyl and single-photon $SU(1,1)$ algebras. *J. Phys.* A (2002) **35** 6755–6766.

[525] Trifonov, D.A., Completeness and geometry of Schrödinger minimum uncertainty states. *J. Math. Phys.* (1993) **34** 100–110.

[526] Trifonov, D.A., Robertson intelligent states. *J. Phys.* A (1997) **30** 5941–5957.

[527] Yu, D. and Hillery, M., Minimum uncertainty states for amplitude-squared squeezing: general solutions. *Quant. Opt.* (1994) **6** 37–56.

[528] Puri, R.R., Minimum-uncertainty states for noncanonical operators. *Phys. Rev.* A (1994) **49** 2178–2180.

[529] Gerry, C.C. and Grobe, R., Two-mode intelligent $SU(1,1)$ states. *Phys. Rev.* A (1995) **51** 4123–4131.

[530] Puri R.R. and Agarwal, G.S., $SU(1,1)$ coherent states defined via a minimum-uncertainty product and an equality of quadrature variances. *Phys. Rev.* A (1996) **53** 1786–1790.

[531] Weigert, S., Landscape of uncertainty in Hilbert space for one-particle states. *Phys. Rev.* A (1996) **53** 2084–2088.

[532] Gerry, C.C. and Grobe, R., Intelligent photon states associated with the Holstein-Primakoff realisation of the $SU(1,1)$ Lie algebra. *Quant. Semiclass. Opt.* (1997) **9** 59–67.

[533] Arvieu, R. and Rozmej, P., Geometrical properties of intelligent spin states and time evolution of coherent states. *J. Phys.* A (1999) **32** 2645–2652.
[534] Peřinová, V., Lukš, A., and Křepelka, J., Intelligent states in $SU(2)$ and $SU(1,1)$ interferometry. *J. Opt.* B (2000) **2** 81–89.
[535] Fan, H.Y. and Sun, Z.H., Minimum uncertainty $SU(1,1)$ coherent states for number difference: Two-mode phase squeezing. *Mod. Phys. Lett.* (2000) **14** 157–166.
[536] Liu, N.L., Sun, Z.H., and Huang, L.S., Intelligent states of the quantized radiation field associated with the Holstein-Primakoff realisation of $su(2)$. *J. Phys.* A (2000) **33** 3347–3360.
[537] Sa, N.B.E., Uncertainty for spin systems. *J. Math. Phys.* (2001) **42** 981–990.
[538] Hoffman, D.K. and Kouri, D.J., Hierarchy of local minimum solutions of Heisenberg's uncertainty principle. *Phys. Rev.* A (2002) **65** 052106.
[539] Alvarez, M.N. and Hussin, V., Generalized coherent and squeezed states based on the $h(1) \otimes su(2)$ algebra. *J. Math. Phys.* (2002) **43** 2063–2096.
[540] El Kinani, A.H. and Daoud, M., Generalized intelligent states for an arbitrary quantum system. *J. Phys.* A (2001) **34** 5373–5387.
[541] El Kinani, A.H. and Daoud, M., Generalized coherent and intelligent states for exact solvable quantum systems. *J. Math. Phys.* (2002) **43** 714–733.
[542] Marhic, M.E. and Kumar, P., Squeezed states with a thermal photon distribution. *Opt. Commun.* (1990) **76** 143–146.
[543] Vourdas, A., $SU(2)$ and $SU(1,1)$ phase states. *Phys. Rev.* A (1990) **41** 1653–1661.
[544] Shapiro, J.H. and Shepard, S.R., Quantum phase measurement: a system-theory perspective. *Phys. Rev.* A (1991) **43** 3795–3818.
[545] Chizhov, A.V. and Paris, M.G.A., Phase squeezed states. *Acta Phys. Slovaca* (1998) **48** 343–348.
[546] Summy, G.S. and Pegg, D.T., Phase optimized quantum states of light. *Opt. Commun.* (1990) **77** 75–79.
[547] Bandilla, A., Paul, H., and Ritze, H.-H., Realistic quantum states of light with minimum phase uncertainty. *Quant. Opt.* (1991) **3** 267–282.
[548] Bandilla, A., How to realize phase optimized quantum states. *Opt. Commun.* (1992) **94** 273–280.
[549] Szabo, S., Adam, P., Karpati, A., and Janszky, J., Phase optimized light in complex optical systems. *Fortschr. Phys.* (2001) **49** 1109–1116.
[550] Vourdas, A., Analytic representations in the unit disk and applications to phase state and squeezing. *Phys. Rev.* A (1992) **45** 1943–1950.
[551] Sudarshan, E.C.G., Diagonal harmonious state representation. *Int. J. Theor. Phys.* (1993) **32** 1069–1076.
[552] Brif, C. and Ben-Aryeh, Y., Phase-state representation in quantum optics. *Phys. Rev.* A (1994) **50** 3505–3516.
[553] Vourdas, A., Brif, C., and Mann, A., Factorization of analytic representations in the unit disc and number–phase statistics of a quantum harmonic oscillator. *J. Phys.* A (1996) **29** 5887–5898.
[554] Brif, C., Photon states associated with the Holstein-Primakoff realisation of the $SU(1,1)$ Lie algebra. *Quant. Semiclass. Opt.* (1995) **7** 803–834.
[555] Luo, S.L., Minimum uncertainty states for Dirac's number-phase pair. *Phys. Lett.* A (2000) **275** 165–168.
[556] Dodonov, V.V. and Mizrahi, S.S., Uniform nonlinear evolution equations for pure and mixed quantum states. *Ann. Phys.* (NY) (1995) **237** 226–268.
[557] Chaturvedi, S., Kapoor, A.K., Sandhya, R., Srinivasan, V., and Simon, R., Gener-

alized commutation relations for a single-mode oscillator. *Phys. Rev.* A (1991) **43** 4555–4557.

[558] Agarwal, G.S., Infinite statistics and the relation to a phase operator in quantum optics. *Phys. Rev.* A (1991) **44** 8398–8399.

[559] Lukš, A., Peřinová, V., and Křepelka, J., Special states of the plane rotator relevant to the light field. *Phys. Rev.* A (1992) **46** 489–498.

[560] Hall, M.J.W., Phase resolution and coherent phase states. *J. Mod. Opt.* (1993) **40** 809–824.

[561] Gantsog, T., Joshi, A, and Tanaś, R., Phase properties of binomial and negative binomial states. *Quant. Opt.* (1994) **6** 517–526.

[562] Opatrný, T., Number–phase uncertainty relations. *J. Phys.* A (1995) **28** 6961–6975.

[563] Royer, A., Phase states and phase operators for the quantum harmonic oscillator. *Phys. Rev.* A (1996) **53** 70–108.

[564] Baseia, B., Dantas, C.M.A., and Moussa, M.H.Y., Pure states having thermal photon distribution revisited: Generation and phase-optimization. *Physica* A (1998) **258** 203–210.

[565] Peřinová, V., Lukš, A., and Křepelka, J., Quasidistributions for noncommuting cosine and sine operators. *Eur. Phys. J.* D (1999) **5** 417–432.

[566] Wünsche, A., Realizations of $SU(1,1)$ by boson operators with application to phase states. *Acta Phys. Slovaca* (1999) **49** 771–782.

[567] Wünsche, A., A class of phase-like states. *J. Opt.* B (2001) **3** 206–218.

[568] Bužek, V., $SU(1,1)$ squeezing of $SU(1,1)$ generalized coherent states. *J. Mod. Opt.* (1990) **37** 303–316.

[569] Ban, M., Lie-algebra methods in quantum optics: the Liouville-space formulation. *Phys. Rev.* A (1993) **47** 5093–5119.

[570] Wünsche, A., Symplectic groups in quantum optics. *J. Opt.* B (2000) **2** 73–80.

[571] Gerry, C.C., Correlated two-mode $SU(1,1)$ coherent states: nonclassical properties. *J. Opt. Soc. Am.* B (1991) **8** 685–690.

[572] Basu, D., The Barut–Girardello coherent states. *J. Math. Phys.* (1992) **33** 114–121.

[573] Leonhardt, U., Quantum statistics of a two-mode $SU(1,1)$ interferometer. *Phys. Rev.* A (1994) **49** 1231–1242.

[574] Brif, C. and Ben-Aryeh, Y., Subcoherent P-representation for nonclassical photon states. *Quant. Opt.* (1994) **6** 391–396.

[575] Nieto, M.M. and Truax, D.R., Squeezed states for general systems. *Phys. Rev. Lett.* (1993) **71** 2843–2846.

[576] Prakash, G.S. and Agarwal, G.S., Mixed excitation- and deexcitation-operator coherent states for the $SU(1,1)$ group. *Phys. Rev.* A (1994) **50** 4258–4263.

[577] Marian, P., Second-order squeezed states. *Phys. Rev.* A (1997) **55** 3051–3058 [Erratum: *Phys. Rev.* A (1999) **59** 3141].

[578] Prakash, G.S. and Agarwal, G.S., Pair excitation-deexcitation coherent states. *Phys. Rev.* A (1995) **52** 2335–2341.

[579] Liu, X.M., Even and odd charge coherent states and their non-classical properties. *Phys. Lett.* A (2001) **279** 123–132.

[580] Liu, X.M., k-Component charge coherent states and their non-classical properties. *Phys. Lett.* A (2001) **292** 23–35.

[581] Várilly, J.C. and Gracia-Bondía, J.M., The Moyal representation for spin. *Ann. Phys.* (NY) (1989) **190** 107–148.

[582] Weiner, B., Deumens, E., and Öhrn, Y., Spin projection of fermion coherent states. *J. Math. Phys.* (1991) **32** 2413–2426.

[583] Öhrn, Y. and Goscinski, O., Analysis of the Thouless coherent state using the $1/K$ expansion. *Phys. Rev.* A (1993) **48** 1093–1097.

[584] Macomber, J.D. and Lynch, R., Squeezed spin states. *J. Chem. Phys.* (1985) **83** 6514–6519.

[585] Kitagawa, M. and Ueda, M., Squeezed spin states. *Phys. Rev.* A (1993) **47** 5138–5143.

[586] Isham, C.J. and Klauder, J.R., Coherent states for n-dimensional Euclidean groups $E(n)$ and their application. *J. Math. Phys.* (1991) **32** 607–620.

[587] Gitman, D.M. and Shelepin, A.L., Coherent states of $SU(N)$ groups. *J. Phys.* A (1993) **26** 313–327.

[588] Puri, R.R., $SU(m,n)$ coherent states in the bosonic representation and their generation in optical parametric processes. *Phys. Rev.* A (1994) **50** 5309–5316.

[589] Nemoto, K., Generalized coherent states for $SU(n)$ systems. *J. Phys.* A (2000) **33** 3493–3506.

[590] Cao, C.Q. and Haake, F., Coherent states and holomorphic representations for multilevel atoms. *Phys. Rev.* A (1995) **51** 4203–4210.

[591] Gerry, C.C. and Grobe, R., Statistical properties of squeezed Kerr states. *Phys. Rev.* A (1994) **49** 2033–2039.

[592] Fu, H.C. and Sasaki, R., Exponential and Laguerre squeezed states for $su(1,1)$ algebra and the Calogero–Sutherland model. *Phys. Rev.* A (1996) **53** 3836–3844.

[593] Brif, C., Vourdas, A., and Mann, A., Analytic representation based on SU(1,1) coherent states and their applications. *J. Phys.* A (1996) **29** 5873–5885.

[594] Gnutzmann, S and Kus, M, Coherent states and the classical limit on irreducible $SU(3)$ representations. *J. Phys.* A (1998) **31** 9871–9896.

[595] Mathur, M. and Sen, D., Coherent states for $SU(3)$. *J. Math. Phys.* (2001) **42** 4181–4196.

[596] Daubechies, I., Klauder, J.R., and Paul, T., Wiener measures for path integrals with affine kinematic variables. *J. Math. Phys.* (1987) **28** 85–102.

[597] Watson, G. and Klauder, J.R., Generalized affine coherent states: A natural framework for the quantization of metric-like variables. *J. Math. Phys.* (2000) **41** 8072–8082.

[598] Pris, I.E. and Tolkachev, E.A., Dyogen atom as a constrained 4–dimensional isotropic singular oscillator. *Sov. J. Nucl. Phys.* (1991) **54** 582–584.

[599] Zlatev, I., Zhang, W.-M., and Feng, D.H., Possibility that Schrödinger's conjecture for the hydrogen atom coherent states is not attainable. *Phys. Rev.* A (1994) **50** R1973–R1975.

[600] Thomas L.E. and Villegas-Blas, C., Asymptotics of Rydberg states for the hydrogen atom. *Commun. Math. Phys.* (1997) **187** 623–645.

[601] Ghosh, G., Generalized annihilation operator coherent states. *J. Math. Phys.* (1998) **39** 1366–1372.

[602] Bellomo, P. and Stroud, C.R., Classical evolution of quantum elliptic states. *Phys. Rev.* A (1999) **59** 2139–2145.

[603] Toyoda, T. and Wakayama, S., Coherent states for the Kepler motion. *Phys. Rev.* A (1999) **59** 1021–1024.

[604] Nouri, S., Generalized coherent states for the d-dimensional Coulomb problem. *Phys. Rev.* A (1999) **60** 1702–1705.

[605] Crawford, M.G.A., Temporally stable coherent states in energy-degenerate systems: The hydrogen atom. *Phys. Rev.* A (2000) **62** 012104.

[606] Toyoda, T. and Wakayama, S., Coherent states for the Kepler motion. II. *Phys. Rev.* A (2001) **64** 032110.

[607] Pol'shin, S.A., Coherent states for the hydrogen atom: discrete and continuous spectra. *J. Phys.* A (2001) **34** 11083–11094.

[608] Drăgănescu, G.E., Messina, A., and Napoli, A., Radial coherent states for Dirac hydrogen-like atom. *J. Opt.* B (2002) **4** 240–244.

[609] Alber, G. and Zoller, P., Laser excitation of electronic wave packets in Rydberg atoms. *Phys. Rep.* (1991) **199** 232–280.

[610] Bialynicki-Birula, I., Kaliński, M., and Eberly, J.H., Lagrange equilibrium points in celestial mechanics and nonspreading wave packets for strongly driven Rydberg electrons. *Phys. Rev. Lett.* (1994) **73** 1777–1780.

[611] Nieto, M.M., Rydberg wave packets are squeezed states. *Quant. Opt.* (1994) **6** 9–14.

[612] Bluhm, R., Kostelecky, V.A., and Tudose, B., Elliptical squeezed states and Rydberg wave packets. *Phys. Rev.* A (1995) **52** 2234–2244.

[613] Cerjan, C., Lee, E., Farrelly, D., and Uzer, T., Coherent states in a Rydberg atom: Quantum mechanics. *Phys. Rev.* A (1997) **55** 2222–2231.

[614] Hagedorn, G.A. and Robinson, S.L., Approximate Rydberg states of the hydrogen atom that are concentrated near Kepler orbits. *Helv. Phys. Acta* (1999) **72** 316–340.

[615] Michel, L. and Zhilinskii, B.I., Rydberg states of atoms and molecules. Basic group theoretical and topological analysis. *Phys. Rep.* (2001) **341** 173–264.

[616] Verlet, J.R.R. and Fielding, H.H., Manipulating electron wave packets. *Int. Rev. Phys. Chem.* (2001) **20** 283–312.

[617] Karasev, M.V. and Novikova, E.M., Representation of exact and semiclassical eigenfunctions via coherent states. Hydrogen atom in a magnetic field. *Teor. Mat. Fiz.* (1996) **108** 339–387 [*Theor. & Math. Phys.* (1996) **108** 1119–1159].

[618] Cooper, I.L., A simple algebraic approach to coherent states for the Morse oscillator. *J. Phys.* A (1992) **25** 1671–1683.

[619] Drăgănescu, G.E. and Avram, N.M., Creation and annihilation operators for the Morse oscillator and the coherent states. *Canad. J. Phys.* (1998) **76** 273–281.

[620] Benedict, M.G. and Molnár, B., Algebraic construction of the coherent states of the Morse potential based on supersymmetric quantum mechanics. *Phys. Rev.* A (1999) **60** R1737–R1740.

[621] Avram, N.M., Drăgănescu, G.E., and Avram, C.N., Vibrational coherent states for Morse oscillator. *J. Opt.* B (2000) **2** 214–219.

[622] Molnár, B., Benedict, M.G., and Bertrand, J., Coherent states and the role of the affine group in the quantum mechanics of the Morse potential. *J. Phys.* A (2001) **34** 3139–3151.

[623] Dong, S.H., The $SU(2)$ realization for the Morse potential and its coherent states. *Canad. J. Phys.* (2002) **80** 129–139.

[624] Roy, B. and Roy, P., Gazeau–Klauder coherent state for the Morse potential and some of its properties. *Phys. Lett.* A (2002) **296** 187–191.

[625] Crawford, M.G.A. and Vrscay, E.R., Generalized coherent states for the Pöschl–Teller potential and a classical limit. *Phys. Rev.* A (1998) **57** 106–113.

[626] El Kinani, A.H. and Daoud, M., Coherent states à la Klauder-Perelomov for the Pöschl–Teller potentials. *Phys. Lett.* A (2001) **283** 291–299.

[627] Agarwal, G.S. and Chaturvedi, S., Calogero–Sutherland oscillator: Classical behavior and coherent states. *J. Phys.* A (1995) **28** 5747–5755.

[628] Maamache, M., Invariant-angle coherent states for the singular oscillator and geometrical phases and angles. *Ann. Phys.* (NY) (1997) **254** 1–10.

[629] Dodonov, V.V., Man'ko, V.I., and Rosa, L., Quantum singular oscillator as a model of a two-ion trap: An amplification of transition probabilities due to small time

variations of the binding potential. *Phys. Rev.* A (1998) **57** 2851–2858.

[630] Wang, J.-S., Liu, T.-K, and Zhan, M.-S., Nonclassical properties of even and odd generalized coherent states for an isotonic oscillator. *J. Opt.* B (2000) **2** 758–763.

[631] Popov, D., Barut–Girardello coherent states of the pseudoharmonic oscillator. *J. Phys.* A (2001) **34** 5283–5296.

[632] Samsonov, B.F., Coherent states of potentials of soliton origin. *J. Exp. Theor. Phys.* (1998) **87** 1046–1052.

[633] Kowalski, K., Rembieliński, J., and Papaloucas, L.C., Coherent states for a quantum particle on a circle. *J. Phys.* A (1996) **29** 4149–4167.

[634] González, J.A. and del Olmo, M.A., Coherent states on the circle. *J. Phys.* A (1998) **31** 8841–8857.

[635] Kowalski, K. and Rembieliński, J., Quantum mechanics on a sphere and coherent states. *J. Phys.* A (2000) **33** 6035–6048.

[636] Kowalski, K. and Rembieliński, J., On the uncertainty relations and squeezed states for the quantum mechanics on a circle. *J. Phys.* A (2002) **35** 1405–1414.

[637] Hall, B.C. and Mitchell, J.J., Coherent states on spheres. *J. Math. Phys.* (2002) **43** 1211-1236.

[638] Fu, H.C. and Sasaki, R., Probability distributions and coherent states of B_r, C_r and D_r algebras. *J. Phys.* A (1998) **31** 901–925.

[639] Rozmej, P. and Arvieu, R., Clones and other interference effects in the evolution of angular-momentum coherent states. *Phys. Rev.* A (1998) **58** 4314–4329.

[640] Basu, D., The coherent states of the $SU(1,1)$ group and a class of associated integral transforms. *Proc. Roy. Soc. Lond.* A (1999) **455** 975–989.

[641] Solomon, A.I., Quantum optics: Group and non-group methods. *Int. J. Mod. Phys.* B (1999) **13** 3021–3038.

[642] Sunilkumar, V., Bambah, B.A., Jagannathan, R., Panigrahi, P.K., and Srinivasan, V., Coherent states of nonlinear algebras: applications to quantum optics. *J. Opt.* B (2000) **2** 126–132.

[643] Atakishiev, N.M. and Suslov, S.K., Difference analogs of the harmonic oscillator. *Teor. Mat. Fiz.* (1990) **85** 64–73 [*Theor. & Math. Phys.* (1990) **85** 1055–1062].

[644] Solomon, A.I. and Katriel, J., On q-squeezed states. *J. Phys.* A (1990) **23** L1209–L1212.

[645] Katriel, J. and Solomon, A.I., Generalised q-bosons and their squeezed states. *J. Phys.* A (1991) **24** 2093–2105.

[646] Celeghini, E., Rasetti, M., and Vitiello, G., Squeezing and quantum groups. *Phys. Rev. Lett.* (1991) **66** 2056–2059.

[647] Chaturvedi, S. and Srinivasan, V., Aspects of q-oscillator quantum mechanics. *Phys. Rev.* A (1991) **44** 8020–8023.

[648] Chaturvedi, S. and Srinivasan, V., Para-Bose oscillator as a deformed Bose oscillator. *Phys. Rev.* A (1991) **44** 8024–8026.

[649] Chiu, S.-H., Gray, R.W., and Nelson, C.A., The q-analog quantized radiation field and its uncertainty relations. *Phys. Lett.* A (1992) **164** 237–242.

[650] McDermott, R.J. and Solomon, A.I., Squeezed states parametrized by elements of noncommutative algebras. *Czechosl. J. Phys.* (1996) **46** 235–241.

[651] Osland, P. and Zhang, J.Z., Critical phenomenon of a consistent q-deformed squeezed state. *Ann. Phys.* (NY) (2001) **290** 45–52.

[652] Fivel, D.I., Quasi-coherent states and the spectral resolution of the q-Bose field operator. *J. Phys.* A (1991) **24** 3575–3586.

[653] Chakrabarti, R. and Jagannathan, R., A (p,q)-oscillator realisation of two-parameter quantum algebras. *J. Phys.* A (1991) **24** L711–L718.

[654] Arik, M., Demircan, E., Turgut, T., Ekinci, L., and Mungan, M., Fibonacci oscillators. *Z. Phys.* C (1992) **55** 89–95.

[655] Ellinas, D., Quantum phase and a q-deformed quantum oscillator. *Phys. Rev.* A (1992) **45** 3358–3361.

[656] Wang, F.B. and Kuang, L.M., Even and odd q-coherent state representations of the quantum Heisenberg-Weyl algebra. *Phys. Lett.* A (1992) **169** 225–228.

[657] Jing, S.C. and Fan, H.Y., q-deformed binomial state. *Phys. Rev.* A (1994) **49** 2277–2279.

[658] Solomon, A.I. and Katriel, J., Multi-mode q-coherent states. *J. Phys.* A (1993) **26** 5443–5447.

[659] Jurčo, B., On coherent states for the simplest quantum groups. *Lett. Math. Phys.* (1991) **21** 51–58.

[660] Quesne, C., Coherent states, K-matrix theory and q-boson realizations of the quantum algebra $su_q(2)$. *Phys. Lett.* A (1991) **153** 303–307.

[661] Zhedanov, A.S., Nonlinear shift of q-Bose operators and q-coherent states. *J. Phys.* A (1991) **24** L1129–L1132.

[662] Chang, Z., Generalized Holstein-Primakoff realizations and quantum group-theoretic coherent states, *J. Math. Phys.* (1992) **33** 3172–3179.

[663] Wang, F.B. and Kuang, L.M., Even and odd q-coherent states and their optical statistics properties. *J. Phys.* A (1993) **26** 293–300.

[664] Kuang, L.M. and Wang, F.B., The $su_q(1,1)$ q-coherent states and their nonclassical properties. *Phys. Lett.* A (1993) **173** 221–227.

[665] Campos, R.A., Interpolation between the wave and particle properties of bosons and fermions. *Phys. Lett.* A (1994) **184** 173–178.

[666] Codriansky, S., Localized states in deformed quantum mechanics. *Phys. Lett.* A (1994) **184** 381–384.

[667] Aref'eva, I.Y., Parthasarthy, R., Viswanathan, K.S., and Volovich, I.V., Coherent states, dynamics and semiclassical limit of quantum groups. *Mod. Phys. Lett.* A (1994) **9** 689–703.

[668] Celeghini, E., De Martino, S., De Siena, S., Rasetti, M., and Vitiello, G., Quantum groups, coherent states, squeezing and lattice quantum mechanics. *Ann. Phys.* (NY) (1995) **241** 50–67.

[669] Nelson, C.A. and Fields, M.H., Number and phase uncertainties of the q-analog quantized field. *Phys. Rev.* A (1995) **51** 2410–2429.

[670] Spiridonov, V., Universal superpositions of coherent states and self-similar potentials. *Phys. Rev.* A (1995) **52** 1909–1935.

[671] Spiridonov, V., Coherent states of the q-Weyl algebra. *Lett. Math. Phys.* (1995) **35** 179–185.

[672] Mann, A. and Parthasarathy, R., Minimum uncertainty states for the quantum group $SU_q(2)$ and quantum Wigner d-functions. *J. Phys.* A (1996) **29** 427–435.

[673] Atakishiyev, N.M. and Feinsilver, P., On the coherent states for the q-Hermite polynomials and related Fourier transformation. *J. Phys.* A (1996) **29** 1659–1664.

[674] Park, S.U., Equivalence of q-bosons using the exponential phase operator. *J. Phys.* A (1996) **29** 3683–3696.

[675] Chung, W.S., Coherent states of $su_q(n)$-covariant oscillators. *Prog. Theor. Phys.* (1998) **100** 657–663.

[676] Irac-Astaud, M. and Rideau, G., Bargmann representations for deformed harmonic oscillators. *Rev. Math. Phys.* (1998) **10** 1061–1078.

[677] Bonatsos, D. and Daskaloyannis, C., Quantum groups and their applications in nuclear physics. *Prog. Particle & Nucl. Phys.* (1999) **43** 537–618.

[678] Aniello, P., Man'ko, V., Marmo, G., Solimeno, S., and Zaccaria, F., On the coherent states, displacement operators and quasidistributions associated with deformed quantum oscillators. *J. Opt.* B (2000) **2** 718–725.

[679] Becker, J., Debergh, N., and Szafraniec, F.H., A proposal of new sets of squeezed states. *Phys. Lett.* A (1998) **243** 256–260.

[680] Becker, J., Debergh, N., and Szafraniec, F.H., Oscillator-like Hamiltonians and squeezing. *Int. J. Theor. Phys.* (2000) **39** 1515–1527.

[681] Beckers, J., Debergh, N., Cariñena, J.F., and Marmo, G., Non-Hermitian oscillator-like Hamiltonians and λ-coherent states revisited. *Mod. Phys. Lett.* A (2001) **16** 91–98.

[682] Quesne, C., Spectrum generating algebra of the C_λ-extended oscillator and multiphoton coherent states. *Phys. Lett.* A (2000) **272** 313–325.

[683] Quesne, C., Generalized coherent states associated with the C_λ-extended oscillator. *Ann. Phys.* (NY) (2001) **293** 147–188.

[684] Saxena, G.M. and Mehta, C.L., Para-squeezed states. *J. Math. Phys.* (1991) **32** 783–786.

[685] Saxena, G.M., Mehta, C.L., and Mathur, B.S., Eigenstates of bilinears in paraboson operators and their inverses. *J. Math. Phys.* (1993) **34** 2875–2892.

[686] Shanta, P., Chaturvedi, S, Srinivasan, V., and Jagannathan, R., Unified approach to the analogues of single-photon and multiphoton coherent states for generalized bosonic oscillators. *J. Phys.* A (1994) **27** 6433–6442.

[687] Bagchi, B. and Bhaumik, D., Squeezed states for parabosons. *Mod. Phys. Lett.* A (1998) **13** 623–630.

[688] Jing, S.C. and Nelson, C.A., Eigenstates of paraparticle creation operators. *J. Phys.* A (1999) **32** 401–409.

[689] Fukui, T. and Aizawa, N., Shape-invariant potentials and an associated coherent state. *Phys. Lett.* A (1993) **180** 308–313.

[690] Fukui, T., Self-similar potentials and q-coherent states. *Phys. Lett.* A (1994) **189** 7–10.

[691] Fernández, D.J., Hussin, V., and Nieto, L.M., Coherent states for isospectral oscillator Hamiltonians. *J. Phys.* A (1994) **27** 3547–3564.

[692] Fernández, D.J., Nieto, L.M., and Rosas-Ortiz, O., Distorted Heisenberg algebra and coherent states for isospectral oscillator Hamiltonians. *J. Phys.* A (1995) **28** 2693–2708.

[693] Bagrov, V.G. and Samsonov, B.F., Coherent states for anharmonic oscillator Hamiltonians with equidistant and quasi-equidistant spectra. *J. Phys.* A (1996) **29** 1011–1023.

[694] Fu, H.-C. and Sasaki, R., Generally deformed oscillator, isospectral oscillator system and Hermitian phase operator. *J. Phys.* A (1996) **29** 4049–4064.

[695] Samsonov, B.F., Distortion of a phase space under the Darboux transformation. *J. Math. Phys.* (1998) **39** 967–975.

[696] Fernández, D.J. and Hussin, V., Higher-order SUSY, linearized nonlinear Heisenberg algebras and coherent states. *J. Phys.* A (1999) **32** 3603–3619.

[697] Junker, G. and Roy, P., Non-linear coherent states associated with conditionally exactly solvable problems. *Phys. Lett.* A (1999) **257** 113–119.

[698] Samsonov, B.F., Coherent states for transparent potentials. *J. Phys.* A (2000) **33** 591–605.

[699] Fatyga, B.W., Kostelecky, V.A., Nieto, M.M., and Truax, D.R., Supercoherent states. *Phys. Rev.* D (1991) **43** 1403–1412.

[700] Kostelecky, V.A., Man'ko, V.I., Nieto, M.M., and Truax, D.R., Supersymmetry

and a time-dependent Landau system. *Phys. Rev.* A (1993) **48** 951–963.

[701] Kostelecky, V.A., Nieto, M.M., and Truax, D.R., Supersqueezed states. *Phys. Rev.* A (1993) **48** 1045–1054.

[702] Bérubé-Lauzière, Y. and Hussin, V., Comments of the definitions of coherent states for the SUSY harmonic oscillator. *J. Phys.* A (1993) **26** 6271–6275.

[703] Bluhm, R. and Kostelecky, V.A., Atomic supersymmetry, Rydberg wave-packets, and radial squeezed states. *Phys. Rev.* A (1994) **49** 4628–4640.

[704] Bergeron, H. and Valance, A., Overcomplete basis for one-dimensional Hamiltonians. *J. Math. Phys.* (1995) **36** 1572–1592.

[705] Jayaraman, J., Rodrigues, R.D., and Vaidya, A.N., A SUSY formulation a la Witten for the SUSY isotonic oscillator canonical supercoherent states. *J. Phys.* A (1999) **32** 6643–6652.

[706] Bagchi, B. and Quesne, C., Creation and annihilation operators and coherent states for the PT-symmetric oscillator. *Mod. Phys. Lett.* A (2001) **16** 2449–2455.

[707] Nieto, M.M., Coherent states for unusual potentials. *Mod. Phys. Lett.* A (2001) **16** 2305–2311.

[708] Man'ko, V.I., Marmo, G., Solimeno, S., and Zaccaria, F., Physical nonlinear aspects of classical and quantum q-oscillators. *Int. J. Mod. Phys.* A (1993) **8** 3577–3597.

[709] Man'ko, V.I., Marmo, G., Solimeno, S., and Zaccaria, F., Correlation functions of quantum q-oscillators. *Phys. Lett.* A (1993) **176** 173–175.

[710] García-Fernandez, P. and Bermejo, F.J., Sub-Poissonian photon statistics and higher-order squeezing behavior of Bernouilli states in the linear amplifier. *J. Opt. Soc. Am.* (1987) **4** 1737–1741.

[711] Pegg, D.T. and Barnett, S.M., Optical state measurement by information transfer. *J. Mod. Opt.* (1999) **46** 1657–1667.

[712] Wódkiewicz, K., Knight, P.L., Buckle, S.J., and Barnett, S.M., Squeezing and superposition states. *Phys. Rev.* A (1987) **35** 2567–2577.

[713] Lee, H.W., Superposition phases and squeezing. *J. Mod. Opt.* (1993) **40** 1081–1089.

[714] Sanders, B.C., Superposition of two squeezed vacuum states and interference effects. *Phys. Rev.* A (1989) **39** 4284–4287.

[715] Janszky, J. and Vinogradov, An.V., Squeezing via one-dimensional distribution of coherent states. *Phys. Rev. Lett.* (1990) **64** 2771–2774.

[716] Bužek, V. and Knight, P., The origin of squeezing in a superposition of coherent states. *Opt. Commun.* (1991) **81** 331–336.

[717] Vogel, K. and Risken, H., Determinantion of quasiprobability distribution in terms of probability distributions for the rotated quadrature phase. *Phys. Rev.* A (1989) **40** 2847–2849.

[718] Schleich, W., Pernigo, M., and Kien, F.L., Nonclassical state from from two pseudoclassical states. *Phys. Rev.* A (1991) **44** 2172–2187.

[719] Sherman, B. and Kurizki, G., Preparation and detection of macroscopic quantum superpositions by two-photon field–atom interactions. *Phys. Rev.* A (1992) **45** R7674–R7677.

[720] Schaufler, S., Freyberger, M., Schleich, W.P., The birth of a phase-cat. *J. Mod. Opt.* (1994) **41** 1765-1779.

[721] Dodonov, V.V., Kalmykov, S.Y., and Man'ko, V.I., Statistical properties of Schrödinger real and imaginary cat states. *Phys. Lett.* A (1995) **199** 123–130.

[722] Lee, C.T., Theorem on nonclassical states. *Phys. Rev.* A (1995) **52** 3374–3376.

[723] Sanders, B.C., Entangled coherent states. *Phys. Rev.* A (1992) **45** 6811–6815.

[724] Chai, C.L., Two-mode nonclassical state via superpositions of two-mode coherent states. *Phys. Rev.* A (1992) **46** 7187–7191.

[725] Kim, M.S. and Bužek, V., Photon statistics of superposition states in phase-sensitive reservoirs. *Phys. Rev.* A (1993) **47** 610–619.

[726] Ban, M., Superpositions of the $su(1,1)$ coherent states. *Phys. Lett.* A (1994) **193** 121–125.

[727] Moya-Cessa, H., Generation and properties of superpositions of displaced Fock states. *J. Mod. Opt.* (1995) **42** 1741-1754.

[728] Kilin, S.Y. and Shatokhin, V.N., Complex quantum structure of nonclassical superposition states and quantum instability in resonance fluorescence. *Phys. Rev. Lett.* (1996) **76** 1051–1054.

[729] Kryuchkyan, G.Y. and Kheruntsyan, K.V., Exact quantum theory of a parametrically driven dissipative anharmonic oscillator. *Opt. Commun.* (1996) **127** 230–236.

[730] Bandilla A, Drobny G and Jex I 1996 Nondegenerate parametric interactions and nonclassical effects *Phys. Rev.* A **53** 507-516

[731] de Toledo Piza, A.F.R. and Zagury, N., Class of feasible quantum superpositions of squeezed and sub-Poissonian states. *Phys. Rev.* A (1996) **54** 3338–3346.

[732] Obada, A.S.F. and Omar, Z.M, Properties of superposition of squeezed states. *Phys. Lett.* A (1997) **227** 349–356.

[733] Dantas, C.A.M., Queroz, J.R., and Baseia, B., Superposition of displaced number states and interference effects. *J. Mod. Opt.* (1998) **45** 1085–1096.

[734] Ragi, R., Baseia, B., and Bagnato, V.S., Generalized superposition of two coherent states and interference effects. *Int. J. Mod. Phys.* B (1998) **12** 1495–1529.

[735] Obada, A.S.F. and Abd Al-Kader, G.M., Superpositions of squeezed displaced Fock states: properties and generation. *J. Mod. Opt.* (1999) **46** 263–278.

[736] Barbosa, Y.A., Marques, G.C. and Baseia, B., Generalized superposition of two squeezed states: generation and statistical properties. *Physica* A (2000) **280** 346–361.

[737] Liao, J., Wang, X, Wu, L.-A., and Pan, S.H., Superpositions of negative binomial states. *J. Opt.* B (2000) **2** 541–544.

[738] Nemoto, K. and Sanders, B.C., Superpositions of $SU(3)$ coherent states via a nonlinear evolution. *J. Phys.* A (2001) **34** 2051–2062.

[739] Marchiolli, M.A., da Silva, L.F., Melo, P.S., and Dantas, C.A.M., Quantum-interference effects on the superposition of N displaced number states. *Physica* A (2001) **291** 449–466.

[740] Miranowicz, A., Bajer, J., Wahiddin, M.R.B., and Imoto, N., Wehrl information entropy and phase distributions of Schrödinger cat and cat-like states. *J. Phys.* A (2001) **34** 3887–3896.

[741] Mancini, S. and Man'ko, V.I., The survival of quantum coherence in deformed-states superposition. *Europhys. Lett.* (2001) **54** 586–591.

[742] Obada, A.S.F. and Abd Al-Kader, G.M., Generation and properties of a superposition of four displaced Fock states. *Int. J. Theor. Phys.* (2001) **40** 1715–1735.

[743] Wang, X.G., Coherence and squeezing in superpositions of spin coherent states. *Opt. Commun.* (2001) **200** 277–282.

[744] Fu, H.C. and Solomon, A.I., Kerr cat states from the four-photon Jaynes–Cummings model. *J. Mod. Opt.* (2002) **49** 259–268.

[745] Adam, P., Janszky, J. and Vinogradov, An.V., Gaussian coherent state expansion of the squeezed states. *Opt. Commun.* (1990) **80** 155–158.

[746] Adam, P., Janszky, J. and Vinogradov, An.V., Amplitude squeezed and number-phase intelligent states via coherent states superposition. *Phys. Lett.* A (1991) **160**

506–510.

[747] Domokos, P., Adam, P., and Janszky, J., One-dimensional coherent-state representation on a circle in phase space. *Phys. Rev.* A (1994) **50** 4293–4297.

[748] Szabó, S., Kis, Z., Adam, P., and Janszky, J., Number-phase uncertainty properties of the Gaussian arc distribution state. *Quant. Opt.* (1994) **6** 527–539.

[749] D'Ariano, G.M., Amplitude squeezing through photon fractioning. *Phys. Rev.* A (1990) **41** 2636–2644.

[750] Bužek, V., Jex, I., and Quang, T., k-photon coherent states. *J. Mod. Opt.* (1990) **37** 159–163.

[751] Bužek, V. and Jex, I., Multiphoton states and amplitude k-th power squeezing. *Nuovo Cim.* B (1991) **106** 147–157.

[752] Sun, J., Wang, J., and Wang, C., Generation of orthonormalized eigenstates of the operator a^k (for $k \geq 3$) from coherent states and their higher-order squeezing. *Phys. Rev.* A (1992) **46** 1700–1702.

[753] Jex, I. and Bužek, V., Multiphoton coherent states and the linear superposition principle. *J. Mod. Opt.* (1993) **40** 771–783.

[754] Gerry, C.C., Nonclassical properties of even and odd coherent states. *J. Mod. Opt.* A (1993) **40** 1053–1071.

[755] Fan, H. and Zhang, Z., Higher-order squeezing for even- and odd-displaced squeezed states. *Quant. Opt.* (1994) **6** 411–416.

[756] Gerry, C.C. and Grobe, R., Nonclassical properties of correlated two-mode Schrödinger cat states. *Phys. Rev.* A (1995) **51** 1698–1701.

[757] Gerry, C.C. and Grobe, R., Two-mode SU(2) and SU(1,1) Schrödinger-cat states. *J. Mod. Opt.* (1997) **44** 41–53.

[758] Marian, P. and Marian, T.A., Generalized characteristic functions for a single-mode radiation field. *Phys. Lett.* A (1997) **230** 276–282.

[759] Trifonov, D.A., Barut–Girardello coherent states for $u(p, q)$ and $sp(N, R)$ and their macroscopic superpositions. *J. Phys.* A (1998) **31** 5673–5696.

[760] Liu, Y.-X., Atomic odd-even coherent state. *Nuovo Cim.* B (1999) **114** 543–553.

[761] Bužek, V., Vidiella-Barranco, A., and Knight, P.L., Superpositions of coherent states: Squeezing and dissipation. *Phys. Rev. A* (1992) **45** 6570–6585.

[762] Sun, J., Wang, J., and Wang, C., Orthonormalized eigenstates of cubic and higher powers of the annihilation operator. *Phys. Rev.* A (1991) **44** 3369–3372.

[763] Janszky, J., Kis. Z., and Adam, P., Star states. *Fortschr. Phys.* (1998) **46** 829–835.

[764] Banaszek, K. and Knight, P.L., Quantum interference in three-photon down conversion. *Phys. Rev.* A (1997) **55** 2368–2375.

[765] Felbinger, T., Schiller, S., and Mlynek, J., Oscillation and generation of nonclassical states in three-photon down-conversion. *Phys. Rev. Lett.* (1998) **80** 492–495.

[766] Gevorkyan, S.T. and Chaltykyan, V.O., Formation of nonclassical states of light in media with a three-photon resonance. *Opt. Spektrosk.* (2000) **89** 617–622 [*Opt. Spectrosc.* (2000) **89** 563–568].

[767] Hach III, E.E. and Gerry, C.C., Four photon coherent states. Properties and generation. *J. Mod. Opt.* (1992) **39** 2501–2517.

[768] Lynch, R., Simultaneous fourth-order squeezing of both quadrature components. *Phys. Rev.* A (1994) **49** 2800–2805.

[769] Souza Silva, A.L., Mizrahi, S.S., and Dodonov, V.V., Effect of phase-sensitive reservoir on the decoherence of pair-cat coherent states. *J. Russ. Laser Research* (2001) **22** 534–544.

[770] Paprzycka, M. and Tanaś, R., Discrete superpositions of coherent states and phase properties of the m-photon anharmonic oscillator. *Quant. Opt.* (1992) **4** 331–342.

[771] Tara, K., Agarwal, G.S., and Chaturvedi, S., Production of Schrödinger macroscopic quantum-superposition states in a Kerr medium. *Phys. Rev.* A (1993) **47** 5024–5029.

[772] Shanta, P., Chaturvedi, S, Srinivasan, V., Agarwal, G.S., and Mehta, C.L., Unified approach to multiphoton coherent states. *Phys. Rev. Lett.* (1994) **72** 1447–1450.

[773] Nieto, M.M. and Truax, D.R., Arbitrary-order Hermite generating functions for obtaining arbitrary-order coherent and squeezed states. *Phys. Lett.* A (1995) **208** 8–16.

[774] Nieto, M.M. and Truax, D.R., Holstein-Primakoff/Bogoliubov transformations and the multiboson system. *Fortschr. Phys.* (1997) **45** 145–156.

[775] Janszky, J., Domokos, P., and Adam, P., Coherent states on a circle and quantum interference. *Phys. Rev.* A (1993) **48** 2213–2219.

[776] Gagen, M.J., Phase-space interference approaches to quantum superposition states. *Phys. Rev.* A (1995) **51** 2715–2725.

[777] Xin, Z.Z., Wang, D.B., Hirayama, M., and Matumoto, K., Even and odd two-photon coherent states of the radiation field. *Phys. Rev.* A (1994) **50** 2865–2869.

[778] Castaños, O., López-Peña, R., and Man'ko, V.I., Crystallized Schrödinger cat states. *J. Russ. Laser Research* (1995) **16** 477–525.

[779] Nagel, B., Higher power squeezed states, Jacobi matrices, and the Hamburger moment problem. In: *Fifth International Conference on Squeezed States and Uncertainty Relations, Balatonfured, 1997* (D. Han, J. Janszky, Y.S. Kim, and V.I. Man'ko, eds.), NASA Conf. Publ. NASA/CP-1998–206855, pp. 43–48. Goddard Space Flight Center, Greenbelt, 1998.

[780] Jex, I., Törmä, P., and Stenholm, S., Multimode coherent states. *J. Mod. Opt.* (1995) **42** 1377–1386.

[781] An, N.B. and Duc, T.M., Trio coherent states. *J. Opt.* B (2002) **4** 80–85.

[782] Baseia, B., Moussa, M.H.Y., and Bagnato, V.S., Hole burning in Fock space. *Phys. Lett.* A (1998) **240** 277–281.

[783] Napoli, A, and Messina, A., An application of the arithmetic Euler function to the construction of nonclassical states of a quantum harmonic oscillator. *Rep. Math. Phys.* (2001) **48** 159-166.

[784] Chountasis, S. and Vourdas, A., Weyl functions and their use in the study of quantum interference. *Phys. Rev.* A (1998) **58** 848-855.

[785] Napoli, A, and Messina, A., Generalized even and odd coherent states of a single bosonic mode. *Eur. Phys. J.* D (1999) **5** 441–445.

[786] Nieto, M.M. and Truax, D.R., Higher-power coherent and squeezed states. *Opt. Commun.* (2000) **179** 197–213.

[787] Ragi, R., Baseia, B., and Mizrahi, S.S., Non-classical properties of even circular states. *J. Opt.* B (2000) **2** 299–305.

[788] José, W.D. and Mizrahi, S.S., Generation of circular states and Fock states in a trapped ion. *J. Opt.* B (2000) **2** 306–314.

[789] Souza Silva, A.L., José, W.D., Dodonov, V.V., and Mizrahi, S.S., Production of two-Fock states superpositions from even circular states and their decoherence. *Phys. Lett.* A (2001) **282** 235–244.

[790] Dattoli, G., Lorenzutta, S., and Sacchetti, D., Arbitrary-order coherent states and pseudo-hyperbolic functions. *Nuovo Cim.* B (2001) **116** 719–726.

[791] De Siena, S., Di Lisi, A., and Illuminati, F., Quadrature-dependent Bogoliubov transformations and multiphoton squeezed states. *Phys. Rev.* A (2001) **64** 063803.

[792] Wu, Y. and Côté, R., Quadrature-dependent Bogoliubov transformations and multiphoton squeezed states. *Phys. Rev.* A (2002) **66** 025801.

[793] Karassiov, V.P. and Puzyrevsky, V.I., Generalized coherent states of multimode light and biphotons. *J. Sov. Laser Research* (1989) **10** 229–240.
[794] Karassiov, V.P., Polarization structure of quantum light fields – a new insight. 1. General outlook. *J. Phys.* A (1993) **26** 4345–4354.
[795] Karassiov, V.P., Polarization squeezing and new states of light in quantum optics. *Phys. Lett.* A (1994) **190** 387–392.
[796] Korolkova, N.V. and Chirkin, A.S., Formation and conversion of the polarization-squeezed light. *J. Mod. Opt.* (1996) **43** 869–878.
[797] Lehner, J., Leonhardt, U., and Paul, H., Unpolarized light: Classical and quantum states. *Phys. Rev.* A (1996) **53** 2727–2735.
[798] Lehner, J., Paul, H., and Agarwal, G.S., Generation and physical properties of a new form of unpolarized light. *Opt. Commun.* (1997) **139** 262–269.
[799] Klyshko, D.N., Polarization of light: fourth-order effects and polarization-squeezed states. *JETP* (1997) **84** 1065–1079.
[800] Alodjants, A.P., Arakelian, S.M., and Chirkin, A.S., Polarization quantum states of light in nonlinear distributed feedback systems; quantum nondemolition measurements of the Stokes parameters of light and atomic angular momentum. *Appl. Phys.* B (1998) **66** 53–65.
[801] Alodjants, A.P. and Arakelian, S.M., Quantum phase measurements and nonclassical polarization states of light. *J. Mod. Opt.* (1999) **46** 475–507.
[802] Karassiov, V.P., Symmetry approach to reveal hidden coherent structures in quantum optics. General outlook and examples. *J. Russ. Laser Research* (2000) **21** 370–410.
[803] Karassiov, V.P. and Masalov, A.V., Quantum interference of light polarization states via polarization quasiprobability functions. *J. Opt.* B (2002) **4** S366–S371.
[804] Luis, A., Degree of polarization in quantum optics. *Phys. Rev.* A (2002) **66** 013806.
[805] Bowen, W.P., Schnabel, R., Bachor, H.-A., and Lam, P.K., Polarization squeezing of continuous variable Stokes parameters. *Phys. Rev. Lett.* (2002) **88** 093601.
[806] Klauder, J.R., Coherent states without groups: quantization on nonhomogeneous manifolds. *Mod. Phys. Lett.* A (1993) **8** 1735–1738.
[807] Klauder, J.R., Quantization without quantization. *Ann. Phys.* (NY) (1995) **237** 147–160.
[808] Gazeau, J.P. and Klauder, J.R., Coherent states for systems with discrete and continuous spectrum. *J. Phys.* A (1999) **32** 123–132.
[809] Klauder, J.R., Coherent states for the hydrogen atom. *J. Phys.* A (1996) **29** L293–L298.
[810] Sixdeniers, J.M., Penson, K.A., and Solomon, A.I., Mittag-Leffler coherent states. *J. Phys.* A (1999) **32** 7543–7563.
[811] Fox, R.F., Generalized coherent states. *Phys. Rev.* A (1999) **59** 3241–3255.
[812] Fox, R.F. and Choi, M.H., Generalized coherent states and quantum-classical correspondence. *Phys. Rev.* A (2000) **61** 032107.
[813] Fox, R.F. and Choi, M.H., Generalized coherent states for systems with degenerate energy spectra. *Phys. Rev.* A (2001) **64** 042104.
[814] Antoine, J.P., Gazeau, J.P., Monceau, P., Klauder, J.R., and Penson, K.A., Temporally stable coherent states for infinite well and Pöschl–Teller potentials. *J. Math. Phys.* (2001) **42** 2349–2387.
[815] Klauder, J.R., Penson, K.A., and Sixdeniers, J.M., Constructing coherent states through solutions of Stieltjes and Hausdorff moment problems. *Phys. Rev.* A (2001) **64** 013817.
[816] Hollingworth, J.M., Konstadopoulou, A., Chountasis, S., Vourdas, A., and Back-

house, N.B., Gazeau–Klauder coherent states in one-mode systems with periodic potential. *J. Phys.* A (2001) **34** 9463–9474.

[817] Mir-Kasimov, R.M., $SU_q(1,1)$ and the relativistic oscillator *J. Phys.* A (1991) **24** 4283–4302.

[818] Dodonov, V.V. and Mizrahi, S.S., Strict lower bound for the spatial spreading of a relativistic particle. *Phys. Lett.* A (1993) **177** 394–398.

[819] Moshinsky, M. and Szczepaniak, A., The Dirac oscillator. *J. Phys.* A (1989) **22** L817–L819.

[820] Itô, D., Mori, K., and Carriere, E., An example of dynamical systems with linear trajectory. *Nuovo Cim.* A (1967) **51** 1119–1121.

[821] Swamy, N.V.V.J., Exact solution of the Dirac equation with an equivalent oscillator potential. *Phys. Rev.* (1969) **180** 1225–1226.

[822] Cook, P.A., Relativistic harmonic oscillators with intrinsic spin structure. *Lett. Nuovo Cim.* (1971) **1** 419–426.

[823] Nogami, Y. and Toyama, F.M., Coherent state of the Dirac oscillator. *Canad. J. Phys.* (1996) **74** 114–121.

[824] Rozmej, P. and Arvieu, R., The Dirac oscillator. A relativistic version of the Jaynes-Cummings model. *J. Phys.* A (1999) **32** 5367–5382.

[825] Aldaya, V. and Guerrero, J., Canonical coherent states for the relativistic harmonic oscillator. *J. Math. Phys.* (1995) **36** 3191–3199.

[826] Tang, J., Coherent states and squeezed states of massless and massive relativistic harmonic oscillators. *Phys. Lett.* A (1996) **219** 33–40.

[827] Lev, B.I., Semenov, A.A., and Usenko, C.V., Behaviour of $\pi^{\pm}$ mesons and synchrotron radiation in a strong magnetic field. *Phys. Lett.* A (1997) **230** 261–268.

[828] Lev, B.I., Semenov, A.A., Usenko, C.V., and Klauder, J.R., Relativistic coherent states and charge structure of the coordinate and momentum operators. *Phys. Rev.* A (2002) **66** 022115.

[829] Larsen, A.L. and Sanchez, N., Quantum coherent string states in AdS_3 and the $SL(2,R)$ Wess–Zumino–Witten–Novikov model. *Phys. Rev.* D (2000) **62** 046003.

[830] Larsen, A.L. and Sanchez, N., New coherent string states and minimal uncertainty in WZWN models. *Nucl. Phys.* B (2001) **618** 301–311.

[831] Thiemann, T., Gauge field theory coherent states (GCS): I. General properties. *Class. Quant. Gravity* (2001) **18** 2025–2064.

[832] Hall, B.C., Coherent states and the quantization of one-dimensional Yang–Mills theory. *Rev. Math. Phys.* (2001) **13** 1281-1305.

[833] Agarwal, G.S. and Tara, K., Nonclassical properties of states generated by the excitations on a coherent state. *Phys. Rev.* A (1991) **43** 492–497.

[834] Zhang, Z. and Fan, H., Properties of states generated by excitations on a squeezed vacuum state. *Phys. Lett.* A (1992) **165** 14–18.

[835] Kis, Z., Adam, P., and Janszky, J., Properties of states generated by excitations on the amplitude squeezed states. *Phys. Lett.* A (1994) **188** 16–20.

[836] Xin, Z.Z., Duan, Y.B., Zhang, H.M., Hirayama, M., and Matumoto, K.I., Excited two-photon coherent state of the radiation field. *J. Phys.* B (1996) **29** 4493–4506.

[837] Xin, Z.Z., Duan, Y.B., Zhang, H.M., Qian, W.J., Hirayama, M., and Matumoto, K.I., Excited even and odd coherent states of the radiation field. *J. Phys.* B (1996) **29** 2597-2606.

[838] Dodonov, V.V., Korennoy, Y.A., Man'ko, V.I., and Moukhin, Y.A., Nonclassical properties of states generated by the excitation of even/odd coherent states of light. *Quant. Semiclass. Opt.* (1996) **8** 413–427.

[839] Agarwal, G.S. and Tara, K., Nonclassical character of states exhibiting no squeez-

ing or sub-Poissonian statistics. *Phys. Rev.* A (1992) **46** 485–488.

[840] Agarwal, G.S. and Tara, K., Transformations of the nonclassical states by an optical amplifier. *Phys. Rev.* A (1993) **47** 3160–3166.

[841] Jones, G.N., Haight, J., and Lee, C.T., Nonclassical effects in the photon-added thermal states. *Quant. Semiclass. Opt.* (1997) **9** 411–418.

[842] Dakna, M., Knöll, L., and Welsch, D.-G., Photon-added state preparation via conditional measurement on a beam splitter. *Opt. Commun.* (1998) **145** 309–321.

[843] Dakna, M., Knöll, L., and Welsch, D.-G., Quantum state engineering using conditional measurement on a beam splitter. *Europ. Phys. J.* D (1998) **3** 295–308.

[844] Mehta, C.L., Roy, A.K., and Saxena, G.M., Eigenstates of two-photon annihilation operators. *Phys. Rev.* A (1992) **46** 1565–1572.

[845] Fan, H.Y., Inverse in Fock space studied via a coherent-state approach. *Phys. Rev.* A (1993) **47** 4521–4523.

[846] Arvind, Dutta, B., Mehta, C.L., and Mukunda, N. Squeezed states, metaplectic group, and operator Möbius transformations. *Phys. Rev.* A (1994) **50** 39–61.

[847] Roy, A.K. and Mehta, C.L., Boson inverse operators and associated coherent states. *Quant. Semiclass. Opt.* (1995) **7** 877–888.

[848] Moya-Cessa, H., Chavez-Cerda, S., and Vogel, W., Adding and subtracting energy quanta of the harmonic oscillator. *J. Mod. Opt.* (1999) **46** 1641–1656.

[849] Dodonov, V.V., Marchiolli, M.A., Korennoy, Y.A., Man'ko, V.I., and Moukhin, Y.A., Dynamical squeezing of photon-added coherent states. *Phys. Rev.* A (1998) **58** 4087–4094.

[850] Lu, H., Statistical properties of photon-added and photon-subtracted two-mode squeezed vacuum state. *Phys. Lett.* A (1999) **264** 265–269.

[851] Wei, L.F., Wang, S.J., and Xi, D.P., Inverse q-boson operators and their relation to photon-added and photon-depleted states. *J. Opt.* B (1999) **1** 619–623.

[852] Sixdeniers, J.M. and Penson, K.A., On the completeness of photon-added coherent states. *J. Phys.* A (2001) **34** 2859–2866.

[853] Wang, X.B., Kwek, L.C., Liu, Y., and Oh, C.H., Non-classical effects of two-mode photon-added displaced squeezed states. *J. Phys.* B (2001) **34** 1059–1078.

[854] Quesne, C., Completeness of photon-added squeezed vacuum and one-photon states and of photon-added coherent states on a circle. *Phys. Lett.* A (2001) **288** 241–250.

[855] Popov, D., Photon-added Barut–Girardello coherent states of the pseudoharmonic oscillator. *J. Phys.* A (2002) **35** 7205–7223.

[856] Dodonov, V.V. and Klimov, A.B., Exactly solvable model of oscillator with non-polynomial interaction and zeros of Bessel functions. In: *Group Theoretical Methods in Physics. Proceedings of the XIX International Colloquium, Salamanca, Spain, 1992* (M.A. Del Olmo, M. Santander, and J. Mateos Guilarte, eds.), vol. 1, pp. 321–324. CIFMAT, Madrid, 1993.

[857] Bergou, J.A., Hillery, M., and Yu, D., Minimum uncertainty states for amplitude-squared squeezing: Hermite polynomial states. *Phys. Rev.* A (1991) **43** 515–520.

[858] Datta, S. and D'Souza, R., Generalised quasiprobability distribution for Hermite polynomial squeezed states. *Phys. Lett.* A (1996) **215** 149–153.

[859] Fan, H.Y. and Ye, X., Hermite polynomial states in two-mode Fock space. *Phys. Lett.* A (1993) **175** 387–390.

[860] Fan, H.Y., Ye, X.O., and Xu, Z.H., Laguerre polynomial states in single-mode Fock space. *Phys. Lett.* A (1995) **199** 131–136.

[861] Wang, X.G., Coherent states, displaced number states and Laguerre polynomial states for $su(1,1)$ Lie algebra. *Int. J. Mod. Phys.* B (2000) **14** 1093–1103.

[862] Bialynicki-Birula, I., Freyberger, M., and Schleich, W., Various measures of quantum phase uncertainty: A comparative study. *Phys. Scripta* (1993) **T48** 113–118.

[863] Fu, H.C., Pólya states of quantized radiation fields, their algebraic characterization and non-classical properties. *J. Phys.* A (1997) **30** L83–L89.

[864] Fu, H.C. and Sasaki, R., Hypergeometric states and their nonclassical properties. *J. Math. Phys.* (1997) **38** 2154–2166.

[865] Fan, H.Y. and Liu, N.L., Negative hypergeometric states of the quantized radiation field. *Phys. Lett.* A (1998) **250** 88–92.

[866] Liu, N.L., Algebraic structure and nonclassical properties of the negative hypergeometric state. *J. Phys.* A (1999) **32** 6063–6078.

[867] Sixdeniers, J.M., Penson, K.A., and Klauder, J.R., Tricomi coherent states. *Int. J. Mod. Phys.* B (2001) **15** 4231–4243.

[868] Roy, P. and Roy, B., A generalized nonclassical state of the radiation field and some of its properties. *J. Phys.* A (1997) **30** L719–L723.

[869] Atakishiyev, N.M., Jafarov, E.I., Nagiyev, S.M., and Wolf, K.B., Meixner oscillators. *Rev. Mex. Fís.* (1998) **44** 235–244.

[870] Penson, K.A. and Solomon, A.I., New generalized coherent states. *J. Math. Phys.* (1999) **40** 2354–2363.

[871] Shepherd, T.J., A model for photodetection of single-mode cavity radiation. *Optica Acta* (1981) **28** 567–583.

[872] Shepherd, T.J. and Jakeman, E., Statistical analysis of an incoherently coupled, steady-state optical amplifier. *J. Opt. Soc. Am.* B (1987) **4** 1860–1866.

[873] Chaturvedi, S., Sandhya, R., Srinivasan, V., and Simon, R., Thermal counterparts of nonclassical states in quantum optics. *Phys. Rev.* A (1990) **41** 3969–3974.

[874] Baseia, B., Gomes, A.R., and Bagnato, V.S., Intermediate pure mixed state of the quantized radiation field. *Braz. J. Phys.* (1997) **27** 276–284.

[875] Dantas, C.A.M. and Baseia, B., Noncoherent states having Poissonian statistics. *Physica* A (1999) **265** 176–185.

[876] Bekenstein, J.D. and Schiffer, M., Universality in grey-body radiance: extending Kirchhoff's law to the statistics of quanta. *Phys. Rev. Lett.* (1994) **72** 2512–2515.

[877] Lee, C.T., Nonclassical effects in the grey-body state. *Phys. Rev.* A (1995) **52** 1594–1600.

[878] Herzog, U. and Bergou, J., Nonclassical maximum-entropy states. *Phys. Rev.* A (1997) **56** 1658–1661.

[879] Dodonov, V.V. and Mizrahi, S.S., Stationary states in saturated two-photon processes and generation of phase-averaged mixtures of even and odd quantum states. *Acta Phys. Slovaca* (1998) **48** 349–360.

[880] Lee, C.T., Shadow states and shaded states. *Quant. Semiclass. Optics* (1996) **8** 849–860.

[881] Srinivasan, R. and Lee, C.T., Shadowed negative binomial state. *Phys. Lett.* A (1996) **218** 151–156.

[882] Lee, C.T., Application of Klyshko's criterion for nonclassical states to the micromaser pumped by ultracold atoms. *Phys. Rev.* A (1997) **55** 4449–4453.

[883] Scully, M.O., Meyer, G.M., and Walther, H., Induced emission due to the quantized motion of ultracold atoms passing through a micromaser cavity. *Phys. Rev. Lett.* (1996) **76** 4144–4147.

[884] Pegg, D.T. and Barnett, S.M., Unitary phase operator in quantum mechanics. *Europhys. Lett.* (1988) **6** 483–487.

[885] Barnett, S.M. and Pegg, D.T., On the Hermitian optical phase operator. *J. Mod. Opt.* (1989) **36** 7–19.

[886] Shapiro, J.H., Shepard, S.R., and Wong, N.C., Ultimate quantum limits on phase measurement. *Phys. Rev. Lett.* (1989) **62** 2377–2380.

[887] Schleich, W.P., Dowling, J.P., and Horowicz, R.J., Exponential decrease in phase uncertainty. *Phys. Rev.* A (1991) **44** 3365–3368.

[888] Nath, R. and Kumar, P., Quasi-photon phase states. *J. Mod. Opt.* (1991) **38** 263–268.

[889] Bužek, V., Wilson-Gordon, A.D., Knight, P.L., and Lai, W.K., Coherent states in a finite-dimensional basis: Their phase properties and relationship to coherent states of light. *Phys. Rev.* A (1992) **45** 8079–8094.

[890] Kuang, L.-M., Wang, F.B., and Zhou, Y.G., Coherent states of a harmonic oscillator in a finite-dimensional Hilbert space and their squeezing properties. *J. Mod. Opt.* (1994) **41** 1307–1318.

[891] Kuang, L.-M. and Chen, X., Phase-coherent states and their squeezing properties. *Phys. Rev.* A (1994) **50** 4228–4236.

[892] Galetti, D. and Marchiolli, M.A., Discrete coherent states and probability distributions in finite-dimensional spaces. *Ann. Phys.* (NY) (1996) **249** 454–480.

[893] Obada, A.-S.F, Hassan, S.S., Puri, R.R., and Abdalla, M.S., Variation from number-state to chaotic-state fields: a generalized geometric state. *Phys. Rev.* A (1993) **48** 3174–3185.

[894] Obada A.-S.F., Yassin, O.M., and Barnett, S.M., Phase properties of coherent phase and generalized geometric state. *J. Mod. Opt.* (1997) **44** 149–161.

[895] Kuang, L.-M., A q-deformed harmonic oscillator in a finite-dimensional Hilbert space. *J. Phys.* A (1993) **26** L1079–L1083.

[896] Figurny, P., Orlowski, A., and Wódkiewicz, K., Squeezed fluctuations of truncated photon operators. *Phys. Rev.* A (1993) **47** 5151–5157.

[897] Baseia, B., de Lima, A.F., and Marques, G.C., Intermediate number-phase states of the quantized radiation field. *Phys. Lett.* A (1995) **204** 1–6.

[898] Opatrný, T., Miranowicz, A., and Bajer, J., Coherent states in finite-dimensional Hilbert space and their Wigner representation. *J. Mod. Opt.* (1996) **43** 417–432.

[899] Leoński, W., Finite-dimensional coherent-state generation and quantum-optical nonlinear oscillator model. *Phys. Rev.* A (1997) **55** 3874–3878.

[900] Lobo, A.C. and Nemes, M.C., The reference state for finite coherent states. *Physica* A (1997) **241** 637–648.

[901] Roy, B. and Roy, P., Coherent states, even and odd coherent states in a finite-dimensional Hilbert space and their properties. *J. Phys.* A (1998) **31** 1307–1317.

[902] Zhang, Y.Z., Fu, H.C., and Solomon, A.I., Intermediate coherent-phase(PB) states of radiation fields and their nonclassical properties. *Phys. Lett.* A (1999) **263** 257–262.

[903] Wang, X, Ladder operator formalisms and generally deformed oscillator algebraic structures of quantum states in Fock space. *J. Opt.* B (2000) **2** 534–540.

[904] Baseia, B., Duarte, S.B., and Malbouisson, J.M.C., Interpolation from number states to chaotic states of the electromagnetic field. *J. Opt.* B (2001) **3** 152–162.

[905] Barnett, S.M. and Pegg, D.T., Phase measurement by projection synthesis. *Phys. Rev. Lett.* (1996) **76** 4148–4150.

[906] Pegg, D.T., Barnett, S.M., and Phillips, L.S., Quantum phase distribution by projection synthesis. *J. Mod. Opt.* (1997) **44** 2135–2148.

[907] Moussa, M.H.Y. and Baseia, B., Generation of the reciprocal-binomial state. *Phys. Lett.* A (1998) **238** 223–226.

[908] Baseia, B., de Lima, A.F., and da Silva, A.J, Intermediate number-squeezed state of the quantized radiation field. *Mod. Phys. Lett.* B (1995) **9** 1673–1683.

[909] Roy, B., Nonclassical properties of the even and odd intermediate number squeezed states. *Mod. Phys. Lett.* B (1998) **12** 23–33.

[910] Baseia, B., Granja, S.C.G., and Marques, G.C., Intermediate number coherent state of the quantized radiation field. *Phys. Scripta* (1997) **55** 719–723.

[911] Fu, H.C. and Sasaki, R., Negative binomial and multinomial states: probability distribution and coherent state. *J. Math. Phys.* (1997) **38** 3968–3987.

[912] Barnett, S.M., Negative binomial states of the quantized radiation field. *J.Mod.Opt.* (1998) **45** 2201–2205.

[913] Wang, X.G., Pan, S.H., and Yang, G.Z., Nonclassical properties and algebraic characteristics of negative binomial states in quantized radiation fields. *Eur. Phys. J.* D (2000) **10** 415–422.

[914] Fan, H.Y. and Liu, N.L., New generalized binomial states of the quantized radiation field. *Phys. Lett.* A (1999) **264** 154–161.

[915] Fu, H., Feng, Y., and Solomon, A.I., States interpolating between number and coherent states and their interaction with atomic systems. *J. Phys.* A (2000) **33** 2231–2249.

[916] Sixdeniers, J.M. and Penson, K.A., On the completeness of coherent states generated by binomial distribution. *J. Phys.* A (2000) **33** 2907–2916.

[917] Wang, X.G. and Fu, H., Superposition of the λ-parametrized squeezed states. *Mod. Phys. Lett.* B (2000) **14** 243–250.

[918] Joshi, A. and Lawande, S.V., Properties of squeezed binomial states and squeezed negative binomial states. *J. Mod. Opt.* (1991) **38** 2009–2022.

[919] Agarwal, G.S., Negative binomial states of the field-operator representation and production by state reduction in optical processes. *Phys. Rev.* A (1992) **45** 1787–1792.

[920] Fu, H.C. and Sasaki, R., Generalized binomial states: ladder operator approach. *J. Phys.* A (1996) **29** 5637–5644.

[921] Joshi, A. and Obada, A.-S.F., Some statistical properties of the even and the odd negative binomial states. *J. Phys.* A (1997) **30** 81–97.

[922] Dantas, C.M.A., Baseia, B., and Bagnato, V.S., Alternative strategy to the binomial state. *Phys. Scripta* (1998) **58** 145–148.

[923] El-Orany, F.A.A., Mahran, M.H., Obada, A.S.F., and Abdalla, M.S., Statistical properties of the odd binomial states with dynamical applications. *Int. J. Theor. Phys.* (1999) **38** 1493–1520.

[924] Fu, H.C., Wang, X.G., Li, C., and Wang, J.G., $su(2)$ and $su(1,1)$ displaced number states and their nonclassical properties. *Mod. Phys. Lett.* B (2000) **14** 1099–1108.

[925] El-Orany, F.A.A., Abdalla, M.S., Obada, A.S.F., and Abd Al-Kader, G.M., Influence of squeezing operator on the quantum properties of various binomial states. *Int. J. Mod. Phys.* B (2001) **15** 75–100.

[926] Fan, H.Y. and Jing, S.C., Even- and odd-binomial states as eigenstates of a nonlinear combination of radiation field operators. *Mod. Phys. Lett.* B (2001) **15** 1047–1052

[927] Liao, J., Wang, X.G., Wu, L.A., and Pan, S.H., Real and imaginary negative binomial states. *J. Opt.* B (2001) **3** 302–307

[928] Matos Filho, R.L. and Vogel, W., Nonlinear coherent states. *Phys. Rev.* A (1996) **54** 4560–4563.

[929] Man'ko, V.I., Marmo, G., Sudarshan, E.C.G, and Zaccaria, F., f-oscillators and nonlinear coherent states. *Phys. Scripta* (1997) **55** 528–541.

[930] Sivakumar, S., Photon-added coherent states as nonlinear coherent states. *J. Phys.*

A (1999) **32** 3441–3447.

[931] Mancini, S., Even and odd nonlinear coherent states. *Phys. Lett.* A (1997) **233** 291–296.

[932] Sivakumar, S., Even and odd nonlinear coherent states. *Phys. Lett.* A (1998) **250** 257–262.

[933] Roy, B. and Roy, P., Phase properties of even and odd nonlinear coherent states. *Phys. Lett.* A (1999) **257** 264–268.

[934] Roy, B. and Roy, P., Time dependent nonclassical properties of even and odd nonlinear coherent states. *Phys. Lett.* A (1999) **263** 48–52.

[935] Sivakumar, S., Generation of even and odd nonlinear coherent states. *J. Phys.* A (2000) **33** 2289–2297.

[936] Manko, O.V., Symplectic tomography of nonlinear coherent states of a trapped ion. *Phys. Lett.* A (1997) **228** 29–35.

[937] Man'ko, V., Marmo, G., Porzio, A., Solimeno, S., and Zaccaria, F., Trapped ions in laser fields: A benchmark for deformed quantum oscillators. *Phys. Rev.* A (2000) **62** 053407.

[938] Roy, B., Nonclassical properties of the real and imaginary nonlinear Schrödinger cat states. *Phys. Lett.* A (1998) **249** 25–29.

[939] Wang, X.G. and Fu, H.C., Negative binomial states of the radiation field and their excitations are nonlinear coherent states. *Mod. Phys. Lett.* B (1999) **13** 617–623.

[940] Liu, X.M., Orthonormalized eigenstates of the operator $(\hat{a}f(\hat{n}))^k$ $(k \geq 1)$ and their generation. *J. Phys.* A (1999) **32** 8685–8689.

[941] Wang, X.G., Two-mode nonlinear coherent states. *Opt. Commun.* (2000) **178** 365–369.

[942] Sivakumar, S., Studies on nonlinear coherent states. *J. Opt.* B (2000) **2** R61–R75.

[943] Wang, X.G., Spin squeezing in nonlinear spin-coherent states. *J. Opt.* B (2001) **3** 93–96.

[944] Wang, X.G., Photon-added one-photon and two-photon nonlinear coherent states. *Canad. J. Phys.* (2001) **79** 833–840.

[945] Fan, H.Y. and Yu, G.C., Nonlinear $SU(3)$ charged and hypercharged coherent states. *J. Phys.* A (2001) **34** 5995–6001.

[946] Kis, Z., Vogel, W., and Davidovich, L., Nonlinear coherent states of trapped-atom motion. *Phys. Rev.* A (2001) **64** 033401.

[947] An, N.B. and Duc, T.M., Excited K-quantum nonlinear coherent states. *J. Phys.* A (2002) **35** 4749–4754.

[948] Fan, H. and Cheng, H., Nonlinear entangled state representation in quantum mechanics. *Phys. Lett.* A (2002) **295** 65–73.

[949] Vogel, K., Akulin, V.M., and Schleich, W.P., Quantum state engineering of the radiation field. *Phys. Rev. Lett.* (1993) **71** 1816–1819.

[950] Barbosa, G.A. and Monken, C.H., Enhancing the sub-Poissonian character of a light beam from the down-conversion luminescence. *Phys. Rev. Lett.* (1991) **67** 3372–3375.

[951] Parkins, A.S., Marte, P., Zoller, P., and Kimble, H.J., Synthesis of arbitrary quantum states via adiabatic transfer of Zeeman coherence. *Phys. Rev. Lett.* (1993) **71** 3095–3098.

[952] Domokos, P., Janszky, J., and Adam, P., Single-atom interference method for generating Fock states. *Phys. Rev.* A (1994) **50** 3340–3344.

[953] Law, C.K. and Eberly, J.H., Arbitrary control of a quantum electromagnetic field. *Phys. Rev. Lett.* (1996) **76** 1055–1058.

[954] Walther, H., Generation and detection of Fock states of the radiation field. *J. Opt.*

B (2002) **4** S418–S425.

[955] Saito, H. and Ueda, M., Squeezed few-photon states of the field generated from squeezed atoms. *Phys. Rev.* A (1999) **59** 3959–3974.

[956] Andreoni, A., Bondani, M., A'Ariano, G.M., and Paris, M.G.A., Dichromatic squeezing generation. *Eur. Phys. J.* D (2001) **13** 415-421.

[957] Leonhardt, U., Quantum statistics of a lossless beam splitter: $SU(2)$ symmetry in phase space. *Phys. Rev.* A (1993) **48** 3265-3277.

[958] Mattle, K., Michler, M., Weinfurter, H, Zeilinger, A., and Zukowski, M., Nonclassical statistics at multiport beam splitters. *Appl. Phys.* B (1995) **60** S111–S117.

[959] Ban, M., Photon statistics of conditional output states of lossless beam splitter. *J. Mod. Opt.* (1996) **43** 1281–1303.

[960] Dakna, M., Anhut, T., Opatrný, T., Knöll, L., and Welsch, D.-G., Generating Schrödinger-cat-like states by means of conditional measurements on a beam splitter. *Phys. Rev.* A (1997) **55** 3184–3194.

[961] Lvovsky, A.I. and Mlynek, J., Quantum-optical catalysis: generating nonclassical states of light by means of linear optics. *Phys. Rev. Lett.* (2002) **88** 250401.

[962] Kim, M.S, Son, W., Bužek, V., and Knight, P.L., Entanglement by a beam splitter: Nonclassicality as a prerequisite for entanglement. *Phys. Rev.* A (2002) **65** 032323.

[963] Pegg, D.T., Phillips, L.S., and Barnett, S.M., Optical state truncation by projection synthesis. *Phys. Rev. Lett.* (1998) **81** 1604–1606.

[964] Villas-Boas, C.J., Guimarães, Y., Moussa, M.H.Y., and Baseia, B., Recurrence formula for generalized optical state truncation by projection synthesis. *Phys. Rev.* A (2001) **63** 055801.

[965] Özdemir, S.K., Miranowicz, A., Koashi, M., and Imoto, N., Quantum-scissors device for optical state truncation: A proposal for practical realization. *Phys. Rev.* A (2001) **64** 063818.

[966] Mancini, S., Man'ko, V.I., and Tombesi, P., Ponderomotive control of quantum macroscopic coherence. *Phys. Rev.* A (1997) **55** 3042–3050.

[967] Bose, S., Jacobs, K., and Knight, P.L., Preparation of nonclassical states in cavities with a moving mirror. *Phys. Rev.* A (1997) **56** 4175–4186.

[968] Zheng, S.-B., Preparation of even and odd coherent states in the motion of a cavity mirror. *Quant. Semiclass. Opt.* (1998) **10** 657–660.

[969] *Quantum Measurements in Optics. NATO ASI series, B: Physics, vol. 282* (P. Tombesi and D.F. Walls, eds.). Plenum, New York, 1992.

[970] *Fundamental Systems in Quantum Optics* (W. Schleich and G. Rempe, eds.), *Appl. Phys.* B (1995) **60** N 2/3 (supplement) S1–S265.

[971] Sherman, B., Kofman, A.G., and Kurizki, G., Preparation of non-classical field states by resonance fluorescence in photonic band structures. *Appl. Phys.* B (1995) **60** S99–S105.

[972] Davidovich, L., Sub-Poissonian processes in quantum optics. *Rev. Mod. Phys.* (1996) **68** 127–173.

[973] *Quantum state preparation and measurement* (W.P. Schleich and M.G. Raymer, eds.), special issue of *J. Mod. Opt.* (1997) **44** N 11/12.

[974] *Modern Studies of Basic Quantum Concepts and Phenomena* (E.B. Karlsson and E. Brändas, eds.), *Phys. Scripta* (1998) **T76** 1–232.

[975] Malbouisson, J.M.C. and Baseia, B., Higher-generation Schrödinger cat states in cavity QED. *J. Mod. Opt.* (1999) **46** 2015–2041.

[976] Lutterbach, L.G. and Davidovich, L., 2000 Production and detection of highly squeezed states in cavity QED. *Phys. Rev.* A (2000) **61** 023813.

[977] Raimond, J.M., Brune, M., and Haroche, S., Manipulating quantum entanglement

with atoms and photons in a cavity. *Rev. Mod. Phys.* (2001) **73** 565–582.

[978] Meekhof, D.M., Monroe, C., King, B.E., Itano, W.M., and Wineland, D.J., Generation of nonclassical motional states of a trapped atom. *Phys. Rev. Lett.* (1996) **76** 1796–1799.

[979] Monroe, C., Meekhof, D.M., King, B.E. and Wineland, D.J., A "Schrödinger cat" superposition state of an atom. *Science* (1996) **272** 1131–1136.

[980] Cirac, J.I., Parkins, A.S., Blatt, R., and Zoller, P. Nonclassical states of motion in ion traps. *Adv. At. Mol. Opt. Phys.* (1996) **37** 237–296.

[981] Retamal, J.C. and Zagury, N., Generation of nonclassical states of the center-of-mass motion of ions by dispersive coupling. *Phys. Rev.* A (1997) **55** 2387–2396.

[982] Wineland, D.J., Monroe, C., Itano, W.M., Leibfried, D., King, B.E., and Meekhof, D.M., Experimental issues in coherent quantum-state manipulation of trapped atomic ions. *J. Res. Nat. Inst. Stand. Technol.* (1998) **103** 259–328.

[983] Massoni, E. and Orszag, M., Phonon-photon translator. *Opt. Commun.* (2000) **179** 315–321.

[984] Zou, X.B., Kim, J., and Lee, H.W., Generation of two-mode nonclassical motional states and a Fredkin gate operation in a two-dimensional ion trap. *Phys. Rev.* A (2001) **63** 065801.

[985] Hillery, M., Detection of nonclassical states using a Kerr medium. *Phys. Rev.* A (1991) **44** 4578–4581.

[986] Leonhardt, U. and Paul, H., Measuring the quantum state of light. *Prog. Quant. Electron.* (1995) **19** 89–130.

[987] Bužek, V., Adam, G., and Drobný, G., Reconstruction of Wigner functions on different observation levels. *Ann. Phys.* (NY) (1996) **245** 37–97.

[988] Schiller, S., Breitenbach, G., Pereira, S.F., Müller, T., and Mlynek, J., Quantum statistics of the squeezed vacuum by measurement of the density matrix in the number state representation. *Phys. Rev. Lett.* (1996) **77** 2933–2936.

[989] Gardiner, S.A., Cirac, J.I., and Zoller, P., Nonclassical states and measurement of general motional observables of a trapped ion. *Phys. Rev.* A (1997) **55** 1683–1694.

[990] Breitenbach, G. and Schiller, S., Homodyne tomography of classical and nonclassical light. *J. Mod. Opt.* (1997) **44** 2207–2225.

[991] *Quantum Nondemolition Measurements* (J. Mlynek, G. Rempe, S. Schiller, and M. Wilkens, eds.), special issue of *Appl. Phys.* B (1997) **64** N 2.

[992] Welsch, D.-G., Vogel, W., and Opatrný, T., Homodyne detection and quantum state reconstruction. In: *Progress in Optics, vol. XXXIX* (E. Wolf, ed.), pp. 63–211. North Holland, Amsterdam, 1999.

[993] Brif, C. and Mann, A., Phase-space formulation of quantum mechanics and quantum-state reconstruction for physical systems with Lie-group symmetries. *Phys. Rev.* A (1999) **59** 971–987.

[994] Bužek, V., Drobný, G., Derka, R., Adam, G., and Wiedemann, H., Quantum state reconstruction from incomplete data. *Chaos Solitons Fract.* (1999) **10** 981–1074.

[995] Santos, M.F., Lutterbach, L.G., Dutra, S.M., Zagury, N., and Davidovich, L., Reconstruction of the state of the radiation field in a cavity through measurements of the outgoing field. *Phys. Rev.* A (2001) **63** 033813.

[996] Agarwal, G.S. and Banerji, J., Reconstruction of $SU(1,1)$ states. *Phys. Rev.* A (2001) **64** 023815.

[997] Garraway, B.M., Sherman, B., Moya-Cessa, H., Knight, P.L., and Kurizki, G., Generation and detection of nonclassical field states by conditional measurements following two-photon resonant interactions. *Phys. Rev.* A (1994) **49** 535-547.

[998] Ghosh, H. and Gerry, C.C., Measurement-induced nonclassical states of the

Jaynes–Cummings model. *J. Opt. Soc. Am.* B (1997) **14** 2782–2787.

[999] Napoli, A. and Messina, A., Conditional generation of non-classical states in a non-degenerate two-photon micromaser: I. Equal-intensity pair-Fock states preparation. II. Single-mode Fock states preparation. *J. Mod. Opt.* (1997) **44** 2075–2091, 2093–2103.

[1000] Peřinová, V. and Lukš, A., Continuous measurements in quantum optics. In: *Progress in Optics, vol. XL* (E. Wolf, ed.), pp. 115–269. North Holland, Amsterdam, 2000.

[1001] Bambah, B.A. and Satyanarayana, M.V., Squeezed coherent states and hadronic multiplicity distributions. *Prog. Theor. Phys. Suppl.* (1986) **86** 377–382.

[1002] Shih, C.C., Sub-Poissonian distribution in hadronic processes. *Phys. Rev.* D (1986) **34** 2720–2726.

[1003] Bambah, B.A. and Satyanarayana, M.V., Scaling and correlations of squeezed coherent distributions: Application to hadronic multiplicities. *Phys. Rev.* D (1988) **38** 2202–2208.

[1004] Vourdas, A. and Weiner, R.M., Multiplicity distributions and Bose-Einstein correlations in high-energy multiparticle production in the presence of squeezed coherent states. *Phys. Rev.* D (1988) **38** 2209–2217.

[1005] Ruijgrok, T.W., Squeezing and $SU(3)$-invariance in multiparticle production. *Acta Phys. Pol.* B (1992) **23** 629–635.

[1006] Dremin, I.M. and Hwa, R.C., Multiplicity distributions of squeezed isospin states. *Phys. Rev.* D (1996) **53** 1216–1223.

[1007] Dremin, I.M. and Man'ko, V.I., Particles and nuclei as quantum slings. *Nuovo Cim.* A (1998) **111** 439–444.

[1008] Dodonov, V.V., Dremin, I.M., Man'ko, O.V., Man'ko, V.I., and Polynkin, P.G., Nonclassical field states in quantum optics and particle physics. *J. Russ. Laser Research* (1998) **19** 427–463.

[1009] Kuvshinov, V. and Shaporov, V., Gluon squeezed states in QCD jet. *Acta Phys. Pol.* B (1999) **30** 59–68.

[1010] Volya, A., Pratt, S., and Zelevinsky, V., Multiple pion production from an oriented chiral condensate. *Nucl. Phys.* A (2000) **671** 617–643.

[1011] Tsue, Y., Koike, A., and Ikezi, N., Time-evolution of a collective meson field by use of a squeezed state. *Prog. Theor. Phys.* (2001) **106** 807–822.

[1012] Kuvshinov, V.I. and Shaporov, V.A., Fluctuations and correlations of soft gluons at the nonperturbative stage of evolution of QCD jets. *Phys. Atom. Nucl.* (2002) **65** 309–314.

[1013] Kostelecky, V.A. and Potting, R., Analytical construction of a nonperturbative vacuum for the open bosonic string. *Phys. Rev.* D (2001) **63** 046007.

[1014] Sidorov, Y.V., Quantum state of gravitons in expanding Universe. *Europhys. Lett.* (1989) **10** 415–418.

[1015] Grishchuk, L.P. and Sidorov, Y.V., Squeezed quantum states of relic gravitons and primordial density fluctuations. *Phys. Rev.* D (1990) **42** 3413–3421.

[1016] Grishchuk, L.P., Haus, H.A., and Bergman, K., Generation of squeezed radiation from vacuum in the cosmos and the laboratory. *Phys. Rev.* D (1992) **46** 1440–1449.

[1017] Gasperini, M. and Giovannini, M., Entropy production in the cosmological amplification of vacuum fluctuations. *Phys. Lett.* B (1993) **301** 334-338.

[1018] González-Díaz, P.F., Beyond the single universe. *Nuovo Cim.* B (1993) **108** 1197–1225.

[1019] Albrecht, A., Ferreira, P., Joyce, M., and Prokopec, T., Inflation and squeezed quantum states. *Phys. Rev.* D (1994) **50** 4807–4820.

[1020] Hu, B.L., Kang, G., and Matacz, A., Squeezed vacua and the quantum statistics of cosmological particle creation. *Int. J. Mod. Phys.* A (1994) **9** 991–1007.

[1021] Suresh, P.K. and Kuriakose, V.C., Squeezed states, black holes and entropy generation. *Mod. Phys. Lett.* A (1997) **12** 1435–1445.

[1022] Finelli, F., Vacca, G.P., and Venturi, G., Chaotic inflation from a scalar field in nonclassical states. *Phys. Rev.* D (1998) **58** 103514.

[1023] Giovannini, M., Backgrounds of squeezed relic photons and their spatial correlations. *Phys. Rev.* D (2000) **61** 087306.

[1024] Suresh, P.K., Thermal squeezing and density fluctuations in semiclassical theory of gravity. *Mod. Phys. Lett.* A (2001) **16** 707–717.

[1025] Jacobson, D.L., Werner, S.A., and Rauch, H., Spectral modulation and squeezing at high-order neutron interferences. *Phys. Rev.* A (1994) **49** 3196–3200.

[1026] Badurek, G., Rauch, H., Suda, M., and Weinfurter, H., Identification of nonclassical states in neutron spin precession experiments. *Opt. Comm.* (2000) **179** 13–18.

[1027] Spiller, T.P., Clark, T.D., Prance, R.J., Prance, H., and Poulton, D.A., Macroscopic superposition in superconducting circuits. *Int. J. Mod. Phys.* B (1990) **4** 1423–1435.

[1028] Dodonov, V.V., Man'ko, O.V., and Man'ko, V.I., Correlated states in quantum electronics (resonant circuit). *J. Sov. Laser Research* (1989) **10** 413–420.

[1029] Pavlov, S.T. and Prokhorov, A.V., Correlated and compressed states in a parametrized Josephson junction. *Sov. Phys. - Solid State* (1991) **33** 1384–1386.

[1030] Pavlov, S.T. and Prokhorov, A.V., Time-dependent theory of the single-contact quantum interferometer. Correlated and squeezed states. *Sov. Phys. - Solid State* (1992) **34** 50–53.

[1031] Man'ko, O.V., Correlated squeezed states of a Josephson junction. *J. Kor. Phys. Soc.* (1994) **27** 1–4.

[1032] Vourdas, A. and Spiller, T.P., Quantum theory of the interaction of Josephson junctions with non-classical microwaves. *Z. Phys.* B (1997) **102** 43–54.

[1033] Zou, J. and Shao, B., Superpositions of coherent states and squeezing effects in a mesoscopic Josephson junction. *Int. J. Mod. Phys.* B (1999) **13** 917–924.

[1034] Zheng, H., Variational ground state for the periodic Anderson model with an indirect hybridization. *Phys. Rev.* B (1987) **36** 8736–8751.

[1035] Zheng, H., Reconsideration of a simple model for bipolarons. *Solid State Commun.* (1988) **65** 731–734.

[1036] Feinberg, D., Ciuchi, S., and de Pasquale, F., Squeezing phenomena in interacting electron-phonon systems. *Int. J. Mod. Phys.* B (1990) **4** 1317–1367.

[1037] An, N.B., Squeezed excitons in semiconductors. *Mod. Phys. Lett.* B (1991) **5** 587–591.

[1038] Artoni, M. and Birman, J.L., Quantum optical properties of polariton waves. *Phys. Rev.* B (1991) **44** 3736–3756.

[1039] De Melo, C.A.R.S., Squeezed boson states in condensed matter. *Phys. Rev.* B (1991) **44** 11911–11917.

[1040] Peřina, J., Kárská, M., and Křepelka, J., Stimulated Raman scattering of nonclassical light by squeezed phonons. *Acta Phys. Polon.* (1991) **79** 817–828.

[1041] Janszky, J. and Vinogradov, An.V., Phonon squeezing. In: *Mol. Cryst. Liq. Cryst. Sci. Technol. – Sec. B: Nonlinear Optics, vol.* 2, p. 317–329. Gordon & Breach, London, 1992.

[1042] An, N.B., Squeezed state of biexcitons in excited semiconductors. *Int. J. Mod. Phys.* B (1992) **6** 395–407.

[1043] Lo, C.F. and Sollie, R., Correlated squeezed phonon states. *Phys. Lett.* A (1992)

169 91–98.

[1044] Artoni, M. and Birman, J.L., Non-classical states in solids and detection. *Opt. Commun.* (1994) **104** 319–324.

[1045] Sonnek, M. and Wagner, M., Squeezed oscillatory states in extended exciton-phonon systems. *Phys. Rev.* B (1996) **53** 3190–3202.

[1046] Hu, X.D. and Nori, F., Squeezed phonon states: modulating quantum fluctuations of atomic displacements. *Phys. Rev. Lett.* (1996) **76** 2294–2297.

[1047] Garrett, G.A., Rojo, A.G., Sood, A.K., Whitaker, J.F., and Merlin, R., Vacuum squeezing of solids; macroscopic quantum states driven by light pulses. *Science* (1997) **275** 1638–1640.

[1048] Chai, J.-H. and Guo, G-C., Preparation of squeezed-state phonons using the Raman-induced Kerr effect. *Quant. Semiclass. Opt.* (1997) **9** 921–927.

[1049] Artoni, M., Detecting phonon vacuum squeezing. *J. Nonlin. Opt. Phys. Mater.* (1998) **7** 241–254.

[1050] Zou, J. and Shao, B., Nonclassical states evolving from classical states in a polariton system at low temperature. *Int. J. Mod. Phys.* B (1999) **13** 2371–2385.

[1051] Frahm, H. and Holyst, J.A., On spin-squeezed states and their application to semi-classical kink dynamics in magnetic chains. *J. Phys.: Condens. Mat.* (1989) **1** 3083–3094.

[1052] Solomon, A.I. and Penson, K.A., Coherent pairing states for the Hubbard model. *J. Phys.* A (1998) **31** L355–L360.

[1053] Altanhan, T. and Bilge, S., Squeezed spin states and Heisenberg interaction. *J. Phys.* A (1999) **32** 115–121.

[1054] Sorensen, A.S. and Mølmer, K., Entanglement and extreme spin squeezing. *Phys. Rev. Lett.* (2001) **86** 4431–4434.

[1055] Wineland, D.J., Bollinger, J.J., Itano, W.M., and Heinzen, D.J., Squeezed atomic states and projection noise in spectroscopy. *Phys. Rev.* A (1994) **50** 67–88.

[1056] Vinogradov, A.V. and Janszky, J., New mechanism of molecule vibrations squeezing during Franck–Condon transition. *Acta Phys. Polon.* A (1990) **78** 231–237.

[1057] Vinogradov, A.V. and Janszky, J., Excitation of squeezed vibrational wave packets associated with Franck–Condon transitions in molecules. *Zhurn. Eksp. Teor. Fiz.* (1991) **100** 386–399 [*Sov. Phys. – JETP* (1991) **73** 211–217].

[1058] Averbukh, I. and Shapiro, M., Optimal squeezing of molecular wave packets. *Phys. Rev.* A (1993) **47** 5086–5092.

[1059] Janszky, J., Vinogradov, A.V., Walmsley, I.A., and Mostowski, J., Competition between geometrical and dynamical squeezing during a Franck–Condon transition. *Phys. Rev.* A (1994) **50** 732–740.

[1060] Janszky, J., Vinogradov, A.V., Kobayashi, T., and Kis, Z., Vibrational Schrödinger-cat states. *Phys. Rev.* A (1994) **50** 1777–1784.

[1061] Davidovich, L., Orszag, M., and Zagury, N., Quantum diagnosis of molecules: A method for measuring directly the Wigner function of a molecular vibrational state. *Phys. Rev.* A (1998) **57** 2544–2549.

[1062] Irac-Astaud, M., Molecular-Coherent-States and Molecular-Fundamental-States. *Rev. Math. Phys.* (2001) **13** 1437–1457.

[1063] Cirac, J.I., Lewenstein, M., Mølmer, K., and Zoller, P., Quantum superposition states of Bose-Einstein condensates. *Phys. Rev.* A (1998) **57** 1208–1218.

[1064] Horak, P. and Ritsch, H., Manipulating a Bose–Einstein condensate with a single photon. *Europ. Phys. J.* D (2001) **13** 279–287.

[1065] Orzel, C., Tuchman, A.K., Fenselau, M.L., Yasuda, M., and Kasevich, M.A., Squeezed states in a Bose–Einstein condensate. *Science* (2001) **291** 2386-2389.

[1066] Dunningham, J.A. and Burnett, K., Proposals for creating Schrodinger cat states in Bose–Einstein condensates. *J. Mod. Opt.* (2001) **48** 1837–1853.

[1067] Rogel-Salazar, J., Choi, S., New, G.H.C., and Burnett, K., Characterisation of the dynamical quantum state of a zero temperature Bose-Einstein condensate. *Phys. Lett.* A (2002) **299** 476–482.

[1068] Cirac, J.I., Parkins, A.S., Blatt, R., and Zoller, P., Dark squeezed states of the motion of a trapped ion. *Phys. Rev. Lett.* (1993) **70** 556–559.

[1069] Arimondo, E., Coherent population trapping in laser spectroscopy. In: *Progress in Optics, vol. XXXV* (E. Wolf, ed.), pp. 257–354. North Holland, Amsterdam, 1996.

[1070] Wynands, R. and Nagel, A., Precision spectroscopy with coherent dark states. *Appl. Phys.* B (1999) **68** 1–25.

[1071] Kulin, S., Castin, Y., Ol'shanii, M., Peik, E., Saubaméa, B., Leduc, M., and Cohen-Tannoudji, C., Exotic quantum dark states. *Eur. Phys. J.* D (1999) **7** 279–284.

[1072] Kis, Z., Vogel, W., Davidovich, L., and Zagury, N., Dark $SU(2)$ states of the motion of a trapped ion. *Phys. Rev.* A (2001) **63** 053410.

[1073] Einstein, A., Podolsky, B., and Rosen, N., Can quantum-mechanical description of physical reality be considered complete? *Phys. Rev.* (1935) **47** 777–780.

[1074] Bell, J.S., On the Einstein Podolsky Rosen paradox. *Physics* (1964) **1** 195–200.

[1075] Greenberger, D.M., Horne, M., and Zeilingler, A., Going beyond Bell's theorem. In: *Bell's Theorem, Quantum Theory, and Conceptions of the Universe* (M. Kafatos, ed.), pp. 73–76. Kluwer, Dordrecht, 1989.

[1076] Greenberger, D.M., Horne, M.A., Shimony, A., and Zeilingler, A., Bell's theorem without inequalities. *Am. J. Phys.* (1990) **58** 1131–1143.

[1077] Englert, B.-G. and Walther, H., Preparing a GHZ state, or an EPR state, with the one-atom maser. *Opt. Commun.* (2000) **179** 283–288.

[1078] Mann, A., Revzen, M., and Schleich, W., Unique Bell state. *Phys. Rev.* A (1992) **46** 5363–5366.

[1079] Brif, C., Mann, A., and Revzen, M., Generalized coherent states are unique Bell states of quantum systems with Lie-group symmetries. *Phys. Rev.* A (1998) **57** 742–745.

[1080] Agarwal, G.S., Puri, R.R., and Singh, R.P., Vortex states for the quantized radiation field. *Phys. Rev.* A (1997) **56** 4207–4215.

[1081] Arvieu, R., Rozmej, P., and Berej, W., Time-dependent partial waves and vortex rings in the dynamics of wavepackets. *J. Phys.* A (1997) **30** 5381–5392.

[1082] Bialynicki-Birula, I., Bialynicka-Birula, Z., and Sliwa, C., Motion of vortex lines in quantum mechanics. *Phys. Rev.* A (2000) **61** 032110.

[1083] Drobný, G. and Jex, I., Nondegenerate two-photon down-conversion: coherent inputs and nonclassical effects. *Czechosl. J. Phys.* (1994) **44** 827–842.

[1084] Sotskii, B.A. and Glazachev, B.I., Entropy properties of superrandom fields. *Opt. Spektrosk.* (1981) **50** 1057–1061 [*Opt. Spectrosc.* (1981) **50** 582–584].

[1085] Ritze, H.H. and Bandilla, A., Squeezing and first-order coherence. *J. Opt. Soc. Am.* B (1987) **4** 1641–1644.

[1086] Lukš, A., Peřinová, V., and Hradil, Z., Principal squeezing. *Acta Phys. Polon.* A (1988) **74** 713–721.

[1087] Loudon, R., Graphical representation of squeezed-state variances. *Opt. Commun.* (1989) **70** 109–114.

[1088] Hillery, M., Squeezing and photon number in Jaynes-Cummings model. *Phys. Rev.* A (1989) **39** 1556–1557.

[1089] Arvind, $U(2)$ invariant squeezing properties of pair coherent states. *Phys. Lett.* A (2002) **299** 461–468.

[1090] Peřinová, V., Křepelka,J., and Peřina, J., Photon statistics of nonclassical fields. *Opt. Acta* (1986) **33** 1263–1278.

[1091] Lee, C.T., Generalised Q parameters and their evolution under continuous photodetection. *Quant. Opt.* (1994) **6** 27–36.

[1092] Klyshko, D.N., Observable signs of nonclassical light. *Phys. Lett.* A (1996) **213** 7–15.

[1093] Lee, C.T., Simple criterion for nonclassical two-mode states. *J. Opt. Soc. Am.* B (1998) **15** 1187–1191.

[1094] Vyas, R. and Singh, S., Higher-order nonclassical effects in a parametric oscillator. *Phys. Rev.* A (2000) **62** 033803.

[1095] Lee, C.T., Higher-order criteria for nonclassical effects in photon statistics. *Phys. Rev.* A (1990) **41** 1721–1723.

[1096] Lee, C.T., Measure of nonclassicality of nonclassical states. *Phys. Rev.* A (1991) **44** R2775–R2778.

[1097] Lee, C.T., Moments of P functions and nonclassical depths of quantum states. *Phys. Rev.* A (1992) **45** 6586–6595.

[1098] Marchiolli, M.A., Bagnato, V.S., Guimarães, Y., and Baseia, B., Nonclassical depth of the phase state. *Phys. Lett.* A (2001) **279** 294–304.

[1099] Arvind, Mukunda, N., and Simon, R., Gaussian-Wigner distributions and hierarchies of nonclassical states in quantum optics: The single-mode case. *Phys. Rev.* A (1997) **56** 5042–5052.

[1100] Arvind, Mukunda, N., and Simon, R., Characterizations of classical and nonclassical states of quantized radiation. *J. Phys.* A (1998) **31** 565–583.

[1101] Puri, R.R., A quasiprobability based criterion for classifying the states of N spin-$\frac{1}{2}$s as classical or non-classical. *J. Phys.* A (1996) **29** 5719–5726.

[1102] Hillery, M., Nonclassical distance in quantum optics. *Phys. Rev.* A (1987) **35** 725–732.

[1103] Hillery, M., Total noise and nonclassical states. *Phys. Rev.* A (1989) **39** 2994–3002.

[1104] Dodonov,V.V., Man'ko, O.V., Man'ko, V.I., and Wünsche, A., Energy-sensitive and "classical-like" distances between quantum states. *Phys. Scripta* (1999) **59** 81–89.

[1105] Vogel, W., Nonclassical states: an observable criterion. *Phys. Rev. Lett.* (2000) **84** 1849–1852.

[1106] Honegger, R. and Rieckers, A., Construction of classical and non-classical coherent photon states. *Ann. Phys.* (NY) (2001) **289** 213–231.

[1107] Wünsche, A., Dodonov, V.V., Man'ko, O.V., and Man'ko, V.I., Nonclassicality of states in quantum optics. *Fortschr. Phys.* (2001) **49** 1117–1122.

[1108] Lvovsky, A.I. and Shapiro, J.H., Nonclassical character of statistical mixtures of the single-photon and vacuum optical states. *Phys. Rev.* A (2002) **65** 033830.

[1109] Marian, P., Marian, T.A., and Scutaru, H., Quantifying nonclassicality of one-mode Gaussian states of the radiation field. *Phys. Rev. Lett.* (2002) **88** 153601.

[1110] Werner, R.F., Quantum states with Einstein–Podolsky–Rosen correlations admitting a hidden-variable model. *Phys. Rev.* A (1989) **40** 4277–4281.

[1111] Peres, A., Collective tests for quantum nonlocality. *Phys. Rev.* A (1996) **54** 2685–2689.

[1112] Caves, C.M. and Milburn, G.J., Qutrit entanglement. *Opt. Commun.* (2000) **179** 439–446.

[1113] Hiroshima, T. and Ishizaka, S., Local and nonlocal properties of Werner states. *Phys. Rev.* A (2000) **62** 044302.

[1114] Shor, P.W., Smolin, J.A., and Terhal, B.M., Nonadditivity of bipartite distillable entanglement follows from a conjecture on bound entangled Werner states. *Phys. Rev. Lett.* (2001) **86** 2681–2684.

[1115] Schrödinger, E., Discussion of probability relations between separated systems. *Proc. Camb. Phil. Soc.* (1935) **31** 555–563.

[1116] Barnett, S.M. and Phoenix, S.J.D., Entropy as a measure of quantum optical correlation. *Phys. Rev.* A (1989) **40** 2404–2409.

[1117] Barnett, S.M. and Phoenix, S.J.D., Information theory, squeezing and quantum correlations. *Phys. Rev.* A (1991) **44** 535–545.

[1118] Mann, A., Sanders, B.C., and Munro, W.J., Bell's inequality for an entanglement of nonorthogonal states. *Phys. Rev.* A (1995) **51** 989–991.

[1119] Bennett, C.H., Bernstein, H.J., Popescu, S., and Schumacher, B., Concentrating partial entanglement by local operations. *Phys. Rev.* A (1996) **53** 2046–2052.

[1120] Vedral, V. and Plenio, M.B., Entanglement measures and purification procedures. *Phys. Rev.* A (1998) **57** 1619–1633.

[1121] Englert, B.-G., Löffler, M., Benson, O., Varcoe, B., Weidinger, M., and Walther, H., Entangled atoms in micromaser physics. *Fortschr. Phys.* (1998) **46** 897–926.

[1122] Horodecki, M., Horodecki, P., and Horodecki, R., Limits for entanglement measures. *Phys. Rev. Lett.* (2000) **84** 2014–2017.

[1123] Parker, S., Bose, S., and Plenio, M.B., Entanglement quantification and purification in continuous-variable systems. *Phys. Rev.* A (2000) **61** 032305.

[1124] Bennett, C.H., Popescu, S., Rohrlich, D., Smolin, J.A., and Thapliyal, A.V., Exact and asymptotic measures of multipartite pure-state entanglement. *Phys. Rev.* A (2001) **63** 012307.

[1125] Piątek, K. and Leoński, W., Wehrl's entropy and a measure of intermode correlations in phase space. *J. Phys.* A (2001) **34** 4951–4967.

[1126] Audenaert, K., Eisert, J., Jané, E., Plenio, M.B., Virmani, S., and De Moor, B., Asymptotic relative entropy of entanglement. *Phys. Rev. Lett.* (2001) **87** 217902.

[1127] Witte, C. and Trucks, M., A new entanglement measure induced by the Hilbert–Schmidt norm. *Phys. Lett.* A (1999) **257** 14–20.

[1128] Ozawa, M., Entanglement measures and the Hilbert–Schmidt distance. *Phys. Lett.* A (2000) **268** 158–160.

[1129] Verstraete, F., Dehaene, J., and De Moor, B., On the geometry of entangled states. *J. Mod. Opt.* (2002) **49** 1277–1287.

[1130] Man'ko, V.I., Marmo, G., Sudarshan, E.C.G., and Zaccaria, F., Interference and entanglement: an intrinsic approach. *J. Phys.* A (2002) **35** 7137–7157.

[1131] Furuya, K., Nemes, M.C., and Pellegrino, G.Q., Quantum dynamical manifestation of chaotic behavior in the process of entanglement. *Phys. Rev. Lett.* (1998) **80** 5524–5527.

[1132] Życzkowski, K., Horodecki, P., Sanpera, A., and Lewenstein, M., Volume of the set of separable states. *Phys. Rev.* A (1998) **58** 883–892.

[1133] Munro, W.J., James, D.F.V., White, A.G., and Kwiat, P.G., Maximizing the entanglement of two mixed qubits. *Phys. Rev.* A (2001) **64** 030302.

[1134] Vidal, G. and Werner, R.F., Computable measure of entanglement. *Phys. Rev.* A (2002) **65** 032314.

[1135] Coffman, V., Kundu, J., and Wootters, W.K., Distributed entanglement. *Phys. Rev.* A (2000) **61** 052306.

[1136] Badziag, P., Deuar, P., Horodecki, M., Horodecki, P., and Horodecki, R., Concurrence in arbitrary dimensions. *J. Mod. Opt.* (2002) **49** 1289–1297.

[1137] Dodonov, V.V., Castro, A.S.M., and Mizrahi, S.S., Covariance entanglement measure for two-mode continuous variable systems. *Phys. Lett.* A (2002) **296** 73–81.

Chapter 2

Squeezed states

Alfred Wünsche

1 Introduction

The history of squeezing as an operation and of squeezed states as a result of this operation is closely related to the development of quantum mechanics from its beginning, although not under the name "squeezing". The name is connected with the representation of states in the quantum-mechanical phase space by quasiprobabilities and with their transformations. The first phase-space distribution function in quantum mechanics was introduced in 1932 by Wigner [1] and is now called Wigner quasiprobability but its broad application began only in the 1950s after the work of Moyal [2]. Squeezed states were first considered by Kennard [3] (see [4] for this moment of history). The special case of squeezing of coherent states in axes directions is contained in the *Handbuch* article of Pauli [5] as states with minimal uncertainty product. However, a systematic investigation in this field had to be prepared by preceding fundamental researches into coherent states.

Coherent states were introduced as displaced ground states of the harmonic oscillator by Schrödinger [6] in 1926 and were used sporadically in the following three decades. The intense use of coherent states as displaced vacuum states and as the right-hand eigenstates of the annihilation operator in quantum optics began only in the 1960s when their modern name was introduced [7,8] (see also [9,10]). The next step was to consider the canonical transformations of operators or unitary transformations of states by operators with quadratic combinations of the boson operators in the exponent that leads to squeezing. Essential steps in this direction were made by Stoler [11,12] considering the canonical transformations of the boson annihilation and creation operators in connection with states of minimal uncertainty product and by Yuen [13,14] with the introduction of two-photon coherent states (now squeezed coherent states). The name "squeeze" operator was introduced by Hollenhorst [15]. Some basic papers relating to squeezing are republished in [16]. Squeezing has now become a major topic of quantum optics and there are review articles [17–22] (in [21,22] squeezed states are discussed along with generalized coherent and other nonclassical states), special issues of journals [23,24], collections of selected articles [25] and, furthermore, specialized conferences such as the series "International Conference on Squeezed States and

Uncertainty Relations" and its predecessors [26–31]. Every new monograph in the field of quantum optics now discusses squeezed states [32–46].

A rational treatment of squeezing has to involve some group theory, in particular, for symplectic groups. Group theory was applied in quantum mechanics from its beginning in the 1920s, mainly for the treatment of atoms, molecules, solids and elementary particles. The basis of the transformation theory of quantum mechanics, in analogy to canonical transformations in classical mechanics, was mainly laid by Weyl [47], who introduced the name "symplectic group" [48]. The noncommutative group of displacement operators in quantum mechanics is now called the "Heisenberg–Weyl group".

A broad introduction of group-theoretical aspects into quantum optics took place in the 1970s by realization of the Lie group of linear canonical transformations and of its Lie algebra by general quadratic combinations of boson operators (see, e.g., [49–53] for the general theory of Lie groups and symplectic transformations, [54, 55] for group-coherent states, [56, 57] for symplectic geometry and [58] as well as [33, 41, 44] for more specifics to quantum optics). The realization of the unitary groups $U(n)$ and $SU(n)$ by bilinear combinations of boson annihilation and creation operators (Jordan–Schwinger realization) and the Holstein–Primakoff realization of $SU(2)$ for a one-mode system (e.g., [41]) can be considered as the first steps in this direction.

Squeezing is understood in this chapter in its narrow sense as an operation in the phase space leading to some changes in the shape and position of the quasiprobabilities. The aim of the chapter is to give a theoretical description of the properties of squeezed states, in particular, of squeezed coherent and squeezed thermal states as the most general pure and mixed states with Gaussian quasiprobabilities and of the operation of squeezing in the frame of one mode of the electromagnetic field. Two modes are considered very briefly.

We discuss the relation of the squeezing operation to the symplectic group $Sp(2, \mathrm{R})$ which is isomorphic to the group $SU(1,1)$. Squeezing within one mode in its common sense is connected with a special realization of the abstract Lie algebra $sp(2, \mathrm{R}) \sim su(1,1)$ by quadratic combination of boson operators. There are other realizations of this algebra by fractal combinations of boson operators for one mode and by boson operators for two and more modes. The relations derived from the abstract Lie algebra are applicable for all these realizations. For example, there is a relation of this Lie algebra to coherent phase states. In the next section we derive the fundamental irreducible representation of $SU(1,1)$ and discuss the problem of operator ordering. Our main aim is to connect the group-theoretical aspects with the special derivations of the properties of squeezed states.

We do not give a representation of the experimental side of squeezing or of the generation of squeezed states (see [45] for a detailed representation and also [36] and reprinted articles in [16]) or of the application of squeezed states in different processes. The basic possibilities for generation can be seen from the different representations of squeezed states. One main process is the generation of squeezing by the time-dependent harmonic oscillator (e.g., [59]). There are parametric interactions of two and more modes which lead to squeezing. In the simplest parametric approximation, the dynamics described by the time-independent quadratic

Hamiltonian in the boson operators of the modes, the time evolution selects one-parameter subgroups of the quantum-mechanical symplectic group $Sp(2;\mathrm{R})$ with time as the additive parameter. For open systems, one has master equations for density operators (e.g, [34, 46, 60]) instead of the von Neumann equation and the dynamics is described by quantum-mechanical Liouville operators acting in spaces of linear operators instead of the usual operators acting in the Hilbert space of states.

The higher symplectic groups $Sp(2n,\mathrm{R})$ involve $n(2n+1)$ independent real parameters and not only describe squeezing operations (different $Sp(2,\mathrm{R})$ subgroups) but contain many other subgroups as, for example, the unitary groups $U(n)$ and special unitary groups $SU(n)$ and $SU(k,l)$. For example, in $Sp(4,\mathrm{R})$ of a two-mode system one has different subgroups $Sp(2,\mathrm{R}) \sim SU(1,1)$ (squeezing) and $SU(2)$ and devices that realize them may be called $SU(1,1)$ and $SU(2)$ interferometers [44,61]. The group $SU(2)$ describes, for example, light polarization and beam splitting [62]. The systematic treatment of these basic operations on the group-theoretical basis seems to be very attractive but is not possible in the frame of this chapter.

2 Symplectic group of squeezing operators

2.1 Lie algebra to symplectic group in boson realization

In this section, we treat a few selected problems of the representation of the Lie algebra $sp(2,\mathrm{C})$ and Lie group $Sp(2;\mathrm{C})$ of squeezing operators (e.g., [33,48–54]) which prepare a concise discussion of squeezed states beginning from the next section on. We first consider the quadratic combinations of the canonical operators (Q,P) or boson operators $(a,a^\dagger)$ of a single mode

$$
\begin{aligned}
K_1 &\equiv \left(Q^2 - P^2\right)/(4\hbar) = \left(a^2 + a^{\dagger 2}\right)/4,\\
K_2 &\equiv -\left(QP + PQ\right)/(4\hbar) = \mathrm{i}\left(a^2 - a^{\dagger 2}\right)/4,\\
K_0 &\equiv \left(Q^2 + P^2\right)/(4\hbar) = \left(aa^\dagger + a^\dagger a\right)/4 = (2N+I)/4. \qquad (1)
\end{aligned}
$$

They satisfy the commutation relations

$$
[K_1, K_2] = -\mathrm{i}K_0, \quad [K_2, K_0] = \mathrm{i}K_1, \quad [K_0, K_1] = \mathrm{i}K_2, \qquad (2)
$$

following from the commutators $[Q,P] = \mathrm{i}\hbar I$ or $[a,a^\dagger] = I$ by using the identity $[AB,C] = A[B,C] + [A,C]B$. The operators K_1 and K_2 can be considered as some quadratic analogues of the canonical operators Q and P. Their expectation values play a role in the definition of second-order squeezing (section 4).

The operators (K_1, K_2, K_0) in the realization (1) act in the Hilbert space $\mathcal{H}$ of a single boson mode which is the Fock space spanned by the eigenstates $|n\rangle$, $(n = 0, 1, \ldots)$ of the number operator $N = a^\dagger a$ and they are Hermitian operators in this realization. Since the commutation relations are closed with regard to these operators, they form a basis of a three-dimensional Lie algebra $sp(2,\mathrm{R})$

or $sp(2,\mathrm{C})$ in its complex extension. A related important basis of this Lie algebra with the operators K_- and K_+ instead of K_1 and K_2 can be defined by

$$K_- = K_1 - \mathrm{i}K_2 = K_+^\dagger, \quad K_+ = K_1 + \mathrm{i}K_2 = K_-^\dagger, \tag{3}$$

or explicitly in the realization (1) by boson operators

$$\begin{aligned} K_- &= \left\{Q^2 - P^2 + \mathrm{i}(QP + PQ)\right\}/(4\hbar) = a^2/2, \\ K_+ &= \left\{Q^2 - P^2 - \mathrm{i}(QP + PQ)\right\}/(4\hbar) = a^{\dagger 2}/2. \end{aligned} \tag{4}$$

In this realization, the operators K_- and K_+ are complex adjoint to each other. They obey the commutation relations

$$[K_0, K_-] = -K_-, \quad [K_0, K_+] = +K_+, \quad [K_-, K_+] = 2K_0. \tag{5}$$

One can look to the first two of these equations as to the two solutions of the root equation for the operator K_0. The (here scalar) roots of the operator K_0 are -1 and $+1$ to the root operators K_- and K_+ and the (Abelian) Cartan subalgebra is spanned by the only operator K_0. The Cartan subalgebra together with the solution of the root equations for this subalgebra is a very important mean to clarify the structure of a Lie group and to derive its irreducible representations.

Due to the low dimension of the considered Lie algebra, there exist some isomorphisms to initial members of different series of Lie algebras. The real Lie algebra with commutation relations (2) or (5) is isomorphic to the Lie algebras $su(1,1) \sim sp(2,\mathrm{R}) \sim sl(2,\mathrm{R}) \sim so(2,1)$ belonging to the Lie groups $SU(1,1) \sim Sp(2,\mathrm{R}) \sim SL(2,\mathrm{R}) \sim SO(2,1)$. These are the special unitary group $SU(1,1)$, the two-dimensional real symplectic group $Sp(2,\mathrm{R})$, the two-dimensional real special linear (or unimodular) group $SL(2,\mathrm{R})$ and the three-dimensional proper pseudo-orthogonal group SO(2,1) leaving invariant the scalar product in a pseudo-Euclidian space with signature $(++-)$. Operators K_- and K_+ alone are not members of the real Lie algebra $su(1,1)$ but only their combinations $K_-x^- + K_+x^{-*}$. It is useful to embed these real Lie algebras into corresponding complex algebras $sp(2,\mathrm{C}) \sim sl(2,\mathrm{C}) \sim so(3,\mathrm{C})$ by supposing that the coefficients can take on arbitrary complex numbers. The corresponding complex Lie groups are denoted by $Sp(2,\mathrm{C}) \equiv C_1$ (two-dimensional complex symplectic group), by $SL(2,\mathrm{C}) \equiv A_1$ (two-dimensional complex unimodular group) and by $SO(3,\mathrm{C}) \equiv B_1$ (proper three-dimensional complex orthogonal group). The group $Sp(2,\mathrm{C})$ is also homomorphic ((2–1)-correspondence) to the proper Lorentz group. It contains the group $SU(2) \sim SO(3,\mathrm{R})$ as a subgroup. There are also differences in the global structure of the finite-dimensional irreducible representations of $Sp(2,\mathrm{R})$ and of the unitary irreducible representations in infinite-dimensional Hilbert spaces (metaplectic groups).

2.2 Killing form and Casimir operator

If K_k, $(k = 1,\ldots,r)$ are r basis operators of an r-dimensional Lie algebra, then its general element x can be written in this basis as $x = K_k x^k$ (sum convention). The commutators of the basis operators can be written $[K_k, K_l] = K_j c^j_{kl}$, where $c^j_{kl} = -c^j_{lk}$ are the structure coefficients. Due to the Jacobi identity following identically for commutators, the structure coefficients satisfy additional conditions. The commutator of two arbitrary elements $x = K_k x^k$ and $y = K_l y^l$ is a new element $z = K_j z^j$ with $z^j = c^j_{kl} x^k y^l$. From the structure coefficients c^j_{kl}, one can determine a covariant metric tensor $g_{kl} = (1/2) c^n_{km} c^m_{ln}$ which is symmetric ($g_{kl} = g_{lk}$) and the corresponding contravariant tensor g^{kl} according to $g_{kl} g^{lm} = \delta^m_k$. By means of the metric tensors g_{kl} and g^{kl}, one can determine two important quantities, first, the Killing form $(x, y) \equiv g_{kl} x^k y^l$ of two elements x and y of the Lie algebra which has all the properties of an Euclidian (or pseudo-Euclidian) scalar product and, second, the Casimir operator $C \equiv g^{kl} K_k K_l$ which commutes with all operators of the Lie algebra, which means $[C, x] = 0$.

We now specialize this to the Lie algebra $sp(2, \mathrm{C})$ introduced with two convenient bases in (1) and (4) and write the general element x of this Lie algebra in the following convenient form for our purposes

$$x = -\mathrm{i}\xi K_- + 2\eta K_0 + \mathrm{i}\zeta K_+, \quad (\xi, \eta, \zeta \in \mathrm{C}), \tag{6}$$

where (ξ, η, ζ) are arbitrary complex parameters. For $\xi = \zeta^*$ and $\eta = \eta^*$ the elements x become Hermitian operators in realizations (1) and (4) of the abstract Lie algebra. The Killing form (x, x') for two general elements x and x' of $sp(2, \mathrm{C})$ is given by

$$(x, x') \equiv 4\eta\eta' - 2(\xi\zeta' + \zeta\xi'), \quad (x, x) \equiv 4(\eta^2 - \xi\zeta) \equiv -4\varepsilon^2. \tag{7}$$

The value of this form does not depend on the chosen basis. Therefore, (x, x) is a basic invariant of the considered element x and we have introduced a special notation ε for a quantity closely related to it. The Killing form corresponds for $sp(2, \mathrm{R})$ to a pseudo-Euclidian scalar product with signature $(+--)$ of the space $(g_{00} = 1, g_{11} = g_{22} = -1)$. For the Casimir operator C (with appropriately chosen factor), one obtains explicitly $(g^{00} = 1, g^{11} = g^{22} = -1)$

$$C = K_0^2 - K_1^2 - K_2^2 = K_0^2 - \tfrac{1}{2}\left(K_- K_+ + K_+ K_-\right). \tag{8}$$

The Casimir operator C is the only independent operator which commutes with all operators of the Lie algebra $sp(2, \mathrm{C})$. In the boson realization (1) it is proportional to the unit operator I of the Fock space $\mathcal{H}$ according to

$$C = \frac{1}{16}\left(aa^\dagger + a^\dagger a\right)^2 - \frac{1}{8}\left(a^2 a^{\dagger 2} + a^{\dagger 2} a^2\right) = -\frac{3}{16} I \equiv k(k-1)I, \tag{9}$$

but the partial spaces spanned by even Fock states and by odd Fock states belong to different unitary irreducible representations of $Sp(2, \mathrm{R})$ with labels $k = \frac{1}{4}$ and $k = \frac{3}{4}$, respectively [54].

The Killing form (x, x) in (7) is indefinite for real (x^0, x^1, x^2) (this is typical for noncompact groups) and allows the following classification:

$$\begin{aligned} (x, x) &> 0, && \text{rotation-like,} \\ (x, x) &= 0, && \text{degenerate or cone-like,} \\ (x, x) &< 0, && \text{squeezing-like.} \end{aligned} \tag{10}$$

In the case of complex (x^0, x^1, x^2) the Killing form (x, x) takes on complex values and the classification (10) becomes inapplicable.

The special form of the operators of the Lie algebra $su(1,1)$ in (1) or (4) is called a realization of the abstract Lie algebra by quadratic boson operators. There exist many other realizations of $sp(2, \mathrm{C})$ by fractal expressions with boson operators, for example, connected with phase states [54, 63–71] and two-mode realizations (section 4). The abstract form (6) of the general element of the Lie algebra $sp(2, \mathrm{C})$ takes on the following concrete form in the considered realization by boson operators

$$x = -\mathrm{i}\left\{\frac{\xi}{2}a^2 + \mathrm{i}\frac{\eta}{2}(aa^\dagger + a^\dagger a) - \frac{\varsigma}{2}a^{\dagger 2}\right\}. \tag{11}$$

For the general element of the corresponding Lie group $Sp(2, \mathrm{C})$ obtained by exponential mapping $X = \exp(\mathrm{i}x)$ of the Lie algebra onto the Lie group, we introduce the special notation $S(\xi, \eta, \varsigma)$ according to

$$\begin{aligned} S(\xi, \eta, \varsigma) &\equiv \exp\left\{\frac{\xi}{2}a^2 + \mathrm{i}\frac{\eta}{2}\left(aa^\dagger + a^\dagger a\right) - \frac{\varsigma}{2}a^{\dagger 2}\right\} \\ &= \exp\left(\xi K_- + \mathrm{i}2\eta K_0 - \varsigma K_+\right) = \exp(\mathrm{i}x), \end{aligned} \tag{12}$$

and call $S(\xi, \eta, \varsigma)$ the squeezing operator. Its special case $S(0, \eta, 0))$ means a pure rotation. The operator $S(\xi, \eta, \varsigma)$ possesses the property

$$S^{-1}(\xi, \eta, \varsigma) = S(-\xi, -\eta, -\varsigma) = S^\dagger(\varsigma^*, \eta^*, \xi^*). \tag{13}$$

The operator $S(\xi, \eta, \varsigma)$ becomes unitary for real $\eta = \eta^*$ and for $\xi = \varsigma^*$:

$$S(\varsigma^*, \eta = \eta^*, \varsigma) = \exp\left\{\frac{\varsigma^*}{2}a^2 + \mathrm{i}\frac{\eta}{2}\left(aa^\dagger + a^\dagger a\right) - \frac{\varsigma}{2}a^{\dagger 2}\right\}. \tag{14}$$

However, nonunitary squeezing operators, such as, for example, $S(0, 0, \varsigma)$ and $S(\xi, 0, 0)$, also possess importance in the theory of squeezed states (section 3).

2.3 Fundamental two-dimensional representation

We now construct the fundamental two-dimensional irreducible representation of the Lie algebra and Lie group of squeezing operators in their complex extensions $sp(2, \mathrm{C})$ and $Sp(2, \mathrm{C})$ which is the faithful irreducible representation of lowest dimension. The realization (1) of the corresponding abstract Lie algebra is important for this purpose because it provides a basis of this representation in a natural way, which is not the case if we consider only the abstract commutation relations (5). The background is the following commutation relations of (K_-, K_0, K_+) in realization (1) with the boson operators $(a, a^\dagger)$

$$\begin{aligned} &[K_-, a] = 0, \quad [K_0, a] = -\tfrac{1}{2}a, \quad [K_+, a] = -a^\dagger, \\ &[K_-, a^\dagger] = a, \quad [K_0, a^\dagger] = \tfrac{1}{2}a^\dagger, \quad [K_+, a^\dagger] = 0. \end{aligned} \tag{15}$$

The operators $(K_-, K_0, K_+, a, a^\dagger, I)$ form a six-dimensional Lie algebra which is the unification of the Lie algebra $sp(2,\mathrm{C})$ with the Heisenberg–Weyl algebra $w(1,\mathrm{C})$ (inhomogeneous symplectic Lie algebra denoted by $isp(2,\mathrm{C})$ [36]). The commutation relations (15) lead again to superpositions of the boson operators $(a, a^\dagger)$ and these operators are therefore appropriate as a basis for a representation. This means that the symplectic Lie algebra $sp(2,\mathrm{C})$ is an outer automorphism of the Heisenberg–Weyl algebra $w(1,\mathrm{C})$ of one mode. We construct the two-dimensional fundamental representation of the Lie algebra $sp(2,\mathrm{C})$ by using $(a, a^\dagger)$ as basis operators. By applying a generally possible construction of a representation of a Lie algebra in a given basis of this representation (here $(a, a^\dagger)$) to the basis operators (K_-, K_0, K_+) of the Lie algebra, we find

$$
\begin{aligned}
([K_-, a], [K_-, a^\dagger]) &= (0, a) = (a, a^\dagger) \begin{Vmatrix} 0 & 1 \\ 0 & 0 \end{Vmatrix}, \\
([K_0, a], [K_0, a^\dagger]) &= \tfrac{1}{2}(-a, a^\dagger) = (a, a^\dagger) \begin{Vmatrix} -\frac{1}{2} & 0 \\ 0 & \frac{1}{2} \end{Vmatrix}, \\
([K_+, a], [K_+, a^\dagger]) &= (-a^\dagger, 0) = (a, a^\dagger) \begin{Vmatrix} 0 & 0 \\ -1 & 0 \end{Vmatrix}.
\end{aligned} \tag{16}
$$

This corresponds to the mapping

$$
K_- \rightarrow \begin{Vmatrix} 0 & 1 \\ 0 & 0 \end{Vmatrix}, \quad K_0 \rightarrow \frac{1}{2} \begin{Vmatrix} -1 & 0 \\ 0 & 1 \end{Vmatrix}, \quad K_+ \rightarrow \begin{Vmatrix} 0 & 0 \\ -1 & 0 \end{Vmatrix}. \tag{17}
$$

An arbitrary operator $x = -\mathrm{i}(\xi K_- + \mathrm{i}2\eta K_0 - \zeta K_-)$ of the Lie algebra is mapped according to

$$
x \rightarrow \hat{x} \equiv \begin{Vmatrix} -\eta & -\mathrm{i}\xi \\ -\mathrm{i}\zeta & \eta \end{Vmatrix}. \tag{18}
$$

The trace $\langle \hat{x} \rangle$ of the representation matrices $\hat{x}$ is vanishing for arbitrary operators x and together with the determinant $[\hat{x}]$ of $\hat{x}$, one obtains

$$
\langle \hat{x} \rangle \equiv \mathrm{Tr}(\hat{x}) = 0, \quad [\hat{x}] \equiv \det(\hat{x}) = \xi\zeta - \eta^2 = -\tfrac{1}{4}(x, x) \equiv \varepsilon^2. \tag{19}
$$

This means that the determinant of the two-dimensional representation is the only nontrivial scalar invariant of this representation due to vanishing of the trace, and it is proportional to the Killing form (x, x) defined in (7). The matrix corresponding to K_0 is a Hermitian matrix, whereas the matrices corresponding to K_1 and K_2 are anti-Hermitian matrices and therefore the matrices corresponding to (K_-, K_+) do not form a pair of Hermitian-adjoint matrices in this representation.

The connection between elements x of the Lie algebra and elements X of the Lie group is established by exponential mapping according to

$$
x \rightarrow X \equiv \exp(\mathrm{i}x) \quad \Leftrightarrow \quad \hat{x} \rightarrow \hat{X} \equiv \exp(\mathrm{i}\hat{x}) = \sum_{n=0}^{\infty} \frac{\mathrm{i}^n}{n!} \hat{x}^n. \tag{20}
$$

If x is a Hermitian operator then X is a unitary operator and the same is true for the representations, i.e. if x is represented by a Hermitian matrix then X is represented by a unitary matrix; however, according to (18) there is not a unique correspondence of Hermitian operators x to Hermitian matrices $\hat{x}$.

We now map the fundamental two-dimensional representation of the Lie algebra $sp(2,C)$ into a representation of the Lie group $Sp(2,\mathrm{C})$ via exponential mapping $\hat{x} \to \hat{X} = \exp(\mathrm{i}\hat{x}) = \hat{S}(\xi,\eta,\zeta)$ in the basis $(a,a^\dagger)$

$$S(\xi,\eta,\zeta)(a,a^\dagger)S^{-1}(\xi,\eta,\zeta) = (a,a^\dagger)\hat{S}(\xi,\eta,\zeta). \tag{21}$$

By applying the Hamilton–Cayley identity $A^2 - \langle A\rangle A + [A]I_2 = 0$ to the two-dimensional representation matrices $\hat{x}$ in (18) with vanishing trace $\langle\hat{x}\rangle$, one obtains for $\hat{X} \equiv \exp(\mathrm{i}\hat{x})$ specialized with regard to the basis $(a,a^\dagger)$ [72] (special case $\xi = \zeta^*, \eta = 0$, e.g., [11])

$$\hat{X} \equiv \left\| \begin{matrix} \kappa & \lambda \\ \mu & \nu \end{matrix} \right\| = \left\| \begin{matrix} \mathrm{ch}\varepsilon - \mathrm{i}\eta\dfrac{\mathrm{sh}\varepsilon}{\varepsilon} & \xi\dfrac{\mathrm{sh}\varepsilon}{\varepsilon} \\ \zeta\dfrac{\mathrm{sh}\varepsilon}{\varepsilon} & \mathrm{ch}\varepsilon + \mathrm{i}\eta\dfrac{\mathrm{sh}\varepsilon}{\varepsilon} \end{matrix} \right\|, \tag{22}$$

where $\varepsilon \equiv \sqrt{\xi\zeta - \eta^2}$. The traces $\langle\hat{X}\rangle$ which form the character of the representation and which are very important in group theory for the identification of representations and the determinant $[\hat{X}]$ of the matrices $\hat{X}$ are

$$\langle\hat{X}\rangle = \kappa + \nu = 2\,\mathrm{ch}\varepsilon, \quad [\hat{X}] = \kappa\nu - \lambda\mu = 1. \tag{23}$$

The matrix elements (κ,λ,μ,ν) of $\hat{X}$ are not independent from each other since the determinant is equal to 1. This means that the matrices $\hat{X}$ are unimodular matrices but they are, in general, not unitary matrices and the fundamental representation is not a unitary one.

The basic linear transformations of the boson operators defined according to (21)–(23) can be written in the form

$$S(\xi,\eta,\zeta)(a,a^\dagger)S^{-1}(\xi,\eta,\zeta) = \left(\kappa a + \mu a^\dagger, \lambda a + \nu a^\dagger\right), \tag{24}$$

which is sometimes called the Bogolyubov transformation. For the nonunitary operator $S(\xi,\eta,\zeta)$, the transformed pair $\left(\kappa a + \mu a^\dagger, \lambda a + \nu a^\dagger\right)$ is not Hermitian adjoint to each other but due to the unimodularity of the matrix involved it preserves the commutation relation

$$\left[\kappa a + \mu a^\dagger, \lambda a + \nu a^\dagger\right] = \left[a, a^\dagger\right] = I. \tag{25}$$

Equations (24) and (25) reflect the basic meaning of the two-dimensional fundamental representation. The corresponding transformation of the, in general, nonunitary displacement operator $D(\alpha,\beta) \equiv \exp\left(\alpha a^\dagger - \beta a\right)$ can be represented in the form (note that we have changed here, for later application, the positions of $S(\xi,\eta,\zeta)$ and its inverse)

$$S^{-1}(\xi,\eta,\zeta)D(\alpha,\beta)S(\xi,\eta,\zeta) = D(\kappa\alpha + \mu\beta, \lambda\alpha + \nu\beta). \tag{26}$$

The constructed matrices $\hat{X}$ realize the fundamental two-dimensional representation of the complex symplectic group $Sp(2,\mathrm{C})$.

The special case of unitary squeezing operators $S(\zeta^*, \eta = \eta^*, \zeta)$ leads to the specialization $\kappa = \nu^*, \lambda = \mu^*$ in the matrix of the fundamental representation (22) and related formulae. The invariant $(x, x) = -4\varepsilon^2$ then becomes real positive or negative (rotation-like or squeezing-like; see (10)) and ε is a real or an imaginary number. The basic linear transformations of pairs of Hermitian adjoint boson operators $(a, a^\dagger)$ according to (24) lead in this case again to pairs of Hermitian adjoint operators and are the quantum-mechanical analogue of classical linear canonical (symplectic) transformations [33, 56, 57, 72–74]. The complex parameter ζ and the real parameter $\eta = \eta^*$ correspond to 3 real parameters, that is, equal to the number of real parameters involved in the complex parameters (ν, μ) satisfying the condition $\nu\nu^* - \mu\mu^* = 1$. We mention that the eigenvalue problem for the matrix $\hat{X}$ has two solutions $\exp(-\varepsilon)$ and $\exp(\varepsilon)$ that underline the invariance of ε with regard to any applied basis.

The construction of the fundamental representation of the group $Sp(2, \mathrm{C})$ with the operators of the Heisenberg–Weyl group $W(1, \mathrm{C})$ as the basis shows that the symplectic group $Sp(2, \mathrm{C})$ is an outer automorphism group to the Heisenberg–Weyl group $W(1, \mathrm{C})$, and its real subgroup $Sp(2, \mathrm{R}) \sim SU(1,1)$ is an outer automorphism group to the real Heisenberg–Weyl group $W(1, \mathrm{R})$. This property is important for the eigenvalue problem of arbitrary linear combinations of the operators of the Heisenberg–Weyl group (section 3). The transition from the single-mode case to the n-mode case corresponds to the transition from the group $Sp(2, \mathrm{R})$ to the group $Sp(2n, \mathrm{R})$ as the group of outer automorphisms to the Heisenberg–Weyl group $W(n, \mathrm{R})$ and to underline this aspect, it is didactic to refer to $Sp(2, \mathrm{R})$ and to its complexification $Sp(2, \mathrm{C})$ from the mentioned isomorphisms and homomorphisms.

2.4 Inversion of the fundamental representation

The problem of inversion of the fundamental representation (22) consists of two partial problems. The first problem is for given matrices $\hat{X}$ with the elements $(\kappa, \lambda, \mu, \nu)$ satisfying the unimodularity condition $\kappa\nu - \lambda\mu = 1$ to determine the matrix $\hat{x}$ with the parameters (ξ, η, ζ) that provides a multiplicity of solutions. The second problem is to form the operators $S(\xi, \eta, \zeta)$ for the multiplicity of obtained solutions (ξ, η, ζ) and to establish the degree of nonuniqueness of these operators.

The solution of the first problem can be written in the form [72]

$$\xi = \lambda \frac{\mathrm{Arsh}\,\vartheta}{\vartheta}, \quad \eta = \mathrm{i}\frac{\kappa - \nu}{2}\frac{\mathrm{Arsh}\,\vartheta}{\vartheta}, \quad \zeta = \mu \frac{\mathrm{Arsh}\,\vartheta}{\vartheta},$$
$$\vartheta \equiv \pm\sqrt{(\kappa + \nu)^2/4 - 1} = \pm\sqrt{(\kappa - \nu)^2/4 + \lambda\mu} = \pm\mathrm{sh}\varepsilon. \quad (27)$$

The two possible signs of the square root in $\vartheta = \pm\mathrm{sh}\varepsilon \equiv \mathrm{sh}\sqrt{[\hat{x}]}$ are irrelevant in the combination $\mathrm{Arsh}\,\vartheta/\vartheta = \varepsilon/\mathrm{sh}\varepsilon = \sqrt{[\hat{x}]}/\mathrm{sh}\sqrt{[\hat{x}]}$. The problem, however, is that $\mathrm{Arsh}\,\vartheta$ is not a unique function in the complex domain. Therefore, one has to find the whole set of possible solutions for $\hat{x}$ in dependence on $\hat{X}$. From (22), one finds that $\hat{X}$ remains unchanged by the transformations $\hat{x} \to \hat{x}'$

$$\hat{x}' = \left(1 + \mathrm{i}2\pi k/\sqrt{[\hat{x}]}\right)\hat{x}, \quad \Rightarrow \quad [\hat{x}'] = \left(\sqrt{[\hat{x}]} + \mathrm{i}2\pi k\right)^2, \quad (28)$$

or in the variables (ξ,η,ζ)

$$(\xi',\eta',\zeta') = \left(1 + \frac{\mathrm{i}2\pi k}{\sqrt{\xi\zeta - \eta^2}}\right)(\xi,\eta,\zeta), \quad k = 0, \pm 1, \pm 2, \dots . \tag{29}$$

However, only in the case of imaginary $\varepsilon = \sqrt{[\hat{x}]}$, that is, for rotation-like operators, are the multiple solutions $\hat{x}$ and $\hat{x}'$ related by real factors $1 + \mathrm{i}2\pi k/\varepsilon$ for arbitrary integer k. The multiplicity of (ξ,η,ζ) contributes to a possible nonuniqueness of $S(\xi,\eta,\zeta)$ of the form

$$S(\xi',\eta',\zeta') = \exp\left[\frac{\mathrm{i}\pi k}{\sqrt{\xi\zeta - \eta^2}}\left(\xi a^2 + \mathrm{i}\eta(a^\dagger a + a a^\dagger) - \zeta a^{\dagger 2}\right)\right] S(\xi,\eta,\zeta). \tag{30}$$

One can prove the following identity

$$\exp\left[\frac{\mathrm{i}\pi k}{\sqrt{\xi\zeta - \eta^2}}\left(\xi a^2 + \mathrm{i}\eta(a a^\dagger + a^\dagger a) - \zeta a^{\dagger 2}\right)\right] = (-1)^k I. \tag{31}$$

The proof of this relation is easy for $\xi = \zeta = 0$. In this case one finds

$$\begin{aligned} S(0, 2\pi k, 0) &= \exp\left[\mathrm{i}\pi k(a a^\dagger + a^\dagger a)\right] \\ &= \exp(\mathrm{i}\pi k)\exp\left(\mathrm{i}2\pi k N\right) = (-1)^k I. \end{aligned} \tag{32}$$

In the more general case, the proof can be made by decomposition of the operator in (31) into a normally ordered product. The derived relations show that each pair of operators $\pm S(\xi,\eta,\zeta)$ leads to the same matrix $\hat{X}$ in the fundamental representation and we have a 2–1 correspondence between these representations. Therefore, the quantum-mechanical group of operators $S(\xi,\eta,\zeta)$ possesses a global structure which is different from the global structure of its two-dimensional fundamental representation (and of other finite-dimensional representations) and the quantum-mechanical group or unitary representation is sometimes called the metaplectic group $Mp(2,\mathrm{C})$ [57, 75]. These global relations become more complex if one considers other possible quantum-mechanical realizations of the symplectic group $Sp(2,\mathrm{C})$ by fractal combinations of boson operators and among them, one finds realizations with $s-1$ correspondences to $Sp(2,\mathrm{C})$ with arbitrary integer $s \geq 1$.

2.5 Fundamental representation and canonical operators

One can use instead of $(a, a^\dagger)$ linear combinations of them as a basis of the fundamental representation of the group $Sp(2,\mathrm{C})$, for example, the canonical operators (Q,P). Since the corresponding matrices play some role in our further considerations, we will give them here. The exponential mapping of x onto $X = \exp(\mathrm{i}x)$ in analogy to (20) leads to the following two-dimensional representation matrices with regard to the basis (Q,P)

$$\hat{X} \equiv \left\| \begin{matrix} \alpha & \beta \\ \gamma & \delta \end{matrix} \right\| = \left\| \begin{matrix} \mathrm{ch}\varepsilon + \dfrac{\xi+\zeta}{2}\dfrac{\mathrm{sh}\varepsilon}{\varepsilon} & \left(\mathrm{i}\dfrac{\xi-\zeta}{2} - \eta\right)\dfrac{\mathrm{sh}\varepsilon}{\varepsilon} \\ \left(\mathrm{i}\dfrac{\xi-\zeta}{2} + \eta\right)\dfrac{\mathrm{sh}\varepsilon}{\varepsilon} & \mathrm{ch}\varepsilon - \dfrac{\xi+\zeta}{2}\dfrac{\mathrm{sh}\varepsilon}{\varepsilon} \end{matrix} \right\| . \tag{33}$$

Trace and determinant of the matrices remain the same as in (23). The transformation of the basis operators (Q, P) possesses the form

$$\exp(\mathrm{i}x)(Q,P)\exp(-\mathrm{i}x) = (Q,P)\exp(\mathrm{i}\hat{x}) = (Q,P)\hat{X}. \tag{34}$$

The connection between the matrix elements in (22) and (33) is

$$\begin{aligned} \alpha &= (\kappa+\lambda+\mu+\nu)/2, \quad \beta = -\mathrm{i}(\kappa-\lambda+\mu-\nu)/2, \\ \gamma &= \mathrm{i}(\kappa+\lambda-\mu-\nu)/2, \quad \delta = (\kappa-\lambda-\mu+\nu)/2, \end{aligned} \tag{35}$$

$$\begin{aligned} \kappa &= [\alpha+\mathrm{i}(\beta-\gamma)+\delta]/2, \quad \lambda = [\alpha-\mathrm{i}(\beta+\gamma)-\delta]/2, \\ \mu &= [\alpha+\mathrm{i}(\beta+\gamma)-\delta]/2, \quad \nu = [\alpha-\mathrm{i}(\beta-\gamma)+\delta]/2. \end{aligned} \tag{36}$$

One can construct higher-dimensional irreducible representations of $Sp(2,\mathrm{C})$ by using the symmetrized n-th powers of the boson operators as a basis that leads to $(n+1)$-dimensional irreducible representations. In particular, one can construct the regular (or adjoint or associated) representation by means of the basis operators of the Lie algebra itself, which leads in our case to a three-dimensional irreducible representation. This representation is important for considering the inner automorphisms of the group $Sp(2,\mathrm{C})$ which preserve the Killing form and therefore the property to belong to one of the classes (10) in the real case. The regular representation can be mapped onto the fundamental representation by stereographic projection of a two-sheet unit hyperboloid of revolution from the south-pole onto a horizontal plane. The transformations induced by group elements on the hyperboloid are then mapped into fractional linear or Möbius transformations on the complex plane described by the matrices (22). Furthermore, for each representation there exists a corepresentation by transition to the inverse of the adjoint operators which in our case are not equivalent. We do not treat this here.

2.6 Ordering problems of squeezing operators

The derived fundamental representation allows us to solve composition and decomposition problems of squeezing operators. Composition problems can be solved by multiplication of the corresponding matrices in the fundamental representation and by inverse transition from the product matrix to the corresponding squeezing operators. It is very important for applications to make decompositions of the general squeezing operator $S(\xi,\eta,\zeta)$ into ordered products of simpler operators. From (22), one finds the correspondence of special squeezing and rotation operators to unimodular matrices

$$S(\xi,0,0) \rightarrow \left\| \begin{matrix} 1 & \xi \\ 0 & 1 \end{matrix} \right\|, \quad S(0,0,\zeta) \rightarrow \left\| \begin{matrix} 1 & 0 \\ \zeta & 1 \end{matrix} \right\|, \quad S(0,\eta,0) \rightarrow \left\| \begin{matrix} \mathrm{e}^{-\mathrm{i}\eta} & 0 \\ 0 & \mathrm{e}^{\mathrm{i}\eta} \end{matrix} \right\|. \tag{37}$$

By using the unimodularity $\kappa\nu - \lambda\mu = 1$, one can check the following decompositions of the general unimodular representation matrices in products corresponding to the special unimodular ones in (37)

$$\begin{aligned}
\begin{Vmatrix} \kappa & \lambda \\ \mu & \nu \end{Vmatrix} &= \begin{Vmatrix} 1 & 0 \\ \mu/\kappa & 1 \end{Vmatrix} \begin{Vmatrix} 1 & \lambda\kappa \\ 0 & 1 \end{Vmatrix} \begin{Vmatrix} \kappa & 0 \\ 0 & 1/\kappa \end{Vmatrix} \\
&= \begin{Vmatrix} 1 & 0 \\ \mu/\kappa & 1 \end{Vmatrix} \begin{Vmatrix} \kappa & 0 \\ 0 & 1/\kappa \end{Vmatrix} \begin{Vmatrix} 1 & \lambda/\kappa \\ 0 & 1 \end{Vmatrix} \\
&= \begin{Vmatrix} \kappa & 0 \\ 0 & 1/\kappa \end{Vmatrix} \begin{Vmatrix} 1 & 0 \\ \mu\kappa & 1 \end{Vmatrix} \begin{Vmatrix} 1 & \lambda/\kappa \\ 0 & 1 \end{Vmatrix} \\
&= \begin{Vmatrix} 1/\nu & 0 \\ 0 & \nu \end{Vmatrix} \begin{Vmatrix} 1 & \lambda\nu \\ 0 & 1 \end{Vmatrix} \begin{Vmatrix} 1 & 0 \\ \mu/\nu & 1 \end{Vmatrix} \\
&= \begin{Vmatrix} 1 & \lambda/\nu \\ 0 & 1 \end{Vmatrix} \begin{Vmatrix} 1/\nu & 0 \\ 0 & \nu \end{Vmatrix} \begin{Vmatrix} 1 & 0 \\ \mu/\nu & 1 \end{Vmatrix} \\
&= \begin{Vmatrix} 1 & \lambda/\nu \\ 0 & 1 \end{Vmatrix} \begin{Vmatrix} 1 & 0 \\ \mu\nu & 1 \end{Vmatrix} \begin{Vmatrix} 1/\nu & 0 \\ 0 & \nu \end{Vmatrix} . \qquad (38)
\end{aligned}$$

The second and the fifth of these decompositions correspond to the Gauss decomposition of a matrix into a product of three factors with triangular matrices of opposite kind and units in their diagonals to the left and to the right and a diagonal matrix in the centre [49, 54]. The resulting decompositions of the squeezing operators are (particular cases, e.g., [76–79])

$$\begin{aligned}
S(\xi, \eta, \varsigma) &= \exp\Big[-\frac{\mu}{\kappa}K_+\Big] \exp\Big[\lambda\kappa K_-\Big] \exp\Big[-(\log\kappa)\, 2K_0\Big] \\
&= \exp\Big[-\frac{\mu}{\kappa}K_+\Big] \exp\Big[-(\log\kappa)\, 2K_0\Big] \exp\Big[\frac{\lambda}{\kappa}K_-\Big] \\
&= \exp\Big[-(\log\kappa)\, 2K_0\Big] \exp\Big[-\mu\kappa K_+\Big] \exp\Big[\frac{\lambda}{\kappa}K_-\Big] \\
&= \exp\Big[(\log\nu)\, 2K_0\Big] \exp\Big[\lambda\nu K_-\Big] \exp\Big[-\frac{\mu}{\nu}K_+\Big] \\
&= \exp\Big[\frac{\lambda}{\nu}K_-\Big] \exp\Big[(\log\nu)\, 2K_0\Big] \exp\Big[-\frac{\mu}{\nu}K_+\Big] \\
&= \exp\Big[\frac{\lambda}{\nu}K_-\Big] \exp\Big[-\mu\nu K_+\Big] \exp\Big[(\log\nu)\, 2K_0\Big], \qquad (39)
\end{aligned}$$

where the unimodular matrix with the elements $(\kappa, \lambda, \mu, \nu)$ in dependence on (ξ, η, ς) is given in (22). The operators proportional to K_0 in the exponent depend only on the principal ordering of the factors with K_- and K_+ in the exponent and can be represented by

$$\exp\Big[-(\log\kappa)\, 2K_0\Big] = \kappa^{-2K_0}, \quad \exp\Big[(\log\nu)\, 2K_0\Big] = \nu^{2K_0}. \qquad (40)$$

Formulae (39) are not only true for the realization of $sp(2, \mathrm{C})$ according to (1) by boson operators but also for any other realizations, because its derivation uses only the structural relations of the Lie algebra.

Another kind of decomposition is very important for the separation of rotations from special squeezing operations. In analogy to (38), one can make the following decomposition of the general matrix $\hat{X}$ into a product of two special matrices

$$\left\| \begin{matrix} \kappa & \lambda \\ \mu & \nu \end{matrix} \right\| = \left\| \begin{matrix} \sqrt{\kappa\nu} & \lambda\sqrt{\kappa/\nu} \\ \mu\sqrt{\nu/\kappa} & \sqrt{\kappa\nu} \end{matrix} \right\| \left\| \begin{matrix} \sqrt{\kappa/\nu} & 0 \\ 0 & \sqrt{\nu/\kappa} \end{matrix} \right\|$$
$$= \left\| \begin{matrix} \sqrt{\kappa/\nu} & 0 \\ 0 & \sqrt{\nu/\kappa} \end{matrix} \right\| \left\| \begin{matrix} \sqrt{\kappa\nu} & \lambda\sqrt{\nu/\kappa} \\ \mu\sqrt{\kappa/\nu} & \sqrt{\kappa\nu} \end{matrix} \right\| . \tag{41}$$

The diagonal matrix corresponds to a rotation (at least, for $|\nu/\kappa| = 1$)), whereas the other matrix is of a special form belonging to $\eta = 0$. By using this, one obtains from these relations together with (38) the following decomposition of the general squeezing operator into products of partial operators

$$S(\xi,\eta,\zeta) = \exp\left[\frac{\mathrm{Arsh}\sqrt{\lambda\mu}}{\sqrt{\lambda\mu}}\left(\lambda\sqrt{\frac{\kappa}{\nu}}K_- - \mu\sqrt{\frac{\nu}{\kappa}}K_+\right)\right]\left(\frac{\nu}{\kappa}\right)^{K_0}$$
$$= \left(\frac{\nu}{\kappa}\right)^{K_0}\exp\left[\frac{\mathrm{Arsh}\sqrt{\lambda\mu}}{\sqrt{\lambda\mu}}\left(\lambda\sqrt{\frac{\nu}{\kappa}}K_- - \mu\sqrt{\frac{\kappa}{\nu}}K_+\right)\right]. \tag{42}$$

In the case of the general unitary squeezing operator $S(\zeta^*, \eta = \eta^*, \zeta)$, we have $\kappa = \nu^*$ and $\lambda = \mu^*$ and $\nu/\kappa = \nu/\nu^*$ is a phase factor which we can set equal to $\mathrm{e}^{\mathrm{i}2\varphi}$. In this way it follows from (42)

$$S(\zeta^*, \eta{=}\eta^*, \zeta) = \exp\left[\frac{\mathrm{Arsh}\,(|\zeta|\mathrm{sh}\,\varepsilon/\varepsilon)}{|\zeta|}\left(\zeta^*\mathrm{e}^{-\mathrm{i}\varphi}K_- - \zeta\mathrm{e}^{\mathrm{i}\varphi}K_+\right)\right]\mathrm{e}^{\mathrm{i}2\varphi K_0}$$
$$= \mathrm{e}^{\mathrm{i}2\varphi K_0}\exp\left[\frac{\mathrm{Arsh}\,(|\zeta|\mathrm{sh}\,\varepsilon/\varepsilon)}{|\zeta|}\left(\zeta^*\mathrm{e}^{\mathrm{i}\varphi}K_- - \zeta\mathrm{e}^{-\mathrm{i}\varphi}K_+\right)\right], \tag{43}$$

$$\varepsilon \equiv \sqrt{|\zeta|^2 - \eta^2}, \quad \mathrm{e}^{\mathrm{i}2\varphi} \equiv \frac{\nu}{\nu^*} = \frac{\varepsilon\,\mathrm{ch}\varepsilon + \mathrm{i}\eta\,\mathrm{sh}\varepsilon}{\varepsilon\,\mathrm{ch}\varepsilon - \mathrm{i}\eta\,\mathrm{sh}\varepsilon}.$$

Relation (43) gives the decomposition of the general unitary squeezing operator $S(\zeta^*, \eta = \eta^*, \zeta)$ into the product of a special unitary squeezing operator $S(\zeta'^*, 0, \zeta')$ and a rotation operator $S(0, \eta' = \eta'^*, 0)$ in the two possible orderings. The rotation operator is a stable part of this decomposition, whereas the special squeezing operator depends on the ordering.

We now consider the decompositions in the particular case $\eta = 0$. From (39) in connection with (22), one finds, by inserting the explicit form of the matrix elements $(\kappa, \lambda, \mu, \nu)$, a decomposition (cf. [4])

$$\exp\left[\frac{\mathrm{Arth}\sqrt{\xi\zeta}}{\sqrt{\xi\zeta}}(\xi K_- - \zeta K_+)\right]$$
$$= \exp(-\zeta K_+)\exp\left(\frac{\xi}{1-\xi\zeta}K_-\right)(1-\xi\zeta)^{K_0}$$
$$= \exp(-\zeta K_+)(1-\xi\zeta)^{K_0}\exp(\xi K_-)$$

$$
\begin{aligned}
&= (1-\xi\zeta)^{K_0} \exp\left(-\frac{\zeta}{1-\xi\zeta}K_+\right) \exp(\xi K_-) \\
&= (1-\xi\zeta)^{-K_0} \exp\left(\frac{\xi}{1-\xi\zeta}K_-\right) \exp(-\zeta K_+) \\
&= \exp(\xi K_-)\,(1-\xi\zeta)^{-K_0} \exp(-\zeta K_+) \\
&= \exp(\xi K_-) \exp\left(-\frac{\zeta}{1-\xi\zeta}K_+\right) (1-\xi\zeta)^{-K_0} . \qquad (44)
\end{aligned}
$$

Relations (43) and (44) form the basis for using different parameters and for the distinction of the unitary and the nonunitary approaches to squeezing.

3 Representations of squeezed states

3.1 Squeezed coherent states in unitary approach

We define in this section squeezed coherent states in two different approaches, the unitary and the nonunitary. The unitary approach provides these states in the normalized form, but not with the most convenient squeezing parameter. Furthermore, there are two different possibilities of representation with regard to the order of displacement and squeezing operators. The nonunitary approach provides the squeezed coherent states in an analytic form of both involved complex variables of squeezing and displacement and allows us to introduce the most convenient squeezing parameter in a natural way. It is related to $SU(1,1)$ coherent states and to the extension of $SU(1,1)$ by the Heisenberg–Weyl group $W(1,\mathrm{R})$.

Coherent states $|\alpha\rangle$ are defined as displaced vacuum states $|0\rangle$ [6,7]. Squeezed coherent states are obtained by applying squeezing and displacement operations to the vacuum state. Since displacement and squeezing are noncommutative operations, one can define them in two different orderings of these operations. For the expression of the properties of squeezed coherent states (e.g., moments) it is more favourable to make first the squeezing of the vacuum state and then the displacement of the obtained squeezed vacuum state. In this ordering, we introduce our basic notation of squeezed coherent states in the unitary approach according to

$$
\begin{aligned}
|\bar{a},\zeta\rangle &\equiv D(\bar{a},\bar{a}^*)S(\zeta'^*,0,\zeta')|0\rangle \\
&= \exp\left(\bar{a}a^\dagger - \bar{a}^*a\right) \exp\left[\frac{\mathrm{Arth}|\zeta|}{|\zeta|}\left(\frac{\zeta^*}{2}a^2 - \frac{\zeta}{2}a^{\dagger 2}\right)\right]|0\rangle, \qquad (45)
\end{aligned}
$$

where $D(\alpha,\alpha^*) \equiv \exp(\alpha a^\dagger - \alpha^* a)$ is the unitary displacement operator. We use here $(\bar{a},\bar{a}^*)$ instead of (α,α^*) as a pair of complex conjugated parameters. This has the advantage that it is near to the notation $(\overline{a},\overline{a^*})$ of the expectation values of the operators $(a,a^\dagger)$ which in this order of squeezing and displacement become equal for squeezed coherent states and, therefore, possess a clear physical meaning, and it makes (α,α^*) free for the quasiprobabilities considered later. Furthermore, we emphasize that in our notation of the states, we do not use the complex squeezing parameter ζ' which appears basically in the notation of the applied unitary squeezing operator $S(\zeta'^*, 0, \zeta')$ but a complex parameter ζ which is related to ζ' in the following way (cf., e.g., [11–14, 17, 80–82])

$$\zeta = \zeta' \frac{\mathrm{th}|\zeta'|}{|\zeta'|}, \quad \zeta' = \zeta \frac{\mathrm{Arth}|\zeta|}{|\zeta|}, \quad |\zeta| = \mathrm{th}|\zeta'|, \quad |\zeta'| = \mathrm{Arth}|\zeta|,$$
$$\frac{\zeta}{|\zeta|} = \frac{\zeta'}{|\zeta'|}, \quad \frac{1}{\sqrt{1-|\zeta|^2}} = \mathrm{ch}|\zeta'|, \quad \frac{\zeta}{\sqrt{1-|\zeta|^2}} = \frac{\zeta'}{|\zeta'|}\mathrm{sh}|\zeta'|,$$
$$\sqrt{\frac{1+|\zeta|}{1-|\zeta|}} = \exp(|\zeta'|), \qquad 0 \le |\zeta| < 1, \quad 0 \le |\zeta'| < \infty. \tag{46}$$

The reason for this is that almost all formulae for the properties of squeezed coherent states take on a simpler form by using the squeezing parameter ζ in comparison to ζ'. The matrix of the fundamental representation (22), corresponding to the special unitary squeezing operators $S(\zeta'^*, 0, \zeta')$, takes on the following equivalent forms

$$\left\| \begin{matrix} \nu^* & \mu^* \\ \mu & \nu \end{matrix} \right\| = \left\| \begin{matrix} \mathrm{ch}|\zeta'| & \zeta'^* \frac{\mathrm{sh}|\zeta'|}{|\zeta'|} \\ \zeta' \frac{\mathrm{sh}|\zeta'|}{|\zeta'|} & \mathrm{ch}|\zeta'| \end{matrix} \right\| = \frac{1}{\sqrt{1-|\zeta|^2}} \left\| \begin{matrix} 1 & \zeta^* \\ \zeta & 1 \end{matrix} \right\|, \tag{47}$$

and, according to (44), $S(\zeta'^*, 0, \zeta')$ has the following almost normally ordered decomposition (the central factor is not yet normally ordered)

$$\begin{aligned} S(\zeta'^*, 0, \zeta') &\equiv \exp\left[\frac{\mathrm{Arth}|\zeta|}{|\zeta|}\left(\frac{\zeta^*}{2}a^2 - \frac{\zeta}{2}a^{\dagger 2}\right)\right] \\ &= \exp\left(-\frac{\zeta}{2}a^{\dagger 2}\right)\left(\sqrt{1-|\zeta|^2}\right)^{\frac{1}{2}(aa^\dagger + a^\dagger a)} \exp\left(\frac{\zeta^*}{2}a^2\right) \\ &= S(0,0,\zeta)S\left(0, -\mathrm{i}\log\sqrt{1-|\zeta|^2}, 0\right)S(\zeta^*, 0, 0). \end{aligned} \tag{48}$$

This relation shows that with the parameter ζ, one has a simple form of the normally ordered expression for the squeezing operator that becomes more complicated by using ζ' in this ordering. This is the main advantage of using the parameter ζ. If necessary, one can make the transition to the squeezing parameter ζ' by using (46) or (47). The complex squeezing parameter ζ' (or sometimes $-\zeta'$) is often written by real variables as $r\mathrm{e}^{\mathrm{i}\Theta}$ or similar and the exact correspondences can be established in each case by comparison of equivalent formulae and by using (46). The complex squeezing parameter ζ is restricted for normalizable states to the interior of the unit circle, whereas ζ' does not possess such a restriction and stretches the regions of ζ for $|\zeta|$ in the neighbourhood of 1 to the infinity of ζ'.

The squeezing operator $S(\zeta'^*, 0, \zeta')$ in the definition (45) is the stable part when changing the order of the operations of squeezing and displacement. Due to

$$S^\dagger(\zeta'^*, 0, \zeta')D(\bar{a}, \bar{a}^*)S(\zeta'^*, 0, \zeta') = D\left(\frac{\bar{a} + \zeta\bar{a}^*}{\sqrt{1-|\zeta|^2}}, \frac{\bar{a}^* + \zeta^*\bar{a}}{\sqrt{1-|\zeta|^2}}\right), \tag{49}$$

following from (24) with the substitutions $\xi \rightarrow \zeta'^*, \eta \rightarrow 0, \zeta \rightarrow \zeta'$ and with $\nu = \nu^*$, one can write the definition (45) in two equivalent forms

$$|\bar{a}, \zeta\rangle = S(\zeta'^*, 0, \zeta') D\left(\frac{\bar{a} + \zeta\bar{a}^*}{\sqrt{1-|\zeta|^2}}, \frac{\bar{a}^* + \zeta^*\bar{a}}{\sqrt{1-|\zeta|^2}}\right)|0\rangle, \tag{50}$$

$$\left|\frac{\gamma - \zeta\gamma^*}{\sqrt{1-|\zeta|^2}}, \zeta\right\rangle = S(\zeta'^*, 0, \zeta') D(\gamma, \gamma^*)|0\rangle \equiv |\zeta, \gamma\rangle. \tag{51}$$

This is connected with the following relations and their inversion

$$\gamma = \frac{\bar{a} + \zeta\bar{a}^*}{\sqrt{1-|\zeta|^2}}, \quad \bar{a} = \frac{\gamma - \zeta\gamma^*}{\sqrt{1-|\zeta|^2}}. \tag{52}$$

Relations (50)–(52) clarify the transformations of the displacement parameters under changing the order of squeezing and displacement.

The special case of vanishing displacement parameters $(\bar{a}, \bar{a}^*)$ in (45) leads to one of the basic definitions of squeezed vacuum states

$$|0, \zeta\rangle \equiv S(\zeta'^*, 0, \zeta')|0\rangle \equiv \exp\left[\frac{\mathrm{Arth}|\zeta|}{|\zeta|}\left(\frac{\zeta^*}{2}a^2 - \frac{\zeta}{2}a^{\dagger 2}\right)\right]|0\rangle. \tag{53}$$

The special case of vanishing squeezing parameters (ζ, ζ^*) leads to coherent states in the new notation

$$|\bar{a}, 0\rangle \equiv D(\bar{a}, \bar{a}^*)|0\rangle = \exp\left(\bar{a}a^\dagger - \bar{a}^*a\right)|0\rangle \equiv |\bar{a}\rangle. \tag{54}$$

The forms (50) or (51) justify the name squeezed coherent states for the considered states $|\bar{a}, \zeta\rangle$ because they are obtained there by the squeezing of coherent states $|\gamma\rangle$. The primary definition (45) implies the equivalent name of displaced squeezed vacuum states because they are obtained there by the displacement of squeezed vacuum states $|0, \zeta\rangle$.

3.2 Squeezed coherent states in nonunitary approach

We now consider the nonunitary approach to squeezed coherent states. For this purpose, we use the ordered decomposition (48) of the unitary squeezing operator $S(\zeta'^*, 0, \zeta')$ into a product of 3 nonunitary squeezing operators and apply this to the vacuum state $|0\rangle$

$$\begin{aligned} |0, \zeta\rangle &= \exp\left(-\frac{\zeta}{2}a^{\dagger 2}\right)\left(\sqrt{1-|\zeta|^2}\right)^{\frac{1}{2}(aa^\dagger + a^\dagger a)} \exp\left(\frac{\zeta^*}{2}a^2\right)|0\rangle \\ &= \left(1-|\zeta|^2\right)^{\frac{1}{4}} \exp\left(-\frac{\zeta}{2}a^{\dagger 2}\right)|0\rangle. \end{aligned} \tag{55}$$

In application to $|0\rangle$, the right factor $S(\zeta^*, 0, 0)$ leaves it unchanged, whereas the central factor leads to the multiplication with a normalization factor. This means that the squeezed vacuum states $|0, \zeta\rangle$ can be obtained by application of the nonunitary operator $S(0,0,\zeta)$ to the vacuum state $|0\rangle$ and by multiplication with a normalization factor. We call this the nonunitary approach to squeezed vacuum states.

Now we apply the displacement operator $D(\bar{a},\bar{a}^*)$ to the squeezed vacuum states in the form (55). By using the commutation relation

$$\exp\left(\frac{\zeta}{2}a^{\dagger 2}\right)(a,a^\dagger)\exp\left(-\frac{\zeta}{2}a^{\dagger 2}\right) = (a-\zeta a^\dagger, a^\dagger), \tag{56}$$

we find the commutation with the displacement operator in the form

$$\exp\left(\frac{\zeta}{2}a^{\dagger 2}\right)D(\bar{a},\bar{a}^*)\exp\left(-\frac{\zeta}{2}a^{\dagger 2}\right) = D(\bar{a}+\zeta\bar{a}^*,\bar{a}^*). \tag{57}$$

The operator $D(\bar{a}+\zeta\bar{a}^*,\bar{a}^*)$ on the right-hand side is a nonunitary operator which can be represented in normal ordering as follows

$$D(\bar{a}+\zeta\bar{a}^*,\bar{a}^*) = \exp\left(-\frac{(\bar{a}+\zeta\bar{a}^*)\bar{a}^*}{2}\right)D(\bar{a}+\zeta\bar{a}^*,0)D(0,\bar{a}^*). \tag{58}$$

If we now apply the displacement operator $D(\bar{a},\bar{a}^*)$ to squeezed vacuum states in the form (55), then by using the commutation (57) and the decomposition (58), we arrive at the following representation of squeezed coherent states

$$|\bar{a},\zeta\rangle = \left[1-|\zeta|^2\right]^{\frac{1}{4}}\exp\left[-\frac{(\bar{a}+\zeta\bar{a}^*)\bar{a}^*}{2}\right]\exp\left[(\bar{a}+\zeta\bar{a}^*)\,a^\dagger - \frac{\zeta}{2}a^{\dagger 2}\right]|0\rangle. \tag{59}$$

This can also be written in the form

$$\left|\frac{\beta-\zeta\beta^*}{1-|\zeta|^2},\zeta\right\rangle = \left[1-|\zeta|^2\right]^{\frac{1}{4}}\exp\left[-\frac{\beta(\beta^*-\zeta^*\beta)}{2(1-|\zeta|^2}\right]\exp\left[\beta a^\dagger - \frac{\zeta}{2}a^{\dagger 2}\right]|0\rangle, \tag{60}$$

with the following pair of transformation and inverse transformation

$$\beta = \bar{a}+\zeta\bar{a}^*, \quad \bar{a} = \frac{\beta-\zeta\beta^*}{1-|\zeta|^2}\,. \tag{61}$$

The relations between the parameters (β,β^*) and (γ,γ^*) in the representation (51) of squeezed coherent states (where first the displacement operator $D(\gamma,\gamma^*)$ is applied to the vacuum state and then the squeezing operator $S(\zeta'^*,0,\zeta')$ to the result) are

$$\beta = \sqrt{1-|\zeta|^2}\,\gamma, \quad \beta^* = \sqrt{1-|\zeta|^2}\,\gamma^*. \tag{62}$$

The operator $\exp\left(\beta a^\dagger - (\zeta/2)a^{\dagger 2}\right)$ is a nonunitary operator and the factor in front of this operator in (59) or (60) gives the normalization factor of squeezed coherent states. We call this the nonunitary approach to the definition of squeezed coherent states [72, 83, 84]. It is favourable to introduce a special symbol for the result of the action of the operator $\exp[\beta a^\dagger - (\zeta/2)a^{\dagger 2}]$ onto the vacuum state $|0\rangle$ as follows

$$|(\beta,\zeta)\rangle \equiv \exp\left[\beta a^\dagger - \frac{\zeta}{2}a^{\dagger 2}\right]|0\rangle, \quad \langle(\beta^*,\zeta^*)| \equiv \langle 0|\exp\left[\beta^* a - \frac{\zeta^*}{2}a^2\right]. \tag{63}$$

The states $|(\beta,\zeta)\rangle$ are for $|\zeta| < 1$ squeezed coherent states in a nonnormalized but analytic form in the two complex variables (β,ζ) that is shown in our notation by including them within the circular brackets. The adjoint states $\langle(\beta^*,\zeta^*)|$ are nonnormalized costates (or 'bra's) which depend analytically on the variables (β^*,ζ^*) that are (deviating from convention in Dirac's notation) explicitly shown in our notation.

The nonunitary approach has some advantages as an alternative to the unitary one. The only operator $\exp\left(\beta a^\dagger - (\zeta/2)a^{\dagger 2}\right)$ which is involved in the nonunitary approach can be decomposed into two commuting factors

$$\exp\left(\beta a^\dagger - \frac{\zeta}{2}a^{\dagger 2}\right) = D(\beta,0)S(0,0,\zeta) = S(0,0,\zeta)D(\beta,0), \qquad (64)$$

and it provides in a natural way the complex squeezing parameter ζ which we consider as the most appropriate one for representations of formulae connected with squeezed states. The operators $D(\beta,0)S(0,0,\zeta)$ form a complex Abelian 2-parameter group with the additive complex parameters β and ζ.

The squeezing parameter ζ can be extended in relations of the nonunitary approach to values with $|\zeta| \geq 1$. With regard to normalizability, we distinguish three classes of states [85]:

1. *Normalizable states*

$$|\zeta| < 1. \qquad (65)$$

The states $|(\beta,\zeta)\rangle$ are regular states of the Hilbert space $\mathcal{H}$.

2. *Weakly nonnormalizable states*

$$|\zeta| = 1. \qquad (66)$$

The states $|(\beta,\zeta)\rangle$ do not belong to the Hilbert space $\mathcal{H}$ but are states of a *rigged* Hilbert space $\mathcal{K}'$. This is the space of (anti-)linear functionals over some narrower space $\mathcal{K}$ than the Hilbert space $\mathcal{H}$ that leads to a wider space of (anti-)linear functionals $\mathcal{K}'$ than $\mathcal{H}' = \mathcal{H}$ expressed by the enclosure relations $\mathcal{K} \subset \mathcal{H} = \mathcal{H}' \subset \mathcal{K}'$ (Gel'fand triplets of spaces [86]). The states of this class can also be considered as limiting cases of sequences of normalizable states with $|\zeta| < 1$.

3. *Strongly nonnormalizable states*

$$|\zeta| > 1. \qquad (67)$$

The states $|(\beta,\zeta)\rangle$ belong to more general rigged Hilbert spaces as in the case of $|\zeta| = 1$. Such states play an important role for the formulation of new kinds of orthogonality and completeness relations [72, 87–89]. The extension $|\zeta| \to \infty$ leads, after some renormalization, to eigenstates of the boson creation operator $a^\dagger$. We do not treat this subject here.

3.3 Eigenvalue equations to squeezed coherent states

One of the important properties of squeezed coherent states $|(\beta,\zeta)\rangle$ is to satisfy the eigenvalue equation

$$(a + \zeta a^\dagger)|(\beta,\zeta)\rangle = \beta|(\beta,\zeta)\rangle, \quad (\beta,\zeta \in C). \qquad (68)$$

This follows immediately from the operator identity

$$\exp\left(-\beta a^\dagger + \frac{\zeta}{2}a^{\dagger 2}\right)(a + \zeta a^\dagger)\exp\left(\beta a^\dagger - \frac{\zeta}{2}a^{\dagger 2}\right) = a + \beta I, \tag{69}$$

by applying it to the vacuum state $|0\rangle$

$$(a + \zeta a^\dagger)\exp\left(\beta a^\dagger - \frac{\zeta}{2}a^{\dagger 2}\right)|0\rangle = \beta\exp\left(\beta a^\dagger - \frac{\zeta}{2}a^{\dagger 2}\right)|0\rangle, \tag{70}$$

and may be used as a possible starting point to the introduction of squeezed coherent states $|(\beta,\zeta)\rangle$ in the nonunitary approach [72].

There is another operator to which the states $|(\beta,\zeta)\rangle$ are eigenstates, however, to eigenvalues ζ. This is for squeezed vacuum states $|(0,\zeta)\rangle$ the operator $-(N+I)^{-1}a^2$ [71] and it is connected with the property of $|(0,\zeta)\rangle$ to be $SU(1,1)$-coherent states in the sense of Perelomov [54,90] (contrary to that of Barut and Girardello [91]) and can be generalized to $|(\beta,\zeta)\rangle$ as follows

$$-(N+I)^{-1}\,a\,(a-\beta I)\,|(\beta,\zeta)\rangle = \zeta|(\beta,\zeta)\rangle, \tag{71}$$

corresponding to group-coherent states in the unification of $SU(1,1)$ with the Heisenberg–Weyl group $W(1,\mathrm{R})$.

To obtain the eigenvalue equation in the unitary approach, we start from the well-known property of coherent states $|\gamma\rangle$ to be eigenstates of the annihilation operator a to eigenvalues γ and apply the general unitary squeezing transformation (24) of the annihilation operator

$$\begin{aligned}(\nu^* a + \mu a^\dagger)\,S(\zeta^*,\eta{=}\eta^*,\zeta)D(\gamma,\gamma^*)|0\rangle \;&=\; S(\zeta^*,\eta{=}\eta^*,\zeta)a|\gamma\rangle \\ = \gamma S(\zeta^*,\eta{=}\eta^*,\zeta)D(\gamma,\gamma^*)|0\rangle.&\end{aligned} \tag{72}$$

Now, by using the special unitary squeezing operator $S(\zeta'^*,0,\zeta')$ with special (μ,ν) given in (47) and by bringing the factor $\sqrt{1-|\zeta|^2}$ to the right-hand side, one obtains (compare (52) and (61))

$$(a+\zeta a^\dagger)\,|\bar{a},\zeta\rangle = \sqrt{1-|\zeta|^2}\,\gamma|\bar{a},\zeta\rangle = (\bar{a}+\zeta\bar{a}^*)\,|\bar{a},\zeta\rangle. \tag{73}$$

This eigenvalue problem for linear combinations of boson annihilation and creation operators can serve as a starting point for the introduction of squeezed coherent states [13,72].

In the form $|(\beta,\zeta)\rangle$, the squeezed coherent states are analytic functions of both complex variables β and ζ, which means

$$\frac{\partial}{\partial\beta^*}|(\beta,\zeta)\rangle = \frac{\partial}{\partial\zeta^*}|(\beta,\zeta)\rangle = \frac{\partial}{\partial\beta}\langle(\beta^*,\zeta^*)| = \frac{\partial}{\partial\zeta}\langle(\beta^*,\zeta^*)| = 0. \tag{74}$$

This is favourable for the treatment of some analytic problems, for example, of contour and path representations of states by coherent or squeezed states [89]. The derivatives of the states with regard to β and ζ are

$$\frac{\partial}{\partial\beta}|(\beta,\zeta)\rangle = a^\dagger|(\beta,\zeta)\rangle, \quad \frac{\partial}{\partial\zeta}|(\beta,\zeta)\rangle = -\frac{1}{2}a^{\dagger 2}|(\beta,\zeta)\rangle. \tag{75}$$

Therefore, the action of the boson annihilation and creation operator onto the states $|(\beta,\varsigma)\rangle$ can be substituted in connection with the eigenvalue equation (68) by the following differentiation and multiplication operators

$$a^{\dagger}|(\beta,\varsigma)\rangle = \frac{\partial}{\partial\beta}|(\beta,\varsigma)\rangle, \quad a|(\beta,\varsigma)\rangle = \left(\beta - \varsigma\frac{\partial}{\partial\beta}\right)|(\beta,\varsigma)\rangle. \tag{76}$$

From (75) together with (63), one finds the relation between analytic squeezed and coherent states

$$|(\beta,\varsigma)\rangle = \exp\left(-\frac{\varsigma}{2}\frac{\partial^2}{\partial\beta^2}\right)|(\beta,0)\rangle, \tag{77}$$

and therefore

$$\begin{aligned} |(\beta,\varsigma)\rangle\langle(\beta^*,\varsigma^*)| &= \exp\left(-\frac{\varsigma}{2}\frac{\partial^2}{\partial\beta^2} - \frac{\varsigma^*}{2}\frac{\partial^2}{\partial\beta^{*2}}\right)|(\beta,0)\rangle\langle(\beta^*,0)| \\ &= \frac{1}{\pi\sqrt{-\varsigma\varsigma^*}}\exp\left(\frac{\beta^2}{2\varsigma} + \frac{\beta^{*2}}{2\varsigma^*}\right) * |(\beta,0)\rangle\langle(\beta^*,0)|. \end{aligned} \tag{78}$$

This is a two-dimensional convolution (notation $'*'$) in the sense of generalized functions. Since the analytic form $|(\beta,\varsigma)\rangle$ is nonnormalized, the expectation value $\overline{A}$ of an arbitrary operator A has to be calculated as

$$\overline{A} = \frac{\langle(\beta^*,\varsigma^*)|A|(\beta,\varsigma)\rangle}{\langle(\beta^*,\varsigma^*)|(\beta,\varsigma)\rangle}. \tag{79}$$

By using (74)–(76), one finds the expectation values of normally and antinormally ordered powers of the annihilation and creation operators:

$$\begin{aligned} \overline{a^{\dagger k}a^l} &= \frac{1}{\langle(\beta^*,\varsigma^*)|(\beta,\varsigma)\rangle}\left(\beta^* - \varsigma^*\frac{\partial}{\partial\beta^*}\right)^k\left(\beta - \varsigma\frac{\partial}{\partial\beta}\right)^l\langle(\beta^*,\varsigma^*)|(\beta,\varsigma)\rangle, \\ \overline{a^l a^{\dagger k}} &= \frac{1}{\langle(\beta^*,\varsigma^*)|(\beta,\varsigma)\rangle}\frac{\partial^{k+l}}{\partial\beta^k\partial\beta^{*l}}\langle(\beta^*,\varsigma^*)|(\beta,\varsigma)\rangle. \end{aligned} \tag{80}$$

The explicit form of $\langle(\beta^*,\varsigma^*)|(\beta,\varsigma)\rangle$, which we later derive, proves to be sufficiently simple to carry out the necessary differentiations.

The expectation values of the operators a and $a^{\dagger}$ can be calculated in the following simple way. One forms the scalar product from the eigenvalue equation (45) with $\langle(\beta^*,\varsigma^*)|$ and by a following division of the obtained equation by $\langle(\beta^*,\varsigma^*)|(\beta,\varsigma)\rangle$, one finds

$$\beta = \overline{a} + \varsigma\,\overline{a}^*, \qquad \beta^* = \overline{a}^* + \varsigma^*\,\overline{a}. \tag{81}$$

From these equations one obtains by inversion

$$\overline{a} = \frac{\beta - \varsigma\beta^*}{1 - \varsigma\varsigma^*} = \bar{a}, \qquad \overline{a}^* = \frac{\beta^* - \varsigma^*\beta}{1 - \varsigma\varsigma^*} = \overline{a^{\dagger}} = \bar{a}^*. \tag{82}$$

These relations show that the mean value $\overline{a}$ is expressed by a linear combination of the parameters β and β^* but depends also on the complex squeezing parameter ς and only for $\varsigma = 0$ are both equal. The expectation value of a is identical with the displacement parameter $\bar{a}$, introduced earlier for squeezed coherent states in the unitary approach.

3.4 Fock-state expansion, scalar products and distances

The Fock-state (or number) representation of the states $|(\beta,\zeta)\rangle$ can be obtained from the generating function of the Hermite polynomials by applying it to the exponential operator function in (63)

$$\begin{aligned} |(\beta,\zeta)\rangle &= \sum_{n=0}^{\infty} H_n\left(\frac{\beta}{\sqrt{2\zeta}}\right)\frac{1}{n!}\left(\frac{\sqrt{2\zeta}}{2}\,a^\dagger\right)^n|0\rangle \\ &= \sum_{n=0}^{\infty}\frac{\left(\sqrt{2\zeta}\,\right)^n}{2^n\sqrt{n!}}H_n\left(\frac{\beta}{\sqrt{2\zeta}}\right)|n\rangle, \end{aligned} \tag{83}$$

where the well-known generation of Fock states from the vacuum states is used. The sign of the complex roots $\sqrt{2\zeta}$ in this expression may be chosen arbitrarily but it has to be the same in the argument of the Hermite polynomials and in the corresponding power functions. In a more explicit form this number representation can be written

$$|(\beta,\zeta)\rangle = \sum_{n=0}^{\infty}\left[\sum_{k=0}^{[n/2]}\frac{(-1)^k n!}{k!(n-2k)!}\left(\frac{\zeta}{2\beta^2}\right)^k\right]\frac{\beta^n}{\sqrt{n!}}|n\rangle. \tag{84}$$

The ambiguity of $\sqrt{2\zeta}$ in (83) disappeared in this representation. The factor in braces shows the modification of the number representation of $|(\beta,\zeta)\rangle$ in comparison to analytic coherent states $|(\beta,0)\rangle$. In this special case and for analytic squeezed vacuum states, one finds

$$|(\beta,0)\rangle = \sum_{n=0}^{\infty}\frac{\beta^n}{\sqrt{n!}}|n\rangle, \quad |(0,\zeta)\rangle = \sum_{m=0}^{\infty}\frac{(-1)^m\sqrt{(2m)!}}{2^m m!}\zeta^m|2m\rangle. \tag{85}$$

Only the even Fock states $|2m\rangle$ are occupied in squeezed vacuum states.

The Fock-state expansion of the squeezed coherent states $|\bar{a},\zeta\rangle$ in the unitary approach can be obtained from (83) in connection with (59) and (63) and possesses the form (cf., e.g., [13])

$$\begin{aligned} |\bar{a},\zeta\rangle &= (1-|\zeta|^2)^{\frac{1}{4}}\exp\left(-\frac{(\bar{a}+\zeta\bar{a}^*)\,\bar{a}^*}{2}\right) \\ &\quad\times\sum_{n=0}^{\infty}\frac{\left(\sqrt{2\zeta}\,\right)^n}{2^n\sqrt{n!}}H_n\left(\frac{\bar{a}+\zeta\bar{a}^*}{\sqrt{2\zeta}}\right)|n\rangle. \end{aligned} \tag{86}$$

The resulting photon statistics $p_n \equiv \langle n|\bar{a},\zeta\rangle\,\langle\bar{a},\zeta|n\rangle$, $(\sum_{n=0}^{\infty}p_n = 1)$, in dependence on the complex displacement parameter $\bar{a}$ and on the complex squeezing parameter ζ, is

$$\begin{aligned} p_n &= \sqrt{1-|\zeta|^2}\exp\left[-\left(\bar{a}\bar{a}^* + \frac{\zeta^*}{2}\bar{a}^2 + \frac{\zeta}{2}\bar{a}^{*2}\right)\right] \\ &\quad\times\frac{|\zeta|^n}{2^n n!}H_n\left(\frac{\bar{a}+\zeta\bar{a}^*}{\sqrt{2\zeta}}\right)H_n\left(\frac{\bar{a}^*+\zeta^*\bar{a}}{\sqrt{2\zeta^*}}\right). \end{aligned} \tag{87}$$

Due to the presence of the Hermite polynomials which possess zeros for real values of the argument, this photon statistics in dependence on n can show oscillatory character [92–94], in particular, if the argument of the Hermite polynomials is real. These oscillations are easily visible if many zeros coincide with regions of n where the envelope of the distribution is not nearly vanishing. The argument of the Hermite polynomials in (87) becomes real for real $\bar{a}/\sqrt{\zeta}$ which corresponds to amplitude squeezing as discussed later. For imaginary $\bar{a}/\sqrt{\zeta}$ which corresponds to phase squeezing, the argument of the Hermite polynomials becomes imaginary and this case is far from showing oscillations in the photon distribution.

The states $|(\beta,\zeta)\rangle$ are nonorthogonal to each other for arbitrary values of the parameters β and ζ. The scalar product of two such states can be calculated by using the following formula of Mehler [95] (formula (10.13.22) there)

$$\sum_{n=0}^{\infty} \frac{t^n}{2^n n!} H_n(x) H_n(y) = \frac{1}{\sqrt{1-t^2}} \exp\left[\frac{2xyt - (x^2+y^2)t^2}{1-t^2}\right]. \tag{88}$$

Hence from the number representation of the states in (62), one obtains for the scalar product

$$\begin{aligned}\langle(\beta_2^*,\zeta_2^*)|(\beta_1,\zeta_1)\rangle &= \frac{1}{\sqrt{1-\zeta_1\zeta_2^*}} \exp\left[\frac{2\beta_1\beta_2^* - (\zeta_2^*\beta_1^2 + \zeta_1\beta_2^{*2})}{2(1-\zeta_1\zeta_2^*)}\right], \\ \langle(\beta^*,\zeta^*)|(\beta,\zeta)\rangle &= \frac{1}{\sqrt{1-\zeta\zeta^*}} \exp\left[\frac{2\beta\beta^* - (\zeta^*\beta^2 + \zeta\beta^{*2})}{2(1-\zeta\zeta^*)}\right].\end{aligned} \tag{89}$$

This shows that the squeezed states are normalizable only for $|\zeta| < 1$. The square root of $\langle(\beta^*,\zeta^*)|(\beta,\zeta)\rangle$ gives a possible normalization factor but relation (89) shows that the connection to the states $|\bar{a},\zeta\rangle$ involves an additional phase factor according to

$$\begin{aligned}\left|\frac{\beta-\zeta\beta^*}{1-|\zeta|^2},\zeta\right\rangle &= \left(1-|\zeta|^2\right)^{1/4} \exp\left(-\frac{\beta(\beta^*-\zeta^*\beta)}{2(1-|\zeta|^2)}\right)|(\beta,\zeta)\rangle \\ &= \exp\left(\frac{\zeta^*\beta^2 - \zeta\beta^{*2}}{4(1-|\zeta|^2)}\right)\frac{|(\beta,\zeta)\rangle}{\sqrt{\langle(\beta^*,\zeta^*)|(\beta,\zeta)\rangle}},\end{aligned} \tag{90}$$

or in variables $(\bar{a},\bar{a}^*)$

$$\begin{aligned}|\bar{a},\zeta\rangle &= \left(1-|\zeta|^2\right)^{1/4} \exp\left(-\frac{(\bar{a}+\zeta\bar{a}^*)\bar{a}^*}{2}\right)|(\bar{a}+\zeta\bar{a}^*,\zeta)\rangle \\ &= \exp\left(\frac{\zeta^*\bar{a}^2 - \zeta\bar{a}^{*2}}{4}\right)\frac{|(\bar{a}+\zeta\bar{a}^*,\zeta)\rangle}{\sqrt{\langle(\bar{a}^*+\zeta^*\bar{a},\zeta^*)|(\bar{a}+\zeta\bar{a}^*,\zeta)\rangle}}.\end{aligned} \tag{91}$$

Relations (90) and (91) are important for the transition from the nonunitary to the unitary approach to squeezed coherent states.

The scalar product of two squeezed coherent states $|\bar{a}_1, \zeta_1\rangle$ and $|\bar{a}_2, \zeta_2\rangle$ can be obtained from (90) and (91) in the following form:

$$\langle \bar{a}_2, \zeta_2 | \bar{a}_1, \zeta_1 \rangle = \left(\frac{(1 - |\zeta_1|^2)\,(1 - |\zeta_2|^2)}{(1 - \zeta_1 \zeta_2^*)^2} \right)^{1/4}$$
$$\times \exp\left\{ -\frac{1}{2(1 - \zeta_1 \zeta_2^*)} \Big((\bar{a}_1 + \zeta_1 \bar{a}_1^*)(\bar{a}_1^* + \zeta_2^* \bar{a}_1) \right.$$
$$\left. -2(\bar{a}_1 + \zeta_1 \bar{a}_1^*)(\bar{a}_2^* + \zeta_2^* \bar{a}_2) + (\bar{a}_2 + \zeta_1 \bar{a}_2^*)(\bar{a}_2^* + \zeta_2^* \bar{a}_2) \Big) \right\}. \tag{92}$$

It possesses the structure of the special scalar product $\langle 0, \zeta_2 | \bar{a}_1 - \bar{a}_2, \zeta_1 \rangle$ multiplied by a phase factor dependent on the displacement parameters:

$$\begin{aligned} \langle \bar{a}_2, \zeta_2 | \bar{a}_1, \zeta_1 \rangle &= \langle 0, \zeta_2 | (D(\bar{a}_2, \bar{a}_2^*))^\dagger D(\bar{a}_1, \bar{a}_1^*) | 0, \zeta_1 \rangle \\ &= \exp\left[\tfrac{1}{2}\left(\bar{a}_1 \bar{a}_2^* - \bar{a}_2 \bar{a}_1^*\right)\right] \langle 0, \zeta_2 | \bar{a}_1 - \bar{a}_2, \zeta_1 \rangle. \end{aligned} \tag{93}$$

The scalar product (92) and its consequences are very important in different specializations. For example, the specialization $\zeta_1 = 0, \bar{a}_1 \equiv \alpha$ is appropriate for the calculation of the coherent-state quasiprobability $Q(\alpha, \alpha^*)$ for squeezed coherent states $|\bar{a}, \zeta\rangle$.

Another application is found for these scalar products in the calculation of the Hilbert-Schmidt distances between squeezed coherent states. The Hilbert-Schmidt distance $d(\varrho_1, \varrho_2)$ of two states with density operators ϱ_1 and ϱ_2 is defined as the Hilbert-Schmidt norm of the difference operator

$$d(\varrho_1, \varrho_2) \equiv \sqrt{\langle (\varrho_1 - \varrho_2)^2 \rangle}. \tag{94}$$

For the distance of two squeezed coherent states, it takes on the form

$$d(|\bar{a}_1, \zeta_1\rangle\langle \bar{a}_1, \zeta_1|, \bar{a}_2, \zeta_2\rangle\langle \bar{a}_2, \zeta_2|) = \sqrt{2(1 - |\langle \bar{a}_1, \zeta_1 | \bar{a}_2, \zeta_2 \rangle|^2)}, \tag{95}$$

and one can calculate the explicit form by using (92) and (93). The maximum distance between pure states is $\sqrt{2}$; it is obtained if the states are orthogonal to each other. The distances depend, apart from the complex squeezing parameters, only on the difference of the complex displacement parameters. For example, the distance of two squeezed vacuum states is

$$d(|0, \zeta_1\rangle\langle 0, \zeta_1|, |0, \zeta_2\rangle\langle 0, \zeta_2|) = \frac{\sqrt{2}\,|\zeta_1 - \zeta_2|}{|1 - \zeta_1 \zeta_2^*| \sqrt{1 + |\langle 0, \zeta_1 | 0, \zeta_2 \rangle|^2}}$$
$$= |\zeta_1 - \zeta_2| \left(1 + \frac{1}{8}\left(|\zeta_1|^2 + 3\zeta_1 \zeta_2^* + 3\zeta_2 \zeta_1^* + |\zeta_2|^2 \right) + \ldots \right). \tag{96}$$

This is also the distance of two squeezed coherent states with equal displacement parameters. It depends for small squeezing parameters $|\zeta_{1,2}| \ll 1$ only on the distance of the squeezing parameters $|\zeta_1 - \zeta_2|$. The special case $\zeta_1 \equiv \zeta$, $\zeta_2 = 0$

determines the distance of squeezed vacuum states to the vacuum state

$$d(|0,\zeta\rangle\langle 0,\zeta|,|0\rangle\langle 0|) = \frac{\sqrt{2}\,|\zeta|}{\sqrt{1+\sqrt{1-|\zeta|^2}}} = |\zeta|\Big(1+\frac{1}{8}|\zeta|^2+\dots\Big). \tag{97}$$

This is at once the minimal distance of a squeezed coherent state with squeezing parameter ζ to a coherent state. It was proposed to use the smallest distance of a pure state to the set of coherent states as a measure of the nonclassicality of pure states [96–98]. We see from (97) that it depends for squeezed coherent states only on the modulus $|\zeta|$ of the squeezing parameter but not on the position of the squeezing axis in relation to the coordinate axes.

3.5 Bargmann representation and Husimi function

The Bargmann representation [99] is a representation of states by analytic functions of a complex variable α or α^* with their origin from analytic coherent states [9, 10]. It is (for normalizable states) an entire function of this variable, and this leads in a natural way to a generalization of states which can be called squeezed-state excitations [100] or photon-added states [101–103]. The simplest cases are squeezed Fock states and displaced squeezed Fock states [104, 105]. This generalization can be taken from the well-known form of entire functions with a finite number of given zeros in the complex plane. The generalization for an infinite number of zeros without accumulation points in the complex plane can be made by using the theorems of Weierstrass and Hadamard. This is at least one of the possible approaches to squeezed state excitations and the discussed nonunitary approach can be generalized to it. The Bargmann representation of states possesses a close relation to the representation of density operators by the coherent-state quasiprobability $Q(\alpha,\alpha^*)$, now frequently called Husimi quasiprobability (section 4).

The Bargmann representation of states is the mapping of arbitrary states $|\psi\rangle$ onto analytic functions $\psi(\alpha^*)$ of the complex variable α^* which can be obtained by forming the scalar product of $|\psi\rangle$ with the analytic coherent states $\langle(\alpha^*,0)|$ as follows:

$$\psi(\alpha^*) \equiv \langle(\alpha^*,0)|\psi\rangle, \quad \Leftrightarrow \quad \psi^*(\alpha) \equiv \langle\psi|(\alpha,0)\rangle = (\psi(\alpha^*))^*. \tag{98}$$

The scalar product of two states $|\psi\rangle$ and $|\chi\rangle$ is given by the integral

$$\langle\chi|\psi\rangle = \frac{1}{\pi}\int \frac{\mathrm{i}}{2} d\alpha \wedge d\alpha^* \exp(-\alpha\alpha^*)\chi^*(\alpha)\psi(\alpha^*), \tag{99}$$

and is established by the well-known completeness relation for the coherent states [7, 106]. From this form of the scalar product it follows that for normalizable states $|\psi\rangle$ the analytic functions $\langle(\alpha^*,0)|\psi\rangle$ cannot possess singularities because any singularity destroys the normalizability. Therefore, $\langle(\alpha^*,0)|\psi\rangle$ have to be entire functions of the complex variable α^* or $\langle\psi|(\alpha,0)\rangle$ of α. From the relations

$$\langle(\alpha^*,0)|a|\psi\rangle = \frac{\partial}{\partial\alpha^*}\langle(\alpha^*,0)|\psi\rangle, \qquad \langle(\alpha^*,0)|a^\dagger|\psi\rangle = \alpha^*\langle(\alpha^*,0)|\psi\rangle, \tag{100}$$

it follows for the representation of the basic boson operators a and $a^\dagger$

$$a \to \frac{\partial}{\partial \alpha^*}, \qquad a^\dagger \to \alpha^*, \qquad [a, a^\dagger] = I \to \left[\frac{\partial}{\partial \alpha^*}, \alpha^*\right] = 1. \tag{101}$$

The boson operators $(a, a^\dagger)$ act as operators of differentiation and multiplication with α^* in the space of entire functions $\langle(\alpha^*, 0)|\psi\rangle$.

The Bargmann representation of the analytic squeezed coherent states $|(\beta, \zeta)\rangle$ can immediately be obtained from (63)

$$\langle(\alpha^*,0)|(\beta,\zeta)\rangle = \langle(\alpha^*,0)| \exp\left[\beta a^\dagger - \frac{\zeta}{2} a^{\dagger 2}\right]|0\rangle = \exp\left[\beta\alpha^* - \frac{\zeta}{2}\alpha^{*2}\right]. \tag{102}$$

The coherent-state representation $\langle\alpha|(\beta, \zeta)\rangle$ is related to the Bargmann representation by the additional normalization factor $\exp(-\alpha\alpha^*/2)$ in $\langle\alpha|$ which destroys the analyticity in α^*. By means of the connection (91) of $|(\beta, \zeta)\rangle$ to $|\bar{a}, \zeta\rangle$ or directly from the general scalar product (92) by setting $\bar{a}_1 \equiv \bar{a}$, $\zeta_1 \equiv \zeta$, $\bar{a}_2 = \alpha$, and $\zeta_2 = 0$, one finds

$$\exp\left[\frac{\alpha\alpha^*}{2}\right] \langle\alpha|\bar{a}, \zeta\rangle = (1 - |\zeta|^2)^{\frac{1}{4}} \exp\left[\bar{a}\alpha^* - \frac{\bar{a}\bar{a}^*}{2} - \frac{\zeta}{2}(\alpha^* - \bar{a}^*)^2\right]. \tag{103}$$

By using (103) in connection with the completeness of the coherent states, one obtains a representation of squeezed states by an integral over coherent states. Due to overcompleteness of coherent states in the phase plane, one can find representations of squeezed and other states by integrals over coherent states on contours or paths in the complex plane [72, 87–89, 107–110].

As already mentioned, the Bargmann representation of states is closely related to the coherent-state (Husimi) quasiprobability $Q(\alpha, \alpha^*)$ which is defined for pure normalized states $|\psi\rangle$ by

$$Q(\alpha, \alpha^*) \equiv \frac{\langle\alpha|\psi\rangle\langle\psi|\alpha\rangle}{\pi} = \frac{\exp(-\alpha\alpha^*)}{\pi}\langle(\alpha^*, 0|\psi\rangle\langle\psi|(\alpha, 0)\rangle, \tag{104}$$

By using the scalar product in (103), one immediately finds its explicit form for squeezed coherent states $|\bar{a}, \zeta\rangle$ ($\alpha' \equiv \alpha - \bar{a}$, $\alpha'^* \equiv \alpha^* - \bar{a}^*$)

$$Q(\alpha, \alpha^*) = \frac{1}{\pi}\sqrt{1 - |\zeta|^2} \exp\left[-\alpha'\alpha'^* - \frac{\zeta^*}{2}\alpha'^2 - \frac{\zeta}{2}\alpha'^{*2}\right]. \tag{105}$$

We see here that the displacement of a state by the unitary displacement operator $D(\bar{a}, \bar{a}^*)$ leads only to a displacement of the whole figure of the quasiprobability by the complex displacement parameter $\bar{a}$ in the complex plane. This is the same for all quasiprobabilities due to the structure of quasiprobabilities which is a basic requirement. We give here the explicit form of two further very important quasiprobabilities for squeezed coherent states, the Wigner quasiprobability

$$W(\alpha, \alpha^*) = \frac{2}{\pi} \exp\left[-2\frac{(\alpha' + \zeta\alpha'^*)(\alpha'^* + \zeta^*\alpha')}{1 - |\zeta|^2}\right], \tag{106}$$

and the Glauber–Sudarshan quasiprobability

$$P(\alpha,\alpha^*) = \frac{1}{\pi}\sqrt{-\frac{1-|\zeta|^2}{|\zeta|^2}}\exp\left[\alpha'\alpha'^* + \frac{\alpha'^2}{2\zeta} + \frac{\alpha'^{*2}}{2\zeta^*}\right]. \tag{107}$$

All three quasiprobabilities are normalized Gaussian functions over the phase space; however, the last one only in the sense of a generalized function because the quadratic form in the exponent is not negative definite and has one positive and one negative eigenvalue and there are directions in the phase space where this Gaussian function increases in infinity. Squeezed coherent states are the most general pure states with Gaussian Wigner quasiprobability [111, 112]. The given quasiprobabilities can be embedded into a three-parameter class of quasiprobabilities $F_{(r_1,r_2,r_3)}(\alpha,\alpha^*)$ with the additive vector parameter $\boldsymbol{r} \equiv (r_1,r_2,r_3)$ [113]. For convenience, we give here the explicit form of $F_{(0,0,r)}(\alpha,\alpha^*)$ for squeezed coherent states [85] (s-parametrized quasiprobabilities [114]; $r=-s$)

$$F_{(0,0,r)}(\alpha,\alpha^*) = \frac{2}{\pi}\sqrt{\frac{1-|\zeta|^2}{(1+r)^2(1-|\zeta|^2)+4r|\zeta|^2}}$$
$$\times \exp\left[-2\frac{(\alpha'+\zeta\alpha'^*)(\alpha'^*+\zeta^*\alpha')+r(1-|\zeta|^2)\alpha'\alpha'^*}{(1+r)^2(1-|\zeta|^2)+4r|\zeta|^2}\right]. \tag{108}$$

Quasiprobabilities $Q(\alpha,\alpha^*), W(\alpha,\alpha^*), P(\alpha,\alpha^*)$ can be obtained by specialization of r to $1, 0, -1$ in this order. If one goes with the parameter r from the Wigner quasiprobability $W(\alpha,\alpha^*)$ (which is a regular function) to the Glauber–Sudarshan quasiprobability $P(\alpha,\alpha^*)$ corresponding to $r=-1$, then at a certain value r_{sing} the quasiprobability $F_{(0,0,r)}(\alpha,\alpha^*)$ becomes singular and the quadratic form in the exponent becomes indefinite for $r < r_{\text{sing}}$ with a positive and a negative eigenvalue. This value of r is determined from (108) for squeezed coherent states by

$$r_{\text{sing}} = -\frac{1-|\zeta|}{1+|\zeta|}, \quad \Leftrightarrow \quad 1+r_{\text{sing}} = \frac{2|\zeta|}{1+|\zeta|}. \tag{109}$$

This, or better the value $1+r_{\text{sing}}$, is sometimes considered as a measure of nonclassicality of the states [112]. In principle, such a definition can be extended to arbitrary pure (and also mixed) states and it is in agreement with intuition and with other proposed measures of nonclassicality, for example, with the nearest distance to coherent states [96–98], but the exact relations between these two measures are not yet established and it is often very difficult to calculate this value.

3.6 Position and momentum representations

The position and momentum representations of a state are the representations by the complete sets of eigenstates $|q\rangle$ and $|p\rangle$ of the canonical operators Q and P with real eigenvalues q and p. The position and momentum representations of squeezed coherent states $|(\beta,\zeta)\rangle$ can be obtained from their Fock-state expansion together with the position and momentum representations of the Fock states $|n\rangle$

according to

$$\langle q|(\beta,\zeta)\rangle = \sum_{n=0}^{\infty}\langle q|n\rangle\langle n|(\beta,\zeta)\rangle, \qquad \langle p|(\beta,\zeta)\rangle = \sum_{n=0}^{\infty}\langle p|n\rangle\langle n|(\beta,\zeta)\rangle, \quad (110)$$

or via the derived Bargmann representation of the states $|(\beta,\zeta)\rangle$ and of $|q\rangle$ and $|p\rangle$ by evaluation of the phase-space integrals

$$\langle q|(\beta,\zeta)\rangle = \frac{\mathrm{i}}{2\pi}\int d\alpha\wedge d\alpha^* \exp(-\alpha\alpha^*)\langle q|(0,\alpha)\rangle\langle(\alpha^*,0)|(\beta,\zeta)\rangle, \quad (111)$$

and analogously $\langle p|(\beta,\zeta)\rangle$. By using the formula of Mehler (88), one finds from (83) and known $\langle q|n\rangle$ and $\langle p|n\rangle$ in the nonunitary approach

$$\begin{aligned}\langle q|(\beta,\zeta)\rangle &= \frac{1}{(\hbar\pi)^{1/4}\sqrt{1-\zeta}}\exp\left[-\frac{(1+\zeta)q^2 - 2\sqrt{2\hbar}\beta q + \hbar\beta^2}{2\hbar(1-\zeta)}\right],\\ \langle p|(\beta,\zeta)\rangle &= \frac{1}{(\hbar\pi)^{1/4}\sqrt{1+\zeta}}\exp\left[-\frac{(1-\zeta)p^2 + \mathrm{i}2\sqrt{2\hbar}\beta p - \hbar\beta^2}{2\hbar(1+\zeta)}\right].\end{aligned} \quad (112)$$

These are analytic functions of β and ζ with ζ in the unit disk $|\zeta| < 1$.

The position and momentum representation of squeezed coherent states in the unitary approach can be obtained from (112) via the transition (91) from $|(\beta,\zeta)\rangle$ to $|\bar{a},\zeta\rangle$ or directly from their Fock-state expansion (86). By using the abbreviations

$$\bar{q} = \sqrt{\hbar/2}\,(\bar{a}+\bar{a}^*), \qquad \bar{p} = -\mathrm{i}\sqrt{\hbar/2}\,(\bar{a}-\bar{a}^*), \qquad \bar{a} = \frac{\bar{q}+\mathrm{i}\bar{p}}{\sqrt{2\hbar}}, \quad (113)$$

the results can be represented in the following form (cf. [115])

$$\begin{aligned}\langle q|\bar{a},\zeta\rangle &= \frac{(1-|\zeta|^2)^{1/4}}{(\hbar\pi)^{1/4}\sqrt{1-\zeta}}\exp\left[-\frac{1+\zeta}{1-\zeta}\frac{(q-\bar{q})^2}{2\hbar} + \mathrm{i}\frac{\bar{p}}{\hbar}\left(q-\frac{\bar{q}}{2}\right)\right],\\ \langle p|\bar{a},\zeta\rangle &= \frac{(1-|\zeta|^2)^{1/4}}{(\hbar\pi)^{1/4}\sqrt{1+\zeta}}\exp\left[-\frac{1-\zeta}{1+\zeta}\frac{(p-\bar{p})^2}{2\hbar} - \mathrm{i}\frac{\bar{q}}{\hbar}\left(p-\frac{\bar{p}}{2}\right)\right].\end{aligned} \quad (114)$$

From this it follows for the normalized squared amplitudes

$$\begin{aligned}\langle q|\bar{a},\zeta\rangle\langle\bar{a},\zeta|q\rangle &= \sqrt{\frac{1-|\zeta|^2}{|1-\zeta|^2\,\hbar\pi}}\exp\left[-\frac{1-|\zeta|^2}{|1-\zeta|^2}\frac{(q-\bar{q})^2}{\hbar}\right],\\ \langle p|\bar{a},\zeta\rangle\langle\bar{a},\zeta|p\rangle &= \sqrt{\frac{1-|\zeta|^2}{|1+\zeta|^2\,\hbar\pi}}\exp\left[-\frac{1-|\zeta|^2}{|1+\zeta|^2}\frac{(p-\bar{p})^2}{\hbar}\right].\end{aligned} \quad (115)$$

The corresponding functions (114) are the wave functions of squeezed coherent states in the position and momentum representation.

The widths of the Gaussian distributions in (115) are proportional either to $|1-\zeta|/\sqrt{1-|\zeta|^2}$ (in q-direction) or to $|1+\zeta|/\sqrt{1-|\zeta|^2}$ (in p-direction). In case of real $\zeta = \zeta^*$, one has $1-|\zeta|^2 = (1-\zeta)(1+\zeta)$. Taking this into account, one finds that one has squeezing in q-direction and stretching in p-direction for positive $\zeta > 0$ and in the opposite way for negative $\zeta < 0$.

3.7 Limiting cases of squeezed coherent states: eigenstates of rotated canonical operators

We now consider the limiting case $|\zeta| = 1$ of weakly nonnormalizable states $|(\beta,\zeta)\rangle$. In this case the eigenvalue problem in (68) is equivalent to the following eigenvalue problem for the Hermitian operator $\sqrt{\zeta}^* a + \sqrt{\zeta} a^\dagger$ to eigenstates $\beta/\sqrt{\zeta} = \sqrt{\zeta}^* \beta$

$$\left(\sqrt{\zeta}^* a + \sqrt{\zeta} a^\dagger\right) |(\beta,\zeta)\rangle = \sqrt{\zeta}^* \beta |(\beta,\zeta)\rangle, \tag{116}$$

or with the substitution $\zeta \equiv |\zeta| \mathrm{e}^{\mathrm{i}2\chi}$ taking into account $|\zeta| = 1$

$$\left(\mathrm{e}^{-\mathrm{i}\chi} a + \mathrm{e}^{\mathrm{i}\chi} a^\dagger\right) \big|(\beta, \mathrm{e}^{\mathrm{i}2\chi})\big\rangle = \mathrm{e}^{-\mathrm{i}\chi} \beta \big|(\beta, \mathrm{e}^{\mathrm{i}2\chi})\big\rangle. \tag{117}$$

Since β is an arbitrary complex number, the eigenvalue equation is solved in the limiting case $|\zeta| = 1$ for arbitrary complex eigenvalues. The eigenvalue problem (116) is identical with the eigenvalue problem for the operator $Q(\varphi)$ which is the rotated canonical operator Q. We introduce the rotated canonical operators $(Q(\varphi), P(\varphi))$ by

$$\begin{aligned}
Q(\varphi) &\equiv R(\varphi) Q \left(R(\varphi)\right)^\dagger = Q\cos\varphi + P\sin\varphi = \sqrt{\frac{\hbar}{2}} \left(a\mathrm{e}^{-\mathrm{i}\varphi} + a^\dagger \mathrm{e}^{\mathrm{i}\varphi}\right), \\
P(\varphi) &\equiv R(\varphi) P \left(R(\varphi)\right)^\dagger = P\cos\varphi - Q\sin\varphi = \frac{1}{\mathrm{i}}\sqrt{\frac{\hbar}{2}} \left(a\mathrm{e}^{-\mathrm{i}\varphi} - a^\dagger \mathrm{e}^{\mathrm{i}\varphi}\right), \\
R(\varphi) &\equiv \exp(\mathrm{i}\varphi a^\dagger a), \qquad (Q(\varphi), P(\varphi)) = (Q, P)\hat{R}(\varphi).
\end{aligned} \tag{118}$$

The eigenvalue value problem for the operator $Q(\varphi)$ is now related to the eigenvalue problem for the operator $Q(0) \equiv Q$ by

$$Q(\varphi)|q; \mathrm{e}^{\mathrm{i}\varphi}\rangle = q|q; \mathrm{e}^{\mathrm{i}\varphi}\rangle, \qquad Q|q\rangle = q|q\rangle, \qquad |q; \mathrm{e}^{\mathrm{i}\varphi}\rangle \equiv R(\varphi)|q\rangle, \tag{119}$$

By comparison with (68) one finds that the states $|(\beta, \mathrm{e}^{\mathrm{i}2\chi})\rangle$ have to be proportional to the eigenstates $|q; \mathrm{e}^{\mathrm{i}\varphi}\rangle$ with the correspondence $\beta = \sqrt{\hbar/2}\, q\, \mathrm{e}^{\mathrm{i}\varphi}$ and $\chi = \varphi$. It becomes an equality if we set

$$\begin{aligned}
|q; \mathrm{e}^{\mathrm{i}\varphi}\rangle &= (\hbar\pi)^{-1/4} \exp\left(-q^2/2\hbar\right) \left|\left(\sqrt{2/\hbar}\, q\, \mathrm{e}^{\mathrm{i}\varphi}, \mathrm{e}^{\mathrm{i}2\varphi}\right)\right\rangle, \\
\left|\left(\sqrt{2/\hbar}\, q\, \mathrm{e}^{\mathrm{i}\varphi}, \mathrm{e}^{\mathrm{i}2\varphi}\right)\right\rangle &\equiv \exp\left(\sqrt{2/\hbar}\, q\, \mathrm{e}^{\mathrm{i}\varphi} a^\dagger - \tfrac{1}{2} \mathrm{e}^{\mathrm{i}2\varphi} a^{\dagger 2}\right) |0\rangle.
\end{aligned} \tag{120}$$

This is seen by comparing the Fock-state matrix elements $\langle n|q; \mathrm{e}^{\mathrm{i}\varphi}\rangle = \mathrm{e}^{\mathrm{i}n\varphi}\langle n|q\rangle$ with $\langle n|(\beta,\zeta)\rangle$ taken from (83). By using the scalar product of the squeezed coherent states in (89), one obtains

$$\langle q'; \mathrm{e}^{\mathrm{i}\varphi'}|q; \mathrm{e}^{\mathrm{i}\varphi}\rangle = \frac{\exp\left[-\mathrm{i}\dfrac{(q^2 + q'^2)\cos(\varphi - \varphi') - 2qq'}{2\hbar\sin(\varphi - \varphi')}\right]}{\left[\hbar\pi\left(1 - \mathrm{e}^{\mathrm{i}2(\varphi - \varphi')}\right)\right]^{1/2}}. \tag{121}$$

For $\varphi' = \varphi$ by a limiting procedure and for $\varphi' = \varphi + \pi/2$, this yields

$$\begin{aligned} \langle q'; e^{i\varphi}|q; e^{i\varphi}\rangle &= \delta(q - q'), \\ \langle p; ie^{i\varphi}|q; e^{i\varphi}\rangle &= (2\hbar\pi)^{-1/2} \exp(-iqp/\hbar). \end{aligned} \tag{122}$$

The states $|q; e^{i\varphi}\rangle$ obey the completeness relation

$$\int_{-\infty}^{+\infty} dq|q; e^{i\varphi}\rangle\langle q; e^{i\varphi}| = R(\varphi) \int_{-\infty}^{+\infty} dq|q\rangle\langle q| \,(R(\varphi))^\dagger = I, \tag{123}$$

which follows immediately from the completeness relation for the states $|q\rangle$ and the definition of $|q; e^{i\varphi}\rangle$ in (120).

In the special cases $\varphi = 0$ and $\varphi = \pi/2$, one obtains from (120) the eigenstates of the canonical operators Q and P

$$\begin{aligned} |q\rangle \equiv |q; 1\rangle &= (\hbar\pi)^{-1/4} \exp\left(-\frac{q^2}{2\hbar}\right) \exp\left(\sqrt{\frac{2}{\hbar}}\, qa^\dagger - \frac{1}{2}a^{\dagger 2}\right)|0\rangle, \\ |p\rangle \equiv |p; i\rangle &= (\hbar\pi)^{-1/4} \exp\left(-\frac{p^2}{2\hbar}\right) \exp\left(i\sqrt{\frac{2}{\hbar}}\, pa^\dagger + \frac{1}{2}a^{\dagger 2}\right)|0\rangle. \end{aligned} \tag{124}$$

These relations and, more generally, (120) show that the eigenstates $|q; e^{i\varphi}\rangle$ of the rotated canonical operator $Q(\varphi)$ can be obtained from the vacuum state $|0\rangle$ by applying a nonunitary squeezing operator and by transition to the limiting case $|\zeta| = 1$. The states $|q\rangle, |p\rangle$ and, more generally, $|q; e^{i\varphi}\rangle$ can also be expressed by the states $|\bar{a}, \zeta\rangle$ in the unitary approach. However, since $|\bar{a}, \zeta\rangle$ are normalized and are defined as such states for $|\zeta| < 1$, this can be expressed only by a limiting procedure. By using the squeezed coherent states in the nonunitary approach $|(\beta, \zeta)\rangle$ and their relation to the states $|q\rangle$ and $|p\rangle$ given in (120) and the connection between unitary and nonunitary approach in (91), one finds the relation to $|\bar{a}, \zeta\rangle$

$$|q; e^{i\varphi}\rangle = \lim_{\epsilon \to +0} (2\hbar\pi\epsilon)^{-1/4} \left| e^{i\varphi} q/\sqrt{2\hbar},\, e^{i2\varphi}(1-\epsilon)\right\rangle, \tag{125}$$

and in the special cases $\varphi = 0$ and $\varphi = \pi/2$

$$\begin{aligned} |q\rangle \equiv |q; 1\rangle &= \lim_{\epsilon \to +0} (2\hbar\pi\epsilon)^{-1/4} \left| q/\sqrt{2\hbar},\, 1-\epsilon\right\rangle, \\ |p\rangle \equiv |p; i\rangle &= \lim_{\epsilon \to +0} (2\hbar\pi\epsilon)^{-1/4} \left| ip/\sqrt{2\hbar},\, -(1-\epsilon)\right\rangle. \end{aligned} \tag{126}$$

We now calculate more general wave functions $\langle q; e^{i\varphi}|(\beta, \zeta)\rangle$ and $\langle q; e^{i\varphi}|\bar{a}, \zeta\rangle$. From the definition of $|q; e^{i\varphi}\rangle$ in (119), one obtains

$$\begin{aligned} \langle q; e^{i\varphi}|(\beta, \zeta)\rangle &= \langle q| \,(R(\varphi))^\dagger \exp\left(\beta a^\dagger - \frac{\zeta}{2}a^{\dagger 2}\right)|0\rangle \\ &= \langle q| \left(e^{-i\varphi}\beta, e^{-i2\varphi}\zeta\right)\rangle, \end{aligned} \tag{127}$$

where $\langle q|\left(\mathrm{e}^{-\mathrm{i}\varphi}\beta, \mathrm{e}^{-\mathrm{i}2\varphi}\zeta\right)\rangle$ can be taken from (112). The corresponding scalar product in the unitary approach is

$$\langle q; \mathrm{e}^{\mathrm{i}\varphi}|\bar{a}, \zeta\rangle = (\hbar\pi)^{-1/4}\left(1-|\zeta|^2\right)^{1/4}\left(1-\mathrm{e}^{-\mathrm{i}2\varphi}\zeta\right)^{-1/2} \times \exp\left[-\frac{1+\mathrm{e}^{-\mathrm{i}2\varphi}\zeta}{1-\mathrm{e}^{-\mathrm{i}2\varphi}\zeta}\frac{\left(q-\overline{Q(\varphi)}\right)^2}{2\hbar} + \frac{\mathrm{i}}{\hbar}\overline{P(\varphi)}\left(q-\frac{\overline{Q(\varphi)}}{2}\right)\right]. \tag{128}$$

The rotated q-distribution of the state $|\bar{a}, \zeta\rangle$ is

$$\left|\langle q; \mathrm{e}^{\mathrm{i}\varphi}|\bar{a}, \zeta\rangle\right|^2 = \sqrt{\frac{1-|\zeta|^2}{\left|1-\mathrm{e}^{-\mathrm{i}2\varphi}\zeta\right|^2\hbar\pi}}\exp\left[-\frac{\left(1-|\zeta|^2\right)\left(q-\overline{Q(\varphi)}\right)^2}{\left|1-\mathrm{e}^{-\mathrm{i}2\varphi}\zeta\right|^2\hbar}\right]. \tag{129}$$

In the next section we show that the expression on the left-hand side of (129) is the reduced form of the Radon transform of the Wigner quasiprobability which is important in quantum tomography.

Squeezed coherent states $|(\beta, \zeta)\rangle$ in the limit case $|\zeta| = 1$ solve the eigenvalue problem for the displacement operator $D(\alpha, \alpha^*) \equiv \exp\left(\alpha a^\dagger - \alpha^* a\right)$ according to [72]

$$D(\alpha, \alpha^*)\,|(\beta, -\alpha/\alpha^*)\rangle = \exp\left(-\alpha^*\beta\right)|(\beta, -\alpha/\alpha^*)\rangle. \tag{130}$$

The eigenstates of the displacement operator are limiting cases of squeezed coherent states which are infinitely stretched in the direction of the displacement parameter α in the phase plane and infinitely squeezed in the direction perpendicular to α, for example, in the case of displacement in the direction of the q-axis of the eigenstates of the operator P.

3.8 Moments of boson and canonical operators

The scalar products (89) allow to calculate the moments of the canonical operators $Q(\varphi)$ by applying formulae (80). The explicit form of the scalar product $\langle(\beta^*, \zeta^*)|(\beta, \zeta)\rangle$ given in (91) allows us to carry out the calculations with the following result for the normally ordered moments (cf. [116, 117])

$$\begin{aligned}\overline{a^{\dagger k}a^l} &= \left(\frac{\beta^*-\zeta^*\beta}{1-\zeta\zeta^*} - \zeta^*\frac{\partial}{\partial\beta^*}\right)^k\left(\frac{\beta-\zeta\beta^*}{1-\zeta\zeta^*} - \zeta\frac{\partial}{\partial\beta}\right)^l 1 \\ &= \frac{(\zeta^*)^{k/2}\zeta^{l/2}k!l!}{\left[2(1-|\zeta|^2)\right]^{(k+l)/2}}\sum_{j=0}^{\min\{k,l\}}\frac{(2|\zeta|)^j}{j!(k-j)!(l-j)!} \\ &\quad\times H_{k-j}\left(\sqrt{\frac{1-|\zeta|^2}{2\zeta^*}}\,\bar{a}^*\right)H_{l-j}\left(\sqrt{\frac{1-|\zeta|^2}{2\zeta}}\,\bar{a}\right).\end{aligned} \tag{131}$$

The ordered moments $\overline{\mathcal{O}_{(0,0,r)}[a^{\dagger k}a^l]}$ can be obtained from (131) by means of the substitution $(2|\zeta|)^j \rightarrow \left(\left[1+r+(1-r)|\zeta|^2\right]/|\zeta|\right)^j$, in particular $(2|\zeta|)^j \rightarrow (2/|\zeta|)^j$

for $r = 1$ (antinormal ordering) and $(2|\zeta|)^j \to \left([1+|\zeta|^2]/|\zeta|\right)^j$ for $r = 0$ (symmetrical ordering). This can be proved by the transition relation from normal ordering $(0, 0, -1)$ to arbitrary ordering $(0, 0, r)$ (e.g., [114, 118–120]). In an analogous way, one can obtain antinormally ordered moments.

As already derived, one has $\overline{a} = \bar{a}$, $\overline{a^\dagger} = \bar{a}^*$. The second-order normally ordered moments obtained from (131) are (cf. [13])

$$\overline{a^2} = \bar{a}^2 - \frac{\zeta}{1-|\zeta|^2}, \qquad \overline{a^\dagger a} = \bar{a}\bar{a}^* + \frac{|\zeta|^2}{1-|\zeta|^2}, \qquad \overline{a^{\dagger 2}} = \overline{a^2}^{\,*}. \tag{132}$$

The first central moments of powers of a are ($\Delta a \equiv a - \bar{a}I$)

$$\overline{(\Delta a)^2} = -\frac{\zeta}{1-|\zeta|^2}, \qquad \overline{(\Delta a)^3} = 0, \qquad \overline{(\Delta a)^4} = \frac{3\zeta^2}{(1-|\zeta|^2)^2}. \tag{133}$$

All central moments depend only on the squeezing parameter and are the same as for corresponding squeezed vacuum states. The vanishing of the third central moment $\overline{(\Delta a)^3}$ and of all other odd central moments is due to the inversion symmetry of squeezed vacuum states.

The moments of the canonical operators (Q, P) can be obtained from the moments of the boson operators. However, the knowledge of the wave function $\langle q; \mathrm{e}^{\mathrm{i}\varphi}|\bar{a}, \zeta\rangle$ allows their direct calculation in two different ordering

$$\overline{(Q(\varphi))^m (P(\varphi))^n} \quad \text{and} \quad \overline{(P(\varphi))^n (Q(\varphi))^m}.$$

In particular, one finds for the variances of $Q(\varphi)$ and $P(\varphi)$

$$\overline{(\Delta Q(\varphi))^2} = \frac{\hbar}{2}\frac{\left|1-\mathrm{e}^{-\mathrm{i}2\varphi}\zeta\right|^2}{1-\zeta\zeta^*}, \qquad \overline{(\Delta P(\varphi))^2} = \frac{\hbar}{2}\frac{\left|1+\mathrm{e}^{-\mathrm{i}2\varphi}\zeta\right|^2}{1-\zeta\zeta^*}, \tag{134}$$

and for the uncertainty correlations

$$\begin{aligned} \overline{\Delta Q(\varphi)\Delta P(\varphi)} &= \mathrm{i}\frac{\hbar}{2}\frac{(1+\mathrm{e}^{-\mathrm{i}2\varphi}\zeta)(1-\mathrm{e}^{\mathrm{i}2\varphi}\zeta^*)}{1-\zeta\zeta^*} = \overline{\Delta P(\varphi)\Delta Q(\varphi)}^{\,*}, \\ \frac{1}{2}\left(\overline{\Delta Q(\varphi)\Delta P(\varphi)} + \overline{\Delta P(\varphi)\Delta Q(\varphi)}\right) &= \mathrm{i}\frac{\hbar}{2}\frac{\mathrm{e}^{-\mathrm{i}2\varphi}\zeta - \mathrm{e}^{\mathrm{i}2\varphi}\zeta^*}{1-\zeta\zeta^*}, \end{aligned} \tag{135}$$

with the abbreviations $\Delta Q(\varphi) \equiv Q(\varphi) - \overline{Q(\varphi)}I$, $\Delta P(\varphi) \equiv P(\varphi) - \overline{P(\varphi)}I$. From (134) and (135), one obtains for the uncertainty sum

$$\overline{(\Delta Q(\varphi))^2} + \overline{(\Delta P(\varphi))^2} = \hbar\frac{1+|\zeta|^2}{1-|\zeta|^2}, \tag{136}$$

and for the uncertainty product from which we subtract the squared symmetrical uncertainty correlation

$$\overline{(\Delta Q(\varphi))^2}\,\overline{(\Delta P(\varphi))^2} - \frac{1}{4}\left(\overline{\Delta Q(\varphi)\Delta P(\varphi)} + \overline{\Delta P(\varphi)\Delta Q(\varphi)}\right)^2 = \frac{\hbar^2}{4}. \tag{137}$$

This shows that the squeezed coherent states are states with minimal uncertainty product with the modification by subtraction of the squared uncertainty correlation. Without this modification by the uncertainty correlation, only squeezed coherent states with squeezing axes parallel to the coordinate axes are minimum uncertainty states. Such states were considered in Pauli's *Handbuch* article [5].

One finds from (134) that the squeezed coherent states possess reduced uncertainty of the canonical coordinates in one direction and enlarged uncertainty in the perpendicular direction. The minimum and maximum of these uncertainties are obtained by the relations

$$\begin{aligned} e^{i2\varphi_{\min}} &= \frac{\zeta^*}{|\zeta|}, \qquad \Rightarrow \qquad \overline{(\Delta Q(\varphi_{\min}))^2} = \frac{\hbar}{2}\frac{1-|\zeta|}{1+|\zeta|}, \\ e^{i2\varphi_{\max}} &= -\frac{\zeta^*}{|\zeta|}, \qquad \Rightarrow \qquad \overline{(\Delta Q(\varphi_{\max}))^2} = \frac{\hbar}{2}\frac{1+|\zeta|}{1-|\zeta|}, \end{aligned} \tag{138}$$

with $\varphi_{\max} = \varphi_{\min} \pm \pi/2$. These relations reveal the name "squeezed coherent" states. In comparison to coherent states with equal uncertainties (fluctuations) of the canonical coordinates in arbitrary direction, the squeezed coherent states are squeezed by a factor in one direction and stretched by the inverse factor in the perpendicular direction. By using (134), the squared modulus of the wave function (129) can be represented as

$$|\langle q; e^{i\varphi}|\overline{a}, \zeta\rangle|^2 = \left[2\pi\overline{(\Delta Q(\varphi))^2}\right]^{-1/2} \exp\left[-\frac{\left(q - \overline{Q(\varphi)}\right)^2}{2\overline{(\Delta Q(\varphi))^2}}\right]. \tag{139}$$

This is the standard form of a one-dimensional Gaussian distribution represented by its displacement $\overline{Q(\varphi)}$ and by its dispersion $\overline{(\Delta Q(\varphi))^2}$. It depends here on the angle φ as a parameter.

The symmetrical uncertainty correlation given in (135) vanishes for real ζ. In this case, the dispersions (134) take on their extreme values in correspondence with the coordinate axes. One can distinguish the following cases:

Positive squeezing parameter ζ (q-squeezing)

$$\zeta = \zeta^* > 0, \ \Rightarrow \ \overline{(\Delta Q)^2} = \frac{\hbar}{2}\frac{1-|\zeta|}{1+|\zeta|}, \qquad \overline{(\Delta P)^2} = \frac{\hbar}{2}\frac{1+|\zeta|}{1-|\zeta|}. \tag{140}$$

The states are squeezed in the direction of the coordinate q and stretched in the direction of the coordinate p.

Negative squeezing parameter ζ (p-squeezing)

$$\zeta = \zeta^* < 0, \ \Rightarrow \ \overline{(\Delta Q)^2} = \frac{\hbar}{2}\frac{1+|\zeta|}{1-|\zeta|}, \qquad \overline{(\Delta P)^2} = \frac{\hbar}{2}\frac{1-|\zeta|}{1+|\zeta|}. \tag{141}$$

The states are squeezed in the direction of the coordinate p and stretched in the direction of the coordinate q.

The wave function of the vacuum state $|0\rangle$

$$\langle q; e^{i\varphi}|0\rangle = \langle q|\left(R(\varphi)\right)^{\dagger}|0\rangle = (\hbar\pi)^{-1/4} \exp\left[-q^2/(2\hbar)\right] \tag{142}$$

is independent of φ, meaning that the vacuum state is rotation-invariant. By splitting off the vacuum part from the wave function (127) and by using the eigenvalue equation (119) for the operator $Q(\varphi)$, one can represent this equation in the following way in the nonunitary approach

$$
\begin{aligned}
|(\beta,\zeta)\rangle &= \exp\left[-\frac{2\mathrm{e}^{-\mathrm{i}2\varphi}\zeta(Q(\varphi))^2 - 2\sqrt{2\hbar}\mathrm{e}^{-\mathrm{i}\varphi}\beta Q(\varphi) + \hbar\mathrm{e}^{-\mathrm{i}2\varphi}\beta^2 I}{2\hbar(1-\mathrm{e}^{-\mathrm{i}2\varphi}\zeta)}\right] \\
&\quad \times \left(1-\mathrm{e}^{-\mathrm{i}2\varphi}\zeta\right)^{-1/2}|0\rangle, \qquad (143)
\end{aligned}
$$

where φ is a parameter. By scalar multiplication of this identity with $\langle q;\mathrm{e}^{\mathrm{i}\varphi}|$ and by using (119), one comes back to (127). From (143), one finds the corresponding expressions for squeezed vacuum states $|(0,\zeta)\rangle$ and for coherent states $|(\beta,0)\rangle$ in the nonunitary approach which we do not write down explicitly. These equations show that squeezed coherent states can be generated from the vacuum state $|0\rangle$ by a class of nonunitary operators which are quadratic in $Q(\varphi)$ in the exponent [121]. The same can be made in the unitary approach where one obtains from (128) with arbitrary angle φ

$$
\begin{aligned}
|\bar{a},\zeta\rangle &= \exp\left\{\sqrt{\frac{2}{\hbar}}\mathrm{e}^{-\mathrm{i}\varphi}\bar{a}\left(Q(\varphi)-\frac{1}{2}\overline{Q(\varphi)}I\right) - \frac{\mathrm{e}^{-\mathrm{i}2\varphi}\zeta\left(\Delta Q(\varphi)\right)^2}{\hbar\left(1-\mathrm{e}^{-\mathrm{i}2\varphi}\zeta\right)}\right\} \\
&\quad \times\left(1-|\zeta|^2\right)^{1/4}\left(1-\mathrm{e}^{-\mathrm{i}2\varphi}\zeta\right)^{-1/2}|0\rangle. \qquad (144)
\end{aligned}
$$

A more direct proof of relations (143) and (144) can be obtained by normally ordered expansions of the operators on the right-hand sides.

3.9 Photon statistics of squeezed coherent states

The number distribution $p_n = \langle n|\bar{a},\zeta\rangle\,\langle\bar{a},\zeta|n\rangle$ for squeezed coherent states was given in general form in (87) and it involves Hermite polynomials with the displacement parameters $(\bar{a},\bar{a}^*)$ and squeezing parameters (ζ,ζ^*) as substantial parts of the arguments. We can use these formulae for the calculation of the moments of the number operator of not too high order. The other possibility is to calculate, for example, the normally ordered moments $\overline{a^{\dagger k}a^k}$ that is possible by carrying out the differentiations in (80) and is contained in (131). The moments of N depend, apart from the modulus $|\bar{a}|$ of the displacement parameter and the modulus $|\zeta|$ of the squeezing parameter, also from the position of the squeezing axes in relation to the radius to the centre of the distribution. This can be determined by one angle which we denote by χ as follows

$$
\zeta^*\bar{a}^2 = |\zeta||\bar{a}|^2\exp(\mathrm{i}2\chi). \qquad (145)
$$

The angle $\chi = 0$ corresponds to amplitude squeezing where the squeezing axis is perpendicular to the radial direction and $\chi = \pi/2$ to phase squeezing where the squeezing axis is parallel to the radial direction.

One finds for the first two moments of the number operator

$$\begin{aligned}\overline{N} &= |\bar{a}|^2 + \frac{|\zeta|^2}{1-|\zeta|^2}, \quad \Leftrightarrow \quad |\bar{a}|^2 = \overline{N} - \frac{|\zeta|^2}{1-|\zeta|^2},\\ \overline{N^2} &= |\bar{a}|^4 + \frac{1+3|\zeta|^2 - 2|\zeta|\cos(2\chi)}{1-|\zeta|^2}|\bar{a}|^2 + \frac{|\zeta|^2(2+|\zeta|^2)}{(1-|\zeta|^2)^2}. \qquad (146)\end{aligned}$$

With the abbreviation $\Delta N \equiv N - \overline{N}I$ it follows for the first of the nonvanishing central moments $\overline{(\Delta N)^k}$, $(k=2)$ of the number operator

$$\overline{(\Delta N)^2} = \overline{N} + \frac{|\zeta|}{1-|\zeta|^2}\left(2(|\zeta| - \cos(2\chi))|\bar{a}|^2 + \frac{|\zeta|(1+|\zeta|^2)}{1-|\zeta|^2}\right). \qquad (147)$$

The quantity $\overline{(\Delta N)^2}$ is the dispersion of the number operator. The next central moment $\overline{(\Delta N)^3}$, which we did not give in explicit form, is one of the simplest definitions of a skewness. For coherent states $(\zeta = 0)$ both central moments $\overline{(\Delta N)^2}$ and $\overline{(\Delta N)^3}$ are equal to $\overline{N}$. This means that for $\overline{N} \to \infty$ the relative skewness is very small. We have separated from $\overline{(\Delta N)^2}$ in (146) the value $\overline{N}$ for coherent states. This shows that the remaining part can become positive as well as negative. Positive values for $\overline{(\Delta N)^2} - \overline{N}$ are called super-Poissonian and negative values sub-Poissonian statistics. In order to get sub-Poissonian statistics for squeezed coherent states, it is necessary to have more than a minimal displacement $\bar{a}$ in dependence on $|\zeta|$ with the angle $\chi = 0$ as the most favourable condition (amplitude squeezing). For $\chi = \mp\pi/2$ (phase squeezing), the squeezed coherent states possess super-Poissonian statistics. If one fixes $|\zeta|$ and $\bar{a}$ in cases where for $\chi = 0$ one has sub-Poissonian statistics and rotate the squeezing ellipse then for a certain angle χ one obtains a Poissonian-like statistics with $\overline{(\Delta N)^2} - \overline{N} = 0$ which, however, is not a genuine Poissonian statistics.

4 Quasiprobabilities and uncertainties

4.1 Wigner quasiprobability, Fourier and Radon transforms

The Wigner quasiprobability [1] plays a central role in modern quantum optics. Its properties were discussed in detail, e.g., in [2,7,8,33,122–124]. In representation $W(q,p)$ or $W(\alpha,\alpha^*)$ by real or complex variables, it gives the most analogous description of quantum mechanics in one dimension to classical statistical mechanics of Hamilton systems. The Wigner quasiprobability $W(q,p)$ or $W(\alpha,\alpha^*)$ as a normalized function over the phase space can be defined in the following symmetrical form for a given density operator ϱ [recall $\langle A\rangle \equiv \mathrm{Trace}(A)$]:

$$\begin{aligned}&W(q,p) \equiv \left\langle \varrho \exp\left(-Q\frac{\partial}{\partial q} - P\frac{\partial}{\partial p}\right)\right\rangle \delta(q)\delta(p),\\ &W(\alpha,\alpha^*) \equiv \left\langle \varrho \exp\left(-a\frac{\partial}{\partial \alpha} - a^\dagger\frac{\partial}{\partial \alpha^*}\right)\right\rangle \delta(\alpha,\alpha^*),\\ &\int dq \wedge dp\, W(q,p) = 1, \qquad \int \frac{\mathrm{i}}{2} d\alpha \wedge d\alpha^* W(\alpha,\alpha^*) = 1. \qquad (148)\end{aligned}$$

There exist equivalent representations of the Wigner quasiprobability, among them the original one given by Wigner [1] which is mostly taken as starting definition, a representation by displaced Fock states [114, 125, 126] or by the displaced parity operator [127, 128], and another one by a phase-space integral over coherent states [41, 42, 113, 129].

Besides the Wigner quasiprobability, we introduce its two-dimensional Fourier transforms $\tilde{W}(u,v)$ or $\tilde{W}(\tau,\tau^*)$, respectively, by

$$\begin{aligned}\tilde{W}(u,v) &\equiv \int dq \wedge dp \exp\big(-\mathrm{i}(uq+vp)\big)W(q,p),\\ \tilde{W}(\tau,\tau^*) &\equiv \int \frac{\mathrm{i}}{2} d\alpha \wedge d\alpha^* \exp\big(-\mathrm{i}\,(\tau^*\alpha+\tau\alpha^*)\big)W(\alpha,\alpha^*),\\ \tilde{W}(u=0,v=0) &= 1, \qquad \tilde{W}(\tau=0,\tau^*=0)=1.\end{aligned} \tag{149}$$

Our complex variables are related to the real variables by

$$\alpha = \frac{q+\mathrm{i}p}{\sqrt{2\hbar}}, \qquad \tau = \sqrt{\frac{\hbar}{2}}\,(u+\mathrm{i}v), \qquad \tau^*\alpha+\tau\alpha^* = uq+vp. \tag{150}$$

The Fourier transform of the Wigner quasiprobability (there are different conventions [7, 114]) is called the characteristic function in analogy to Fourier transforms of classical distribution functions. It is the generating function for the symmetrically ordered moments of the canonical or of the boson operators depending on the representation.

The *Radon transform* $\breve{W}(u,v;c)$ of the Wigner quasiprobability $W(q,p)$ is defined by [117, 130, 131] (mathematical aspects [132, 133])

$$\breve{W}(u,v;c) \equiv \int dq \wedge dp\, \delta\,(c-uq-vp)\, W(q,p), \tag{151}$$

or in representation by complex variables $\breve{W}(\tau,\tau^*;c)$

$$\breve{W}(\tau,\tau^*;c) \equiv \int \frac{\mathrm{i}}{2} d\alpha \wedge \alpha^*\, \delta\big(c-(\tau^*\alpha+\tau\alpha^*)\big)W(\alpha,\alpha^*), \tag{152}$$

$$\int_{-\infty}^{+\infty} dc\, \breve{W}(u,v;c) = \int_{-\infty}^{\infty} dc\, \breve{W}(\tau,\tau^*;c) = 1.$$

It is important in quantum tomography for state reconstruction from measured data [43]. In contrast to the Fourier transform, the Radon transform is a real-valued function for real variables (u,v,c). It is a homogeneous function of the variables (u,v,c) (μ arbitrary real number)

$$\breve{W}(u,v;c) = |\mu|\breve{W}(\mu u,\mu v;\mu c), \qquad \mu \neq 0, \tag{153}$$

therefore, it depends effectively only on two independent variables. The Radon transform of the Wigner quasiprobability is related to its Fourier transform by

$$\breve{W}(u,v;c) = \frac{1}{2\pi}\int_{-\infty}^{+\infty} db \exp(\mathrm{i}bc)\tilde{W}(bu,bv), \tag{154}$$

with the inversion

$$\tilde{W}(bu, bv) = \int_{-\infty}^{+\infty} dc \exp(-ibc)\breve{W}(u, v; c), \tag{155}$$

and analogously in complex representation. The full inversion of the Wigner quasiprobability can be made with the intermediate step of the transition from the Radon transform to the Fourier transform and then to its inversion.

The explicit definition (148) of the Wigner quasiprobability $W(q,p)$ in dependence on the density operator ϱ generates corresponding relations for the Fourier and Radon transform. By using partial integration, one obtains the representation of the Fourier transform of the Wigner quasiprobability

$$\tilde{W}(u, v) = \left\langle \varrho \exp\left(-\mathrm{i}(uQ + vP)\right)\right\rangle. \tag{156}$$

In an analogous way, one obtains the Radon transform in the form

$$\breve{W}(u, v; c) = \left\langle \varrho\, \delta\big(cI - (uQ + vP)\big)\right\rangle. \tag{157}$$

The reduced form $\breve{W}(\cos\varphi, \sin\varphi; q)$ is given by

$$\breve{W}(\cos\varphi, \sin\varphi; q) = \langle \varrho\, \delta\left(qI - Q(\varphi)\right)\rangle = \left\langle q; \mathrm{e}^{\mathrm{i}\varphi}\right| \varrho \left|q; \mathrm{e}^{\mathrm{i}\varphi}\right\rangle, \tag{158}$$

where $Q(\varphi) = Q\cos\varphi + P\sin\varphi$ denotes the rotated canonical operator Q according to (118) and $|q; \mathrm{e}^{\mathrm{i}\varphi}\rangle$ its eigenstates to eigenvalues q normalized by means of the delta function. For $\varphi = 0$ and $\varphi = \pi/2$ one obtains the marginal distributions of the Wigner quasiprobability

$$\breve{W}(1, 0; q) = \langle q|\varrho|q\rangle, \quad \breve{W}(0, 1; p) = \langle p|\varrho|p\rangle. \tag{159}$$

The Wigner quasiprobability and its Fourier and Radon transforms contain the complete information of the density operator. Therefore, the density operator can be reconstructed from the Wigner quasiprobability or from its transforms. The reconstruction uses the same operators as necessary for the transition from the density operator to the Wigner quasiprobability:

$$\begin{aligned} \varrho &= 2\hbar\pi \int dq \wedge dp\, W(q, p) \exp\left(-Q\frac{\partial}{\partial q} - P\frac{\partial}{\partial p}\right) \delta(q)\delta(p), \\ \varrho &= \frac{\mathrm{i}\pi}{2} \int d\alpha \wedge d\alpha^* W(\alpha, \alpha^*) \exp\left(-a\frac{\partial}{\partial \alpha} - a^\dagger \frac{\partial}{\partial \alpha^*}\right) \delta(\alpha, \alpha^*). \end{aligned} \tag{160}$$

For other reconstruction formulae see, e.g., [43, 131].

4.2 Transformations of the Wigner quasiprobability

We now discuss the transformation properties of the Wigner quasiprobability and of its Fourier and Radon transforms for given states under squeezing and displacement transformations of these states.

Under unitary displacements of states according to

$$\varrho = D(\bar{q},\bar{p})\varrho_0 D^\dagger(\bar{q},\bar{p}) = D(\bar{a},\bar{a}^*)\varrho_0 D^\dagger(\bar{a},\bar{a}^*), \tag{161}$$

the corresponding Wigner quasiprobabilities in real and complex representation are to transform by argument displacements according to

$$W(q,p) = W_0(q-\bar{q},p-\bar{p}), \qquad W(\alpha,\alpha^*) = W_0(\alpha-\bar{a},\alpha^*-\bar{a}^*). \tag{162}$$

The Fourier transform of the Wigner quasiprobability then transforms

$$\begin{aligned}\tilde{W}(u,v) &= \exp\big(-\mathrm{i}(\bar{q}u+\bar{p}v)\big)\tilde{W}_0(u,v),\\ \tilde{W}(\tau,\tau^*) &= \exp\big(-\mathrm{i}\,(\bar{a}^*\tau+\bar{a}\tau^*)\big)\tilde{W}_0(\tau,\tau^*),\end{aligned} \tag{163}$$

and the Radon transform changes according to

$$\begin{aligned}\breve{W}(u,v;c) &= \breve{W}_0\big(u,v;c-(\bar{q}u+\bar{p}v)\big),\\ \breve{W}(\tau,\tau^*;c) &= \breve{W}_0\big(\tau,\tau^*;c-(\bar{a}^*\tau+\bar{a}\tau^*)\big).\end{aligned} \tag{164}$$

We now consider unitary squeezing transformations

$$\varrho = S(\zeta^*,\eta=\eta^*,\zeta)\varrho_0 S^\dagger(\zeta^*,\eta=\eta^*,\zeta). \tag{165}$$

This causes the transformation of the arguments of the primary Wigner function of the form [131, 134, 135]

$$\begin{aligned}W(q,p) &= W_0(\alpha q+\gamma p,\beta q+\delta p),\\ W(\alpha,\alpha^*) &= W_0(\nu^*\alpha+\mu\alpha^*,\mu^*\alpha+\nu\alpha^*).\end{aligned} \tag{166}$$

In complex representation we have the symplectic (or unimodular) transformation of the variables

$$(\alpha',\alpha'^*) = (\alpha,\alpha^*)\begin{Vmatrix}\nu^* & \mu^*\\ \mu & \nu\end{Vmatrix} = (\nu^*\alpha+\mu\alpha^*,\mu^*\alpha+\nu\alpha^*). \tag{167}$$

The involved matrix of the fundamental representation of $Sp(2,\mathrm{R})$ is given in (22). The matrix with the real elements $(\alpha,\beta,\gamma,\delta)$ is given in (33) in general form and has to be specialized by $\xi=\zeta^*$ and $\eta=\eta^*$ where the elements become real numbers. With the definitions of the Fourier and Radon transforms of the Wigner quasiprobability, one finds their transformations. In particular, one finds for the Fourier transform

$$\begin{aligned}\tilde{W}(u,v) &= \tilde{W}_0(\delta u-\beta v,-\gamma u+\alpha v),\\ \tilde{W}(\tau,\tau^*) &= \tilde{W}(\nu^*\tau-\mu\tau^*,-\mu^*\tau+\nu\tau^*),\end{aligned} \tag{168}$$

and, similarly, for the Radon transform

$$\begin{aligned}\breve{W}(u,v;c) &= \breve{W}_0(\delta u-\beta v,-\gamma u+\alpha v;c),\\ \breve{W}(\tau,\tau^*;c) &= \breve{W}(\nu^*\tau-\mu\tau^*,-\mu^*\tau+\nu\tau^*;c).\end{aligned} \tag{169}$$

Whereas displacements of a state transform the variable c in the Radon transform, the squeezing of a state leads to a symplectic transformation of the variables (u,v) or (τ,τ^*). The involved matrix is here (as in the case of the Fourier transform) the inverse to the matrix involved in the transformation of the arguments of the Wigner quasiprobability.

4.3 Wigner function for displaced squeezed thermal states

We can now calculate the Wigner quasiprobability and its Fourier and Radon transforms for displaced squeezed thermal states with the density operator ϱ defined by (e.g., [117, 136])

$$\begin{aligned} \varrho &= D(\bar{a},\bar{a}^*)S(\zeta'^*,0,\zeta')\,\varrho_0\,(S(\zeta'^*,0,\zeta'))^\dagger(D(\bar{a},\bar{a}^*))^\dagger, \\ \varrho_0 &= \frac{1}{1+\overline{N}_0}\left(\frac{\overline{N}_0}{1+\overline{N}_0}\right)^N = \frac{1}{1+\overline{N}_0}\sum_{n=0}^{\infty}\left(\frac{\overline{N}_0}{1+\overline{N}_0}\right)^n |n\rangle\langle n|. \end{aligned} \tag{170}$$

We have taken the special squeezing operator $S(\zeta'^*,0,\zeta')$ because the initial thermal states are rotational-invariant. As undisplaced and unsqueezed initial states, we consider the thermal states with a mean value $\overline{N}_0$ of the number operator N. Their Wigner quasiprobability is well known:

$$\begin{aligned} W_0(q,p) &= \frac{1}{\pi\,(1+2\overline{N}_0)\,\hbar}\exp\left[-\frac{q^2+p^2}{(1+2\overline{N}_0)\,\hbar}\right], \\ W_0(\alpha,\alpha^*) &= \frac{2}{\pi\,(1+2\overline{N}_0)}\exp\left[-\frac{2\alpha\alpha^*}{1+2\overline{N}_0}\right]. \end{aligned} \tag{171}$$

The Fourier transforms of thermal states possess the forms

$$\begin{aligned} \tilde{W}_0(u,v) &= \exp\left[-(1+2\overline{N}_0)\,\hbar(u^2+v^2)/4\right], \\ \tilde{W}_0(\tau,\tau^*) &= \exp\left[-(1+2\overline{N}_0)\,\tau\tau^*/2\right]. \end{aligned} \tag{172}$$

By using (154), one finds from (172) for the Radon transform of the Wigner quasiprobability for thermal states

$$\begin{aligned} \breve{W}_0(u,v;c) &= \sqrt{\frac{1}{\pi(1+2\overline{N}_0)\hbar(u^2+v^2)}}\exp\left[-\frac{c^2}{(1+2\overline{N}_0)\hbar(u^2+v^2)}\right], \\ \breve{W}_0(\tau,\tau^*;c) &= \sqrt{\frac{1}{2\pi\,(1+2\overline{N}_0)\,\tau\tau^*}}\exp\left[-\frac{c^2}{2\,(1+2\overline{N}_0)\,\tau\tau^*}\right]. \end{aligned} \tag{173}$$

The transition from the vacuum state ($\overline{N}_0 = 0$) to thermal states can be formally made by the substitution $\hbar \to (1+2\overline{N}_0)\hbar$. However, we emphasize that this is not true for other quasiprobabilities. The Wigner quasiprobability for displaced squeezed thermal states (170) obtained from (171) by the transformations (162) and (166) has the form

$$\begin{aligned} W(q,p) &= \frac{1}{\pi\,(1+2\overline{N}_0)\,\hbar}\exp\left[-\frac{|1+\zeta|^2q'^2+|1-\zeta|^2p'^2+4\,\mathrm{Im}(\zeta)q'p'}{(1+2\overline{N}_0)\,\hbar(1-|\zeta|^2)}\right], \\ W(\alpha,\alpha^*) &= \frac{2}{\pi\,(1+2\overline{N}_0)}\exp\left[-\frac{2(\alpha'+\zeta\alpha'^*)(\alpha'^*+\zeta^*\alpha')}{(1+2\overline{N}_0)\,(1-\zeta\zeta^*)}\right], \end{aligned} \tag{174}$$

where phase-space variables with primes are the displaced corresponding variables without primes (recall $\bar{\alpha} = \bar{a}, \bar{\alpha}^* = \bar{a}^*$ in our case)

$$q' \equiv q - \bar{q}, \quad p' \equiv p - \bar{p}, \qquad \alpha' \equiv \alpha - \bar{a}, \quad \alpha'^* \equiv \alpha^* - \bar{a}^*. \tag{175}$$

For the Fourier transforms of the Wigner quasiprobability, one obtains

$$\begin{aligned}\tilde{W}(u,v) &= \exp\left(-\mathrm{i}(\bar{q}u + \bar{p}v)\right)\\ &\quad\times \exp\left[-\frac{\left(1+2\overline{N}_0\right)\hbar\left(|1-\zeta|^2u^2 + |1+\zeta|^2v^2 - 4\,\mathrm{Im}(\zeta)uv\right.}{4(1-|\zeta|^2)}\right],\\ \tilde{W}(\tau,\tau^*) &= \exp\left(-\mathrm{i}(\bar{a}^*\tau + \bar{a}\tau^*)\right)\\ &\quad\times \exp\left[-\frac{\left(1+2\overline{N}_0\right)(\tau-\zeta\tau^*)(\tau^*-\zeta^*\tau)}{2(1-\zeta\zeta^*)}\right],\end{aligned} \tag{176}$$

and for its Radon transform

$$\begin{aligned}\breve{W}(u,v;c) &= \left[\frac{1-|\zeta|^2}{\pi\left(1+2\overline{N}_0\right)\hbar[\,|1-\zeta|^2u^2 + |1+\zeta|^2v^2 - 4\,\mathrm{Im}(\zeta)uv]}\right]^{1/2}\\ &\quad\times \exp\left[-\frac{(1-|\zeta|^2)(c-\bar{q}u-\bar{p}v)^2}{\left(1+2\overline{N}_0\right)\hbar[\,|1-\zeta|^2u^2 + |1+\zeta|^2v^2 - 4\,\mathrm{Im}(\zeta)uv]}\right],\\ \breve{W}(\tau,\tau^*;c) &= \left[\frac{1-\zeta\zeta^*}{2\pi\left(1+2\overline{N}_0\right)(\tau-\zeta\tau^*)(\tau^*-\zeta^*\tau)}\right]^{1/2}\\ &\quad\times \exp\left[-\frac{(1-\zeta\zeta^*)\left(c-(\bar{a}^*\tau+\bar{a}\tau^*)\right)^2}{2\left(1+2\overline{N}_0\right)(\tau-\zeta\tau^*)(\tau^*-\zeta^*\tau)}\right].\end{aligned} \tag{177}$$

By inserting $u = \cos\varphi, v = \sin\varphi, c = q$ in the expression for $\breve{W}(u,v;c)$, one obtains the reduced Radon transform $\breve{W}(\cos\varphi, \sin\varphi; q)$.

4.4 Husimi and Glauber–Sudarshan quasiprobabilities

Besides the Wigner quasiprobability, there are other quasiprobabilities which carry the whole information of the density operator and possess their specific advantages (and disadvantages). All intensively used quasiprobabilities are Gaussian transforms of the Wigner quasiprobability which can be obtained by convolutions with normalized Gaussian functions. This preserves the displacement structure of quasiprobabilities, which means that a displacement of states leads to a displacement of the whole graphics in the phase space by the displacement vector and is a necessary requirement to speak about quasiprobabilities (e.g., Fourier transforms of quasiprobabilities which also carry complete information of the state are not quasiprobabilities).

The Husimi quasiprobability $Q(\alpha,\alpha^*)$ is defined in the following way:

$$Q(\alpha,\alpha^*) \equiv \frac{\langle\alpha|\varrho|\alpha\rangle}{\pi}, \qquad \int \frac{\mathrm{i}}{2} d\alpha \wedge d\alpha^* Q(\alpha,\alpha^*) = 1. \tag{178}$$

It is similar to the structure of genuine probability densities over the phase space with the only difference that the normalized states $|\alpha\rangle$ are not orthogonal to each other for different $|\alpha\rangle$. We mention that the original definition of Husimi [137] comprises a more general class of positive-semidefinite probability functions (or *propensities* in modern terminology [138]) and that the modern definition was given much later in [138–140] (see also [8]).

The Glauber–Sudarshan quasiprobability $P(\alpha,\alpha^*)$ is (indirectly) defined by the relations [7, 106]

$$\varrho = \int \frac{\mathrm{i}}{2} d\alpha \wedge d\alpha^* P(\alpha,\alpha^*)|\alpha\rangle\langle\alpha|, \quad \int \frac{\mathrm{i}}{2} d\alpha \wedge d\alpha^* P(\alpha,\alpha^*) = 1. \tag{179}$$

In principle, this is the reconstruction formula of the density operator ϱ from the Glauber–Sudarshan quasiprobability $P(\alpha,\alpha^*)$.

There is the class of more general quasiprobabilities, as for example r-ordered quasiprobabilities $F_{(0,0,r)}(\alpha,\alpha^*)$ (in the notation of [113])

$$F_{(0,0,r)}(\alpha,\alpha^*) = \exp\left(\frac{r}{2}\frac{\partial^2}{\partial\alpha\partial\alpha^*}\right) W(\alpha,\alpha^*), \tag{180}$$

which is embedded into the Gaussian class of quasiprobabilities $F_{(r_1,r_2,r_3)}(\alpha,\alpha^*)$ with a vector parameter $\boldsymbol{r} = (r_1, r_2, r_3)$ which we do not treat here. By using (174) for the Wigner quasiprobability and the additivity of the vector parameters in convolutions of Gaussian function (see [113]), one obtains explicitly

$$\begin{aligned} F_{(0,0,r)}(\alpha,\alpha^*) = &\frac{2}{\pi}\sqrt{\frac{1-|\zeta|^2}{\left(1+r+2\overline{N}_0\right)^2(1-|\zeta|^2)+4r\left(1+2\overline{N}_0\right)|\zeta|^2}} \\ \times \exp &\left[-\frac{2\left(1+2\overline{N}_0\right)(\alpha'+\zeta\alpha'^*)(\alpha'^*+\zeta^*\alpha')+r\left(1-|\zeta|^2\right)\alpha'\alpha'^*}{\left(1+r+2\overline{N}_0\right)^2(1-|\zeta|^2)+4r\left(1+2\overline{N}_0\right)|\zeta|^2}\right] \end{aligned} \tag{181}$$

(here $\alpha' \equiv \alpha - \bar{\alpha}$, $\alpha'^* \equiv \alpha^* - \bar{\alpha}^*$). This yields the Husimi quasiprobability ($r = 1$)

$$\begin{aligned} Q(\alpha,\alpha^*) = &\frac{1}{\pi}\sqrt{\frac{1-|\zeta|^2}{\left(1+\overline{N}_0\right)^2-\overline{N}_0^2|\zeta|^2}} \\ \times \exp &\left[-\frac{2\left(1+\overline{N}_0+\overline{N}_0|\zeta|^2\right)\alpha'\alpha'^*+\left(1+2\overline{N}_0\right)\left(\zeta^*\alpha'^2+\zeta\alpha'^{*2}\right)}{2\left(\left(1+\overline{N}_0\right)^2-\overline{N}_0^2|\zeta|^2\right)}\right] \end{aligned} \tag{182}$$

and the Glauber–Sudarshan quasiprobability ($r = -1$)

$$\begin{aligned} P(\alpha,\alpha^*) = &\frac{1}{\pi}\sqrt{\frac{1-|\zeta|^2}{\overline{N}_0^2-\left(1+\overline{N}_0\right)^2|\zeta|^2}} \\ \times \exp &\left[-\frac{2\left(|\zeta|^2+\overline{N}_0+\overline{N}_0|\zeta|^2\right)\alpha'\alpha'^*+\left(1+2\overline{N}_0\right)\left(\zeta^*\alpha'^2+\zeta\alpha'^{*2}\right)}{2\left(\overline{N}_0^2-\left(1+\overline{N}_0\right)^2|\zeta|^2\right)}\right]. \end{aligned} \tag{183}$$

For fixed squeezing parameter ζ, the quadratic form in the exponent of (183) remains negatively definite (or regular) for $\overline{N}_0 > \overline{N}_{0,\mathrm{sing}}$ but becomes singular for $\overline{N}_0 = \overline{N}_{0,\mathrm{sing}}$ and, finally, it becomes indefinite for $\overline{N}_0 < \overline{N}_{0,\mathrm{sing}}$. The value $\overline{N}_{0,\mathrm{sing}}$ is determined by

$$\overline{N}_{0,\mathrm{sing}} = |\zeta| \,/\, (1 - |\zeta|). \tag{184}$$

One can look to this value as the minimal number of thermal photons which one has to add to the squeezed coherent state to get a regular function for the Glauber–Sudarshan quasiprobability and can use this as a measure of nonclassicality of the corresponding state without thermal photons (e.g., [142]). The disadvantage of this measure of nonclassicality is that in most cases of other states it is not easy to define the process of adding thermal photons and then to determine its value. In the considered case, this value is for small squeezing parameters proportional to $|\zeta|$ and does not depend on the position of the squeezing axes in relation to the coordinate axes. For small $|\zeta|$ it is proportional to the measure $1 + r_{\mathrm{sing}}$ in (114) and to the shortest distance to a coherent state in (100) (see [98] for discussion).

4.5 Uncertainty relations and variance matrix

The basis for the derivation of uncertainty relations is the following Cauchy–Bunyakovski–Schwarz inequality for, in general, unnormalized states $|\varphi\rangle$ and $|\chi\rangle$ or operators A and B

$$\langle\varphi|\varphi\rangle\langle\chi|\chi\rangle \geq \langle\varphi|\chi\rangle\langle\chi|\varphi\rangle, \qquad \langle A^\dagger A\rangle\langle B^\dagger B\rangle \geq \langle A^\dagger B\rangle\langle B^\dagger A\rangle. \tag{185}$$

Evidently, $\langle A^\dagger B\rangle$ possesses all properties of a Hermitian scalar product in a Hilbert space of linear operators. First, we derive the uncertainty relations for the canonical operators Q and P by taking into account the uncertainty correlation. We make the derivations on the more general level of mixed states by taking into account that the square root $\sqrt{\varrho}$ of the density operator ϱ is well defined if one takes the eigenvalues as the nonnegative square roots of the (nonnegative) eigenvalues of the density operator and then this is a Hermitian operator $\sqrt{\varrho} = \left(\sqrt{\varrho}\right)^\dagger$. With

$$\Delta Q \equiv Q - \overline{Q}I, \quad \Delta P \equiv P - \overline{P}I, \quad \Rightarrow \quad [\Delta Q, \Delta P] = \mathrm{i}\hbar I, \tag{186}$$

where $\overline{Q}$ and $\overline{P}$ are the expectation values of Q and P, one finds by applying the Cauchy–Bunyakovski–Schwarz operator inequality

$$\begin{aligned}
&\langle \varrho\,(\Delta Q)^2\,\rangle\langle \varrho\,(\Delta P)^2\,\rangle = \langle \sqrt{\varrho}\,\Delta Q\Delta Q\sqrt{\varrho}\,\rangle\langle \sqrt{\varrho}\,\Delta P\Delta P\sqrt{\varrho}\,\rangle \\
&\geq \langle \sqrt{\varrho}\,\Delta Q\Delta P\sqrt{\varrho}\,\rangle\langle \sqrt{\varrho}\,\Delta P\Delta Q\sqrt{\varrho}\,\rangle = \langle \varrho\,\Delta Q\Delta P\rangle\langle \varrho\,\Delta P\Delta Q\rangle \\
&= \frac{1}{4}\left|\langle \varrho\,(\Delta Q\Delta P + \Delta P\Delta Q + [\Delta Q, \Delta P]\,)\rangle\right|^2 .
\end{aligned} \tag{187}$$

The corresponding relations for pure states are obtained by the formal "substitutions" $\langle\sqrt{\varrho} \rightarrow \langle\psi|$ and $\sqrt{\varrho}\,\rangle \rightarrow |\psi\rangle$ and, naturally, $\varrho \rightarrow |\psi\rangle\langle\psi|$. This shows that the derivations for mixed states are not more difficult than for pure states but the minimum states again become pure states. If one uses the commutation relation $[Q,P] = \mathrm{i}\hbar I$ and in addition the inequality between geometric and arithmetic mean, one can represent this in the following form of a chain of inequalities

$$\left[\frac{1}{2}\left(\overline{(\Delta Q)^2} + \overline{(\Delta P)^2}\right)\right]^2 \geq \overline{(\Delta Q)^2}\ \overline{(\Delta P)^2}$$
$$\geq \overline{(\Delta Q)^2}\ \overline{(\Delta P)^2} - \frac{1}{4}\left(\overline{\Delta Q \Delta P} + \overline{\Delta P \Delta Q}\right)^2 \geq \frac{\hbar^2}{4}. \qquad (188)$$

The quantities $\overline{(\Delta Q)^2} + \overline{(\Delta P)^2}$ and $\overline{(\Delta Q)^2}\ \overline{(\Delta P)^2}$ are the uncertainty sum and the uncertainty product. The quantity $\frac{1}{2}\left(\overline{\Delta Q \Delta P} + \overline{\Delta P \Delta Q}\right)$ is called the uncertainty correlation and the last part of the chain of inequalities (188) is a modification of the usual uncertainty relation by the uncertainty correlation (e.g., [143-145]).

The uncertainty sum and the uncertainty product modified by the uncertainty correlation are rotational-invariant. By separating the symmetrical part of this matrix and by transition to the expectation values with the density operator ϱ, we now introduce the following symmetrical variance (or dispersion) matrix $S(\varphi)$ (see (118) for $\hat{R}(\varphi)$)

$$S(\varphi) = \left\| \begin{array}{cc} S_{qq}(\varphi) & S_{qp}(\varphi) \\ S_{pq}(\varphi) & S_{pp}(\varphi) \end{array} \right\| = \left(\hat{R}(\varphi)\right)^{-1} S(0) \hat{R}(\varphi),$$
$$S_{qq}(\varphi) \equiv \overline{(\Delta Q(\varphi))^2}, \quad S_{pp}(\varphi) \equiv \overline{(\Delta P(\varphi))^2},$$
$$S_{qp}(\varphi) \equiv \frac{1}{2}\left(\overline{\Delta Q(\varphi)\Delta P(\varphi)} + \overline{\Delta P(\varphi)\Delta Q(\varphi)}\right) = S_{pq}(\varphi). \qquad (189)$$

One finds explicitly the following transformation of the variance matrix

$$S_{qq}(\varphi) = S_{qq}(0)\cos^2\varphi + 2S_{qp}(0)\cos\varphi\sin\varphi + S_{pp}(0)\sin^2\varphi,$$
$$S_{qp}(\varphi) = S_{qp}(0)(\cos^2\varphi - \sin^2\varphi) - (S_{qq}(0) - S_{pp}(0))\cos\varphi\sin\varphi,$$
$$S_{pp}(\varphi) = S_{qq}(0)\sin^2\varphi - 2S_{qp}(0)\cos\varphi\sin\varphi + S_{pp}\cos^2\varphi. \qquad (190)$$

This is a similarity transformation of the matrix $S(0)$ which leaves invariant the trace $\langle S(\varphi)\rangle$ and the determinant $[S(\varphi)]$

$$\langle S(\varphi)\rangle = S_{qq}(\varphi) + S_{pp}(\varphi) = \langle S(0)\rangle \equiv \langle S\rangle,$$
$$[S(\varphi)] = S_{qq}(\varphi)S_{pp}(\varphi) - S_{qp}(\varphi)S_{pq}(\varphi) = [S(0)] \equiv [S]. \qquad (191)$$

The uncertainty relations can now be written

$$\frac{1}{2}\langle S\rangle \geq \sqrt{[S]} \geq \frac{\hbar}{2}. \qquad (192)$$

The quantities $S_{qq}(\varphi) = \overline{(\Delta Q(\varphi))^2}$ and $S_{pp}(\varphi) = \overline{(\Delta P(\varphi))^2}$ are often taken to define (first-order) squeezing. In this sense, a state is squeezed in direction of the angle φ in the phase plane if

$$\overline{(\Delta Q(\varphi))^2} < \frac{\hbar}{2}, \quad \Rightarrow \quad \overline{(\Delta P(\varphi))^2} > \frac{\hbar}{2}. \qquad (193)$$

Often, the corresponding operators Q and P are multiplied by $\sqrt{2/\hbar}$ to get dimensionless operators X_1 and X_2 which are called the quadrature components of the boson operators (or of the field). The definition (193) admits more general "squeezed" states than the squeezed coherent states since they do not require to minimize the uncertainty product taking into account the uncertainty correlation (last part of (188)). The quantity $\overline{(\Delta Q(\varphi))^2}$ can be expressed by means of the Glauber–Sudarshan quasiprobability $P(q,p)$ in the following way

$$\overline{(\Delta Q(\varphi))^2} = \frac{\hbar}{2} + \int dq \wedge dp P(q,p) \left(q\cos\varphi + p\sin\varphi - \overline{Q(\varphi)} \right)^2 . \quad (194)$$

This expression shows that to satisfy (193) it is necessary that the Glauber–Sudarshan quasiprobability possesses regions of negativities in a certain region of the angle φ. On the other side, the only pure states without negativities in the Glauber–Sudarshan quasiprobability are the coherent states where this quasiprobability is a two-dimensional delta function which is a positively definite linear functional and $\overline{(\Delta Q(\varphi))^2}$ and $\overline{(\Delta P(\varphi))^2}$ are equal $\hbar/2$ for arbitrary φ.

Hermitian operators K_1 and K_2 (1) in the considered realization of $sp(2,\mathrm{R}) \sim su(1,1)$ are some quadratic analogues of the canonical operators Q and P. In analogy to (118), one can define angle-dependent operators $K_1(\varphi)$ and $K_2(\varphi)$ with the following transformation properties

$$\begin{aligned} &(K_1(\varphi), K_2(\varphi)) \equiv R(\varphi)(K_1, K_2)(R(\varphi))^\dagger = (K_1, K_2)\hat{R}(-2\varphi), \\ &K_0(\varphi) = R(\varphi)K_0(R(\varphi))^\dagger = K_0, \end{aligned} \quad (195)$$

Operator K_0 is rotation-invariant. Apart from the sign the essential difference is that herein appears the angle 2φ in comparison to φ in the transformation relations for $(Q(\varphi), P(\varphi))$. This means that the operators $K_1(\varphi)$ and $K_2(\varphi)$ are insensitive with regard to the inversion $\varphi \to \varphi + \pi$ of the direction in the phase space and, furthermore, one has $K_1(-\pi/4) = K_2, K_2(\pi/4) = K_1$. The derivation of uncertainty relations according to (187) can be extended to arbitrary pairs of Hermitian operators. By defining with regard to the Hermitian operators K_1 and K_2

$$\begin{aligned} &\Delta K_1 \equiv K_1 - \overline{K_1} I, \qquad \Delta K_2 \equiv K_2 - \overline{K_2} I, \quad \Rightarrow \\ &[\Delta K_1, \Delta K_2] = -\mathrm{i}K_0 = -\mathrm{i}\,(2N + I)\,/4, \end{aligned} \quad (196)$$

one obtains by substitutions in (187) and (188) the uncertainty relations

$$\begin{aligned} &\frac{1}{4}\left[\overline{(\Delta K_1)^2} + \overline{(\Delta K_2)^2}\right]^2 \geq \overline{(\Delta K_1)^2}\;\overline{(\Delta K_2)^2} \\ &\geq \overline{(\Delta K_1)^2}\;\overline{(\Delta K_2)^2} - \frac{1}{4}\left(\overline{\Delta K_1 \Delta K_2} + \overline{\Delta K_2 \Delta K_1}\right)^2 \\ &\geq \overline{K_0}^2/4 \;=\; \left(2\overline{N} + 1\right)^2/64. \end{aligned} \quad (197)$$

The quantities $\overline{(\Delta K_1)^2}$ and $\overline{(\Delta K_2)^2}$ are often used to define second-order squeezing with regard to $K_1(\varphi)$ by [146, 147] and, e.g., [36, 44, 65]

$$\overline{(\Delta K_1(\varphi))^2} < (2\overline{N} + 1)/8, \quad \Rightarrow \quad \overline{(\Delta K_2(\varphi))^2} > (2\overline{N} + 1)/8, \quad (198)$$

and similarly with regard to $K_2(\varphi)$. The uncertainty sum in (197) and the uncertainty product taking into account the uncertainty correlation are rotation-invariant quantities.

Unlike to the usual Heisenberg uncertainty relations (188), on the right-hand side of (197) stands not a constant value but a value which depends on the considered state. In the usual uncertainty relations (188), one speaks about minimum uncertainty states if the inequality becomes an equality. One can consider this in two variants, taking into account the modification by the uncertainty correlation or not taking it into account. If one does not take into account the uncertainty correlation, one has squeezed coherent states with squeezing axes in the coordinate direction as minimum uncertainty states. By taking into account the modification by the uncertainty correlation (last part of (188)), one gets an equality of this last part for all squeezed coherent states independently of the position of the squeezing axes. Finally the uncertainty sum in (188) takes on its minimal value only for coherent states and all parts of (188) then hold with the equality sign. In the same way, one can consider the states which make the different parts of (197) into equalities. The states which make the last part of (197) into an equality are sometimes called $SU(1,1)$-intelligent states [44, 65, 146–148]. These are all eigenstates of arbitrary linear combinations $K_- - \gamma^2 K_+$ of the operators $K_- \equiv K_1 - \mathrm{i}K_2$ and $K_+ \equiv K_1 + \mathrm{i}K_2$ with arbitrary complex γ as a parameter (they are normalizable only for $|\gamma| < 1$). These eigenstates belong to eigenvalues which are twice degenerate and their Bargmann representation and their wave functions can be expressed by functions of the parabolic cylinder. This and the Fock-state representation are difficult to deal with in generality. However, among them are the eigenstates of the operator a^2 (by choosing $\gamma = 0$) which can be taken as the even and odd coherent states $N_\pm(|\alpha\rangle \pm |-\alpha\rangle)$ introduced in [58, 151]. They belong to the large category of finite superpositions of coherent states, sometimes called Schrödinger cat states [152].

4.6 Moments up to second order and variance matrix

We now give the explicit form of the symmetrically ordered moments for displaced squeezed thermal states up to the second order and discuss representations of the Wigner quasiprobability and its transforms by the variance matrix. The first-order moments are already taken into account in the notations of the parameters of the states according to

$$\overline{Q} = \bar{q}, \qquad \overline{P} = \bar{p}, \qquad \overline{a} = \bar{a}, \qquad \overline{a^\dagger} = \bar{a}^*. \tag{199}$$

For the second-order moments of the canonical operators, one finds

$$\begin{aligned}
\overline{Q^2} &= \overline{Q}^2 + \frac{\hbar}{2}\left(1 + 2\overline{N}_0\right)\frac{(1-\zeta)(1-\zeta^*)}{1-|\zeta|^2},\\
\overline{P^2} &= \overline{P}^2 + \frac{\hbar}{2}\left(1 + 2\overline{N}_0\right)\frac{(1+\zeta)(1+\zeta^*)}{1-|\zeta|^2},\\
\frac{1}{2}\left(\overline{QP} + \overline{PQ}\right) &= \overline{Q}\,\overline{P} + \frac{\hbar}{2}\left(1 + 2\overline{N}_0\right)\frac{\mathrm{i}(\zeta-\zeta^*)}{1-|\zeta|^2}.
\end{aligned} \tag{200}$$

The second-order moments of the boson operators are

$$\overline{a^2} = \bar{a}^2 - \frac{(1+2\overline{N}_0)\,\zeta}{1-|\zeta|^2}, \qquad \overline{a^{\dagger 2}} = \bar{a}^{*2} - \frac{(1+2\overline{N}_0)\,\zeta^*}{1-|\zeta|^2},$$
$$\frac{1}{2}\left(\overline{aa^\dagger} + \overline{a^\dagger a}\right) = \bar{a}\,\bar{a}^* + \frac{(1+2\overline{N}_0)\,(1+|\zeta|^2)}{2(1-|\zeta|^2)}. \tag{201}$$

The expectation value of the number operator is

$$\overline{N} \equiv \overline{a^\dagger a} = \bar{a}\,\bar{a}^* + \frac{1+|\zeta|^2}{1-|\zeta|^2}\overline{N}_0 + \frac{|\zeta|^2}{1-|\zeta|^2}. \tag{202}$$

From (201) and (202), one finds for the elements of the matrix $S(\varphi)$

$$S_{qq}(\varphi) = \frac{\hbar}{2}\,(1+2\overline{N}_0)\,\frac{(1-\mathrm{e}^{-\mathrm{i}2\varphi}\zeta)(1-\mathrm{e}^{\mathrm{i}2\varphi}\zeta^*)}{1-|\zeta|^2},$$
$$S_{pp}(\varphi) = \frac{\hbar}{2}\,(1+2\overline{N}_0)\,\frac{(1+\mathrm{e}^{-\mathrm{i}2\varphi}\zeta)(1+\mathrm{e}^{\mathrm{i}2\varphi}\zeta^*)}{1-|\zeta|^2},$$
$$S_{qp}(\varphi) = \frac{\hbar}{2}\,(1+2\overline{N}_0)\,\frac{\mathrm{i}(\mathrm{e}^{-\mathrm{i}2\varphi}\zeta - \mathrm{e}^{\mathrm{i}2\varphi}\zeta^*)}{1-|\zeta|^2}. \tag{203}$$

The uncertainty product depends on the angle φ according to

$$\overline{(\Delta Q(\varphi))^2}\;\overline{(\Delta P(\varphi))^2} = \frac{\hbar^2}{4}\,(1+2\overline{N}_0)^2\left[1 - \left(\frac{\mathrm{e}^{-\mathrm{i}2\varphi}\zeta - \mathrm{e}^{\mathrm{i}2\varphi}\zeta^*}{1-|\zeta|^2}\right)^2\right]. \tag{204}$$

It becomes the minimal one with fixed parameters $(\overline{N}_0, \bar{a}, \bar{a}^*, \zeta, \zeta^*)$ for four angles $\varphi_{\min}$ determined by vanishing of $S_{qp}(\varphi)$, which means

$$\exp(\mathrm{i}4\varphi_{\min}) = \zeta/\zeta^*. \tag{205}$$

The angles $\varphi_{\min}$ determine the directions of the squeezing axes in the phase space. The uncertainty sum and the modified uncertainty product (taking into account the uncertainty correlation) are independent on angle φ:

$$\langle S\rangle = \hbar\,(1+2\overline{N}_0)\,\frac{1+|\zeta|^2}{1-|\zeta|^2}, \qquad [S] = \frac{\hbar^2}{4}\,(1+2\overline{N}_0)^2. \tag{206}$$

The uncertainty relations (188) take on the following specialized form

$$\frac{1}{4}\langle S\rangle^2 \geq \overline{(\Delta Q(\varphi))^2}\;\overline{(\Delta P(\varphi))^2} \geq [S] = \frac{\hbar^2}{4}\,(1+2\overline{N}_0)^2 \geq \frac{\hbar^2}{4}. \tag{207}$$

It is convenient to introduce the following notations for two-dimensional vectors over the phase space and its dual space

$$\mathbf{x} \equiv (q,p), \quad \bar{\mathbf{x}} \equiv (\bar{q},\bar{p}), \quad \mathbf{y} \equiv (u,v). \tag{208}$$

The Wigner quasiprobability and its Fourier and Radon transforms for displaced squeezed thermal states in real representation can now be written in the form [117]

$$W(\mathbf{x}) = \left(2\pi\sqrt{[S]}\right)^{-1} \exp\left[-\frac{1}{2}(\mathbf{x}-\bar{\mathbf{x}})S^{-1}(\mathbf{x}-\bar{\mathbf{x}})\right],$$

$$\tilde{W}(\mathbf{y}) = \exp\left[-\mathrm{i}\mathbf{y}\bar{\mathbf{x}} - \frac{1}{2}\mathbf{y}S\mathbf{y}\right],$$

$$\breve{W}(\mathbf{y};c) = (2\pi\mathbf{y}S\mathbf{y})^{-1/2} \exp\left[-\frac{(c-\mathbf{y}\bar{\mathbf{x}})^2}{2\mathbf{y}S\mathbf{y}}\right], \tag{209}$$

$$\int d^2\mathbf{x} W(\mathbf{x}) = 1, \quad \int_{-\infty}^{\infty} dc\, \breve{W}(\mathbf{y};c) = 1.$$

In more detailed form these functions can be represented by

$$\begin{aligned} W(q,p) &= \left(2\pi\sqrt{S_{qq}S_{pp}-S_{qp}^2}\right)^{-1} \\ &\times \exp\left[-\frac{S_{pp}(q-\bar{q})^2 + S_{qq}(p-\bar{p})^2 - 2S_{qp}(q-\bar{q})(p-\bar{p})}{2(S_{qq}S_{pp}-S_{qp}^2)}\right], \end{aligned}$$

$$\tilde{W}(u,v) = \exp\left[-\mathrm{i}(u\bar{q}+v\bar{p}) - \frac{1}{2}\left(S_{qq}u^2 + S_{pp}v^2 + 2S_{qp}uv\right)\right], \tag{210}$$

$$\begin{aligned} \breve{W}(u,v;c) &= \left[2\pi(S_{qq}u^2 + S_{pp}v^2 + 2S_{qp}uv)\right]^{-1/2} \\ &\times \exp\left[-\frac{(c-u\bar{q}-v\bar{p})^2}{2(S_{qq}u^2 + S_{pp}v^2 + 2S_{qp}uv)}\right]. \end{aligned} \tag{211}$$

The reduced Radon transform $\breve{W}(\cos\varphi, \sin\varphi; q)$ can also be represented by

$$\breve{W}(\cos\varphi, \sin\varphi; q) = \frac{1}{\sqrt{2\pi S_{qq}(\varphi)}} \exp\left[-\frac{(q-\bar{q}\cos\varphi - \bar{p}\sin\varphi)^2}{2S_{qq}(\varphi)}\right].$$

The Fourier transform $\tilde{W}(u,v)$ of the Wigner function $W(q,p)$ for displaced squeezed thermal states possesses the form of Gaussian functions normalized as $\tilde{W}(0,0) = 1$. If we write this Fourier transform in more general cases as

$$\tilde{W}(u,v) = \exp\left(\Phi(u,v)\right), \qquad \tilde{W}(y) = \exp\left(\Phi(y)\right), \tag{212}$$

then its special form (210) can be considered as the Taylor series expansions of the exponent $\Phi(u,v)$ in powers of $-\mathrm{i}u$ and $-\mathrm{i}v$, where this expansion becomes truncated with the quadratic powers. The coefficients in front of the linear terms in (u,v) form the displacement vector and the coefficients in front of the quadratic terms in (u,v) the variance matrix. If we continue this Taylor series expansion to the so-called cumulant expansion, then from cubic terms on, we get the cumulant tensors as coefficients in front of the higher powers of $-\mathrm{i}u$ and $-\mathrm{i}v$. They are exactly vanishing for displaced squeezed thermal states which are the most general states with Gaussian form of the Wigner quasiprobability and of its Fourier transform. One can establish general relations between the symmetrically ordered moments and the cumulants (see [60]). In the considered case of squeezed thermal states all symmetrically ordered moments are determined by the displacement vector $(\bar{q}, \bar{p})$ and by the variance matrix S according to

$$\langle \varrho\, \mathcal{S}\{Q^m P^n\}\rangle = \sum_{j=0}^{\{m,n\}} \frac{m!n!\,(S_{qp})^j}{j!(m-j)!(n-j)!} \left(-\mathrm{i}\sqrt{\frac{S_{qq}}{2}}\right)^{m-j} \left(-\mathrm{i}\sqrt{\frac{S_{pp}}{2}}\right)^{n-j} \times H_{m-j}\left(\frac{\mathrm{i}\bar{q}}{\sqrt{2S_{qq}}}\right) H_{n-j}\left(\frac{\mathrm{i}\bar{p}}{\sqrt{2S_{pp}}}\right). \qquad (213)$$

For $S_{qp} = 0$, the moments $\langle \varrho\, \mathcal{S}\{Q^m P^n\}\rangle$ split into products of the moments $\langle \varrho\, \mathcal{S}\{Q^m\}\rangle$ and $\langle \varrho\, \mathcal{S}\{P^n\}\rangle$.

4.7 Second-order squeezing

We now discuss second-order squeezing in connection with the inequalities (197). The quantities which enter the inequalities possess for displaced squeezed thermal states the following specialized form

$$\begin{aligned}
\overline{(\Delta K_1)^2} &= \frac{1+2\overline{N}_0}{8(1-|\zeta|^2)}\left[2(\bar{a}-\zeta^*\bar{a}^*)(\bar{a}^*-\zeta\bar{a}) + (1+2\overline{N}_0)\,\frac{(1+\zeta^2)\,(1+\zeta^{*2})}{1-|\zeta|^2}\right],\\
\overline{(\Delta K_2)^2} &= \frac{1+2\overline{N}_0}{8(1-|\zeta|^2)}\left[2(\bar{a}+\zeta^*\bar{a}^*)(\bar{a}^*+\zeta\bar{a}) + (1+2\overline{N}_0)\,\frac{(1-\zeta^2)\,(1-\zeta^{*2})}{1-|\zeta|^2}\right],\\
\overline{K}_0 &= \frac{1}{4}\left(2\bar{a}\bar{a}^* + (1+2\overline{N}_0)\,\frac{1+|\zeta|^2}{1-|\zeta|^2}\right).
\end{aligned} \qquad (214)$$

Therefore the generalized uncertainty product

$$(\Delta K)^2 \equiv \overline{(\Delta K_1)^2}\;\overline{(\Delta K_2)^2} - \frac{1}{4}\left(\overline{\Delta K_1 \Delta K_2} + \overline{\Delta K_2 \Delta K_1}\right)^2$$

satisfies the relations

$$\begin{aligned}
(\Delta K)^2 &= \frac{(1+2\overline{N}_0)^2}{64}\left[\left(2|\bar{a}|^2 + (1+2\overline{N}_0)\,\frac{1+|\zeta|^2}{1-|\zeta|^2}\right)^2 + \frac{8\,(1+2\overline{N}_0)\,|\zeta\beta|^2}{(1-|\zeta|^2)^3}\right]\\
&= (1+2\overline{N}_0)^2\left[\frac{1}{4}\overline{K}_0^2 + \frac{(1+2\overline{N}_0)\,|\zeta|^2\beta\beta^*}{8\,(1-|\zeta|^2)^2}\right] \geq \frac{1}{4}\overline{K}_0^2,
\end{aligned} \qquad (215)$$

where β is the eigenvalue parameter introduced in the nonunitary approach (64). This can become an equality only for $\overline{N}_0 = 0$ as a necessary but not sufficient condition. Under $\overline{N}_0 = 0$ it becomes an equality for $\zeta = 0$ (that means for coherent states) or for $\beta = \beta^* = 0$ (that means for squeezed vacuum states). In the last case, the squeezed vacuum states $|(0,\zeta)\rangle$ are eigenstates of the operator $a+\zeta a^\dagger$ to eigenvalue $\beta = 0$. Consequently (in notations of the nonunitary approach),

$$\left(a^2 - \zeta^2 a^{\dagger\,2} + \zeta I\right)|(0,\zeta)\rangle = (a - \zeta a^\dagger)(a + \zeta a^\dagger)|(0,\zeta)\rangle = 0, \tag{216}$$

$$\left(K_- - \zeta^2 K_+\right)|(0,\zeta)\rangle = \left[\left(1-\zeta^2\right)K_1 - \mathrm{i}\left(1+\zeta^2\right)K_2 - \frac{\zeta}{2}\right]|(0,\zeta)\rangle.$$

This shows that the squeezed vacuum states $|(0,\zeta)\rangle$ are eigenstates of a linear combination of the operators K_- and K_+ or K_1 and K_2. More generally, exactly all normalizable eigenstates of arbitrary linear combinations of the operators K_- and K_+ make the last part of the inequalities (197) to an equality. These eigenstates are normalizable only if the part of K_- is larger than the part of K_+ in the linear combinations, similarly to the classification (71)–(73). One can pose a more general eigenvalue problem for arbitrary linear combinations of the three operators K_-, K_+ and K_0 which is a difficult but beautiful problem that cannot be treated here. As already mentioned, in special cases such as, for example, the eigenstates of K_- (coherent states, even and odd coherent states) this problem simplifies considerably. The solutions of the general eigenvalue problem for arbitrary linear combinations of K_-, K_+ and K_0 comprises a very large spectrum of interesting states, in principle, including all intensively used pure states in quantum optics.

4.8 An important two-mode realization of $SU(1,1)$

We cannot consider here the problem of establishing the general (fractal) single-mode and two-mode realizations of $SU(1,1)$ (the last problem seems to be not fully solved) and restrict ourselves to the important case of the most common two-mode realization of $SU(1,1)$. It is determined by the following realization of the basis operators of the Lie algebra $su(1,1)$ by quadratic combinations of the boson operators $(a_1, a_1^\dagger, a_2, a_2^\dagger)$ of the two modes

$$K_- = a_1 a_2, \quad K_0 = \frac{1}{4}\left(a_1 a_1^\dagger + a_1^\dagger a_1 + a_2 a_2^\dagger + a_2^\dagger a_2\right), \quad K_+ = a_1^\dagger a_2^\dagger,$$
$$C = \frac{1}{4}\left[\left(a_1^\dagger a_1 - a_2^\dagger a_2\right)^2 - I\right] = \frac{1}{4}\left[(N_1 - N_2)^2 - I\right]. \tag{217}$$

We have here two different sets of "$SU(1,1)$-vacuum" states $|0,l\rangle$ and $|l,0\rangle$ with $l = 0,1,2,\ldots$ which are annihilated by the operator K_- according to

$$K_-|0,l\rangle = a_1 a_2 |0,l\rangle = 0, \qquad K_-|l,0\rangle = a_1 a_2 |l,0\rangle = 0, \tag{218}$$

and belong to different irreducible representations (irreps) of $SU(1,1)$. Due to

$$K_-|n,n+l\rangle = \sqrt{n(n+l)}\,|n-1,n+l-1\rangle,$$
$$K_+|n,n+l\rangle = \sqrt{(n+1)(n+l+1)}\,|n+1,n+l+1\rangle,$$
$$K_0|n,n+l\rangle = \left[n + \tfrac{1}{2}(l+1)\right]|n,n+l\rangle, \tag{219}$$

the series of states $|n,n+l\rangle$ with fixed l form a basis of an irrep of $SU(1,1)$ with the label $k = \frac{1}{2}(l+1)$ [54]. In full analogy, the states $|n+l,n\rangle$ with fixed l form a basis of an irrep of $SU(1,1)$ with the label $k = \frac{1}{2}(l+1)$. With the exception of $k = \frac{1}{2}$ with the two-mode vacuum state $|0,0\rangle$ as ground state, all irreps with the label $k = \frac{1}{2}(l+1)$, $l = 1, 2, \ldots$ are twofold degenerate within the two-mode Fock states $|n_1, n_2\rangle$ and arbitrary superpositions of $|n, n+l\rangle$ and $|n+l, n\rangle$ belong to the same irrep with the label $k = \frac{1}{2}(l+1)$. The difference of the numbers of the Fock states involved in the two modes remains constant and the Casimir operator C is proportional to the unit operator I of these particular Hilbert spaces according to $C = k(k-1)I$.

One of the basic problems is to establish the transformations of the boson operators by squeezing transformations with the operators

$$S(\xi, \eta, \zeta) \equiv \exp(\xi K_- + \mathrm{i}2\eta K_0 - \zeta K_+)$$

(extension to $Sp(2, \mathrm{C})$). In an analogous way as presented in section 2, this can be written in the following form by a 4×4 matrix $\hat{X}$

$$S(\xi,\eta,\zeta)\big((a_1, a_2^\dagger), (a_2, a_1^\dagger)\big)S^{-1}(\xi,\eta,\zeta) = \big((a_1, a_2^\dagger), (a_2, a_1^\dagger)\big)\hat{X},$$

$$\hat{X} = \begin{pmatrix} \left\| \begin{matrix} \kappa & \lambda \\ \mu & \nu \end{matrix} \right\| & \left\| \begin{matrix} 0 & 0 \\ 0 & 0 \end{matrix} \right\| \\ \left\| \begin{matrix} 0 & 0 \\ 0 & 0 \end{matrix} \right\| & \left\| \begin{matrix} \kappa & \lambda \\ \mu & \nu \end{matrix} \right\| \end{pmatrix}, \tag{220}$$

where 2×2 diagonal matrices are exactly the same as discussed for the single mode in (22). These transformations describe an outer automorphism of the Heisenberg–Weyl algebra $w(2, \mathrm{R})$ which can be written in the form

$$\begin{aligned} S(\xi,\eta,\zeta)\big(a_1, a_2^\dagger\big)\big(S(\xi,\eta,\zeta)\big)^{-1} &= \big(\kappa a_1 + \mu a_2^\dagger, \lambda a_1 + \nu a_2^\dagger\big), \\ S(\xi,\eta,\zeta)\big(a_2, a_1^\dagger\big)\big(S(\xi,\eta,\zeta)\big)^{-1} &= \big(\kappa a_2 + \mu a_1^\dagger, \lambda a_2 + \nu a_1^\dagger\big). \end{aligned} \tag{221}$$

The corresponding relations for the transformation of the canonical operators (Q_1, P_1, Q_2, P_2) are more complicated. They do not split into two independent groups of transformations and involve linear combination of all the 4 operators (Q_1, P_1, Q_2, P_2). Another substructure of the 4×4 matrices by 2×2 matrices is provided if one considers the subgroup $SU(2)$ of $Sp(4, \mathrm{R})$ in the most common two-mode realization [62]. The most general outer automorphism group to the Heisenberg–Weyl group $w(2, \mathrm{R})$ is, however, the symplectic group $Sp(4, \mathrm{R})$ with 10 basis operators and therefore 10 involved parameters and with the general symplectic 4×4 matrix in its fundamental representation. The four-dimensional fundamental representation becomes very complicated and it is not clear whether the inverse determination of the 10 parameters from the 10 independent elements of the representation matrix can or cannot be found in closed form.

Two-mode squeezed coherent states can be defined similarly as in the single-mode case by applying the squeezing and displacement operator to the two-mode vacuum state $|0,0\rangle$ [33, 44, 153–157]. The argument transformations of the Wigner quasiprobability $W(\alpha_1, \alpha_2, \alpha_1^*, \alpha_2^*)$ under displacement and squeezing are in analogy to the single-mode case. In particular, if an initially unsqueezed state described by the density operator ϱ_0 underlies squeezing described by the unitary operator $S(\zeta^*, \eta = \eta^*, \zeta)$ according to

$$\varrho = S(\zeta^*, \eta = \eta^*, \zeta)\varrho_0 S^\dagger(\zeta^*, \eta = \eta^*, \zeta), \tag{222}$$

then the corresponding Wigner quasiprobabilities are related by

$$W(\alpha_1, \alpha_2, \alpha_1^*, \alpha_2^*) = W_0(\nu^*\alpha_1 + \mu\alpha_2^*, \nu^*\alpha_2 + \mu\alpha_1^*, \mu^*\alpha_1 + \nu\alpha_2^*, \mu^*\alpha_2 + \nu\alpha_1^*). \tag{223}$$

This means that the induced argument transformations are similar to the transformations in the case of one-mode squeezing (166). Due to the rotation invariance of the two-mode vacuum state $|0,0\rangle$, two-mode squeezed coherent states with different displacement parameters $\bar{a}_1$ and $\bar{a}_2$ can be defined by means of the special squeezing operator $S(\zeta'^*, 0, \zeta')$ according to

$$|\bar{a}_1, \bar{a}_2; \zeta\rangle \equiv \exp\left(\bar{a}_1 a_1^\dagger - \bar{a}_1^* a_1 + \bar{a}_2 a_2^\dagger - \bar{a}_2^* a_2\right) S(\zeta'^*, 0, \zeta')|0,0\rangle, \tag{224}$$

with the relation between ζ and ζ' given in (46). For vanishing displacement parameters $\bar{a}_1 = \bar{a}_2 = 0$, one obtains two-mode squeezed vacuum states which can be represented in the following form

$$|0,0;\zeta\rangle \equiv S(\zeta'^*, 0, \zeta')|0,0\rangle = \sqrt{1-|\zeta|^2}\exp\left(-\zeta a_1^\dagger a_2^\dagger\right)|0,0\rangle, \tag{225}$$

where the decomposition (44) with $\xi = \zeta^*$ is used together with the special meaning of (K_-, K_0, K_+) in (217).

We now make a transformation of the two modes with labels $(1,2)$ to two new modes with labels $(+,-)$ according to

$$a_+ = (a_1 + a_2)/\sqrt{2}, \qquad a_- = (a_1 - a_2)/\sqrt{2}. \tag{226}$$

This is a special symplectic transformation and, therefore, it preserves the Heisenberg commutation relations. Under this transformation, the realization (217) of the Lie algebra $SU(1,1)$ takes on the following form

$$\begin{aligned} K_- &= \frac{1}{2}\left(a_+^2 - a_-^2\right), & K_0 &= \frac{1}{4}\left(a_+ a_+^\dagger + a_+^\dagger a_+ + a_- a_-^\dagger + a_-^\dagger a_-\right), \\ K_+ &= \frac{1}{2}\left(a_+^{\dagger 2} - a_-^{\dagger 2}\right), & C &= \frac{1}{4}\left[\left(a_+^\dagger a_- + a_-^\dagger a_+\right)^2 - I\right]. \end{aligned} \tag{227}$$

In the new mode picture, the Lie algebra operators (K_-, K_0, K_+) split into the sum of two partial operators which act only in the partial Hilbert spaces of one mode without coupling. An initially disentangled state remains disentangled after application of Lie group operators. However, the operators act at once in both particular spaces in a similar way, which means that squeezing in one mode is accompanied by squeezing in the other mode, but the modes remain disentangled if they were disentangled initially.

There are many problems which may be considered for two-mode systems. For example, one may consider the behaviour of squeezed coherent states at a beam splitter. Here we have some interference between the two groups $SU(1,1)$ of squeezing and of $SU(2)$ which describes the beam splitter and which are both different subgroups of the 10-parameter symplectic group $Sp(4,\mathrm{R})$ which is the maximal linear group preserving the Heisenberg commutation relations of a two-mode system [outer automorphism group to Heisenberg–Weyl group $W(2,\mathrm{R})$]. The treatment of this group and of its unification with the Heisenberg–Weyl group for two modes seems to be very attractive but is also connected with some difficulties, since many necessary basic relations have not yet been explicitly derived (e.g., fundamental four-dimensional representation and its inversion).

5 Conclusion

We have considered squeezing of states mainly in the case of a single boson mode. Since the squeezing operators can be applied to arbitrary states, it was very effective to consider the transformations of quasiprobabilities and of their Fourier and Radon transforms caused by squeezing and, in addition, by displacement of states. The special case of displaced squeezed thermal states is the most general case which leads to Gaussian Wigner quasiprobabilities. By specialization to temperature zero, it leads to squeezed coherent states which form the most general pure states with Gaussian Wigner quasiprobabilities. We have derived in explicit form the Wigner quasiprobability and its Fourier and Radon transforms, furthermore, the ordered moments, and in the case of squeezed coherent states the wave functions in position and momentum representation, the Bargmann representation and photon statistics of the states. However, we could not treat in detail squeezed-state excitations as, for example, displaced squeezed Fock states or photon-added squeezed states.

The volume of the chapter also did not allow us to discuss in detail many interesting problems connected with squeezing, for example, the unitary representations of $SU(1,1)$ and the general theory of $SU(1,1)$-coherent states (e.g., [54,70,71]) and different (fractal) realizations of $SU(1,1)$ by single-mode boson operators with application to coherent phase states. The same is true with the general eigenvalue problem for squeezing operators which is a very complicated problem. It was not possible to proceed to an analogous description of the basic realizations of $Sp(2,\mathrm{R}) \sim SU(1,1)$ for the two-mode case and to embed this group into the more general symplectic group $Sp(4,\mathrm{R})$ and its complexification $Sp(4,\mathrm{C})$ which, in addition to $SU(1,1)$, contains $SU(2)$ as one of its subgroups with interesting applications to polarization of light and to the lossless beam splitter. We also could not provide a section about the representation of the squeezing operators themselves [117]. Moreover, some interesting characteristics such as the quasiprobabilities for coherent phase states lead to great difficulties concerning their calculation. A puzzling paradox connected with convergence problems in normally ordered expansion of vacuum expectation values was shown in [79] in connection with higher-order squeeze operators and a way to its solution was announced in [159]. In the case of two-mode realizations of $SU(1,1)$, we considered the most common example but could not achieve a similar completeness as in the analogous single-mode case. We also did not touch transformations of squeezed states in different processes and devices such as, for example, $SU(2)$-interferometers and, concretely, beam splitters and light polarization. Our aim was to develop reliable recipes for the treatment of squeezing operations starting from representation theory of $Sp(2, \mathrm{R}) \sim SU(1,1)$ and to apply it to the simplest states which, in some regard, are the coherent states and the displaced thermal states, and to derive their most important characteristics.

Bibliography

[1] Wigner, E., On the quantum corrections for thermodynamic equilibrium. *Phys. Rev.* (1932) **40** 749–759.

[2] Moyal, J.E., Quantum mechanics as a statistical theory. *Proc. Camb. Phil. Soc.* (1949) **45** 99–124.

[3] Kennard, E.H., Zur Quantenmechanik einfacher Bewegungstypen. *Z. Phys.* (1927) **44** 326–352.

[4] Nieto, M.M., The discovery of squeezed states – in 1927. In: *Fifth Int. Conf. on Squeezed States and Uncertainty Relations, Balatonfured, 1997* (D. Han, J. Janszky, Y.S. Kim, and V.I. Man'ko, eds.), NASA Conf. Publ. NASA/CP-1998-206855, pp. 175–180. Goddard Space Flight Center, Greenbelt, 1998.

[5] Pauli, W., *Die allgemeinen Prinzipien der Wellenmechanik.* In: *Handbuch der Physik, vol. V, part 1* (S. Flügge, ed.), pp. 1–168. Springer, Berlin, 1958.

[6] Schrödinger, E., Der stetige Übergang von der Mikro- zur Makromechanik. *Naturwissenschaften* (1926) **14** 664–666.

[7] Glauber, R.J., Coherent and incoherent states of the radiation field. *Phys. Rev.* (1963) **131** 2766–2788 (reprinted in [10, 16]).

[8] Glauber, R.J., Optical coherence and statistics of photons. In: *Quantum Optics and Electronics* (C. DeWitt, A. Blandin, and C. Cohen-Tannoudji, eds.), pp. 63–185. Gordon & Breach, New York, 1965.

[9] Klauder, J.R. and Sudarshan, E.C.G., *Fundamentals of Quantum Optics*. W.A. Benjamin, New York, 1968.

[10] *Coherent States* (J.R. Klauder and B.-S. Skagerstam, eds.). World Scientific, Singapore, 1985.

[11] Stoler, D., Equivalence classes of minimum uncertainty packets. I. *Phys. Rev.* D (1970) **1** 3217–3219 (reprinted in [16]).

[12] Stoler, D. Equivalence classes of minimum uncertainty packets. II. *Phys. Rev.* D (1971) **4** 1925–1926.

[13] Yuen, H.P., Two-photon coherent states of the radiation field. *Phys. Rev.* A (1976) **13** 2226–2243 (reprinted in [16]).

[14] Yuen, H.P., Nonclassical light. In: *Photons and Quantum Fluctuations* (E.R. Pike and H. Walther, eds.), pp. 1–9. Adam Hilger, Bristol, 1988.

[15] Hollenhorst, J.N., Quantum limits on resonant-mass gravitational-radiation detectors. *Phys. Rev.* D (1979) **19** 1669–1679 (reprinted in [10]).

[16] *Nonclassical Effects in Quantum Optics* (P. Meystre and D.F. Walls, eds.). American Institute of Physics, New York, 1991.

[17] Walls, D.F., Squeezed states of light. *Nature* (1983) **306** 141–146 (reprinted in [16]).

[18] Loudon, R. and Knight, P.L., Squeezed light. *J. Modern Opt.* (1987) **34** 709–759.

[19] Teich, M.C. and Saleh, B.E.A., Squeezed state of light. *Quant. Opt.* (1989)**1** 153–191.

[20] Zaheer, K. and Zubairy, M.S., Squeezed states of the radiation field. In: *Advances in Atomic, Molecular and Opticsl Physics, vol. 28* (D.R. Bates and B. Bederson, eds.), pp. 143–235. Academic, New York, 1990.
[21] Zhang, W.M., Feng, D.H., and Gilmore, R., Coherent states: Theory and some applications. *Rev. Mod. Phys.* (1990) **62** 867–927.
[22] Klyshko, D.N., Non-classical light. *Uspekhi Fiz. Nauk* (1996) **166** 613–638 [*Physics – Uspekhi* (1996) **39** 573].
[23] *Squeezed States of the Electromagnetic Field*, Special issue of *J. Opt. Soc. Am.* B (1987) **4** 1453–1741.
[24] *Special Issue on Squeezed Light, J. Mod. Opt.* (1987) **34** 707–1020.
[25] *Squeezed and Nonclassical Light* (P. Tombesi and E.R. Pike, eds.). Plenum, New York, 1989.
[26] *Workshop on Squeezed States and Uncertainty Relations, Maryland, USA, 1991* (D. Han, Y.S. Kim, and W.W. Zachary, eds.), NASA Conf. Publ. 3135. Goddard Space Flight Center, Greenbelt, 1992.
[27] *Second International Workshop on Squeezed States and Uncertainty Relations, Moscow, Russia, 1992* (D. Han, Y.S. Kim, and V.I. Man'ko, eds.), NASA Conf. Publ. 3219. Goddard Space Flight Center, Greenbelt, 1993.
[28] *Third International Workshop on Squeezed States and Uncertainty Relations, Baltimore, USA, 1993* (D. Han, Y.S. Kim, M.H. Rubin, Y. Shih, and W.W. Zachary, eds.), NASA Conf. Publ. 3270. Goddard Space Flight Center, Greenbelt, 1994.
[29] *Fourth International Conference on Squeezed States and Uncertainty Relations, Tayuan, P.R. China, 1995* (D. Han, K. Peng, Y.S. Kim, and V.I. Man'ko, eds.), NASA Conf. Publ. 3322. Goddard Space Flight Center, Greenbelt, 1996.
[30] *Fifth International Conference on Squeezed States and Uncertainty Relations, Balatonfured, Hungary, 1997* (D. Han, J. Janszky, Y.S. Kim, and V.I. Man'ko, eds.), NASA Conf. Publ. 206855. Goddard Space Flight Center, Greenbelt, 1998.
[31] *Sixth International Conference on Squeezed States and Uncertainty Relations, Naples, Italy, 1999* (D. Han, Y.S. Kim, and S. Solimeno, eds.), NASA Conf. Publ. 209899. Goddard Space Flight Center, Greenbelt, 2000.
[32] Meystre, P. and Sargent III, M., *Elements of Quantum Optics*. Springer, Berlin, 1990.
[33] Kim, Y.S. and Noz, M.E., *Phase Space Picture of Quantum Mechanics*. World Scientific, Singapore, 1991.
[34] Carmichael, H., *An Open System Approach to Quantum Optics*. Springer, Berlin, 1993.
[35] Walls, D.F. and Milburn, G.J., *Quantum Optics*. Springer, Berlin, 1994.
[36] Peřina, J., Hradil, Z., and Jurčo, B., *Quantum Optics and Fundamental Physics*. Kluwer, Dordrecht, 1994.
[37] Vogel, W. and Welsch, D.-G., *Lectures on Quantum Optics*. Akademie-Verlag, Berlin, 1994.
[38] Mandel, L. and Wolf, E., *Optical Coherence and Quantum Optics*. Cambridge University Press, Cambridge, 1995.
[39] Pike, E.R. and Sarkar, S., *The Quantum Theory of Radiation*. Clarendon Press, Oxford, 1995.
[40] Scully, M.O. and Zubairy, M.S., *Quantum Optics*. Cambridge Univerity Press, Cambridge, 1997.
[41] Fan, H.-Y., *Representation and Transformation Theory in Quantum Mechanics* (in Chinese). Shanghai Scientific and Technical Publishers, Shanghai, 1997.
[42] Barnett, S.M. and Radmore, P.M., *Methods in Theoretical Quantum Optics*. Claren-

don Press, Oxford, 1997.

[43] Leonhardt, U., *Measuring the Quantum State of Light*. Cambridge University Press, Cambridge, 1997.

[44] Peřinová, V., Lukš, A., and Peřina, J., *Phase in Optics*. World Scientific, Singapore, 1998.

[45] Bachor, H.-A., *A Guide to Experiments in Quantum Optics*. Wiley-VCH, Weinheim, 1998.

[46] Carmichael, H.J., *Statistical Methods in Quantum Optics I*. Springer, Berlin, 1999.

[47] Weyl, H., *The Theory of Groups and Quantum Mechanics*. Dover, New York, 1931 [*Gruppentheorie und Quantenmechanik*. Hirzel, Leipzig, 1931].

[48] Weyl, H., *The Classical Groups*. Princeton Univ. Press, Princeton, 1953.

[49] Zhelobenko, D.P., *Compact Lie Groups and Their Application*. Amer. Math. Soc. Transl. of Math. Monographs **40**, 1973 [Russian original: Nauka, Moscow, 1970].

[50] Kirillov, A.A., *Elements of the Theory of Representations*. Springer, Berlin, 1976 [Russian original: Nauka, Moscow, 1972].

[51] Wybourne, B.G., *Classical Groups for Physicists*. Wiley, New York, 1974.

[52] Gilmore, R., *Lie Groups, Lie Algebras and Some of Their Applications*. Wiley, New York, 1974.

[53] Barut, A.O. and Rączka, R., *Theory of Group Representations and Applications*. PWN-Polish Scientific Publishers, Warszawa, 1977.

[54] Perelomov, A., *Generalized Coherent States and Their Applications*. Springer, Berlin, 1986.

[55] Hecht, K.T., *The Vector Coherent State Method and Its Application to Problems of Higher Symmetry*. Springer, Berlin, 1987.

[56] Arnold, V.I., *Mathematical Methods in Classical Mechanics* (2nd Edition). Springer, Berlin, 1989.

[57] Guillemin, V. and Sternberg, S., *Symplectic Techniques in Physics*. Cambridge University Press, Cambridge, 1984.

[58] Malkin, I.A. and Man'ko, V.I., *Dynamical Symmetries and Coherent States of Quantum Systems* (in Russian). Nauka, Moscow, 1979.

[59] Dodonov, V.V. and Man'ko, V.I., Invariants and correlated states of nonstationary quantum systems. In: *Invariants and the Evolution of Nonstationary Quantum Systems. Proc. Lebedev Phys. Inst., vol. 183* (M.A. Markov, ed.), pp. 103–261. Nova Science, Commack, 1989.

[60] Gardiner, C.W., *Handbook of Stochastic Methods*. Springer, Berlin, 1986.

[61] Yurke, B., McCall, S.M., and Klauder, J.R., $SU(2)$ and $SU(1,1)$ interferometers. *Phys. Rev.* A (1986) **33** 4033–4054.

[62] Campos, R.A., Saleh, B.E.A., and Teich, M.C., Quantum-mechanical lossless beam splitter: $su(2)$ symmetry and photon statistics. *Phys. Rev.* A (1989) **40** 1371–1384.

[63] Katriel, J., Solomon, A.I., D'Ariano, G.M. and Rasetti, M., Multiboson Holstein–Primakoff squeezed states for $SU(2)$ and $SU(1,1)$. *Phys. Rev.* D (1986) **34** 2332–2238.

[64] D'Ariano, G.M., Rasetti, M.G., Katriel, J., and Solomon, A.I., Multiphoton and fractional-photon squeezed states. In: *Squeezed and Nonclassical Light* (P. Tombesi and E.R. Pike, eds.), pp. 301–319. Plenum, New York, 1989.

[65] Bužek, V., $SU(1,1)$ squeezing of $SU(1,1)$ generalized coherent states. *J. Modern Opt.* (1990) **37** 303–316.

[66] Vourdas, A., $SU(2)$ and $SU(1,1)$ phase states. *Phys. Rev.* A (1990) **41** 1653–1661.

[67] Vourdas, A., Analytic representations in the unit disk and applications to phase state and squeezing. *Phys. Rev.* A (1992) **45** 1943–1950.

[68] Brif, C., Photon states associated with the Holstein–Primakoff realization of the $SU(1,1)$ Lie algebra. *Quant. Semiclass. Opt.* (1995) **8** 803–834.

[69] Vourdas, A., Resolutions of the identity in terms of SU(2) coherent states and their use for quantum-state engineering. *Phys. Rev.* A (1996) **54** 4544–4552.

[70] Vourdas, A. and Wünsche, A., Resolutions of the identity in terms of line integrals of $SU(1,1)$ coherent states. *J. Phys.* A (1998) **31** 9341–9352.

[71] Wünsche, A., Realizations of $SU(1,1)$ by boson operators with application to phase states. *Acta Phys. Slovaca* (1999) **49** 771–782.

[72] Wünsche, A., Eigenvalue problem for arbitrary linear combinations of a boson annihilation and creation operator. *Ann. Phys.* (Leipzig) (1992) **1** 181–197.

[73] Han, D., Kim, Y.S., and Noz, M.E., Linear canonical transformations of coherent and squeezed states in the Wigner phase space. *Phys. Rev. A* (1988) **37** 807–814.

[74] Han, D., Kim, Y.S., and Noz, M.E., Linear canonical transformations of coherent and squeezed states in the Wigner phase space. II. Quantitative analysis. *Phys. Rev.* A (1989) **40** 902–912.

[75] Arvind, Dutta, B., Mehta, C.L., and Mukunda, N., Squeezed states, metaplectic group, and operator Möbius transformations. *Phys. Rev.* A (1994) **50** 39–61.

[76] Mehta, C.L., Ordering of exponential of a quadratic in boson operators. 1. Single-mode case. *J. Math. Phys.* (1977) **18** 404–407.

[77] Agrawal, G.P., and Mehta, C.L., Ordering of exponential of a quadratic in boson operators. 2. Multimode case. *J. Math. Phys.* (1977) **18** 408–412.

[78] Wódkiewicz, K. and Eberly, J.H., Coherent states, squeezed fluctuations, and the $SU(2)$ and $SU(1,1)$ groups in quantum-optics applications. *J. Opt. Soc. Am.* B (1985) **2** 458–466.

[79] Fisher, R.A., Nieto, M.M., and Sandberg, V.D., Impossibility of naively generalizing squeezed coherent states. *Phys. Rev.* D (1984) **29** 1107–1110.

[80] Caves, C.M., Quantum-mechanical noise in an interferometer. *Phys. Rev.* D (1981) **23** 1693–1708.

[81] Caves, C.M., Quantum limits on noise in linear amplifiers. *Phys. Rev.* D (1982) **26** 1817–1839.

[82] Schleich, W., Horowicz, R.J., and Varro, S. Bifurcation in the phase probability distribution of a highly squeezed state. *Phys. Rev.* A (1989) **40** 7405–7408.

[83] Roy, A.K. and Mehta, C.L., Squeezed states generated by boson creation operator. *J. Mod. Opt.* (1992) **39** 1619–1622.

[84] Wünsche, A., Nonunitary and unitary approach to eigenvalue problem of boson operators and squeezed coherent states. In: *Second International Workshop on Squeezed States and Uncertainty Relations*, Moscow 1992, (see [27], p. 277–281).

[85] Wünsche, A., The total Gaussian class of quasiprobabilities and its relation to squeezed-state excitations. In: *Fourth International Conference on Squeezed States and Uncertainty Relations*, Taiyuan 1995, (see [29], p. 73–82).

[86] Gel'fand, I.M. and Vilenkin, N.Ya., *Generalized Functions, vol. 4, Applications of Harmonic Analysis*. Academic, San Diego, 1964.

[87] Fan, H.-Y., Liu, Z.-W., and Ruan, T.-N., Does the creation operator $a^\dagger$ possess eigenvectors? *Commun. Theor. Phys.* (1984) **3** 175–188.

[88] Fan, H.-Y. and Klauder, J.R., On the common eigenvectors of two-mode creation operators and the change operator. *Mod. Phys. Lett.* A (1994) **9** 1291–1297.

[89] Wünsche, A., The coherent states as basis states on areas contours and paths in the phase space. *Acta Phys. Slovaca* (1996) **46** 505–516.

[90] Perelomov, A.M., Coherent states for arbitrary Lie group. *Commun. Math. Phys.* (1972) **26** 222–236 (reprinted in [10]).

[91] Barut, A.O. and Girardello, L., New "coherent" states associated with noncompact groups. *Commun. Math. Phys.* (1971) **21** 41–55 (reprinted in [10]).

[92] Vourdas, A. and Weiner, R.M., Photon-counting distribution in squeezed states. *Phys. Rev.* A (1987) **36** 5866–5869.

[93] Schleich, W. and Wheeler, J.A., Oscillations in photon distribution of squeezed states and interference in phase space. *Nature* (1987) **326** 574–577.

[94] Schleich, W. and Wheeler, J.A., Oscillations in photon distribution of squeezed states. *J. Opt. Soc. Am.* B (1987) **4** 1715–1722.

[95] *Bateman Manuscript Project, Higher Transcendental Functions, vol. 2* (A. Erdélyi, ed.). McGraw-Hill, New York, 1953.

[96] Wünsche, A., The distance to Poissonian statistics as a supplementary measure in a quantum optics. *Appl. Phys.* B (1995) **60** S119–S122.

[97] Dodonov, V.V., Man'ko, O.V., Man'ko, V.I., and Wünsche, A., Energy-sensitive and "classical-like" distances between quantum states. *Phys. Scripta* (1999) **59** 81–89.

[98] Dodonov, V.V., Man'ko, O.V., Man'ko, V.I., and Wünsche, A., Hilbert-Schmidt distance and nonclassicality of states in quantum optics. *J. Modern Opt.* (2000) **47** 633–654.

[99] Bargmann, V., On a Hilbert space of analytic functions and an associated integral transform. Part I. *Commun. Pure Appl. Math.* (1961) **14** 187–214.

[100] Man'ko, V.I. and Wünsche, A., Properties of squeezed-state excitations. *Quant. Semiclass. Opt.* (1997) **9** 381–409.

[101] Agarwal, G.S. and Tara, K., Nonclassical properties of states generated by the excitations on a coherent state. *Phys. Rev.* A (1991) **43** 492–497.

[102] Jones, G.N., Haight, J., and Lee, C.T., Nonclassical effects in the photon-added thermal states. *Quant. Semiclass. Opt.* (1997) **9** 411–418.

[103] Dakna, M., Anhut, T., Opatrný, T., Knöll, L., and Welsch, D.-G., Generating Schrödinger-cat-like states by means of conditional measurements on a beam splitter. *Phys. Rev.* A (1997) **55** 3184–3194.

[104] Král, P., Displaced and squeezed Fock states. *J. Mod. Opt.* (1990) **37** 889–917.

[105] Král, P., Kerr interaction with displaced and squeezed Fock states. *Phys. Rev.* A (1990) **42** 4177–4192.

[106] Sudarshan, E.C.G., Equivalence of semiclassical and quantum mechanical descriptions of statistical light beams. *Phys. Rev. Lett.* (1963) **10** 277–279.

[107] Vourdas, A. and Bishop, R.F., Dirac's contour representation in thermofield dynamics. *Phys. Rev.* A (1996) **53** R1205–R1208.

[108] Janszky, J., Domokos, P., and Adam, P., Coherent states on a circle and quantum interference. *Phys. Rev.* A (1993) **48** 2213–2219.

[109] Adam, P., Földesi, I., and Janszky, J., Complete basis set via straight-line coherent-state superpositions. *Phys. Rev.* A (1994) **49** 1281–1287.

[110] Kis, Z., Adam, P., and Janszky, J., Properties of states generated by excitations on the amplitude squeezed state. *Phys. Lett.* A (1994) **188** 16–20.

[111] Hudson, P.L., When is the Wigner quasi-probability density non-negative. *Rep. Math. Phys.* (1974) **6** 249–252.

[112] Lütkenhaus, N. and Barnett, S.M., Nonclassical effects in phase space. *Phys. Rev.* A (1995) **51** 3340–3342.

[113] Wünsche, A., The complete Gaussian class of quasiprobabilities and its relation to squeezed states and their discrete excitations. *Quant. Semiclass. Opt.* (1996) **8** 343–379.

[114] Cahill, K.E. and Glauber, R.J., Density operators and quasiprobability distributions. *Phys. Rev.* (1969) **177** 1882–1902.

[115] Nieto, M.M., Displaced and squeezed number states. *Phys. Lett.* A (1997) **229** 135–143.

[116] Adam, G, Density matrix elements and moments for generalized Gaussian state fields. *J. Mod. Opt.* (1995) **42** 1311–1328.

[117] Wünsche, A., Ordered moments and relation to Radon transform of Wigner quasiprobability. *J. Modern Opt.* (2000) **47** 33–56.

[118] Agarwal, G.S. and Wolf, E., Calculus for functions of noncommuting operators and general phase-space methods in quantum mechanics. 1. Mapping theorems and ordering of functions of noncommuting operators. 2. Quantum mechanics in phase space. 3. A generalized Wick theorem and multitime mapping. *Phys. Rev.* D (1970) **2** 2161–2186, 2187–2205, 2206–2225.

[119] Peřina, J., *Coherence of Light.* Van Nostrand Reinhold, London, 1972.

[120] Wünsche, A., Ordered operator expansions and reconstruction from ordered moments *J. Opt.* B (1999) **1** 264–288.

[121] Agarwal, G.S. and Simon, R., A new representation for squeezed states. *Opt. Comm.* (1992) **92** 105–107.

[122] De Groot, S.R. and Suttorp, L.G., *Foundations of Electrodynamics.* North-Holland, Amsterdam, 1972.

[123] Tatarskii, V.I., The Wigner representation of quantum mechanics. *Uspekhi Fiz. Nauk* (1983) **139** 587–619 [Sov. Phys. Usp. (1983) **26** 311–327].

[124] Hillery, M., O'Connell, R.F., Scully, M.O., and Wigner, E.P., Distribution functions in physics: Fundamentals. *Phys. Rep.* (1984) **106** 121–167.

[125] de Oliveira, F.A.M., Kim, M.S., Knight, P.L. and Bužek, V., Properties of displaced number states. *Phys. Rev.* A (1990) **41** 2645–2652.

[126] Wünsche, A., Displaced Fock states and their connection to quasi-probabilities. *Quant. Opt.* (1991) **3** 359–383.

[127] Bishop, R.F. and Vourdas, A., Displaced and squeezed parity operator: its role in classical mappings of quantum theories. *Phys. Rev.* A (1994) **50** 4488–4501.

[128] Czirják, A. and Benedict, M.G., Joint Wigner function for atom-field interactions. *Quant. Semiclass. Opt.* (1996) **8** 975–981.

[129] Fan, H.Y. and Weng, H.-G., Simple approach for discussing the properties of displaced Fock states. *Quant. Opt.* (1992) **4** 265–270.

[130] Bertrand, J. and Bertrand, P., A tomographic approach to Wigner function. *Found. Phys.* (1987) **17** 397–405.

[131] Wünsche, A., Radon transform and pattern functions in quantum tomography. *J. Mod. Opt.* (1997) **44** 2293–2331.

[132] Gel'fand, I.M., Grayev, M.I., and Vilenkin, N.Y., *Generalized Functions, vol. 5, Integral Geometry and Representation Theory.* Academic, New York, 1966.

[133] Barrett, H.H., The Radon transform and its applications. In: *Progress in Optics, vol. XXI* (E. Wolf, ed.), pp. 217–286. North Holland, Amsterdam, 1984.

[134] Ekert, A.K. and Knight, P.L., Canonical transformation and decay into phase-sensitive reservoirs. *Phys. Rev.* A (1990) **42** 487–493.

[135] Ekert, A.K. and Knight, P.L., Relationship between semiclassical and quantum-mechanical input-output theories of optical response. *Phys. Rev.* A (1990) **43** 3934–3938.

[136] Marian, P. and Marian, T.A., Squeezed states with thermal noise. I. Photon-number statistics. *Phys. Rev.* A (1993) **47** 4474–4486.

[137] Husimi, K., Some formal properties of the density matrix. *Proc. Phys. Math. Soc. Japan* (1940) **22** 264–314.

[138] Wünsche, A. and Bužek, V., Reconstruction of quantum states from propensities.

Quant. Semiclass. Opt. (1997) **9** 631–653.

[139] Kano, Y., A new phase-space distribution function in statistical theory of electromagnetic field. *J. Math. Phys.* (1965) **6** 1913–1915.

[140] Mehta, C.L. and Sudarshan, E.C.G., Relation between quantum and semiclassical description of optical coherence. *Phys. Rev.* (1965) **138** B274–B280.

[141] McKenna, J. and Frisch, H.L., High field magnetoresistance of inhomogeneous semiconductors V: Quantum mechanical theory. *Ann. Phys.* (NY) (1965) **33** 156–195.

[142] Janszky, J., Kim, M.G., and Kim, M.S., Quasiprobabilities and nonclassicality of fields. *Phys. Rev. A* (1996) **53** 502–506.

[143] Robertson, H.P., A general formulation of the uncertainty principle and its classical interpretation. *Phys. Rev.* (1930) **35** 667.

[144] Schrödinger, E., Zum Heisenbergschen Unschärfeprinzip. *Ber. Kgl. Akad. Wiss. Berlin* (1930) **24** 296–303.

[145] Dodonov, V.V. and Man'ko, V.I., Generalization of uncertainty relation in quantum mechanics. In: *Invariants and the Evolution of Nonstationary Quantum Systems. Proc. Lebedev Phys. Inst., vol. 183* (M.A. Markov, ed.), pp. 5–70. Nauka, Moscow, 1987 [translated by Nova Science, Commack, 1989, pp. 3–101].

[146] Hillery, M., Squeezing of the square of the field amplitude in second harmonic generation. *Opt. Commun.* (1987) **62** 135–138.

[147] Hillery, M., Amplitude-squared squeezing of the electromagnetic field. *Phys. Rev.* A (1987) **36** 3796–3802.

[148] Trifonov, D.A., Generalized intelligent states and squeezing. *J. Math. Phys.* (1994) **35** 2297–2308.

[149] Brif, C. and Mann, A., Nonclassical interferometry with intelligent light. *Phys. Rev.* A (1996) **54** 4505–4518.

[150] Brif, C. and Mann, A., High-accuracy $SU(1,1)$ interferometers with minimum-uncertainty input states. *Phys. Lett.* A (1996) **219** 257–262.

[151] Dodonov, V.V., Malkin, I.A., and Man'ko, V.I., Even and odd coherent states and excitations of a singular oscillator. *Physica* (1974) **72** 597–615.

[152] Bužek, V. and Knight, P.L., Quantum interference, superposition states of light, and nonclassical effects. In: *Progress in Optics, vol. XXXIV* (E. Wolf, ed.), pp. 1–158. North Holland, Amsterdam, 1995.

[153] Caves, C.M. and Schumaker, B.L., New formalism for two-photon quantum optics. 1. Quadrature phases and squeezed states. *Phys. Rev.* A (1985) **31** 3068–3092.

[154] Schumaker, B.L. and Caves, C.M., New formalism for two-photon quantum optics. 2. Mathematical foundation and compact notation. *Phys. Rev.* A (1985) **31** 3093–3111.

[155] Schumaker, B.L., Quantum-mechanical pure states with Gaussian wave-functions. *Phys. Rep.* (1986) **135** 317–408.

[156] Ekert, A.K. and Knight, P.L., Correlations and squeezing of two-mode oscillations. *Am. J. Phys.* (1989) **57** 692–697.

[157] Schrade, G., Akulin, V.M., Man'ko, V.I., and Schleich, W., Photon statistics of a two-mode squeezed vacuum. *Phys. Rev.* A (1993) **48** 2398–2406.

[158] Wünsche, A., About integration within ordered products in quantum optics. *J. Opt. B: Quant. Semiclass. Opt.* (1999) **1** R11–R21.

[159] Braunstein, S.L. and McLachlan, R.I., Generalized squeezing. *Phys. Rev.* A (1987) **35** 1659–1667.

Chapter 3

Parametric excitation and generation of nonclassical states in linear media

V. V. Dodonov

1 Introduction

This chapter is devoted to the problems related to the evolution of quantum states in the specific but important case when the Hamiltonian of the system is an arbitrary quadratic form with respect to the quadrature operators or the bosonic annihilation and creation operators. In certain sense, the evolution operator can be considered in this case as some sort of generalized multidimensional squeezing operator. A remarkable feature of the systems concerned is the possibility of complete analytical treatment through several different approaches. Our exposition is based on the method of *quantum time-dependent operator invariants* (*integrals of motion*), which enables us to solve the problem in the simplest and most effective way. Besides giving the explicit form of solutions, both for the generic quadratic Hamiltonians and for their most important special cases, we consider some global features of quadratic systems, established in the series of publications by our group in the Lebedev Physics Institute, but still not well known, such as the existence of universal quantum invariants, the problem of factorization of solutions, the description of the most general open systems in terms of the effective Fokker–Planck equations, and some others. The problems discussed in this chapter were the subjects of many hundreds, if not thousands, of publications, especially in the simplest one-mode case. Therefore, although we have tried to give references to the most significant studies, many of them could not be included due to the limitations on the volume.

The single-mode case is equivalent to the model of a one-dimensional generalized oscillator with the time-dependent Hamiltonian

$$\hat{H}(t) = \tfrac{1}{2}\left[\mu(t)\hat{p}^2 + \nu(t)\hat{x}^2 + \rho(t)\left(\hat{x}\hat{p} + \hat{p}\hat{x}\right)\right] + c_1(t)\hat{p} + c_2(t)\hat{x}, \qquad (1)$$

where the *quadrature operators* $\hat{x}$ and $\hat{p}$ satisfy the canonical commutation relation $[\hat{x}, \hat{p}] = i\hbar$. The same Hamiltonian can also be written in terms of the bosonic annihilation and creation operators as

$$\hat{H} = \tfrac{1}{2}\hbar\left[D_0\left(\hat{a}^\dagger\hat{a} + \hat{a}\hat{a}^\dagger\right) + D_1\hat{a}^2 + D_1^*\hat{a}^{\dagger 2}\right] + \hbar\left(g\hat{a} + \hat{a}^\dagger g^*\right), \qquad (2)$$

where $[\hat{a}, \hat{a}^\dagger] = 1$. We assume that the Hamiltonian is Hermitian (although this is not obligatory for the validity of the integrals of motion method), therefore all the coefficients of Hamiltonian (1) are real. In the case of Hamiltonian (2), D_0 is a real coefficient (which may depend on time, as well as the other parameters), and the asterisk means the complex conjugation.

The multidimensional quadratic Hamiltonian can be written as

$$H = \tfrac{1}{2} \sum_{j,k=1}^{2N} B_{jk}(t) q_j q_k + \sum_{j=1}^{2N} C_j(t) q_j = \tfrac{1}{2}\, \mathbf{q} B(t) \mathbf{q} + \mathbf{C}(t)\mathbf{q}. \tag{3}$$

Here $\mathbf{q}$ and $\mathbf{C}$ are $2N$-dimensional vectors, whereas B is a $2N \times 2N$ matrix. They are divided into the N-dimensional blocks as follows,

$$\mathbf{q} = \begin{bmatrix} \mathbf{p} \\ \mathbf{x} \end{bmatrix}, \qquad \mathbf{C} = \begin{bmatrix} \mathbf{c}_1 \\ \mathbf{c}_2 \end{bmatrix}, \qquad B = \left\| \begin{matrix} b_1 & b_2 \\ b_3 & b_4 \end{matrix} \right\| \tag{4}$$

where $\mathbf{x}$ and $\mathbf{p}$ are the N-dimensional vectors of the Cartesian coordinates and conjugated momenta, N being the number of degrees of freedom of the system. The N-dimensional vectors $\mathbf{c}_1$ and $\mathbf{c}_2$, as well as the $N \times N$ matrices b_j, may be arbitrary functions of time ($j = 1, 2, 3, 4$). Evidently, matrix B can always be symmetrized, so we assume that $b_1 = \tilde{b}_1$, $b_4 = \tilde{b}_4$, $b_2 = \tilde{b}_3$; the tilde means the matrix transposition. If the Hamiltonian is Hermitian, then matrix B and vector $\mathbf{C}$ are real, but this property is not significant.

The multimode quadratic Hamiltonian can also be written in terms of the annihilation and creation operators:

$$\hat{H} = \frac{\hbar}{2} \left(\hat{\mathbf{a}}, \hat{\mathbf{a}}^\dagger\right) \left\| \begin{matrix} D_1 & \tilde{D}_0 \\ D_0 & D_1^* \end{matrix} \right\| \begin{pmatrix} \hat{\mathbf{a}} \\ \hat{\mathbf{a}}^\dagger \end{pmatrix} + \hbar \mathbf{g} \hat{\mathbf{a}} + \hbar \hat{\mathbf{a}}^\dagger \mathbf{g}^* \tag{5}$$

where the components of the vector operator $\hat{\mathbf{a}} = (\hat{a}_1, \ldots, \hat{a}_N)$ and its hermitially conjugated partner $\hat{\mathbf{a}}^\dagger$ obey the commutation relations

$$[\hat{a}_j\, , \, \hat{a}_k] = 0, \quad [\hat{a}_j\, , \, \hat{a}_k^\dagger] = \delta_{jk}. \tag{6}$$

The matrices D_1 and D_1^* must be symmetrical, and $D_0 = D_0^\dagger$ if the Hamiltonian is Hermitian.

2 Propagators and evolution operators

The development of a closed quantum system in time is determined by the Hermitian evolution operator $\widehat{U}(t)$, satisfying the Schrödinger equation

$$i\hbar \partial \widehat{U} / \partial t = \widehat{H}(t) \widehat{U}, \quad t > 0 \tag{7}$$

and the initial condition $\widehat{U}(0) = \widehat{\mathbf{1}}$, where $\widehat{\mathbf{1}}$ is the unit operator in the related Hilbert space. In particular, for the state vector $|\psi(t)\rangle$ (pure quantum state) or the statistical operator $\hat{\rho}(t)$ (mixed quantum state) we have

$$|\psi(t)\rangle = \widehat{U}(t)|\psi(0)\rangle, \quad \hat{\rho}(t) = \widehat{U}(t)\hat{\rho}(0)\widehat{U}^\dagger(t). \tag{8}$$

Working in some concrete quantum representation one needs the kernel of the evolution operator, usually called the *propagator* or the *Green function.* For example, in the coordinate representation, the propagator $G(x, x', t)$ connects the current wave function $\psi(x, t)$ and its initial value $\psi(x, 0)$ as follows,

$$\psi(x,t) = \int G(x,x',t)\psi(x',0)\,dx'. \tag{9}$$

The analogous relation for the density matrix (the kernel of the statistical operator) $\rho(x, y, t)$ reads

$$\rho(x,y,t) = \int\int G(x,x',t)\rho(x',y',0)G^*(y',y,t)\,dx'\,dy'. \tag{10}$$

If one knows any complete set of solutions $\{\psi_n(x, t)\}$ to the nonstationary Schrödinger equation, then the propagator can be written as

$$G(x,x',t) = \sum_n \psi_n(x,t)\psi_n^*(x',0), \tag{11}$$

where the asterisk means complex conjugation, and the summation should be replaced by integration in the case of the continuum spectrum. This method of calculating the propagator is rather involved, since first one has to find *all* (in some sense) the solutions to the Schrödinger equation; moreover, one must be able to calculate the corresponding infinite series or integrals.

Even more cumbersome from the point of view of concrete applications is the Feynman path integral method [1]. Its consistent application implies laborious calculations of intermediate N-fold integrals with the subsequent limiting transition $N \to \infty$. True, in the case of quadratic Lagrangians there is a "roundabout" way, since it is known that in this case the propagator can be expressed through the classical action function. However, this is "a step aside" from the first principles. Besides, the calculation of the explicit form of the action function is not a very simple problem, either.

2.1 Propagators and integrals of motion

It is remarkable that there exists a simple and elegant method, which permits us to find the propagator of the quantum system with any quadratic Hamiltonian practically instantaneously, without first calculating the eigenfunctions or complicated sums and integrals. This is the method of *quantum operator time-dependent integrals of motion*, which was actually used (in the simplest cases) at the dawn of quantum mechanics [2,3] (later in [4]) and developed completely in [5–11] (some special cases were also considered in [12–15]).

The essence of the method is as follows. Let $\hat{x}$ be the usual coordinate operator in the Schrödinger picture. Suppose that we know the explicit form of the operator

$$\hat{X} = \hat{U}\hat{x}\hat{U}^{-1}. \tag{12}$$

Multiplying both sides of equation (12) by the operator $\hat{U}$ *from the right* we obtain the equation

$$\hat{X}\hat{U} = \hat{U}\hat{x}. \tag{13}$$

In the propagator language it reads

$$\hat{X}G(x, x', t) = x'G(x, x', t). \tag{14}$$

In other words, the propagator appears to be the *eigenfunction* of operator $\hat{X}$, and the eigenvalue is equal to the "initial" coordinate x'. In equation (14) the operator $\hat{X}$ acts on G as a function of the *first argument* (the "current" coordinate) x, while the second argument x' must be treated as a parameter. To find the total dependence of G on x', we introduce the operator

$$\hat{P} = \hat{U}\hat{p}\hat{U}^{-1}. \tag{15}$$

Then instead of equations (13) and (14) we have

$$\hat{P}\hat{U} = \hat{U}\hat{p}, \tag{16}$$

$$\hat{P}G(x, x', t) = i\hbar\partial G(x, x', t)/\partial x'. \tag{17}$$

Again, the operator $\hat{P}$ in the left-hand side of (17) acts on G as a function of the first argument x. Equations (14) and (17) determine the propagator up to a factor dependent on time only. To obtain this last factor one should take into account the Schrödinger equation

$$\left(i\hbar\partial/\partial t - \widehat{H}\right) G(x, x', t) = 0 \tag{18}$$

(again, operator $\widehat{H}$ acts on G as a function of x) and the initial condition

$$G(x, x', 0) = \delta(x - x'). \tag{19}$$

All we need now are the explicit forms of operators $\hat{X}$ and $\hat{P}$. To find them we notice that the operators (12) and (15) satisfy the equation (due to the definition and equation (7))

$$\left(i\hbar\partial\widehat{I}/\partial t - [\widehat{H}\,,\,\widehat{I}]\right)\psi(t) = 0 \tag{20}$$

for any function $\psi(t)$ which is a solution to the Schrödinger equation (the symbol $[\hat{A}\,,\,\hat{B}]$ means, as usual, the commutator of the operators $\hat{A}$ and $\hat{B}$). Consequently, the operators $\hat{X}$ and $\hat{P}$ are nothing but the *quantum operator integrals of motion.* Moreover, since they coincide, respectively, with the operators $\hat{x}$ and $\hat{p}$ at the initial moment $t = 0$, they have quite clear physical meanings. $\hat{X}$ is the initial coordinate operator while $\hat{P}$ is the initial momentum operator in the phase space.

2.2 Single mode: quadrature representation

Since the equations of motion resulting from the quadratic Hamiltonians are linear with respect to $\hat{x}$ and $\hat{p}$, we look for the integrals of motion in the form of linear combinations of these operators,

$$\hat{P}(t) = \lambda_1(t)\hat{p} + \lambda_2(t)\hat{x} + \delta_1(t), \quad \hat{X}(t) = \lambda_3(t)\hat{p} + \lambda_4(t)\hat{x} + \delta_2(t). \tag{21}$$

Putting expressions (21) into equation (20) we arrive at the set of first-order ordinary differential equations and initial conditions

$$\begin{aligned}
\dot{\lambda}_1 &= \lambda_1\rho - \lambda_2\mu, & \lambda_1(0) &= 1 & (22)\\
\dot{\lambda}_2 &= \lambda_1\nu - \lambda_2\rho, & \lambda_2(0) &= 0 & (23)\\
\dot{\lambda}_3 &= \lambda_3\rho - \lambda_4\mu, & \lambda_3(0) &= 0 & (24)\\
\dot{\lambda}_4 &= \lambda_3\nu - \lambda_4\rho, & \lambda_4(0) &= 1 & (25)\\
\dot{\delta}_1 &= \lambda_1 c_2 - \lambda_2 c_1, & \delta_1(0) &= 0 & (26)\\
\dot{\delta}_2 &= \lambda_3 c_2 - \lambda_4 c_1, & \delta_2(0) &= 0 & (27)
\end{aligned}$$

(the dot means the time derivative). An immediate consequence of equations (22)–(25) and the initial conditions is the identity

$$\lambda_4\lambda_1 - \lambda_3\lambda_2 \equiv 1. \tag{28}$$

Its physical meaning is very simple: the commutation relations between the operators $\hat{x}$ and $\hat{p}$ are the same as between $\hat{X}$ and $\hat{P}$, $[\hat{x}\,,\hat{p}] = [\hat{X}\,,\hat{P}] = i\hbar$, i.e., transformation (21) is *canonical* (and the evolution operator is unitary).

Now equation (14) assumes the form

$$-i\hbar\lambda_3\partial G/\partial x_2 + \lambda_4 x_2 G = (x_1 - \delta_2)\,G. \tag{29}$$

Solving it, we find the dependence of G on the first argument x_2

$$G = \exp\left\{-\frac{i}{2\hbar\lambda_3}\left[\lambda_4 x_2^2 - 2x_2\,(x_1 - \delta_2)\right]\right\} G_1\,(x_1;t)\,, \tag{30}$$

but $G_1\,(x_1;t)$ may still be an arbitrary function. Its explicit form can be found from equation (17), which is reduced after simple algebra to

$$i\hbar\partial G_1/\partial x_1 = \left[\lambda_1\lambda_3^{-1}\,(x_1 - \delta_2) + \delta_1\right] G. \tag{31}$$

The solution to equation (31) is given by

$$G_1\,(x_1;t) = \exp\left\{-\frac{i}{2\hbar\lambda_3}\left[\lambda_1 x_1^2 + 2x_1\,(\lambda_3\delta_1 - \lambda_1\delta_2)\right] + \varphi(t)\right\}. \tag{32}$$

Putting expressions (30) and (32) into the Schrödinger equation (18) one arrives at the ordinary differential equation for the function $\varphi(t)$

$$\dot{\varphi} = \tfrac{1}{2}\left(\lambda_3^{-1}\lambda_4\mu - \rho\right) + \frac{i}{\hbar}\left(c_1\lambda_3^{-1}\delta_2 - \tfrac{1}{2}\,\mu\lambda_3^{-2}\delta_2^2\right). \tag{33}$$

Replacing functions μ, ρ, and c_1 by the combinations of functions λ_j, δ_k and their derivatives, which follow from equations (22)–(27), and using the initial condition (19), one can obtain after some algebra the final result in the form [6,7]

$$\begin{aligned}
G\,(x_2,x_1,t) = (-2\pi i\hbar\lambda_3)^{-1/2}\exp\Bigg\{ &-\frac{i}{2\hbar\lambda_3}\Big[\lambda_4 x_2^2 - 2x_2x_1 + \lambda_1 x_1^2\\
&+2x_2\delta_2 + 2x_1\,(\lambda_3\delta_1 - \lambda_1\delta_2) + \lambda_1\delta_2^2\Big] - \frac{i}{\hbar}\int_0^t \dot{\delta}_1(\tau)\delta_2(\tau)\mathrm{d}\tau\Bigg\}.
\end{aligned} \tag{34}$$

Connection with the classical oscillator. Explicit solutions

Actually, the dynamics of any one-dimensional quantum system with a quadratic Hamiltonian (1) is determined by two independent solutions of the *classical* equation for the oscillator with a time-dependent frequency

$$\ddot{\varepsilon} + \Omega^2(t)\varepsilon = 0, \tag{35}$$

with

$$\Omega^2(t) = \mu\nu - \rho^2 + \frac{\ddot{\mu}}{2\mu} - \frac{3\dot{\mu}^2}{4\mu^2} - \dot{\rho} + \rho\frac{\dot{\mu}}{\mu}. \tag{36}$$

For example,

$$\lambda_1 = \sqrt{\mu}\varepsilon_1, \quad \lambda_3 = \sqrt{\mu}\varepsilon_0, \quad \lambda_4 = \tfrac{1}{2}\,\mu^{-3/2}\left[(2\mu\rho - \dot{\mu})\varepsilon_0 - 2\mu\dot{\varepsilon}_0\right] \tag{37}$$

where the fundamental solutions of equation (35) ε_0 and ε_1 are defined by the initial conditions

$$\varepsilon_0(0) = 0, \quad \dot{\varepsilon}_0(0) = -\sqrt{\mu(0)}, \tag{38}$$

$$\varepsilon_1(0) = \frac{1}{\sqrt{\mu(0)}}, \quad \dot{\varepsilon}_1(0) = \left.\frac{2\mu\rho - \dot{\mu}}{2\mu^{3/2}}\right|_{t=0}. \tag{39}$$

Equation (35) has evident trigonometric solutions

$$\varepsilon_0 = -\frac{\sqrt{\mu_0}}{\Omega_0}\sin(\Omega_0 t), \quad \varepsilon_1 = \frac{\cos(\Omega_0 t)}{\sqrt{\mu_0}} + \frac{\gamma + \rho}{\Omega_0\sqrt{\mu_0}}\sin(\Omega_0 t),$$

if $\mu = \mu_0\exp(-2\gamma t)$ and $\nu = \nu_0\exp(2\gamma t)$, with constant $\mu_0, \nu_0, \rho, \gamma$, when the effective frequency (36) is given by $\Omega_0^2 = \mu_0\nu_0 - (\rho+\gamma)^2 = const$. Taking for simplicity $c_1 = c_2 = 0$, we arrive at the expression

$$\begin{aligned} G(x_2, x_1, t) &= \left[2\pi i\hbar\mu_0 e^{-\gamma t}\sin(\Omega_0 t)/\Omega_0\right]^{-1/2} \\ &\times \exp\left\{\frac{i\Omega_0 e^{\gamma t}}{2\hbar\mu_0\sin(\Omega_0 t)}\left[e^{\gamma t}\left(\cos(\Omega_0 t) - \frac{\gamma+\rho}{\Omega_0}\sin(\Omega_0 t)\right)x_2^2\right.\right. \\ &\left.\left. -2x_2x_1 + e^{-\gamma t}\left(\cos(\Omega_0 t) + \frac{\gamma+\rho}{\Omega_0}\sin(\Omega_0 t)\right)x_1^2\right]\right\}. \end{aligned} \tag{40}$$

If $\mu_0 = m^{-1}$ and $\rho = \gamma = 0$, then equation (40) yields the propagator for the oscillator with constant frequency Ω_0, which is known since Kennard's paper [3]. A lot of papers were devoted to the problem of *forced harmonic oscillator* with constant frequency. We mention here only a few references related to the epoch of 1940s-1960s [16–24].

The case of $\gamma \neq 0$ is known under the name *Caldirola–Kanai model* [25, 26] (originally, with $\rho = 0$). This model was studied in numerous publications, devoted mainly to the problem of damping in quantum mechanics. We can cite here only a small part of relevant papers [22,27–49]; for more detailed lists see, e.g., reviews [9, 10, 50–54]. Different nonexponential dependences of the coefficient $\mu(t)$ (and others) were considered in [55–58] (the so-called "pulsating oscillator"). It is worth noting that the function $\Omega^2(t)$ (or the constant parameter Ω_0^2) in equations (35)–(40) may be not only positive, but negative as well; in the last case we have the so-called "upside-down oscillator" which is sometimes used [59–63] for modelling the tunnelling effects.

For the first time, the propagator of a quantum oscillator with an arbitrary time-dependent frequency ($\mu = const$ and $\rho = 0$) was expressed through the solutions to the classical equation (35) by Fujiwara [64] and in a more explicit form by Husimi [21] (independently, it was rediscovered in [65]). This propagator, together with some examples related to the Caldirola–Kanai model, was discussed in book [66]. During the second half of the 20th century, various approaches to the problem of a generalized nonstationary quantum oscillator (arbitrary functions $\mu(t)$, $\nu(t)$, and $\rho(t)$) were considered in connection with different applications by many authors: see, e.g., [67–85] in addition to the references given above and in Chapter 1.

In many cases it is convenient to use *complex* solutions of equation (35), imposing the additional condition

$$\dot{\varepsilon}\varepsilon^* - \dot{\varepsilon}^*\varepsilon = 2i \tag{41}$$

(the right-hand side of this equation does not depend on time, since it is nothing but the Wronskian of equation (35)). Introducing the amplitude and the phase of ε as $\varepsilon = \rho\exp(i\phi)$, one can easily verify that condition (41) is equivalent to the relation $\phi(t) = \int^t d\tau/\rho^2(\tau)$, and function $\rho(t)$ obeys the nonlinear equation

$$\ddot{\rho} + \Omega^2(t)\rho = \rho^{-3}\,. \tag{42}$$

Equation (42) appeared for the first time, presumably, in the paper by Ermakov [86], who looked for solutions of ordinary differential equations "in quadratures". In the quantum mechanical context this equation was used by Milne [87], who considered equation (35) itself as a stationary Schrödinger equation. Sometimes, equation (42) is called "Pinney's equation", after a short (12 lines of text, including 3 equations) note [88]. During the last two decades, "Ermakov's systems" and "Ermakov's invariants" attracted the attention of many authors: see, e.g., [89–91] and references therein.

The importance of equation (35) for solving the problem of a *classical* charged particle moving in a time-dependent magnetic field was emphasized in [92], where explicit solutions to this equation have been obtained in terms of the Bessel functions in two special cases: $\Omega \sim t^b$ (see also [93]) and $\Omega^2 \sim d^2e^{ct} - p^2$. The existence of exact solutions in these cases, as well as in the case of the dependence $\Omega^2(t) = at^{-1} + bt^{-2}$, was pointed out in [94], where linear integrals of motion were used to find a solution of the Schrödinger equation in the Gaussian form. The systematic study of linear integrals of motion and the related generalized time-dependent coherent states was begun in [95]. Another approach (also based on equation (35)) was used in [96].

Explicit solutions to equations (35) or (42) for different dependences $\Omega(t)$ were also considered, e.g., in [97,98] ($\Omega \sim t^b$, $\mu \sim t^a$; solutions in terms of the Bessel functions for a generic exponent b and in terms of elementary functions if

$b = -1$ or $b = -2$), [99, 100] ($\Omega = a + be^{-\kappa t}$; solutions in terms of the Bessel functions of the orders 0 and 1), [101, 102] ($\Omega^2 = a + bt$; solutions in terms of the Bessel functions of the order 1/3, equivalent to the Airy functions). Periodic delta-kicks of the frequency were considered in [103–107]. Such models were used in the studies of different aspects of quantum chaos [108, 109]. The quantum oscillator driven by a stochastic force was studied in [110], and the case of a delta-kicked force was considered in [111]. Kicked systems with time-dependent Hamiltonians were considered within the framework of the path integral approach in [112]. The evolution of Gaussian wave packets in the case of periodical delta-kicks of the frequency was studied in [113]. The case of piecewise constant function $\Omega(t)$ (with abrupt jumps of the frequency) was studied, e.g., in [114–117]. Solutions in terms of elementary functions (power, logarithmic, and trigonometric) exist for the dependence $\Omega(t) = (a + bt)^{-1}$ (considered for $a = 0$ as far back as in [92, 93]). Periodic modulations of the frequency (parametric resonance) were studied, e.g., in [8, 118–122]. The stochastic variations of parameters were the subjects of studies [118, 123–125]. The case of the frequency going to zero ("quantum sling") has been analyzed in [126, 127].

2.3 Single mode: coherent state representation

The scheme of calculating the propagator given in subsections 2.1 and 2.2 can be generalized to any other quantum representation. Let $\hat{f}$ be some time-independent operator (in the Schrödinger picture), and $\hat{F}(t) = \hat{U}(t)\hat{f}\hat{U}^{-1}(t)$ be the operator integral of motion coinciding with $\hat{f}$ at $t = 0$. Then the evolution operator obeys the equation

$$\hat{F}(t)\hat{U}(t) = \hat{U}(t)\hat{f}. \tag{43}$$

The propagator in some abstract $|z\rangle$-representation is nothing but the matrix element $G(z, z', t) = \langle z|\hat{U}(t)|z'\rangle$. It satisfies the equation

$$\hat{F}(t)_{(1)}G(z, z', t) = \hat{f}^{\mathrm{T}}_{(2)}G(z, z', t) \tag{44}$$

where the indices (1) or (2) show explicitly the argument (the first z or the second z') of the function $G(z, z', t)$, which the operators act on. The symbol $\hat{f}^{\mathrm{T}}$ means the transposed operator: if the kernel of operator $\hat{f}$ equals $f(z, z')$, then the kernel of the transposed operator is $f(z', z)$.

In the holomorphic (coherent state) representation each ket-vector $|\psi\rangle$ is represented by an entire analytic function of the complex argument

$$\psi(\alpha^*) \equiv \langle\alpha^*|\psi\rangle \exp\left(|\alpha|^2/2\right),$$

where $|\alpha\rangle$ is the normalized eigenvector of the annihilation operator $\hat{a}$ with the eigenvalue α. The following relations hold

$$\hat{a}\psi(\alpha^*) = \partial\psi(\alpha^*)/\partial\alpha^*, \qquad \hat{a}^\dagger\psi(\alpha^*) = \alpha^*\psi(\alpha^*). \tag{45}$$

The integrals of motion corresponding to the operators $\hat{a}$ and $\hat{a}^\dagger$ can be written as

$$\hat{A}(t) = \xi(t)\hat{a} + \eta(t)\hat{a}^\dagger + \delta(t), \quad \hat{A}^\dagger(t) = \xi^*(t)\hat{a}^\dagger + \eta^*(t)\hat{a} + \delta^*(t), \tag{46}$$

where the coefficients must satisfy the equations

$$-i\dot{\xi} = \xi D_0 - \eta D_1, \quad -i\dot{\eta} = \xi D_1^* - \eta D_0, \quad -i\dot{\delta} = \xi g^* - \eta g. \tag{47}$$

The unitarity of transformation (46) is equivalent to the identity

$$|\xi(t)|^2 - |\eta(t)|^2 \equiv 1 \tag{48}$$

which is an immediate consequence of equations (47).

Taking $\hat{f} = \hat{a}$ and $\hat{f} = \hat{a}^\dagger$ in equation (44) we obtain two equations for the propagator $G(\alpha^*, \beta, t)$:

$$\left[\xi(t)\frac{\partial}{\partial\alpha^*} + \eta(t)\alpha^* + \delta(t)\right] G(\alpha^*, \beta, t) = \beta G(\alpha^*, \beta, t), \tag{49}$$

$$\left[\eta^*(t)\frac{\partial}{\partial\alpha^*} + \xi^*(t)\alpha^* + \delta^*(t)\right] G(\alpha^*, \beta, t) = \frac{\partial}{\partial\beta} G(\alpha^*, \beta, t). \tag{50}$$

Using the same scheme as in the preceding subsection, we first solve equation (49) and determine the dependence of the propagator on the first argument α^*. Then equation (50) gives the dependence on the second argument β. Finally, equation (18) together with the identity (48) enables us to find the time-dependent phase factor, giving rise to the following expression [6,7,9]:

$$G(\alpha^*, \beta, t) = [\xi(t)]^{-1/2} \exp\left\{\frac{1}{2\xi}\left[2\alpha^*\beta - \eta\alpha^{*2} + \eta^*\beta^2 \right.\right.$$
$$\left.\left. -2\alpha^*\delta + 2\beta\left(\xi\delta^* - \eta^*\delta\right) + \eta^*\delta^2\right] - \int_0^t \delta(t')\dot{\delta}^*(t')dt'\right\}. \tag{51}$$

The initial condition now reads $G(\alpha^*, \beta, 0) = \exp(\alpha^*\beta)$, since this exponential function is the kernel of the unit operator in the holomorphic representation. The action of the propagator (51) on the initial function $f(\beta^*, 0)$ is defined by the relation

$$f(\alpha^*, t) = \pi^{-1} \int G(\alpha^*, \beta, t) f(\beta^*, 0)\, d\,\mathrm{Re}\beta\, d\,\mathrm{Im}\beta. \tag{52}$$

Sometimes it is convenient to transform the first two equations in (47) using the substitutions

$$\xi = \chi \exp[i\Phi(t)], \quad \eta = \psi \exp[-i\Phi(t)], \quad \Phi(t) = \int_0^t D_0(\tau)d\tau. \tag{53}$$

Then we obtain the equations

$$\dot{\chi} = -iD_1 e^{-2i\Phi}\psi, \quad \dot{\psi} = iD_1^* e^{2i\Phi}\chi, \tag{54}$$

which can be easily solved in the important case of the *parametric amplifier*, when $D_1 = \kappa \exp[2i\Phi(t)]$ and $\kappa = const$:

$$\chi(t) = \cosh(|\kappa|t), \quad \psi(t) = \frac{i\kappa^*}{|\kappa|}\sinh(|\kappa|t). \tag{55}$$

This case was studied in numerous papers; see, e.g., [128–131]. There exists a vast literature on different types of amplifiers in quantum mechanics. We refer the reader to studies [132–158], where other references can be found.

2.4 Multimode systems

Quadrature representation

Any quadratic system possesses *linear* vector integrals of motion $\hat{\mathbf{X}}(t)$ and $\hat{\mathbf{P}}(t)$ which can be represented in the following matrix form:

$$\begin{bmatrix} \hat{\mathbf{P}} \\ \hat{\mathbf{X}} \end{bmatrix} = \Lambda(t) \begin{bmatrix} \hat{\mathbf{p}} \\ \hat{\mathbf{x}} \end{bmatrix} + \Delta(t); \quad \Lambda = \left\| \begin{matrix} \lambda_1 & \lambda_2 \\ \lambda_3 & \lambda_4 \end{matrix} \right\|, \quad \Delta = \begin{bmatrix} \delta_1 \\ \delta_2 \end{bmatrix}. \tag{56}$$

$2N \times 2N$ matrix $\Lambda(t)$ and $2N$-dimensional vector $\mathbf{\Delta}(t)$ are solutions to the system of the first-order ordinary differential equations, which are generalizations of similar equations from subsection 2.2 (E_N is $N \times N$ unit matrix):

$$\dot{\Lambda} = \Lambda \Sigma B \equiv \left\| \begin{matrix} \lambda_1 b_3 - \lambda_2 b_1 & \lambda_1 b_4 - \lambda_2 b_2 \\ \lambda_3 b_3 - \lambda_4 b_1 & \lambda_3 b_4 - \lambda_4 b_2 \end{matrix} \right\|, \qquad \Lambda(0) = E_{2N}, \tag{57}$$

$$\dot{\Delta} = \Lambda \Sigma \mathbf{C} \equiv \begin{bmatrix} \lambda_1 \mathbf{c}_2 - \lambda_2 \mathbf{c}_1 \\ \lambda_3 \mathbf{c}_2 - \lambda_4 \mathbf{c}_1 \end{bmatrix}, \qquad \Delta(0) = \mathbf{0}. \tag{58}$$

Antisymmetric matrix Σ is defined as

$$\Sigma = \|\Sigma_{jk}\| = \left\| \begin{matrix} 0 & E_N \\ -E_N & 0 \end{matrix} \right\|, \qquad \Sigma_{jk} = \frac{i}{\hbar} [\hat{q}_j, \hat{q}_k]. \tag{59}$$

A formal solution to equation (57) reads

$$\Lambda(t) = \tilde{\mathrm{T}} \exp \left[\int_0^t \Sigma B(\tau)\, \mathrm{d}\tau \right],$$

where symbol $\tilde{\mathrm{T}}$ means the antichronological ordering of operators. If matrix B does not depend on time, then we have simply

$$\Lambda(t) = \exp(\Sigma B t). \tag{60}$$

Due to equation (57), $\mathrm{d}(\Lambda \Sigma \tilde{\Lambda})/\mathrm{d}t = 0$. Therefore matrix $\Lambda(t)$ is symplectic,

$$\Lambda(t) \Sigma \tilde{\Lambda}(t) \equiv \Sigma. \tag{61}$$

This identity is nothing but the canonicity condition of transformation (56). Its immediate consequence is the relation

$$\Lambda^{-1} = -\Sigma \tilde{\Lambda} \Sigma = \left\| \begin{matrix} \tilde{\lambda}_4 & -\tilde{\lambda}_2 \\ -\tilde{\lambda}_3 & \tilde{\lambda}_1 \end{matrix} \right\|. \tag{62}$$

Identity (61) is equivalent to the set of identities for $N \times N$ matrices λ_j:

$$\lambda_2 \tilde{\lambda}_1 = \lambda_1 \tilde{\lambda}_2, \quad \lambda_3 \tilde{\lambda}_4 = \lambda_4 \tilde{\lambda}_3, \quad \tilde{\lambda}_4 \lambda_2 = \tilde{\lambda}_2 \lambda_4, \quad \tilde{\lambda}_1 \lambda_3 = \tilde{\lambda}_3 \lambda_1, \tag{63}$$

$$\lambda_4 \tilde{\lambda}_1 - \lambda_3 \tilde{\lambda}_2 = E_N, \qquad \tilde{\lambda}_4 \lambda_1 - \tilde{\lambda}_2 \tilde{\lambda}_3 = E_N. \tag{64}$$

Putting the explicit expressions of the operators $\hat{\mathbf{X}}$ and $\hat{\mathbf{P}}$ from (56) into the equations (14) and (17), we arrive at a system of the first order partial differential equations, which can be easily solved. The important phase factor, which depends only on time, is obtained with the aid of equation (18) and the initial condition (19) (with account of the identities (63)–(64), which help to simplify the expressions arising in the intermediate calculations). The details of the calculations can be found in [6,7,9,11]. If det $\lambda_3 \neq 0$, then

$$G(\mathbf{x}_2,\mathbf{x}_1,t) = [\det(-2\pi i\hbar\lambda_3)]^{-1/2}\exp\left\{-\frac{i}{2\hbar}\left[\mathbf{x}_2\lambda_3^{-1}\lambda_4\mathbf{x}_2 -2\mathbf{x}_2\lambda_3^{-1}\mathbf{x}_1 + \mathbf{x}_1\lambda_1\lambda_3^{-1}\mathbf{x}_1 + 2\mathbf{x}_2\lambda_3^{-1}\delta_2 + 2\mathbf{x}_1\left(\delta_1 - \lambda_1\lambda_3^{-1}\delta_2\right) +\delta_2\lambda_1\lambda_3^{-1}\delta_2 - 2\int_0^t \dot{\delta}_1(\tau)\delta_2(\tau)\mathrm{d}\tau\right]\right\}. \tag{65}$$

The explicit expressions in the case of det $\lambda_3 = 0$ were given in [6–9,11]. For the applications of general formulae for the propagators to concrete physical problems see, e.g., [8, 159]. Many explicit expressions in various special cases were given in [9,11].

Coherent state representation

The vector operator integrals of motion $(\hat{\mathbf{A}},\hat{\mathbf{A}}^\dagger)$, coinciding with $(\hat{\mathbf{a}},\hat{\mathbf{a}}^\dagger)$ at the initial moment, can be written as

$$\begin{pmatrix}\hat{\mathbf{A}}\\ \hat{\mathbf{A}}^\dagger\end{pmatrix} = \Lambda(t)\begin{pmatrix}\hat{\mathbf{a}}\\ \hat{\mathbf{a}}^\dagger\end{pmatrix} + \begin{pmatrix}\delta\\ \delta^*\end{pmatrix}, \quad \Lambda = \left\|\begin{matrix}\xi & \eta\\ \eta^* & \xi^*\end{matrix}\right\|. \tag{66}$$

Matrix $\Lambda(t)$ must satisfy the following equation and the initial condition:

$$-i\dot{\Lambda} = \Lambda\Sigma B \equiv \left\|\begin{matrix}\xi D_0 - \eta D_1 & \xi D_1^* - \eta\tilde{D}_0\\ \eta^* D_0 - \xi^* D_1 & \eta^* D_1^* - \xi^*\tilde{D}_0\end{matrix}\right\|, \quad \Lambda(0) = E_{2N} \tag{67}$$

(matrix Σ here is the same as above). Complex vector δ obeys the equations

$$-i\dot{\delta} = \xi\mathbf{g}^* - \eta\mathbf{g}, \quad \delta(0) = 0. \tag{68}$$

The consequences of the symplecticity condition (61) are the identities

$$\xi\tilde{\eta} = \eta\tilde{\xi}, \quad \tilde{\xi}\eta^* = \eta^\dagger\xi, \quad \xi\xi^\dagger - \eta\eta^\dagger = \xi^\dagger\xi - \tilde{\eta}\eta^* = E_N. \tag{69}$$

Solving the vector analogues of equations (49) and (50), one can express the propagator in the coherent state representation as follows [6,7]:

$$G(\alpha^*,\beta,t) = [\det\xi(t)]^{-1/2}\exp\left[\alpha^*\xi^{-1}(\beta-\delta) + \beta\left(\delta^* - \eta^*\xi^{-1}\delta\right) +\tfrac{1}{2}\left(\beta\eta^*\xi^{-1}\beta - \alpha^*\xi^{-1}\eta\alpha^* + \delta\eta^*\xi^{-1}\delta\right) - \int_0^t \delta(\tau)\dot{\delta}^*(\tau)\mathrm{d}\tau\right]. \tag{70}$$

All formulae become extremely simple if $D_1 = 0$, when we have the identity $\eta \equiv 0$, so that any coherent state remains coherent in the process of evolution. This special case was studied for the first time in [160, 161]. For other special cases see [9,11].

2.5 Canonical transformations and evolution operators

Actually, functions (65) and (70) are the kernels of the unitary operators performing *linear canonical transformations* (56) and (66) in accordance with the relations like (12). In this context, these functions were calculated, with the aid of equations similar to those used in this chapter, in [162–167] (and repeated in many more recent papers). However, in the latter case the phase of the preexponential factor (which does not depend on $\mathbf{x}_2$ and $\mathbf{x}_1$, for instance) is not significant, while in the case of the propagator it is very important, since it must be adjusted to the Schrödinger equation. We did not go into details of calculations of this factor in the multidimensional case, but this problem is not quite trivial (it is worth emphasizing that in looking for the kernel of a symplectic transformation, one assumes matrix Λ and vector $\boldsymbol{\Delta}$ to be given beforehand, whereas in the propagator case one has first to find this matrix and vector, starting from the given quadratic Hamiltonian). For example, general linear multidimensional canonical (symplectic) transformations between "in" and "out" operators describing linear amplifiers were considered as far back as in [134]. One of the first studies on *dynamics* of multimode parametric systems with generic quadratic Hamiltonians was performed by Mollow [168] (see also [169], who expressed the solutions of the Heisenberg equations of motion in a form similar to (66) and looked for the kernel of the related canonical transformation in the form equivalent to (70). However, he did not give a closed compact expression for the propagator. Further progress was achieved in [170–172], although some integrals remained in implicit forms. The closed compact expressions given above were obtained for the first time in [6, 7]. Later, formulae analogous to (65) and (70) were derived, e.g., in [173–176]. Linear canonical *supertransformations* (when bosonic operators are coupled with fermionic ones) were studied, e.g., in [9, 177, 178]. The connection between quantum invariants and the Noether theorem was considered in [179].

For the first time (as far as we know), the *evolution operator* for the generalized one-dimensional oscillator with time-dependent parameters was constructed by Fujiwara [64], who developed some "operator calculus" technics to handle operator "T-exponentials". The Schrödinger equation with a generic *multidimensional time-dependent* quadratic Hamiltonian (3) was considered by Kulsrud [180], who looked for the evolution operator in the Gaussian form

$$\hat{U}(t) = \exp\left\{-(i/\hbar)\left[\tfrac{1}{2}\hat{\mathbf{q}}D(t)\hat{\mathbf{q}} + \mathbf{g}(t)\hat{\mathbf{q}} + \phi(t)\right]\right\}.$$

He obtained a *nonlinear* equation for the symmetric matrix $D(t)$ and expressed vector $\mathbf{g}(t)$ and scalar function $\phi(t)$ in terms of some integrals involving elements of matrix $D(t)$. As a matter of fact, the solution can be found in a simple and straightforward manner, if one uses quantum operator integrals of motion. Then one can easily express matrix $D(t)$ in terms of matrix $\Lambda(t)$ as follows:

$$D(t) = \Sigma^{-1} \ln \Lambda(t). \tag{71}$$

To prove this result, let us suppose that we have $2N$ operators $\hat{y}_1, \hat{y}_2, \ldots, \hat{y}_{2N}$, whose commutators form an antisymmetric c-number matrix

$$\Omega = \|\Omega_{ij}\|, \qquad \Omega_{ij} = -\Omega_{ji} = [\hat{y}_i, \hat{y}_j]. \tag{72}$$

We combine these operators in the $2N$-dimensional vector $\hat{\mathbf{y}}$. Consider operator $\hat{U}_\varepsilon \equiv \exp\left(-\frac{\varepsilon}{2}\hat{\mathbf{y}}\mathcal{D}\hat{\mathbf{y}}\right)$, where $\mathcal{D}$ is a symmetric matrix, and ε is a parameter. Introducing operator $\hat{\mathbf{Y}} = \hat{U}_\varepsilon \hat{\mathbf{y}} \hat{U}_\varepsilon^{-1}$, one can easily verify the equation $\partial\hat{\mathbf{Y}}/\partial\varepsilon = \Omega D\hat{\mathbf{Y}}$. Integrating this equation, one arrives at the formula

$$\exp\left(-\frac{\varepsilon}{2}\hat{\mathbf{y}}\mathcal{D}\hat{\mathbf{y}}\right)\hat{\mathbf{y}}\exp\left(\frac{\varepsilon}{2}\hat{\mathbf{y}}\mathcal{D}\hat{\mathbf{y}}\right) = \exp(\Omega\mathcal{D}\varepsilon)\hat{\mathbf{y}}. \qquad (73)$$

If $\hat{U}_\varepsilon$ is the unitary evolution operator, then vector $\hat{\mathbf{Y}}$ is nothing but the operator integral of motion coinciding with $\hat{\mathbf{y}}$ at the initial moment: $\hat{\mathbf{Y}} = \Lambda\hat{\mathbf{y}}$, where $2N \times 2N$ matrix Λ can be found from linear equations of motion given in the preceding sections. Then matrix $\mathcal{D}$, which determines the explicit form of the evolution operator, is proportional to the logarithm of matrix Λ. Choosing the free parameter ε as $\varepsilon = 1$, one obtains $\mathcal{D} = \Omega^{-1}\ln\Lambda$ (we suppose that the commutator matrix Ω is nonsingular). If vector $\hat{\mathbf{y}}$ consists of the usual Cartesian momentum and coordinate operators, as in equation (56), then $\Omega = -i\hbar\Sigma$, where matrix Σ was defined in (59), and we obtain equation (71).

It is well known that Van Vleck's quasiclassical formula [181–183]

$$G(x_2, x_1, t) = \left(\frac{i}{2\pi\hbar}\frac{\partial^2 S}{\partial x_1 \partial x_2}\right)^{1/2} \exp\left[\frac{i}{\hbar}S(x_2, x_1, t)\right], \qquad (74)$$

where $S(x_2, x_1, t)$ is the classical action function (we consider here the one-dimensional case), is exact for systems with quadratic Hamiltonians [184, 185]. Therefore, in many papers the propagator was searched directly in the form (for homogeneous Hamiltonians, for simplicity)

$$G(\mathbf{x}_2, \mathbf{x}_1, t) = \exp\left\{(i/\hbar)\left[\tfrac{1}{2}\hat{\mathbf{x}}_2 a(t)\hat{\mathbf{x}}_2 + \hat{\mathbf{x}}_2 b(t)\hat{\mathbf{x}}_1 + \tfrac{1}{2}\hat{\mathbf{x}}_1 c(t)\hat{\mathbf{x}}_1 + \phi(t)\right]\right\}.$$

However, since matrices a, b, c satisfy *nonlinear* ordinary differential equations, this method is rather complicated, even in the one-dimensional case (Gaussian-like *quasiclassical* propagators were studied, e.g., in [186, 187]).

Many authors used the Feynman path integrals scheme [112, 188–195]. Another method takes into account the Lie-algebraic structure of the quadratic Hamiltonians (for example, Hamiltonian (1) is a linear combination of generators of the $su(1,1)$ algebra). It is based on various algebraic transformations which enable us to factorize the exponential of the sum of N operators, like $\exp(\sum_{k=1}^{N}\xi_k\hat{X}_k)$, to the product of N exponentials of the form $\prod_{k=1}^{N}\exp(\eta_k\hat{X}_k)$. This method, originating from the papers [65, 196, 197], was applied, e.g., in [198–208]. Many more references for path integrals and algebraic approaches can be found in [209–211] ([211] contains a lot of explicit expressions for the propagators). However, although these approaches can be effective in some specific cases, in the generic multidimensional case they are much more complicated from the point of view of calculations, than the simple and elegant method of quantum invariants. Close or similar methods were used, e.g., in [212, 213].

2.6 Wigner–Weyl representation

Perhaps the simplest description of quantum systems whose evolution is governed by quadratic Hamiltonians is achieved in the Wigner–Weyl representation [214, 215] (considered also by Dirac [216]), where the statistical operator $\hat{\rho}$ (introduced for the first time by Landau [217] and a few months later in a more detailed form by von Neumann [218]) is replaced by the *real* function of the vector argument $\mathbf{q} = (\mathbf{p}, \mathbf{x})$ (here ξ is a $2N$-dimensional vector, and matrix Σ is given by (59))

$$W(\mathbf{q}) = \mathrm{Tr}\left(\hat{\rho} \int \exp\left[-\frac{i}{\hbar}\xi\Sigma(\mathbf{q} - \hat{\mathbf{q}})\right] d\xi/(2\pi\hbar)^N\right), \tag{75}$$

The equivalent formula is

$$W(\mathbf{x}, \mathbf{p}) = \int d\mathbf{v}\; e^{i\mathbf{p}\mathbf{v}/\hbar}\langle \mathbf{x} - \mathbf{v}/2|\hat{\rho}|\mathbf{x} + \mathbf{v}/2\rangle, \tag{76}$$

or, for pure quantum states,

$$W_\psi(\mathbf{x}, \mathbf{p}) = \int d\mathbf{v}\; e^{i\mathbf{p}\mathbf{v}/\hbar}\psi(\mathbf{x} - \mathbf{v}/2)\psi^*(\mathbf{x} + \mathbf{v}/2). \tag{77}$$

The inverse relation reads

$$\rho(\mathbf{x}, \mathbf{x}') = \int W\left([\mathbf{x} + \mathbf{x}']/2, \mathbf{p}\right) \exp\left[i(\mathbf{x} - \mathbf{x}')\mathbf{p}/\hbar\right] d\mathbf{p}/(2\pi\hbar)^N. \tag{78}$$

In this chapter, we use the normalization

$$\mathrm{Tr}\hat{\rho} = \int W(\mathbf{x}, \mathbf{p}) d\mathbf{x} d\mathbf{p}/(2\pi\hbar)^N = 1. \tag{79}$$

Properties of the Wigner function were discussed in Chapter 2. There are many studies and reviews devoted to this function and other "quasiprobabilities". In addition to those cited in Chapter 2, it is worth mentioning also [219–233].

There exists the following correspondence between the operators and their symbols in the Wigner representation:

$$\hat{\mathbf{q}}\hat{\rho} \to \left(\mathbf{q} - \frac{i\hbar}{2}\Sigma\frac{\partial}{\partial \mathbf{q}}\right) W(\mathbf{q}), \quad \hat{\rho}\hat{\mathbf{q}} \to \left(\mathbf{q} + \frac{i\hbar}{2}\Sigma\frac{\partial}{\partial \mathbf{q}}\right) W(\mathbf{q}). \tag{80}$$

Therefore the Liouville–von Neumann equation for the statistical operator

$$i\hbar\partial\hat{\rho}/\partial t = [\hat{H}, \hat{\rho}] \tag{81}$$

is transformed in the case of quadratic Hamiltonian (3) to the first-order partial differential equation

$$\frac{\partial W}{\partial t} = \sum_{k=1}^{2n} \frac{\partial}{\partial q_k}\left[(\Sigma B\mathbf{q} + \Sigma\mathbf{C})_k W\right]. \tag{82}$$

The solution to this equation is an arbitrary function of the *integrals of motion* (56). If the initial Wigner function was $W_0(\mathbf{q})$, then [11,234] (see also [235,236])

$$W(\mathbf{q},t) = W_0(\Lambda(t)\mathbf{q} + \mathbf{\Delta}(t)). \tag{83}$$

Explicit expressions of the Wigner functions related to the evolution operators of general quadratic Hamiltonians and their special cases were given, e.g., in [175, 237–240]. For example, the Wigner–Weyl symbol of the propagator (65) has the following symmetric form [238] (here $E \equiv E_{2n}$)

$$\begin{aligned} G(\mathbf{q},t) &= 2^n\left[\det(\Lambda+E)\right]^{-1/2}\exp\left[\frac{i}{\hbar}\mathbf{q}\Lambda\mathbf{q} + \frac{i}{2\hbar}\int_0^t \dot{\Delta}\Sigma\Delta\, d\tau \right.\\ &+ \left.\frac{2i}{\hbar}\mathbf{q}\Sigma(\Lambda+E)^{-1}\Delta + \frac{i}{4\hbar}\Delta\Sigma(\Lambda+E)^{-1}(\Lambda-E)\Delta\right]. \end{aligned} \tag{84}$$

The kernels of linear canonical transformations in the Wigner–Weyl representation were considered, e.g., in [235,241].

Generalized Wigner functions

The structure of the Wigner function becomes even more transparent, if one generalizes it to the case of arbitrary set of coordinates in the phase space, related to the vector $\mathbf{q} = (\mathbf{p},\mathbf{x})$ by means of the *linear* transformation $\mathbf{y} = T\mathbf{q}$, where T is a nonsingular matrix (it may be complex, for example, if one wishes to replace the coordinate and momenta operators by annihilation and creation ones). Then the new fundamental commutator matrix Ω (72) is related to matrix Σ (59) as $\Omega = -i\hbar T\Sigma\tilde{T}$. Making the change of variables in the integral in equation (75) and remembering that the Jacobian of transformation is $|\det T|$, we arrive at the following expressions [11]:

$$W(\mathbf{y}) = \mathrm{Tr}\left(\hat{\rho}\int\frac{\exp\left[\zeta\Omega^{-1}(\mathbf{y}-\hat{\mathbf{y}})\right]d\zeta}{|\det(2\pi\Omega)|^{1/2}}\right), \quad \int\frac{W(\mathbf{y})d\mathbf{y}}{|\det(2\pi\Omega)|^{1/2}} = 1. \tag{85}$$

They clearly show that the factor $\hbar^N$ in the integrals in formulae (75), (78) and (79) is not accidental, because it is related to the determinant of the fundamental commutator matrix.

If the Hamiltonian operator $\hat{H}$ does not depend on time, then the formal substitution $t = -i\beta\hbar$ transforms the Schrödinger equation into the Bloch equation [242] $\partial\widetilde{\rho}_{eq}/\partial\beta + \hat{H}\widetilde{\rho}_{eq} = 0$ for the unnormalized equilibrium statistical operator $\widetilde{\rho}_{eq}(\beta) = \exp(-\beta\hat{H})$, where $\beta = (k_B T)^{-1}$ is the inverse absolute temperature in the energy units. Making the same substitution in equation (84) we immediately obtain the corresponding Wigner function, taking into account formula (60). Then after simple integration we find the statistical sum $\mathcal{Z}(\beta) \equiv \mathrm{Tr}\exp(-\beta\hat{H})$ and the normalized equilibrium Wigner function for an arbitrary N-dimensional quadratic Hamiltonian with *positively definite* real symmetrical matrix B [243]. We give it for an arbitrary c-number commutator matrix (72) (confining ourselves here to the case of homogeneous quadratic forms):

$$\mathcal{Z}(\beta) = 2^{-N} \left|\det \sinh(\beta\Omega B/2)\right|^{-1/2}, \tag{86}$$

$$\begin{aligned} W_{eq}(\mathbf{y};\beta) &= 2^{N} \left|\det\left\{\tanh\left(\beta\Omega B/2\right)\right\}\right|^{1/2} \\ &\times \exp\left[-\mathbf{y}\Omega^{-1}\tanh\left(\beta\Omega B/2\right)\mathbf{y}\right]. \end{aligned} \tag{87}$$

The function (87) corresponds to the operator

$$\hat{\rho}_{eq}(\beta) = [\mathcal{Z}(\beta)]^{-1} \exp\left[-(\beta/2)\hat{\mathbf{y}}B\hat{\mathbf{y}}\right]. \tag{88}$$

3 Evolution of pure states

3.1 Squeezed states as generalized coherent states

The operator integrals of motion $\hat{\mathbf{A}}, \hat{\mathbf{A}}^\dagger$ (66) are unitarily equivalent to the initial operators $\hat{\mathbf{a}}, \hat{\mathbf{a}}^\dagger$. Consequently, they possess the same properties. Thus it is natural to call the eigenstate of operator $\hat{\mathbf{A}}$ the generalized coherent state $|\alpha\rangle$ [95]: $\hat{\mathbf{A}}|\alpha\rangle = \alpha|\alpha\rangle$. Let us designate as $\langle\mathbf{x}|\alpha\rangle$ the normalized wave function of the coherent state in the coordinate representation. It is easy to verify that function $f(\mathbf{x};\alpha) = \langle\mathbf{x}|\alpha\rangle \exp\left(|\alpha|^2/2\right)$ satisfies the equations

$$\hat{\mathbf{A}} f(\mathbf{x};\alpha) = \alpha f(\mathbf{x};\alpha), \quad \hat{\mathbf{A}}^\dagger f(\mathbf{x};\alpha) = \partial f(\mathbf{x};\alpha)/\partial\alpha \tag{89}$$

(here operators $\hat{\mathbf{A}}, \hat{\mathbf{A}}^\dagger$ act on f as a function of $\mathbf{x}$, while α is considered as a parameter in the left-hand sides of the equations). Operator $\hat{\mathbf{A}}$ can be written as a linear combination of the coordinate and momenta operators

$$\hat{\mathbf{A}} = \lambda_p\hat{\mathbf{p}} + \lambda_q\hat{\mathbf{x}} + \delta, \tag{90}$$

where complex matrix coefficients obey the equations

$$\dot{\lambda}_p = \lambda_p b_3 - \lambda_q b_1, \quad \dot{\lambda}_q = \lambda_p b_4 - \lambda_q b_2, \quad \dot{\delta} = \lambda_p \mathbf{c}_2 - \lambda_q \mathbf{c}_1, \tag{91}$$

$$\lambda_p\tilde{\lambda}_q = \lambda_q\tilde{\lambda}_p, \quad \tilde{\lambda}_q\lambda_q^* = \lambda_q^\dagger\lambda_q, \quad \tilde{\lambda}_p\lambda_p^* = \lambda_p^\dagger\lambda_p, \tag{92}$$

$$\lambda_p\lambda_q^\dagger - \lambda_q\lambda_p^\dagger = \lambda_q^\dagger\lambda_p - \tilde{\lambda}_q\lambda_p^* = (i/\hbar)E_N. \tag{93}$$

Using the same strategy as in the previous sections, i.e., solving equations (89) together with the Schrödinger equation, with account of equations (91)–(93), we obtain the explicit expression for the generalized coherent state wave function [11]

$$\begin{aligned} \langle\mathbf{x}|\alpha\rangle = \left(2\pi\hbar^2\right)^{-N/4} (\det\lambda_p)^{-1/2} \exp\Big[&-\frac{i}{2\hbar}\mathbf{x}\lambda_p^{-1}\lambda_q\mathbf{x} + \frac{i}{\hbar}\mathbf{x}\lambda_p^{-1}(\alpha-\delta) \\ +\tfrac{1}{2}(\alpha-\delta)\lambda_p^*\lambda_p^{-1}(\alpha-\delta) + \alpha\delta^* &- \tfrac{1}{2}(|\delta|^2+|\alpha|^2) + i\int_0^t \mathrm{Im}[\dot{\delta}(\tau)\delta^*(\tau)]d\tau\Big]. \end{aligned} \tag{94}$$

Let us introduce the $2N \times 2N$ symmetric covariance matrix

$$\mathcal{M} = \|\overline{q_j q_k}\| = \left\| \begin{matrix} \mathcal{M}_{pp} & \mathcal{M}_{px} \\ \mathcal{M}_{xp} & \mathcal{M}_{xx} \end{matrix} \right\|, \qquad \mathcal{M}_{px} = \tilde{\mathcal{M}}_{xp}, \tag{95}$$

where the (co)variances are defined as

$$\overline{q_j q_k} = \tfrac{1}{2}\langle \hat{q}_j\hat{q}_k + \hat{q}_k\hat{q}_j\rangle - \langle \hat{q}_j\rangle\langle \hat{q}_k\rangle, \tag{96}$$

and the splitting of matrix $\mathcal{M}$ in $N \times N$ blocks is performed in accordance with the structure of the $2N$-dimensional vector $\mathbf{q} = (\mathbf{p}, \mathbf{x})$.

A generic Gaussian pure state is described by means of the wave function

$$\psi(\mathbf{x}) = \mathcal{N}\exp\left(-\mathbf{x}a\mathbf{x} + \mathbf{b}\mathbf{x}\right), \tag{97}$$

where a is a symmetrical complex $N \times N$ matrix with a nonnegatively definite real part, $\mathbf{b}$ is an arbitrary complex vector, and $\mathcal{N}$ is the normalization constant. Calculating the first- and second-order moments with the aid of the Poisson integral

$$\int \exp\left(-\mathbf{x}a\mathbf{x} + \mathbf{b}\mathbf{x}\right) d\mathbf{x} = \pi^{N/2}(\det a)^{-1/2}\exp\left(\tfrac{1}{4}\,\mathbf{b}a^{-1}\mathbf{b}\right) \tag{98}$$

one finds that the variance matrix of the Gaussian state (97) is determined completely by the real and imaginary parts of matrix $a = u + iv$, being independent from vector $\mathbf{b}$:

$$\mathcal{M}_{xx} = (4u)^{-1}, \quad \mathcal{M}_{pp} = \hbar^2\left(u + vu^{-1}v\right), \quad \mathcal{M}_{px} = -(\hbar/2)vu^{-1}. \tag{99}$$

Looking at the expressions above, one can immediately obtain the identities which hold for any pure Gaussian state

$$\mathcal{M}_{xx}\mathcal{M}_{pp} - \mathcal{M}_{xp}^2 = \mathcal{M}_{pp}\mathcal{M}_{xx} - \mathcal{M}_{px}^2 = (\hbar^2/4)E_N. \tag{100}$$

In the specific case of the state (94) we obtain, using (92)–(93) [11],

$$\mathcal{M}_{xx} = \hbar^2\lambda_p^\dagger\lambda_p, \quad \mathcal{M}_{pp} = \hbar^2\lambda_q^\dagger\lambda_q, \quad \mathcal{M}_{px} = -\hbar^2\mathrm{Re}\left(\lambda_q^\dagger\lambda_p\right). \tag{101}$$

Then identities (100) hold due to relations (92)–(93).

The variance matrices $\mathcal{M}_{xx}$ and $\mathcal{M}_{pp}$ are, in general, nondiagonal, and their elements do not coincide with the vacuum state variances. Thus the state (94) can also be named the *multidimensional squeezed state*, because in the process of evolution the variances of some canonical variables can assume values below the ground state level. We illustrate this statement in the one-dimensional example considered in the next section. The general case was considered in [186, 244].

3.2 One-mode case: correlated states and geometric phase

In the one-dimensional case the matrices λ_j become the usual numbers, as well as the variance matrices $\mathcal{M}_{ab}$. In particular, functions $\lambda_p(t)$ and $\lambda_q(t)$ can be expressed as follows [cf. Eq. (37)],

$$\lambda_p = \sqrt{\mu}\varepsilon, \quad \lambda_q = \left[(2\mu\rho - \dot{\mu})\varepsilon - 2\mu\dot{\varepsilon}\right]/(2\mu^{3/2}), \tag{102}$$

where $\varepsilon(t)$ is a *complex* solution to equation (35), satisfying the condition

$$\mathrm{Im}\left(\dot{\varepsilon}\varepsilon^*\right) = (2\hbar)^{-1}. \tag{103}$$

Therefore the most general wave function of the coherent state (94) can be represented in the one-dimensional case as

$$\langle x|\alpha\rangle = \left(2\pi\hbar^2\mu\varepsilon^2\right)^{-1/4}\exp\left\{\frac{i}{2\hbar}\left(\frac{\dot\varepsilon}{\mu\varepsilon}+\frac{\dot\mu}{2\mu^2}-\frac{\rho}{\mu}\right)x^2+\frac{i}{\hbar}\frac{(\alpha-\delta)x}{\sqrt{\mu}\varepsilon}\right.$$
$$\left.+\frac{\varepsilon^*}{2\varepsilon}(\alpha-\delta)^2+\alpha\delta^*-\tfrac{1}{2}\left(|\delta|^2+|\alpha|^2\right)+i\int_0^t \mathrm{Im}\left[\dot\delta(\tau)\delta^*(\tau)\right]d\tau\right\}. \quad (104)$$

For special cases and other examples see, e.g. (in addition to the already cited papers), [131,245–251].

Using the notation σ_{ab} (where $a,b = p,x$) for the variances in the one-mode case, we can express them as

$$\sigma_{xx} = \hbar^2|\lambda_p|^2, \quad \sigma_{pp} = \hbar^2|\lambda_q|^2, \quad \sigma_{px} = -\hbar^2\mathrm{Re}\left(\lambda_q^*\lambda_p\right). \qquad (105)$$

Since functions $\lambda_p(t)$ and $\lambda_q(t)$ satisfy the constraint

$$\mathrm{Im}\left(\lambda_p\lambda_q^*\right) = (2\hbar)^{-1}, \qquad (106)$$

the variances of the Gaussian (pure) state satisfy the identity

$$\Delta \equiv \sigma_{xx}\sigma_{pp} - \sigma_{px}^2 = \hbar^2/4, \qquad (107)$$

minimizing the *Schrödinger–Robertson uncertainty relation* $\Delta \geq \hbar^2/4$ (Chapter 1, page 18). Taking into account relations (105) and (106) one can express the *one-dimensional* generalized coherent state (94) or (97) as the *correlated coherent state* (57) of Chapter 1, where the coefficients σ_{xx}, r, and b are, in general, some functions of time [252–255].

There exists an interesting relation between the correlated states and the Berry (geometric) phase [256] arising in the process of adiabatic evolution of quantum systems. Let us consider, following [253,257], a generic homogeneous quadratic Hamiltonian for the single degree of freedom given by (1). It is sufficient to confine ourselves to the special case when the linear terms in the Hamiltonian are absent. Suppose that the coefficients of the Hamiltonian are *slowly varying in time* functions. Then one can easily find the linear *adiabatic* integral of motion $\hat A$, satisfying the condition $[\hat A, \hat A^\dagger] = 1$, neglecting the time derivatives in the expression for the coefficient λ_q in (102) and taking the approximate solution to equation (35) as $\exp(i\int^t \Omega(\tau)d\tau)$:

$$\hat A(t) = \frac{\exp\left[i\Phi(t)\right]}{\sqrt{2\mu\hbar\omega(t)}}\left\{\mu(t)\hat p - i\left[\omega(t)+i\rho(t)\right]\hat x\right\}, \qquad \omega = \sqrt{\mu\nu-\rho^2}. \quad (108)$$

If the duration of the process has some characteristic time T, then $\omega \sim T^0$, whereas $\dot\rho \sim \dot\mu \sim T^{-1}$, and $\ddot\mu \sim \dot\mu^2 \sim T^{-2}$. Expanding the phase $\Phi(t)$ in the asymptotic series with respect to T^{-1}, one has $\Phi(t) = \Phi_d + \Phi_g + \mathcal{O}(T^{-1})$, where the "dynamical" phase $\Phi_d = \int_0^t \omega(\tau)\,d\tau$ is proportional to T, but the correction

$$\Phi_g = \int_0^t \frac{d\tau}{2\omega(\tau)}\left[\rho(\tau)\frac{\dot\mu(\tau)}{\mu(\tau)} - \dot\rho(\tau)\right] = \int_C \frac{1}{2\omega}\left[\rho\frac{d\mu}{\mu} - d\rho\right] \qquad (109)$$

does not depend on the duration of the process, being determined completely by the contour C in the parameter space (μ, ν, ρ) which these coefficients move along.

The knowledge of linear integrals of motion enables us to construct the quadratic adiabatic invariant

$$\hat{A}^{\dagger}\hat{A} = \hat{H}(t)/[\hbar\omega(t)] - 1/2. \tag{110}$$

The eigenfunctions of this operator (the generalized Fock states) acquire in the process of *adiabatic* evolution the phase factor $\exp[-i(n+1/2)\Phi]$, in the same manner as the eigenfunctions of the number operator $\hat{a}^{\dagger}\hat{a}$ acquire the phase factor $\exp[-i(n+1/2)\omega t]$ in the usual stationary case. The part of the phase, $(n+1/2)\Phi_g$, is just the geometric phase obtained in [256] with the aid of much more complicated calculations. It is easy to verify that the eigenstates of operator (110) possess the *nonzero correlation coefficient*

$$r = \sigma_{px}/\sqrt{\sigma_p\sigma_x} = -\rho/\sqrt{\mu\nu}\,. \tag{111}$$

This coefficient is proportional to the factor $\rho(t)$, which determines the geometric phase (109). Besides, it does not depend on the quantum number n. We see that a nontrivial geometric phase exists only for the correlated states, and for the Hamiltonians possessing the terms such as $\hat{x}\hat{p} + \hat{p}\hat{x}$. Perhaps this observation explains why the quantum mechanical geometric phase was discovered so late, although the initial construction of quantum mechanics was based to a great extent on the theory of adiabatic invariants: for decades, nobody thought that Hamiltonians with terms like $\hat{x}\hat{p}+\hat{p}\hat{x}$ could have physical meanings. Similar approaches to the Berry phase problem, based on quantum invariants, were considered in [258–261]. For other treatments see, e.g., [262,263]. The Berry phase for *fermionic* coherent states was calculated in [264]. Reviews on the geometric phase in quantum mechanics and optics can be found in [265–267].

3.3 Traps and coupled oscillators

An important application of the nonstationary quadratic Hamiltonians is the theory of electromagnetic traps of particles [268–271]. In this context, the motion in nonstationary potentials (in most cases, an oscillator with a time-dependent frequency) was considered, e.g., in [272–278]. Glauber [279] applied the general method of linear time-dependent integrals of motion of [10, 11] to study the motion of quantum charged particles in electromagnetic traps. This approach was developed, e.g., in [280–287]. The *singular oscillator* model of two ions in the Paul trap was considered in [288, 289]. References to the papers devoted to the motion of a charged particle in a homogeneous time-dependent magnetic field and a binding harmonic potential can be found in [11] and in Chapter 1.

The problem of coupled oscillators with *constant* parameters can be reduced, in principle, to the problem of independent oscillators, by means of some linear canonical transformation diagonalizing the quadratic form $\mathbf{q}B\mathbf{q}$ in Hamiltonian (3) [290–294]. This can be always done only for positively definite quadratic forms. Normal forms of arbitrary quadratic Hamiltonians were given by Williamson [295] (see also [296,297]); a detailed discussion of this subject and the related references can be found in [9,11]. The situation is more complicated in the generic *nonstationary* case, when "instantaneous proper frequencies" and coupling constants are *arbitrary* functions of time. Then, applying the *time-dependent* transformation $\psi = \hat{S}(t)\psi'$ to the *time-dependent* Schrödinger equation $i\hbar\dot{\psi} = \hat{H}\psi$, one arrives at the equation $i\hbar\dot{\psi}' = \hat{H}'\psi'$, where the new Hamiltonian is not simply $\hat{S}^{-1}\hat{H}\hat{S}$ (as it would be in the stationary case), but $\hat{H}' = \hat{S}^{-1}\hat{H}\hat{S} - i\hbar\hat{S}^{-1}(\partial\hat{S}/\partial t)$. Consequently, a mere diagonalization of the initial quadratic form does not solve the problem in the generic case. On the other hand, using time-dependent canonical transformations, one can transform the *Schrödinger equation* (not the Hamiltonian itself) with an arbitrary Hamiltonian to an equivalent Schrödinger equation with *any desired another Hamiltonian* [9,11] (provided the Hamiltonians are physically admissible, in the sense that unitary evolution operators do exist; this property holds, in particular, for quadratic Hamiltonians [293] and their special examples, such as generalized time-dependent oscillators [298,299]).

The model of two coupled harmonic oscillators has been studied extensively by many authors, who applied it to various problems of quantum mechanics and quantum optics. For instance, it was used to describe quantum amplifiers [133, 136, 138] and converters in [133,300,301]. Walls [302] applied the coupled oscillator model to the problem of cyclotron resonance. Explicit exact solutions and propagators of the Schrödinger equation, as well as solutions of the Heisenberg equations of motion were considered and applied to different problems in [303–313]. Squeezing, photon statistics and entanglement in the system of two coupled oscillators were studied in [314–325]. As a rule, only specific couplings (mainly via coordinates) were considered in all that papers. The most general bilinear coupling with *time-independent* coefficients was studied in detail in [234,324,326]. The problem of *squeezing exchange* between resonancely coupled modes with different frequencies (and time-dependent coupling coefficients) was considered in some special cases in [315,324,327–331]. It was analyzed in detail for the most general bilinear coupling in [332], where different quantitative *measures of entanglement* between two modes were also calculated. A more general problem of the *quantum state exchange* between two weakly coupled modes was studied in [333–336].

The case of *three* coupled modes was considered in [337–339]. Different systems consisting of an arbitrary number of interacting particles (modes) (in particular, the *chains* of finite or infinite number of quantum oscillators) were considered, e.g., in [340–348]. An emphasis on the problem of *relaxation* of one or a few oscillators interacting with a "bath" or "reservoir" consisting of other oscillators was made in studies [172,234,349–357] (for other references see, e.g., [51,243,358]).

3.4 Generalized Fock states and quantum invariants

Taking different components of the linear vector integral of motion (90) one can construct N Hermitian quadratic integrals of motion $\hat{A}_k^\dagger \hat{A}_k$ (no summation over repeated indices), whose eigenstates $|\mathbf{n}\rangle$ can be called the generalized Fock states:

$$\hat{A}_k^\dagger \hat{A}_k |\mathbf{n}\rangle = n_k |\mathbf{n}\rangle, \quad \mathbf{n} = (n_1, \ldots, n_N), \quad \langle \mathbf{m}|\mathbf{n}\rangle = \delta_{\mathbf{mn}}. \tag{112}$$

These states are related to generalized coherent states (94) as follows

$$|\alpha\rangle = \exp\left(-|\alpha|^2/2\right) \sum_{\mathbf{n}=0}^{\infty} \frac{\alpha_1^{n_1}\cdots\alpha_N^{n_N}}{\sqrt{n_1!\cdots n_N!}} |\mathbf{n}\rangle. \tag{113}$$

Equation (113) has the same structure as the generating function of the multidimensional Hermite polynomials [359]

$$\exp\left(-\tfrac{1}{2}\,\mathbf{a}R\mathbf{a} + \mathbf{a}R\mathbf{y}\right) = \sum_{\mathbf{n}=0}^{\infty} \frac{a_1^{n_1}\cdots a_N^{n_N}}{n_1!\cdots n_N!} H_{\mathbf{n}}^{\{R\}}(\mathbf{y}), \tag{114}$$

where $\mathbf{R}$ is a symmetric matrix. Consequently, we have the formula

$$\langle \mathbf{x}|\mathbf{n}\rangle = \frac{\langle \mathbf{x}|0\rangle}{\sqrt{n_1!\cdots n_N!}} H_{\mathbf{n}}^{\{R\}}\left(-\frac{i}{\hbar}(\lambda_p^{\dagger})^{-1}\left[\mathbf{x} - 2\hbar\mathrm{Im}\left(\lambda_p^{\dagger}\delta\right)\right]\right), \tag{115}$$

where $\langle \mathbf{x}|0\rangle$ is the wave function of the time-dependent ground state (given by (94) with $\alpha = 0$), and the defining matrix R is given by $R = -\lambda_p^{*}\lambda_p^{-1}$. In the one-mode case, we can rewrite (115) in terms of the usual Hermite polynomials defined through the generating function [359]

$$\exp\left(2xz - z^2\right) = \sum_{n=0}^{\infty} \frac{z^n}{n!} H_n(x). \tag{116}$$

$$\langle x|n\rangle = \langle x|0\rangle \frac{i^n}{\sqrt{n!}} \left(\frac{\lambda_p^{*}}{2\lambda_p}\right)^{n/2} H_n\left(\frac{x - 2\hbar\mathrm{Im}\left(\lambda_p^{*}\delta\right)}{\hbar\sqrt{2}|\lambda_p|}\right). \tag{117}$$

The probability density can be expressed as

$$|\langle x|n\rangle|^2 = \frac{\exp\left(-y^2\right) H_n^2(y)}{2^{n+1} n! \pi \sigma_{xx}}, \quad y = \frac{x - \bar{x}_0}{\sqrt{2\sigma_{xx}}}, \tag{118}$$

where $\bar{x}_0 = 2\hbar\mathrm{Im}\left(\lambda_p^{*}\delta\right)$ is the mean value of the coordinate in the time-dependent ground state $\langle x|0\rangle$.

Multidimensional Hermite polynomials can be expressed in terms of the products of the usual ones. In the case of two variables one has [11, 360]

$$H_{mn}^{\{R\}}(x_1, x_2) = m!n! \left(\frac{R_{11}^m R_{22}^n}{2^{m+n}}\right) \sum_{k=0}^{\min(m,n)} [k!(m-k)!(n-k)!]^{-1}$$
$$\times \left(-\frac{2R_{12}}{\sqrt{R_{11}R_{22}}}\right)^k H_{m-k}\left(\frac{y_1}{\sqrt{2R_{11}}}\right) H_{n-k}\left(\frac{y_2}{\sqrt{2R_{22}}}\right), \tag{119}$$

$$y_1 = R_{11}x_1 + R_{12}x_2, \quad y_2 = R_{21}x_1 + R_{22}x_2. \tag{120}$$

In particular, in the case of equation (115), y_1 and y_2 in (119) are the components of vector $\frac{i}{\hbar}\tilde{\lambda}_p^{-1}\left[\mathbf{x} - 2\hbar\mathrm{Im}\left(\lambda_p^{\dagger}\delta\right)\right]$. For the first time, multidimensional Hermite polynomials appeared as solutions to the Schrödinger equation with a time-dependent quadratic (homogeneous) Hamiltonian in the paper by Chernikov [361]. Later, they were used, e.g., in [170, 362–365]. The two-dimensional Hermite polynomials were studied in [11, 366–369].

In one dimension, the linear nonhermitian operator $\hat{A}$ (90) has especially simple form for $\mu = const$ and $\rho = 0$, when $\hat{A} = \sqrt{\mu}\,\varepsilon\hat{p} - (\dot{\varepsilon}/\sqrt{\mu})\hat{x}$, where complex function $\varepsilon(t)$ is determined by equations (35) and (103). Then we have the quadratic invariant

$$\hat{I} \equiv \hat{A}^\dagger\hat{A} = \mu|\varepsilon|^2\hat{p}^2 + (|\dot{\varepsilon}|^2/\mu)\hat{x}^2 - \varepsilon^*\dot{\varepsilon}\hat{p}\hat{x} - \varepsilon\dot{\varepsilon}^*\hat{x}\hat{p}. \tag{121}$$

This integral of motion was discovered in the classical case (for the problem of transverse oscillations of particle beams in accelerators) in [370], so it is known in beam physics under the name *Courant-Snyder invariant*. In the theory of classical and quantum nonstationary oscillators the same invariant is frequently called *Lewis' invariant*, due to the paper [371]. The role of integrals of motion for solving time-dependent Schrödinger equations was emphasized in the paper by Lewis and Riesenfeld [372], which gave birth to a wide stream of research in this direction: see, e.g., [249, 373–377] and references therein (besides already cited papers).

Solutions for generic N-dimensional (but *time independent*) quadratic Hamiltonians and their classifications were given, e.g., in [294, 378–380]. The explicit expression for the *Wigner function* of the Fock state was found by Groenewold [381]. It was generalized in [382–385]. In particular, in the case of generic time-dependent quadratic Hamiltonians one can always construct a set of *completely factorized* solutions to equation (82) in the form [385]

$$W_{\mathbf{lr}} = 2^n \prod_{i=1}^{n} 2^{r_i - l_i}(-1)^{l_i}\left[\frac{l_i!}{r_i!}\right]^{1/2} z_i^{r_i - l_i} e^{-2z_i^* z_i} L_{l_i}^{(r_i - l_i)}\left(4z_i^* z_i\right), \tag{122}$$

where (cf. equation (90) for the *operator* integral of motion)

$$\mathbf{z} = (z_1, z_2, \ldots, z_n) = \lambda_p(t)\mathbf{p} + \lambda_q(t)\mathbf{x} + \delta(t),$$

$\mathbf{l} = (l_1, l_2, \ldots, l_n)$, $\mathbf{r} = (r_1, r_2, \ldots, r_n)$, and $L_m^{(\alpha)}(z)$ is the associated Laguerre polynomial. Function (122) is the Weyl symbol of the operator $\hat{\rho}_{\mathbf{lr}} = |\mathbf{l}\rangle\langle\mathbf{r}|$, where $|\mathbf{l}\rangle$ is the Fock state vector. Taking $n = 1$, $l = r$, and $z^* z = E/(\hbar\omega)$ (where E is the classical energy in the phase space variables), one arrives at Groenewold's result.

One can also obtain a complete set of factorized solutions to equation (82), which are related to the eigenstates of operator integrals of motion with *continuous* spectra. For example, in the one-dimensional case, the eigenstates of the operator $\hat{F} = \frac{1}{2}\hat{P}^2(t) - f\hat{X}(t)$ (where f is an arbitrary real constant; $\hat{X}(t)$ and $\hat{P}(t)$ are the Hermitian operators (12) and (15)), satisfying the equations

$$\hat{F}W_{\xi\eta} = \xi W_{\xi\eta}\,, \qquad W_{\xi\eta}\hat{F} = \eta W_{\xi\eta}$$

with real eigenvalues ξ, η, can be expressed through the Airy function [386, 387]

$$\begin{aligned} W_{\xi\eta} &= 2\pi^{-1/2}(\hbar f)^{-2/3}\exp\left[iv(\xi-\eta)/(\hbar f)\right] \\ &\quad \times \mathrm{Ai}\left(2(\hbar f)^{-2/3}\left[\tfrac{1}{2}v^2 - fu - \tfrac{1}{2}(\xi+\eta)\right]\right). \end{aligned} \tag{123}$$

An appearance of the Airy function in solutions of the quantum-mechanical problem of motion in a uniform potential field is a well-known fact. In this connection, it is worth noting that Berry and Balazs [388] found *unnormalizable* solutions of the Schrödinger equation for a *free* particle in the form $\psi(x,t) \sim \mathrm{Ai}(ax - bt^2)\exp[ict(x - dt^2)]$ ("accelerated packets").

4 Evolution of mixed states

4.1 Gaussian Wigner functions and density matrices

Gaussian states [11,186,243,244,363,389] possess extremely simple Wigner and characteristic functions:

$$W(\mathbf{q}) = \left|\det\left(\mathcal{M}\Omega^{-1}\right)\right|^{-1/2} \exp\left[-\tfrac{1}{2}\left(\mathbf{q} - \langle\mathbf{q}\rangle\right)\mathcal{M}^{-1}\left(\mathbf{q} - \langle\mathbf{q}\rangle\right)\right], \tag{124}$$

$$\chi(\xi) = \langle\exp(\xi\hat{\mathbf{q}})\rangle = \mathrm{Tr}\left[\hat{\rho}\exp(\xi\hat{\mathbf{q}})\right] = \exp\left(\tfrac{1}{2}\,\xi\mathcal{M}\xi + \xi\langle\mathbf{q}\rangle\right). \tag{125}$$

Matrices $\mathcal{M}$ and Ω were defined in (95) and (72), i.e., vector $\mathbf{q}$ is understood in this section as an arbitrary set of independent linear combinations of the coordinate and momentum operators. An immediate consequence of equation (125) is the formula expressing the statistical moments in terms of the Hermite polynomials of several variables [11]

$$\langle q_1^{n_1} q_2^{n_2} \cdots q_{2N}^{n_{2N}}\rangle_W = H^{\{-\mathcal{M}\}}_{n_1 n_2 \cdots n_{2N}}\left(-\mathcal{M}^{-1}\langle\mathbf{q}\rangle\right), \tag{126}$$

where the symbol

$$\langle q_1^{n_1} q_2^{n_2} \cdots q_{2N}^{n_{2N}}\rangle_W \equiv \int \frac{d\mathbf{q}}{(2\pi\hbar)^N} W(\mathbf{q}) q_1^{n_1} q_2^{n_2} \cdots q_{2N}^{n_{2N}} \tag{127}$$

means the "Wigner", or *symmetrized* mean value, i.e., the average value of the sum of all possible different products of n_1 operators $\hat{q}_1$, n_2 operators $\hat{q}_2$, ..., n_{2N} operators $\hat{q}_{2N}$, divided by the number of terms in this sum. For example,

$$\langle p^2 x\rangle_W = \frac{1}{3}\langle\hat{p}^2\hat{x} + \hat{x}\hat{p}^2 + \hat{p}\hat{x}\hat{p}\rangle.$$

The density matrix in the coordinate representation can be obtained with the help of equation (78). Using also the Frobenius formulae for the inversion of the block matrix [390]

$$\left\|\begin{matrix} A & B \\ C & D \end{matrix}\right\|^{-1} = \left\|\begin{matrix} A^{-1} + A^{-1}BH^{-1}CA^{-1} & -A^{-1}BH^{-1} \\ -H^{-1}CA^{-1} & H^{-1} \end{matrix}\right\| \tag{128}$$

$$= \left\|\begin{matrix} K^{-1} & -K^{-1}BD^{-1} \\ -D^{-1}CK^{-1} & D^{-1} + D^{-1}CK^{-1}BD^{-1} \end{matrix}\right\|, \tag{129}$$

$$H = D - CA^{-1}B, \quad K = A - BD^{-1}C,$$

we arrive at the following expression:

$$\rho(\mathbf{x}_2,\mathbf{x}_1) = [\det(2\pi\mathcal{M}_{xx})]^{-1/2} \exp\Big[-\tfrac{1}{2}(\mathbf{x}-\langle\mathbf{x}\rangle)\mathcal{M}_{xx}^{-1}(\mathbf{x}-\langle\mathbf{x}\rangle) + \frac{i}{\hbar}\Delta\mathbf{x}\langle\mathbf{p}\rangle$$
$$+\frac{i}{\hbar}\Delta\mathbf{x}\mathcal{M}_{px}\mathcal{M}_{xx}^{-1}(\mathbf{x}-\langle\mathbf{x}\rangle) - \frac{1}{2\hbar^2}\Delta\mathbf{x}\left(\mathcal{M}_{pp}-\mathcal{M}_{px}\mathcal{M}_{xx}^{-1}\mathcal{M}_{xp}\right)\Delta\mathbf{x}\Big], \quad (130)$$

where $\mathbf{x} = \frac{1}{2}(\mathbf{x}_2+\mathbf{x}_1)$ and $\Delta\mathbf{x} = \mathbf{x}_2-\mathbf{x}_1$.

Entropy and purity

To calculate the von Neumann entropy

$$S = -\mathrm{Tr}\,(\hat{\rho}\ln\hat{\rho}) \equiv -\langle\ln\hat{\rho}\rangle \quad (131)$$

of the Gaussian state we notice [243] that function (124) can be considered formally as the "equilibrium" Wigner function (87) corresponding to the statistical operator (88), if one introduces an "effective Hamiltonian" defined by the matrix

$$B = 2\,(\beta\Omega)^{-1}\tanh^{-1}\left(\Omega\mathcal{M}^{-1}/2\right).$$

This is always possible, because the covariance matrix $\mathcal{M}$ is nondegenerate. Also, since the entropy does not depend on the first-order average values of coordinates and momenta (they can be removed by means of a shift in the phase space performed through a *unitary* transformation, which does not affect the entropy), it is sufficient to consider the case of $\langle\mathbf{q}\rangle = 0$. Then, taking into account the formula $\tanh^{-1}(z) \equiv \frac{1}{2}\ln[(1+z)/(1-z)]$ and the Liouville formula connecting the trace and determinant of any matrix A [390]

$$\det\exp(A) = \exp\mathrm{Tr}(A), \quad (132)$$

one can obtain the following chain of equalities [243]:

$$\begin{aligned} S &= \ln\mathcal{Z}(\beta) + (\beta/2)\langle\hat{q}_\alpha B_{\alpha\beta}\hat{q}_\beta\rangle = \ln\mathcal{Z}(\beta) + (\beta/2)\mathrm{Tr}(\mathcal{M}B) \\ &= \tfrac{1}{4}\ln\det\left(\tfrac{1}{4}E_{2N}-\mathcal{X}^2\right) + \ln\det\exp\left[-\frac{\mathcal{X}}{2}\ln\left(\frac{\mathcal{X}-\frac{1}{2}}{\mathcal{X}+\frac{1}{2}}\right)\right], \end{aligned} \quad (133)$$

where

$$\mathcal{X} = \mathcal{M}\Omega^{-1}, \quad \mathcal{M}_{jk} = \tfrac{1}{2}\langle\hat{q}_j\hat{q}_k+\hat{q}_k\hat{q}_j\rangle, \quad \Omega_{jk} = \langle\hat{q}_j\hat{q}_k-\hat{q}_k\hat{q}_j\rangle. \quad (134)$$

Consequently, the entropy depends only on the eigenvalues of the traceless matrix $\mathcal{X}$, which is, roughly speaking, the "ratio" of the matrix of *symmetrized* second-order (central) moments to the matrix of *antisymmetrized* moments constructed from the components of the $2N$-dimensional operator vector $\hat{\mathbf{q}}$. To find this dependence explicitly, let us consider first the simplest case when vector $\hat{\mathbf{q}}$ has the form $(\hat{p}_1,\hat{x}_1,\ldots,\hat{p}_N,\hat{x}_N)$. Then $\Omega = -i\hbar\Sigma$, where matrix Σ has a block-diagonal form, each block being a 2×2 matrix given in equation (59) with $N = 1$. Now we use the known fact that any *positively definite* matrix $\mathcal{M}$ can be diagonalized by means of a *canonical* (symplectic) linear transformation of vector $\hat{\mathbf{q}}$ preserving the antisymmetric matrix Σ [9, 11, 290–294, 378]: $\mathcal{M} \rightarrow S\mathcal{M}\tilde{S} = \text{diag}\,(\mathcal{M}_1, \ldots, \mathcal{M}_N)$, $S\Sigma\tilde{S} = \Sigma$, where $\mathcal{M}_k$ is a two-dimensional matrix ($k = 1, \ldots, N$). Such a transformation of coordinates in the phase space is equivalent to a *unitary* transformation of the statistical operator, which does not affect the entropy. On the other hand, matrix $\mathcal{X}$ is transformed as $S\mathcal{X}S^{-1}$, i.e., without changing its eigenvalues. Consequently, it is sufficient to find an explicit form of the last expression in (133) for the two-dimensional matrix $\mathcal{X}$.

In the one-mode case, one can easily verify that $\text{Tr}\mathcal{X} = 0$ (due to symmetricity of matrix $\mathcal{M}$ and antisymmetricity of matrix Σ) and $\det \mathcal{X} = -\kappa^2$, where

$$\kappa = \hbar^{-1}\sqrt{\Delta}\,, \qquad \Delta \equiv \sigma_{xx}\sigma_{pp} - \sigma_{px}^2 \geq \hbar^2/4\,. \tag{135}$$

Consequently, the eigenvalues of matrix $\mathcal{X}$ are $\pm\kappa$ (and $\mathcal{X}^2 = \kappa^2 E_2$). Therefore, putting in equation (133) the diagonal form of matrix $\mathcal{X}$, $\text{diag}(\kappa, -\kappa)$, and using the additivity property of the logarithmic function and the multiplicativity property of the determinant, one can arrive after simple algebra to the following expression for the entropy of any N-mode Gaussian state:

$$S_N = \sum_{j=1}^{N} \left(\kappa_j + \tfrac{1}{2}\right) \ln\left(\kappa_j + \tfrac{1}{2}\right) - \left(\kappa_j - \tfrac{1}{2}\right) \ln\left(\kappa_j - \tfrac{1}{2}\right), \tag{136}$$

where $\kappa_j \geq \frac{1}{2}$ ($j = 1, \ldots, N$) are N *positive* eigenvalues of matrix $\mathcal{X}$. In the one-mode case expressions equivalent to formula (136) were found in [243, 354, 391], whereas the generic case was considered in [392] (where the coefficients equivalent to κ_j were given in terms of some average values of the annihilation and creation operators) and [393]. The validity of (136) is not restricted by the specific form of the commutator matrix $\Omega = -i\hbar\Sigma$ used above. Indeed, using transformations $\Omega \rightarrow T\Omega\tilde{T}$ ($\det T \neq 0$), one can transform the initial block-diagonal antisymmetric matrix Ω to any antisymmetric (nonsingular) matrix [9]. Such transformations are accompanied by the transformations $\mathcal{M} \rightarrow T\mathcal{M}\tilde{T}$ and $\mathcal{X} \rightarrow T\mathcal{X}T^{-1}$, which do not affect the eigenvalues of matrix $\mathcal{X}$.

The *quantum purity* of the Gaussian state (124) is given by formulae [243]

$$\mu \equiv \text{Tr}\hat{\rho}^2 = \int \frac{[W(\mathbf{q})]^2 d\mathbf{q}}{|\det(2\pi\Omega)|^{1/2}} = \left|\det(2\mathcal{M}\Omega^{-1}\right|^{-1/2} = \prod_{j=1}^{N} (2\kappa_j)^{-1}\,. \tag{137}$$

Photon distribution in the Gaussian state

The photon (quantum number) distribution function $f(n) \equiv \langle n|\hat{\rho}|n\rangle$ of Gaussian states was calculated in many papers cited in Chapter 1. We give here the results in a compact form close to that obtained in [366, 394]. The information on the photon distribution in some mode is contained in the *generating function*

$$G(z) = \sum_{n=0}^{\infty} f(n) z^n.$$

In particular, the derivatives of $G(z)$ at $z = 1$ generate the factorial moments

$$d^r G(z)/dz^r|_{z=1} = \overline{n(n-1)(n-2)...(n-r+1)} \equiv n^{(r)}. \tag{138}$$

To express $G(z)$ in the most simple form we assume the "coordinate" and "momenta" operator to be dimensionless, i.e., that they are related to the annihilation and creation operators as $\hat{x} = (\hat{a}+\hat{a}^\dagger)/\sqrt{2}$ and $\hat{p} = (\hat{a}-\hat{a}^\dagger)/(i\sqrt{2})$. In this case, $G(z)$ depends on four parameters, which are invariant with respect to rotations in the $x - p$ plane:

$$G(z) = [\mathcal{G}(z)]^{-1/2} \exp\left(\frac{1}{D}\left[\frac{zg_1 - z^2 g_2}{\mathcal{G}(z)} - g_0\right]\right), \tag{139}$$

$$\mathcal{G}(z) = \frac{1}{4}(1+z)^2 + \kappa^2(1-z)^2 + \frac{\tau}{2}\left(1-z^2\right), \qquad D = 1+2\tau+4\kappa^2,$$

$$g_0 = \vartheta+2\gamma, \quad g_1 = 2\gamma(\tau+1) - 2\vartheta\left(\kappa^2-\tfrac{1}{4}\right), \quad g_2 = 2\gamma\tau - 2\vartheta\left(\kappa^2+\tfrac{1}{4}\right),$$

where parameter κ is defined in (135), and other three parameters are as follows,

$$\tau \equiv \sigma_{xx} + \sigma_{pp}, \qquad \vartheta \equiv \langle\hat{x}\rangle^2 + \langle\hat{p}\rangle^2 = 2|\langle\hat{a}\rangle|^2,$$

$$\gamma \equiv \sigma_{pp}\langle\hat{x}\rangle^2 + \sigma_{xx}\langle\hat{p}\rangle^2 - 2\sigma_{xp}\langle\hat{x}\rangle\langle\hat{p}\rangle.$$

The mean number of quanta and their variance are given by the formulae

$$\overline{n} = \tfrac{1}{2}(\tau+\vartheta-1), \qquad \sigma_n \equiv \overline{n^2} - \overline{n}^2 = \tfrac{1}{2}\tau^2-\kappa^2-\tfrac{1}{4} + \tau\vartheta-\gamma. \tag{140}$$

In the generic case $f(n)$ can be expressed in terms of the two-dimensional "diagonal" Hermite polynomials (see section 3.4) or as a sum of the usual Hermite polynomials [394]:

$$f(n) = \frac{2\exp\left(-g_0/D\right)}{n!\sqrt{D}} H_{nn}^{\{\mathcal{R}\}}(\chi,\chi^*) \tag{141}$$

$$= \frac{2n!}{\sqrt{D}} \exp\left(-\frac{g_0}{D}\right)\left(\frac{\sigma}{D}\right)^n \sum_{k=0}^{n}\left(\frac{s}{\sigma}\right)^k \frac{|H_{n-k}(\xi)|^2}{k![(n-k)!]^2}, \tag{142}$$

where

$$\sigma = \sqrt{\tau^2 - 4\kappa^2}, \qquad s = 4\kappa^2 - 1,$$

$$\xi = \frac{(2\sigma_{pp}+1)\langle x\rangle - 2\sigma_{xp}\langle p\rangle + i\left[(1+2\sigma_{xx})\langle p\rangle - 2\sigma_{xp}\langle x\rangle\right]}{\left[2D\left(\sigma_{pp}-\sigma_{xx}-2i\sigma_{xp}\right)\right]^{1/2}},$$

$$\chi = \frac{\sqrt{2}\left\{(2\sigma_{pp}-1)\langle x\rangle - 2\sigma_{xp}\langle p\rangle + i\left[(1-2\sigma_{xx})\langle p\rangle + 2\sigma_{xp}\langle x\rangle\right]\right\}}{2\tau-4\kappa^2-1},$$

and 2×2 symmetric matrix $\mathcal{R}$ has the elements

$$\mathcal{R}_{11} = \mathcal{R}_{22}^* = 2D^{-1}\left(\sigma_{pp}-\sigma_{xx}-2i\sigma_{xp}\right), \quad \mathcal{R}_{12} = \mathcal{R}_{21} = D^{-1}\left(1-4\kappa^2\right).$$

For *pure* (squeezed) Gaussian states $s = 0$, and the only term with $k = 0$ differs from zero in the sum (142), resulting in the formula equivalent to (23) from Chapter 1. The analysis of formula (141) and its asymptotics was done in [367, 394].

If $\langle p\rangle = \langle q\rangle = 0$, then the generating function (139) is reduced to $[\mathcal{G}(z)]^{-1/2}$, which means that $f(n)$ is expressed in terms of the Legendre polynomials:

$$f(n) = \frac{2}{\sqrt{1+2\tau+4\kappa^2}} \left(\frac{1+4\kappa^2-2\tau}{1+4\kappa^2+2\tau}\right)^{n/2} P_n\left(\frac{4\kappa^2-1}{\sqrt{(4\kappa^2+1)^2-4\tau^2}}\right). \quad (143)$$

Note that the argument of the Legendre polynomial in (143) is always outside the interval $(-1,1)$, being equal to 1 only for thermal states with $\tau = 2\kappa$. For highly excited Gaussian states with $\tau \gg 2\kappa$ (and $\gamma = \vartheta = 0$), Mandel's parameter is about twice bigger than for the thermal states: $\mathcal{Q} \equiv \sigma_n/\overline{n} - 1 \approx \tau \approx 2\overline{n}$, which means that such states are "superchaotic" (see Chapter 1, page 43). The asymptotics of $f(n)$ for $n \gg 1$ in this case was found in [395]:

$$f(n) \approx \frac{2\left|1+4\kappa^2-2\tau\right|^{n/2}}{\sqrt{\pi n\tau}\,(1+4\kappa^2+2\tau)^{(n+1)/2}} \times \begin{cases} \cosh\left([n+\frac{1}{2}]\ln|\zeta|\right), & n \text{ even,} \\ \sinh\left([n+\frac{1}{2}]\ln|\zeta|\right), & n \text{ odd,} \end{cases}$$

where

$$\zeta = \frac{4\kappa^2 - 1 + 2\sqrt{\tau^2 - 4\kappa^2}}{\sqrt{(4\kappa^2+1)^2 - 4\tau^2}}.$$

If $\tau \gg \kappa^2$ (then ζ is pure imaginary and small), we have a strongly oscillating distribution (typical for squeezed states) for $n < \tau/(2\kappa^2)$:

$$f(n) \approx \frac{2}{\sqrt{\pi n\tau}} \left(1 - \frac{4\kappa^2+1}{\tau}\right)^{n/2} \times \begin{cases} \cosh\left[n\left(4\kappa^2-1\right)/(2\tau)\right], & n \text{ even,} \\ \sinh\left[n\left(4\kappa^2-1\right)/(2\tau)\right], & n \text{ odd.} \end{cases}$$

The generic case of multimode distributions was considered in [363].

Evolution of the moments

Resolving equation (56) with respect to the operator $\hat{\mathbf{q}} = (\hat{\mathbf{p}}, \hat{\mathbf{x}})$ one obtains the solution to the *Heisenberg equations of motion* for the operator $\hat{\mathbf{q}}(t)$

$$\hat{\mathbf{q}}(t) = \Lambda^{-1}(t)\left[\hat{\mathbf{q}}(0) - \delta(t)\right]. \quad (144)$$

A similar equation holds for the first-order average value $\langle \mathbf{q}\rangle$:

$$\langle \mathbf{q}\rangle(t) = \Lambda^{-1}(t)\left[\langle \mathbf{q}\rangle(0) - \delta(t)\right]. \quad (145)$$

Introducing the "shifted" operators $\hat{r}_k \equiv \hat{q}_k - \langle q_k\rangle$ (so that $\langle \hat{r}_k\rangle \equiv 0$), one immediately obtains a formula for centralized moments of an arbitrary order

$$\langle \hat{r}_\alpha \hat{r}_\beta \cdots \hat{r}_\gamma\rangle(t) = \sum_{mn\ldots k} \mathcal{L}_{\alpha m}\mathcal{L}_{\beta n} \cdots \mathcal{L}_{\gamma k}\langle \hat{r}_m \hat{r}_n \cdots \hat{r}_k\rangle(0), \quad \mathcal{L} \equiv \Lambda^{-1}.$$

For the second-order centralized moments (covariances) this formula can be written in matrix form

$$\mathcal{M}(t) = \Lambda^{-1}(t)\mathcal{M}(0)\tilde{\Lambda}^{-1}(t). \quad (146)$$

Since matrix Λ is symplectic (61), its inverse matrix can be expressed in terms of the transposed one:

$$\Lambda^{-1}(t) = \Sigma\tilde{\Lambda}(t)\Sigma^{-1} = \left\| \begin{array}{cc} \tilde{\lambda}_4 & -\tilde{\lambda}_2 \\ -\tilde{\lambda}_3 & \tilde{\lambda}_1 \end{array} \right\|. \tag{147}$$

It is worth emphasizing that equations (145) and (146) hold for any state (not only Gaussian, and no matter, pure or mixed). This is a specific feature of *quadratic* Hamiltonians. The only requirement is that the quantum system must be *closed*, so that its evolution can be described by means of the Liouville–von Neumann equation (81).

4.2 Evolution of density matrices of linear systems

Consider a quantum system with N degrees of freedom described in terms of $2N$ operators $\hat{z}_1, \hat{z}_2, \ldots \hat{z}_{2N}$, combined in the vector $\hat{\mathbf{z}}$. The commutators between these operators are assumed to be constant complex numbers:

$$[\hat{z}_\alpha\, \hat{z}_\beta] = \Xi_{\alpha\beta} = -\Xi_{\beta\alpha}, \tag{148}$$

so that the coefficients $\Xi_{\alpha\beta}$ form an antisymmetric $2N \times 2N$ matrix Ξ. Another important matrix S describes the transformations of the operators $\hat{z}_j$ under the Hermitian conjugation:

$$\hat{\mathbf{z}}^\dagger = S\hat{\mathbf{z}}, \qquad S^*S = E_{2N}, \quad \Xi^\dagger = S\Xi\tilde{S}.$$

In the most general case the state of the system is described by means of the statistical operator $\hat{\rho}$. The most general *linear* evolution equation for this operator, containing only *quadratic* combinations of the operators $\hat{z}_j$, can be written as

$$\frac{d\hat{\rho}}{dt} = \hat{\mathbf{z}}L\hat{\mathbf{z}}\hat{\rho} + \hat{\rho}\hat{\mathbf{z}}R\hat{\mathbf{z}} + \hat{\mathbf{z}}K\hat{\rho}\hat{\mathbf{z}}, \tag{149}$$

$L = \tilde{L}$, $R = \tilde{R}$, and K being certain $2N \times 2N$ matrices, which must obey the constraints resulting from the properties of the statistical operator. In particular, the hermiticity of $\hat{\rho}$ leads to the relations

$$L = \tilde{S}R^\dagger S, \qquad R = \tilde{S}L^\dagger S, \qquad K = \tilde{S}K^\dagger S. \tag{150}$$

The normalization $\mathrm{Tr}\hat{\rho}(t) \equiv 1$ is preserved under the condition

$$L + R + \tilde{K} = 0. \tag{151}$$

The most important is the third restriction, which must ensure the nonnegative definiteness of the statistical operator for all instants of time. It is equivalent to the condition that the *generalized uncertainty relations* should not be violated in the process of evolution [243, 396, 397]. A simple way to obtain these conditions [398] is to reduce the equation (149) to the "canonical" form of the linear master equation preserving all the properties of the statistical operator (in other words, generating a so-called "quantum dynamical semigroup")

$$\frac{d\hat{\rho}}{dt} = -\frac{i}{\hbar}\left[\hat{H},\hat{\rho}\right] + \sum_j \left(2\hat{\Phi}_j\hat{\rho}\hat{\Phi}_j^\dagger - \hat{\Phi}_j^\dagger\hat{\Phi}_j\hat{\rho} - \hat{\rho}\hat{\Phi}_j^\dagger\hat{\Phi}_j\right). \tag{152}$$

Here $\hat{H}$ is a Hermitian operator, and $\hat{\Phi}_j$ are arbitrary linear operators (their number also may be arbitrary). Sometimes the right-hand side of (152) is called the "Lindblad form", after the study [399], although this general structure was discovered by several authors earlier [400–402]. The general structure of dynamical mappings of density operators preserving positive semidefiniteness was established in [403, 404] and later in [405]. For reviews and details see, e.g., [406–408].

In the case of equation (149) it is natural to suppose that the operators $\hat{\Phi}_j$ must be linear combinations of the operators $\hat{z}_\alpha$, while operator $\hat{H}$ must be a quadratic form of the same operators (the summation over repeated indices is assumed here):

$$\hat{\Phi}_j = \varphi_{j\alpha}\hat{z}_\alpha, \qquad \hat{H} = \tfrac{1}{2}\hat{\mathbf{z}}B\hat{\mathbf{z}}, \quad B = \tilde{B}.$$

Comparing equations (149) and (152) we obtain the relations

$$K = 2\tilde{\Psi}S, \quad L = -\frac{i}{2\hbar}B - \frac{1}{2}\tilde{K}, \quad R = \frac{i}{2\hbar}B - \frac{1}{2}\tilde{K}, \tag{153}$$

where elements of Hermitian matrix Ψ are given by $\Psi_{\alpha\beta} = \sum_j \varphi_{j\alpha}^*\varphi_{j\beta}$. Matrix Ψ is nonnegatively definite, since for any complex vector $\mathbf{x}$ we have $x_\alpha^*\Psi_{\alpha\beta}x_\beta = \sum_j |\varphi_{j\alpha}x_\alpha|^2 \geq 0$. Thus it follows from the first relation in (153) that matrix

$$Q = \tilde{S}^{-1}\tilde{K} = S^\dagger\tilde{K} \tag{154}$$

must be Hermitian and nonnegatively definite. It is easy to see that the conditions (150) and (151) are satisfied automatically due to relations (153) and formula $B = \tilde{S}B^\dagger S$, which means the hermiticity of operator $\hat{H}$. Therefore the evolution equations having the physical sense are determined by an arbitrary Hermitian nonnegatively definite matrix Q and an arbitrary symmetric matrix B satisfying the above formulated condition.

An immediate consequence of equation (149) and condition (151) is the linear equation for the mean values $\langle\hat{\mathbf{z}}\rangle = \mathrm{Tr}(\hat{\rho}\hat{\mathbf{z}})$:

$$d\langle\hat{\mathbf{z}}\rangle/dt = A\langle\hat{\mathbf{z}}\rangle, \tag{155}$$

$$A = \Xi\left(L - \tilde{R}\right) = -\Xi\left[(i/\hbar)B + \tfrac{1}{2}(\tilde{K} - K)\right]. \tag{156}$$

Consequently, equation (149) describes quantum linear systems. Suppose that the equation for the mean values is given, and the problem is to construct an equation for the statistical operator resulting in (155). If matrix $\Xi^{-1}A$ is symmetrical, then one may assume $K = 0$, which means that the system is *Hamiltonian* (closed). But if matrix $\Xi^{-1}A$ has a nonzero antisymmetric part, then one has to use the general equation (149) with $K \neq 0$. However, matrix A defines only the antisymmetrical part of matrix K. The symmetrical part of matrix K enters the equation for the matrix of the second-order moments (here this matrix is not symmetrized) $M = ||M_{\alpha\beta}||$, $M_{\alpha\beta} = \mathrm{Tr}\,(\hat{\rho}\hat{z}_\alpha \hat{z}_\beta)$,

$$\dot{M} = AM + M\tilde{A} + \Xi\tilde{K}\Xi. \tag{157}$$

The physical meaning of matrix K becomes clear, if one takes into account that, besides the description of the quantum system in the Schrödinger picture, there exists an equivalent description in the Heisenberg picture [409–411], based on the operator Langevin equation

$$d\hat{\mathbf{z}}/dt = A\hat{\mathbf{z}} + \hat{\chi}(t), \tag{158}$$

where $\hat{\chi}_\alpha(t)$ ($\alpha = 1, 2, \ldots 2N$) are the noise operators, which are assumed to be delta-correlated in time with zero mean values:

$$\langle \hat{\chi}_\alpha(t) \rangle = 0, \qquad \langle \hat{\chi}_\alpha(t)\hat{\chi}_\beta(t') \rangle = \delta\,(t - t')\,X_{\alpha\beta}. \tag{159}$$

The solution to equation (158) at $\Delta t \to 0$ reads

$$\hat{\mathbf{z}}(t + \Delta t) = \hat{\mathbf{z}}(t) + A(t)\hat{\mathbf{z}}(t)\Delta t + \int_t^{t+\Delta t} \hat{\chi}(\tau)d\tau.$$

Then up to the linear terms with respect to Δt we have

$$M_{\alpha\beta}(t + \Delta t) = M_{\alpha\beta}(t) + \Delta t\left\{A(t)M(t) + M(t)\tilde{A}(t)\right\}_{\alpha\beta} + \Delta t X_{\alpha\beta}(t).$$

Comparing this expression with equation (157) we obtain the relation between the "noise matrix" $X = ||X_{\alpha\beta}||$ and matrix K:

$$X = \Xi\tilde{K}\Xi = S^*\Xi^\dagger Q\Xi, \quad X^\dagger = SX\tilde{S}. \tag{160}$$

Consequently, SX must be a Hermitian nonnegatively definite matrix.

Frequently it is more convenient to solve not the operator equation (149), but an equivalent equation in some fixed representation for the corresponding distribution function ("quasiprobability"). Consider, for example, the Glauber–Sudarshan P-representation of the density operator in terms of the coherent states

$$\hat{\rho} = \pi^{-1} \int P(\alpha)|\alpha\rangle\langle\alpha|\, d\mathrm{Re}\alpha\, d\mathrm{Im}\alpha. \tag{161}$$

Here $P(\alpha)$ is a real function of the complex argument α (and its complex conjugated partner α^*). If $2N$-dimensional vector $\hat{\mathbf{z}}$ combines N-dimensional annihilation and creation operators as $\hat{\mathbf{z}} = (\hat{\mathbf{a}}, \hat{\mathbf{a}}^\dagger)$, then

$$\Xi = \Sigma = \left\| \begin{array}{cc} 0 & E_N \\ -E_N & 0 \end{array} \right\|, \qquad S = \left\| \begin{array}{cc} 0 & E_N \\ E_N & 0 \end{array} \right\|. \tag{162}$$

Using the correspondence rules between the operators $\hat{\mathbf{z}}\hat{\rho}$, $\hat{\rho}\hat{\mathbf{z}}$, and their symbols in the P-representation

$$\begin{array}{ll} \hat{\mathbf{a}}\hat{\rho} \leftrightarrow \alpha P & \hat{\rho}\hat{\mathbf{a}} \leftrightarrow \alpha P - \partial P/\partial\alpha^* \\ \hat{\mathbf{a}}^\dagger\hat{\rho} \leftrightarrow \alpha^* P - \partial P/\partial\alpha & \hat{\rho}\hat{\mathbf{a}}^\dagger \leftrightarrow \alpha^* P \end{array} \tag{163}$$

one can transform equation (149) to the Fokker–Planck equation

$$\frac{\partial P}{\partial t} = -\frac{\partial}{\partial \mathbf{y}}(A\mathbf{y}P) + \frac{1}{2}\frac{\partial}{\partial \mathbf{y}}Y\frac{\partial}{\partial \mathbf{y}}P, \tag{164}$$

where $\mathbf{y} = (\alpha, \alpha^*)$ is the $2N$-dimensional vector, and

$$Y = D + \tfrac{1}{2}\left(AS + S\tilde{A}\right) \tag{165}$$

is the "diffusion matrix". Matrix A was defined in (156), and $D = \frac{1}{2}(X + \tilde{X})$ is the symmetrical part of the "noise matrix" (160). Equation (164) is the most general linear equation describing in the P-representation the evolution of the quantum linear open system possessing the equations for the mean values such as (155). Matrix A may be quite arbitrary, but D must be chosen in such a way that

$$D_* = S\left[D - \tfrac{1}{2}\left(A\Sigma + \Sigma\tilde{A}\right)\right] \tag{166}$$

is Hermitian and nonnegatively definite matrix. For instance, for a single degree of freedom, matrix A can be chosen as

$$A = \left\| \begin{array}{cc} \xi & \eta \\ \eta^* & \xi^* \end{array} \right\|, \tag{167}$$

ξ, η being arbitrary complex numbers. The hermiticity of matrix (166) imposes the following restrictions on the coefficients of matrix D:

$$D = \left\| \begin{array}{cc} D_a & D_n \\ D_n & D_a^* \end{array} \right\|, \quad D_n = D_n^*. \tag{168}$$

To ensure the nonnegative definiteness of matrix (166) it is necessary and sufficient to satisfy the inequalities

$$D_n^2 - |D_a|^2 \geq \tfrac{1}{4}(\mathrm{Tr}\,A)^2 = (\mathrm{Re}\,\xi)^2, \qquad D_n \geq |\mathrm{Re}\,\xi|. \tag{169}$$

Taking $\eta = 0$ and $\xi = -i\omega - \gamma$ we obtain the system which is frequently used as a model of a damped quantum oscillator. In this case the standard choice is

$$D_a = 0, \quad D_n = \gamma\left[1 + 2\langle n(\omega, T)\rangle\right], \quad \langle n(\omega, T)\rangle = \left[\exp\left(\hbar\omega/kT\right) - 1\right]^{-1},$$

where $\langle n(\omega, T)\rangle \equiv \nu$ is the equilibrium mean photon number of the heat bath. The corresponding operator equation

$$\begin{aligned} \frac{d\hat{\rho}}{dt} + i\left[\omega\hat{a}^\dagger\hat{a},\, \hat{\rho}\right] \;&=\; \gamma(1+\nu)\left(2\hat{a}\hat{\rho}\hat{a}^\dagger - \hat{a}^\dagger\hat{a}\hat{\rho} - \hat{\rho}\hat{a}^\dagger\hat{a}\right) \\ &\quad + \gamma\nu\left(2\hat{a}^\dagger\hat{\rho}\hat{a} - \hat{a}\hat{a}^\dagger\hat{\rho} - \hat{\rho}\hat{a}\hat{a}^\dagger\right) \end{aligned} \tag{170}$$

was in fact considered for the first time by Landau in the same paper [217] where the concept of density matrix was introduced. In the mid-1960s this equation was derived in the framework of different approaches, e.g., in [401,412–414], and its Fokker–Planck image in the P-representation (in the interaction picture)

$$\frac{\partial P}{\partial t} = \gamma \left(\frac{\partial}{\partial \alpha}\alpha + \frac{\partial}{\partial \alpha^*}\alpha^* \right) P + 2\gamma \langle n \rangle \frac{\partial^2 P}{\partial \alpha \partial \alpha^*} \tag{171}$$

was considered in [415,416]. A similar equation for Husimi's Q function was solved in [414]. More references can be found in reviews [51,417,418]. The case of time-dependent coefficients was discussed in [419].

For $\hat{\mathbf{z}} = (\hat{\mathbf{p}}, \hat{\mathbf{x}})$, where $\hat{\mathbf{p}}$ and $\hat{\mathbf{x}}$ are the canonical momentum and coordinate operators, it is more convenient to work in the Wigner representation. Then the diffusion matrix in the Fokker–Planck equation depends only on the noise matrix:

$$\frac{\partial W}{\partial t} = -\frac{\partial}{\partial \mathbf{y}}(A\mathbf{y}W) + \frac{\partial}{\partial \mathbf{y}}\mathcal{D}\frac{\partial}{\partial \mathbf{y}}W, \qquad \mathbf{y} = (\mathbf{p}, \mathbf{x}). \tag{172}$$

Matrix A is the same as in (155), whereas the symmetrical real matrix

$$\mathcal{D} = \tfrac{1}{2}\, D = \left\| \begin{array}{cc} \mathcal{D}_{pp} & \mathcal{D}_{pq} \\ \mathcal{D}_{pq} & \mathcal{D}_{qq} \end{array} \right\|$$

must be chosen in such a way that matrix

$$\mathcal{D}_* = \mathcal{D} + \tfrac{1}{4} i\hbar (A\Sigma + \Sigma\tilde{A}) \tag{173}$$

would be Hermitian and nonnegatively definite. Matrix Σ was defined in (162), and $S = E_{2N}$. The condition $\mathcal{D}_* \geq 0$ was derived in [397], and its special case

$$\mathcal{D}_{pp}\mathcal{D}_{qq} - \mathcal{D}_{pq}^2 \geq \tfrac{1}{4}\,\hbar^2\gamma^2 \tag{174}$$

was given for the first time in [396] (see also [420–424]). In the case of one-dimensional systems with an arbitrary real matrix A we have the inequality

$$\mathcal{D}_{pp}\mathcal{D}_{qq} - \mathcal{D}_{pq}^2 \equiv \det \mathcal{D} \geq (\hbar\, \mathrm{Tr} A/4)^2. \tag{175}$$

The concrete forms of the Fokker–Planck equation in other representations (e.g., Glauber's R-representation [425] or the "positive P-representation" [426]) were considered in [398]. We confine ourselves to the equation for the density matrix in the coordinate representation $\rho\,(\mathbf{x}', \mathbf{x}'')$. Using vector $\mathbf{y} = (\mathbf{x}', \mathbf{x}'')$, one can write this equation as

$$\begin{aligned} \frac{\partial \rho}{\partial t} \;=\;& \frac{\partial}{\partial \mathbf{y}}\left[\tilde{T}_+\mathcal{D}T_+ + \frac{i\hbar}{4}\left(\tilde{T}_+AT_- + \tilde{T}_-\tilde{A}T_+\right)\right]\frac{\partial}{\partial \mathbf{y}}\rho(\mathbf{y}) \\ &+\mathbf{y}\left[-\frac{2i}{\hbar}\tilde{T}_-\mathcal{D}T_+ + \frac{1}{2}\left(\tilde{T}_-AT_- - \tilde{T}_+\tilde{A}T_+\right)\right]\frac{\partial}{\partial \mathbf{y}}\rho(\mathbf{y}) \\ &+\left\{\mathbf{y}\left[-\frac{1}{\hbar^2}\tilde{T}_-\mathcal{D}T_- + \frac{i}{4\hbar}\left(\tilde{T}_-AT_+ + \tilde{T}_+\tilde{A}T_-\right)\right]\mathbf{y}\right. \\ &\left.-\tfrac{1}{2}\,\mathrm{Tr}\left(\tilde{T}_+AT_+\right)\right\}\rho(\mathbf{y}), \end{aligned} \tag{176}$$

$$T_+ = \left\| \begin{array}{cc} 0 & 0 \\ E_N & E_N \end{array} \right\|, \qquad T_- = \left\| \begin{array}{cc} E_N & -E_N \\ 0 & 0 \end{array} \right\|.$$

It is clearly seen to what extent the Wigner representation is simpler in the case of linear equations of motion.

The general case of *three* diffusion coefficients (given, e.g., by (168)) in the one-mode master or Fokker–Planck equation was considered for the first time perhaps in [427]. Such a situation is typical for *nonthermal* reservoirs, whose special cases are *squeezed* [143, 149, 428–438], *rigged* [148, 439–441], or *phase-sensitive* [147, 152, 153, 236, 442, 443] reservoirs. In the one-mode case, the master equation describing the "squeezed" reservoir acquires some extra terms in the right-hand side, compared with the form (170) [428]:

$$\delta\mathcal{L}_{sqz}\{\hat{\rho}\} = \gamma\mathcal{M}\left(2a^\dagger\rho a^\dagger - a^\dagger a^\dagger\rho - \rho a^\dagger a^\dagger\right) + h.c., \tag{177}$$

where the complex parameter $\mathcal{M} = |\mathcal{M}|e^{i\phi}$, responsible for the "strength of squeezing" of the reservoir, must satisfy the relation $|\mathcal{M}| \le \sqrt{\nu(\nu+1)}$. Using the linear canonical transformation [236]

$$\hat{b} = u\hat{a} + v\hat{a}^\dagger, \quad u = \sqrt{\nu+1}, \quad v = \sqrt{\nu}e^{i\phi},$$

one can simplify the right-hand of the master equation, reducing it to the "zero temperature" expression $\mathcal{L}_{vac}\hat{\rho} = \gamma\left(2\hat{a}\hat{\rho}\hat{a}^\dagger - \hat{a}^\dagger\hat{a}\hat{\rho} - \hat{\rho}\hat{a}^\dagger\hat{a}\right)$. The "price" for this simplification is the transformation of a simple Hamiltonian $\hat{H} = \hat{a}^\dagger\hat{a}$ in the left-hand side of equation (170) into a more complicated "effective Hamiltonian"

$$\hat{H}_{eff} = (2\nu+1)\hat{a}^\dagger\hat{a} + \mathcal{M}^*\hat{a}^2 + \mathcal{M}\hat{a}^{\dagger 2} + \nu.$$

It should be noted that relatively simple forms of the master equation (149), (152) or (170) can be derived from the "microscopic" models of interaction between the distinguished subsystem and the "rest of Universe" only under strong limitations on the possible form of interactions (in particular, the "rotating-wave approximation" plays an important role): see, e.g., [234]. Therefore, many other forms of "master equations" were considered. For example, Golubev and Sokolov [444] introduced the equation of the type

$$d\hat{\rho}/dt = r\ln(1+\hat{u})\hat{\rho}, \tag{178}$$

which was generalized in [445]. Milburn [446] proposed the equation

$$d\hat{\rho}/dt = \gamma\left\{\exp\left(-i\hat{H}/\hbar\gamma\right)\hat{\rho}\exp\left(i\hat{H}/\hbar\gamma\right) - \hat{\rho}\right\},$$

whose solutions were studied, e.g., in [447–449]. An equation of the same form as (178) but with different physical meaning was considered in [450, 451]. A list of master equations admitting exact solutions can be found in [452].

4.3 Propagators of the Fokker–Planck equation

The Fokker–Planck equation (172) for the Wigner function or equation (176) for the density matrix in the coordinate representation can be considered formally as the "Schrödinger equations" for the "extended wave functions" of $2N$ variables, with some effective quadratic Hamiltonians of the same structure as (3). Consequently, one can solve these evolution equations, applying the methods of the general theory of quadratic systems discussed in the preceding sections, with certain modifications due to the non-Hermiticity of the arising effective Hamiltonians. Using such an approach, the propagator for equation (176) was obtained in [453]. We confine ourselves here to the most symmetrical case of the Wigner function, when the evolution equation (172) can be written as

$$\frac{\partial W}{\partial t} = \hat{\mathcal{H}} W, \quad \hat{\mathcal{H}} = \tfrac{1}{2}\,\hat{\mathbf{z}}\mathcal{B}\hat{\mathbf{z}} + \mathbf{C}\hat{\mathbf{z}} - \tfrac{1}{2}\,\mathrm{Tr}A, \tag{179}$$

$$\hat{\mathbf{z}} = \begin{pmatrix} \partial/\partial\mathbf{q} \\ \mathbf{q} \end{pmatrix}, \quad \mathcal{B} = \left\| \begin{matrix} 2\mathcal{D} & -A \\ -\tilde{A} & 0 \end{matrix} \right\|, \quad \mathbf{C} = \begin{pmatrix} -\mathbf{f} \\ \mathbf{0} \end{pmatrix}. \tag{180}$$

Here $\mathcal{B}$ is the symmetrical $4N \times 4N$ matrix. The propagator of equation (179) can be calculated in the framework of the same method of operator integrals of motion [243, 454] as was used in section 2. We introduce the operator

$$\hat{\mathbf{Z}}(t) = \hat{U}(t)\hat{\mathbf{z}}\hat{U}^{-1}(t) = \Lambda(t)\hat{\mathbf{z}} + \Delta(t), \tag{181}$$

where matrix $\Lambda(t)$ and vector $\Delta(t)$ satisfy the following linear ordinary differential equations and initial conditions:

$$\dot{\Lambda} = -\Lambda\Sigma\mathcal{B}, \qquad \Lambda(0) = E_{4N}, \tag{182}$$

$$\dot{\Delta} = -\Lambda\Sigma\mathbf{C}, \qquad \Delta(0) = \mathbf{0}. \tag{183}$$

(The minus signs on the right-hand sides appear due to the absence of the factor $-i\hbar$ in the definition of the first $2N$ components of the vector operator $\hat{\mathbf{z}}$.) Matrix Σ has the same structure as (59), but its dimension is $4N \times 4N$ now. The blocks of matrix $\Lambda(t)$ (56) satisfy the equations

$$\begin{aligned} \dot{\lambda}_1 &= \lambda_1\tilde{\mathbf{A}} + 2\lambda_2\mathbf{D}, & \dot{\lambda}_2 &= -\lambda_2\mathbf{A}, \\ \dot{\lambda}_3 &= \lambda_3\tilde{\mathbf{A}} + 2\lambda_4\mathbf{D}, & \dot{\lambda}_4 &= -\lambda_4\mathbf{A}, \end{aligned} \tag{184}$$

which are simpler than the analogous equations (57), because the lowest diagonal block of matrix $\mathcal{B}$ is equal to zero identically. In particular, due to the initial condition $\lambda_2(0) = 0$ we have $\lambda_2(t) \equiv 0$. Therefore

$$\Delta(t) = \begin{pmatrix} \mathbf{0} \\ \delta(t) \end{pmatrix}, \qquad \delta(t) = -\int_0^t \lambda_4(\tau)\mathbf{f}(\tau)d\tau,$$

$$\lambda_1(t) = \tilde{\lambda}_4^{-1}(t), \qquad \lambda_3(t) = \left[2\int_0^t \lambda_4(\tau)\mathbf{D}(\tau)\tilde{\lambda}_4(\tau)d\tau\right]\tilde{\lambda}_4^{-1}(t), \tag{185}$$

and the problem is reduced to finding the $2N \times 2N$ matrix $\lambda_4(t)$. The propagator satisfies the first-order equations

$$\lambda_3(t)\partial G/\partial\mathbf{q} + \lambda_4(t)\mathbf{q}G + \delta(t)G = \mathbf{q}'G, \quad \lambda_1(t)\partial G/\partial\mathbf{q} = -\partial G/\partial\mathbf{q}'.$$

Solving these equations together with (179) we obtain [243]

$$G(\mathbf{q};\mathbf{q}';t) = (2\pi)^{-N}\left[\det\left(\lambda_4\lambda_3^{-1}\right)\right]^{1/2}\exp\Big[-\tfrac{1}{2}\mathbf{q}\lambda_3^{-1}\lambda_4\mathbf{q}+\mathbf{q}\lambda_3^{-1}\mathbf{q}' -\tfrac{1}{2}\mathbf{q}'\lambda_1\lambda_3^{-1}\mathbf{q}'-\mathbf{q}\lambda_3^{-1}\delta+\mathbf{q}'\lambda_1\lambda_3^{-1}\delta-\tfrac{1}{2}\delta\lambda_1\lambda_3^{-1}\delta\Big], \quad (186)$$

where the normalization is chosen according to the relation

$$W(\mathbf{q},t) = \int G(\mathbf{q};\mathbf{q}';t)\,W(\mathbf{q}',0)d\mathbf{q}'. \quad (187)$$

The preexponential factor $\det\lambda_4$ appears in (186) due to the term $\mathrm{Tr}A$ in the effective Hamiltonian (180) and due to the Liouville formula [390]

$$\det\Lambda(t) = \exp\left[\int_0^t \mathrm{Tr}\left(\dot{\Lambda}(\tau)\Lambda^{-1}(\tau)\right)\,d\tau\right], \quad (188)$$

There exists an even simpler way to obtain the explicit expression for the propagator. Indeed, since it has the *Gaussian* form, it can be written in the same form (124) as any Gaussian Wigner function:

$$G(\mathbf{q},\mathbf{q}',t) = (2\pi)^{-N}\left[\det\mathcal{M}_*(t)\right]^{-1/2} \times\exp\left\{-\tfrac{1}{2}\left[\mathbf{q}-\mathbf{q}_*(\mathbf{q}',t)\right]\mathcal{M}_*^{-1}\left[\mathbf{q}-\mathbf{q}_*(\mathbf{q}',t)\right]\right\}. \quad (189)$$

Here function $\mathbf{q}_*(\mathbf{q}',t)$ is the mean value of the phase space vector $\mathbf{q}$ at the moment t, provided it coincided with $\mathbf{q}'$ at $t=0$. Analogously, $\mathcal{M}_*(t)$ is the specific (non-physical) variance matrix, which was equal to zero at $t=0$, due to the initial condition $G(\mathbf{q},\mathbf{q}',0) = \delta(\mathbf{q}-\mathbf{q}')$. This condition distinguishes the propagator from all other Gaussians.

The explicit form of $\mathbf{q}_*$ and $\mathcal{M}_*$ can be found by solving the equations (which are immediate consequences of the Fokker–Planck equation)

$$\dot{\mathcal{M}} = A\mathcal{M}+\mathcal{M}\widetilde{A}+2\mathcal{D}, \qquad \dot{\mathbf{q}} = A\mathbf{q}+\mathbf{K}, \quad (190)$$

with the initial conditions $\mathcal{M}_*(0)=0$ and $\mathbf{q}_*(\mathbf{q}',0)=\mathbf{q}'$.

Example: "standard model" of the damped oscillator

The "standard master equation" (170) for the one-dimensional damped oscillator assumes the following form in the Wigner representation ($\hbar=\omega=m=1$):

$$\frac{\partial W}{\partial t} = \frac{\partial}{\partial q}\left([\gamma q-p]W\right)+\frac{\partial}{\partial p}\left([\gamma p+q]W\right)+\gamma\left(\nu+\tfrac{1}{2}\right)\left(\frac{\partial^2 W}{\partial q^2}+\frac{\partial^2 W}{\partial p^2}\right).$$

The equations for the first-order moments and variances read

$$d\bar{q}/dt = \bar{p}-\gamma\bar{q}, \qquad d\bar{p}/dt = -\bar{q}-\gamma\bar{p},$$

$$\dot{\sigma}_q = 2\sigma_{qp}-2\gamma\sigma_q+\gamma(1+2\nu), \qquad \dot{\sigma}_p = -2\sigma_{qp}-2\gamma\sigma_p+\gamma(1+2\nu),$$

$$\dot{\sigma}_{qp} = \sigma_p - \sigma_q - 2\gamma\sigma_{qp}.$$

The solutions are as follows [243]

$$\bar{q}(t) = e^{-\gamma t}\left[\bar{q}_0 \cos t + \bar{p}_0 \sin t\right], \quad \bar{p}(t) = e^{-\gamma t}\left[\bar{p}_0 \cos t - \bar{q}_0 \sin t\right],$$

$$\sigma_q(t) = \sigma_* + e^{-2\gamma t}\left[\left(\sigma_q^{(0)} - \sigma_*\right)\cos^2 t + \left(\sigma_p^{(0)} - \sigma_*\right)\sin^2 t + \sigma_{pq}^{(0)}\sin(2t)\right],$$

$$\sigma_p(t) = \sigma_* + e^{-2\gamma t}\left[\left(\sigma_p^{(0)} - \sigma_*\right)\cos^2 t + \left(\sigma_q^{(0)} - \sigma_*\right)\sin^2 t - \sigma_{pq}^{(0)}\sin(2t)\right],$$

$$\sigma_{qp}(t) = e^{-2\gamma t}\left[\sigma_{qp}^{(0)}\cos(2t) + \tfrac{1}{2}\left(\sigma_p^{(0)} - \sigma_q^{(0)}\right)\sin(2t)\right],$$

where $\sigma_* \equiv \nu + \frac{1}{2}$. The Wigner propagator is given by [146,243]

$$G(q,p;t|q',p',0) = (2\pi\sigma_* u)^{-1}\exp\Big\{-(2\sigma_* u)^{-1}\Big[q_t^2 + p_t^2 + e^{-2\gamma t}\left(q'^2 + p'^2\right) - 2e^{-\gamma t}\left(q' q_t + p' p_t\right)\Big]\Big\}, \tag{191}$$

where $q_t = q\cos t - p\sin t$, $p_t = p\cos t + q\sin t$, and $u(t) \equiv 1 - e^{-2\gamma t}$.

Integrals of motion of the Fokker–Planck equation

Besides the operator integrals of motion, one can introduce the usual ones, i.e., such functions $I(\mathbf{q};t)$ that the integral $\int I(\mathbf{q};t)W(\mathbf{q};t)\,d\mathbf{q}$ does not depend on time for the solutions of the Fokker–Planck equation. These functions must satisfy the equation [243,454]

$$\frac{\partial I}{\partial t} = -\left(A_{\alpha\beta}q_\beta + f_\alpha\right)\frac{\partial I}{\partial q_\alpha} - D_{\alpha\beta}\frac{\partial^2 I}{\partial q_\alpha \partial q_\beta}. \tag{192}$$

Its simplest solution is the linear combination $\mathbf{Q}(\mathbf{q};t) = \lambda_4(t)\mathbf{q} + \delta(t)$, with the same matrix $\lambda_4(t)$ and vector $\delta(t)$ as above. It is worth noting a significant difference between the integrals of motion of Hamiltonian and non-Hamiltonian systems. In the Hamiltonian classical and quantum mechanics, any function of the integral of motion is an integral of motion, too. However, this property does not exist in the non-Hamiltonian case. For example, if $I = \xi^\alpha(t)q_\alpha$ is a linear integral of motion (if $\mathbf{f} = 0$), then the quadratic integral of motion is not I^2, but $I^2 - 2\int \xi^\alpha(\tau)D_{\alpha\beta}\xi^\beta(\tau)d\tau$.

A recipe of constructing polynomial solutions to equation (192) is as follows. Taking as a basis the linear c-number integrals of motion $Q_\alpha(\mathbf{q};t)$, one should act on them by the *operator integrals of motion* related to equation (192). They are obtained from the operator integrals of motion of the original equation (179) (given by equations (181)–(184) simply by changing the sign of matrix $\mathbf{D}$ in the solutions of equations (184). Note that products of the *operator integrals of motion* are again integrals of motion. Therefore the polynomial c-number integrals of motion are linear combinations of the functions

$$\left(\mu_{\alpha\beta}\frac{\partial}{\partial q_\beta} + \lambda_{\alpha\beta}q_\beta + \delta_\alpha\right)\cdots\left(\mu_{\xi\eta}\frac{\partial}{\partial q_\eta} + \lambda_{\xi\eta}q_\eta + \delta_\xi\right)\left(\lambda_{\gamma\nu}q_\nu + \delta_\gamma\right),$$

where the coefficients $\lambda_{\alpha\beta}$ are elements of matrix $\lambda_4(t)$, and the coefficients $\mu_{\alpha\beta}$ are elements of matrix obtained from $\lambda_3(t)$ (185) by means of changing the sign of matrix $\mathbf{D}$. Mean values of all such functions do not depend on time, if the averaging is performed over the Wigner functions satisfying equation (179).

4.4 Discrete (Fock) representation

Due to its specific form (each term in the right-hand side contains pairs of operators $\hat{a}$ and $\hat{a}^\dagger$, but there are no terms with $\hat{a}^2$ or $\hat{a}$ alone), the operator equation (170) takes a rather simple form in the Fock basis, since only matrix elements $\rho_{nm}\langle n|\hat{\rho}|m\rangle$ with the constant difference $m-n$ are connected:

$$\begin{aligned}\dot{\rho}_{nm} &= -i\omega(n-m)\rho_{nm} + \gamma\nu\left[2\sqrt{nm}\rho_{n-1,m-1} - (n+m+2)\rho_{nm}\right] \\ &\quad + \gamma(1+\nu)\left[2\sqrt{(n+1)(m+1)}\rho_{n+1,m+1} - (n+m)\rho_{nm}\right]. \qquad (193)\end{aligned}$$

Then the new functions

$$p_n^{(l)}(t) = \exp[\gamma(2\nu+1)lt - i\omega lt][(n+l)!/n!]^{1/2}\rho_{n,n+l}(t) \qquad (194)$$

(which coincide with the diagonal matrix elements for $l=0$) satisfy the equations with *integral* coefficients (with respect to the index n)

$$\dot{p}_n^{(l)} = 2\gamma(1+\nu)\left[(n+1)p_{n+1}^{(l)} - np_n^{(l)}\right] + 2\gamma\nu\left[(n+l)p_{n-1}^{(l)} - (n+1)p_n^{(l)}\right]. \qquad (195)$$

Introducing the generating function of an auxiliary variable z,

$$G^{(l)}(z;t) \equiv \sum_{n=0}^{\infty} p_n^{(l)}(t)z^n, \qquad (196)$$

one can reduce an infinite set of coupled ordinary differential equations (195) to a simple first-order partial differential equation

$$\frac{\partial G^{(l)}}{\partial t} = 2\gamma(1-z)[1+\nu(1-z)]\frac{\partial G^{(l)}}{\partial z} - 2\gamma\nu[1-z(l+1)]G^{(l)}. \qquad (197)$$

The solution to (197) was found in the generic case in [455,456]:

$$G^{(l)}(z,u) = [1+\nu u(1-z)]^{-l-1}(1-u)^{-l\nu}G_0^{(l)}\left(\frac{z+u(1+\nu)(1-z)}{1+\nu u(1-z)}\right), \qquad (198)$$

where $G_0^{(l)}(z) \equiv G^{(l)}(z,0)$ and $u \equiv 1-\exp(-2\gamma t)$. The special case $l=0$ (diagonal matrix elements) was considered, e.g., in [132,457–459]. Putting $z=1$ in (198) we verify the normalization condition $G^{(0)}(1,t) \equiv 1$. The particular cases of the initial thermal, coherent, and Fock states were also considered in [460]. The role of binomial statistics was emphasized in [461].

A specific feature of equation (195) is that it results in the universal behavior of the mean number of quanta

$$\overline{n}(t) = \overline{n}(0)\exp(-2\gamma t) + \nu\left[1-\exp(-2\gamma t)\right],$$

independently of the initial state. Moreover, the factorial moments of the rth order (138) can be expressed through the initial moments of the orders $k \le r$ (multiplied by some functions of $\exp(-2\gamma t)$ and ν). The simplest expressions exist in the case of zero temperature reservoir ($\nu = 0$): $n^{(r)}(t) = n^{(r)}(0)\exp(-2r\gamma t)$.

For applications of methods and special solutions discussed in this section to various problems of quantum mechanics (in particular, to the problem of *decoherence* [462–465]) see [466–480]. For other approaches see, e.g., [481–483].

5 Universal invariants of quantum systems

For any (normalizable) solution $|\psi(t)\rangle$ of the Schrödinger equation with a *Hermitian* Hamiltonian $\hat{H}(t)$, the average value $I\{\psi\} \equiv \langle\psi(t)|\hat{I}(t)|\psi(t)\rangle$ does not depend on time, if operator $\hat{I}(t)$ satisfies equation (20), which means that it is the operator integral of motion. The concrete value $I\{\psi\}$ obviously depends on the initial state $|\psi(0)\rangle$ and on the explicit form of the Hamiltonian which enters equation (20), in a complete analogy with the classical case. It is known, however, that in classical mechanics there exist invariants of another type, which preserve their values in time simply due to the Hamiltonian structure of the equations of motion, *independently of the concrete form of the Hamiltonian function*. The most famous example is the preservation of the phase volume (the Liouville theorem), while the general construction is known under the name *the universal integral invariants by Poincaré–Cartan* [297]. Similar constructions exist for certain specific classes of quantum systems, as well [11,243,484], in particular, for those described in terms of multidimensional quadratic Hamiltonians considered in this chapter.

5.1 Universal invariants and the second-order moments

Since matrix Σ (59) is nondegenerate, an immediate consequence of equation (61) and the initial condition $\Lambda(0) = E_N$ is the identity $\det\Lambda(t) \equiv 1$ (since $\Lambda(t)$ is a continuous matrix). Evidently, if Λ is a symplectic matrix, then matrix Λ^{-1} is symplectic, too: $\Lambda^{-1}\Sigma\tilde{\Lambda}^{-1} = \Sigma$. Comparing this relation with formula (146), it is easy to verify the identity

$$\mathcal{D}(\mu;t) = \det\left[\mathcal{M}(t) - \mu\Sigma\right] = \sum_{m=0}^{2N} \mathcal{D}_m \mu^m = \mathcal{D}(\mu;0), \qquad (199)$$

which holds for an arbitrary auxiliary parameter μ. The coefficients $\mathcal{D}_m$ are *universal invariants*, since they depend neither on time nor on the functions $B_{\alpha\beta}(t)$ and $C_\alpha(t)$ which determine the concrete form of the quadratic Hamiltonian (3). These coefficients depend only on the initial state of the system and on the commutation relations (defined by matrix Σ) between the operators $\hat{q}_\alpha$. Note that $\mathcal{D}(\mu)$ is an *even* function:

$$\det\left(\mathcal{M} - \mu\Sigma\right) = \det\left(\tilde{\mathcal{M}} - \mu\tilde{\Sigma}\right) = \det\left(\mathcal{M} + \mu\Sigma\right) = \mathcal{D}(-\mu).$$

Moreover, the coefficient $\mathcal{D}_{2N}$ does not depend on $\mathcal{M}$. Therefore the number of different independent invariants equals N. We shall use the notation $\mathcal{D}_j^{(N)}$,

$j = 0, 2, \ldots, 2(N-1)$, in order to emphasize the dependence of the universal invariants on the number of degrees of freedom N. The simplest invariant is $\mathcal{D}_0^{(N)} = \det \mathcal{M}(t)$. In particular, in the case of one degree of freedom we obtain the universal invariant

$$\Delta \equiv \mathcal{D}_0^{(1)} = \sigma_{pp}\sigma_{xx} - \sigma_{px}^2. \tag{200}$$

For $N \geq 2$ we confine ourselves to the explicit form of the invariant $\mathcal{D}_{2N-2}^{(N)}$, which is quadratic with respect to the variances $\mathcal{M}_{\alpha\beta} \equiv \overline{q_\alpha q_\beta}$:

$$\mathcal{D}_{2N-2}^{(N)} = \sum_{i,j=1}^{N} \left(\overline{p_i p_j} \cdot \overline{x_i x_j} - \overline{p_i x_j} \cdot \overline{x_i p_j} \right). \tag{201}$$

Expressions for the higher-order invariants were given in [484]. Special cases of the invariants (200) and (201) have also been discovered in [485,486].

The set of the bosonic annihilation/creation operators (6) possesses the same commutator matrix Σ as above (up to the constant factor $i/\hbar$, which can be removed by the redefinition of the auxiliary parameter μ). In the one-dimensional case the analogue of the invariant (200) reads

$$\tilde{\Delta} = \sigma_{\mathcal{N}}\left(\sigma_{\mathcal{N}} + 1\right) - \left|\sigma_a\right|^2, \quad \sigma_a = \langle \hat{a}^2 \rangle - \langle \hat{a} \rangle^2, \quad \sigma_{\mathcal{N}} = \langle \hat{a}^\dagger \hat{a} \rangle - |\langle \hat{a} \rangle|^2. \tag{202}$$

The analogue of the invariant (201) is as follows:

$$\tilde{\mathcal{D}}_{2N-2}^{(N)} = \sum_{i,j=1}^{N} \left(\overline{a_i a_j^\dagger} \cdot \overline{a_i^\dagger a_j} - \left|\overline{a_i a_j}\right|^2 \right). \tag{203}$$

It is easy to verify that for any natural number m, the quantity

$$\mathcal{L}_m = \mathrm{Tr}\left(\left[\mathcal{M}\Sigma^{-1}\right]^m \right) \tag{204}$$

is also the universal invariant [11,243,484], because it is proportional to the mean value of the *universal integral of motion* [487]

$$\hat{K}_{2n}(t) = -\hat{\mathbf{R}} \left[\Sigma^{-1}\mathcal{M}(t)\right]^{1+2n} \Sigma^{-1}\hat{\mathbf{R}}, \tag{205}$$

where $m = n - 1$ and $\hat{\mathbf{R}} = \hat{\mathbf{q}} - \langle \hat{\mathbf{q}} \rangle$. Due to the symmetry properties of the matrices $\mathcal{M}$ and Σ we have $\mathcal{L}_{2k+1} \equiv 0$. On the other hand, the invariants $\mathcal{L}_m$ with $m > N$ can be expressed in terms of the invariants with indices $m \leq N$, due to the Hamilton–Cayley theorem [390]. For this reason, the independent invariants are given by formula (204) with even indices not exceeding the number $2N$. Each invariant $\mathcal{L}_m$ is some function of the invariants $\mathcal{D}_{2N-j}$, $2 \leq j \leq m$. For example, for $N = 2$ we have $\mathcal{L}_4^{(2)} = 2[\mathcal{D}_2^{(2)}]^2 - 4\mathcal{D}_0^{(2)}$. Other invariants are, obviously, the *eigenvalues* (or their absolute values) of matrix $\mathcal{M}\Sigma^{-1}$.

5.2 Geometric interpretation

To elucidate the meaning of universal quantum invariants, let us consider the Gaussian one-mode state, whose Wigner function reads [252, 394]

$$W(x,p) = \Delta^{-1/2}\hbar \exp\left(-\frac{1}{2\Delta}\left[\sigma_{pp}\tilde{x}^2 + \sigma_{xx}\tilde{p}^2 - 2\sigma_{px}\tilde{x}\tilde{p}\right]\right), \tag{206}$$

with $\tilde{x} = x - \langle\hat{x}(t)\rangle$, $\tilde{p} = p - \langle\hat{p}(t)\rangle$, and Δ given by Eq. (200). For quadratic Hamiltonians, any initial Gaussian state remains Gaussian for any $t > 0$; although the average values and variances of coordinates and momenta change with the course of time, the value of Δ is not changed. The lines of the fixed values of the quasiprobability $W = const$ are the ellipses. Consider the ellipse corresponding to the value of the argument of the exponential in (206) equal to -1. Let $a_\pm$ be the semiaxes of this ellipse. Then (choosing such a system of units that x and p have the same dimensionalities) $a_\pm = \sqrt{T + \sqrt{\Delta}} \pm \sqrt{T - \sqrt{\Delta}}$, where $T = (\sigma_{pp} + \sigma_{xx})/2$. The ellipse area equals $\pi a_+ a_- = 2\pi\sqrt{\Delta}$. Consequently, Eq. (200) means nothing but the conservation of the phase volume (phase area in the one-dimensional case) contained inside the surface of the constant quasiprobability (expressed by the Wigner function), i.e., it can be interpreted as the quantum analogue of the classical Liouville theorem. In the multidimensional case the preservation of the phase volume is related to the preservation of the invariant $\mathcal{D}_0^{(N)} = \det \mathcal{M}$, which is proportional to the square of the volume confined inside the surface of the constant quasiprobability. Actually, the volume of any region in the phase space, confined with the surface $W = const$, remains constant in time for *any* (not only Gaussian) initial Wigner function. This is a consequence of equation (83). Of course, this remarkable property holds only for quadratic Hamiltonians, since in other cases the evolution equation for the Wigner function is no more the classical Liouville equation (82).

The universal invariants of Poincaré were used as far back as by Robertson [488] to illustrate the geometrical meaning of the generalized uncertainty relations, but he did not consider the dynamical aspects of the problem. The relations between the other invariants $\mathcal{D}_{2j}^{(N)}$ and the classical integral universal invariants by Poincaré–Cartan were discussed in [243, 484]. The connections between the *generalized uncertainty relations* and the universal invariants were studied in [484].

5.3 Arbitrary Lie algebras. Higher-order moments

The existence of universal invariants discussed above is the consequence of three factors: the linearity of the Heisenberg equations of motion, the identity (61), and the identity $\det \Lambda = 1$. Linear equations of motion arise for any Hamiltonian which is a linear combination of generators of some Lie algebra:

$$\hat{H} = \sum_{\nu} f^{\nu}(t)\hat{z}_{\nu}, \quad [\hat{z}_\alpha\,,\,\hat{z}_\beta] = i\hbar\sum_{\nu=1}^{N} c^{\nu}_{\alpha\beta}\hat{z}_\nu, \quad c^{\nu}_{\alpha\beta} = -c^{\nu}_{\beta\alpha}, \tag{207}$$

$$\hat{z}_\alpha(t) = \sum_{\beta=1}^{N} \Lambda_{\alpha\beta}(t)\hat{z}_\beta(0), \quad \dot{\Lambda}_{\alpha\beta} = \sum_{\nu,\delta=1}^{N} c_{\alpha\nu}^{\delta} f^\nu(t)\Lambda_{\delta\beta}. \tag{208}$$

However, not all algebras result in the identity $\det \Lambda = 1$. Using the Liouville formula for the matrix determinant (188) and the second equation in (208), one can verify that the necessary and sufficient condition of the unimodularity of matrix $\Lambda(t)$ for arbitrary coefficients $f^\nu(t)$ is the validity of the set of equations

$$\sum_{\alpha=1}^{N} c_{\alpha\beta}^{\alpha} = 0, \qquad \beta = 1, 2, \ldots, N. \tag{209}$$

Instead of (61) now we have the identity

$$\Lambda(t)g\tilde{\Lambda}(t) = g, \qquad \Lambda = \|\Lambda_{\alpha\beta}\|, \quad g = \|g_{\alpha\beta}\|, \tag{210}$$

$g_{\alpha\beta}$ being the elements of the Killing–Cartan tensor,

$$g_{\alpha\beta} = g_{\beta\alpha} = \sum_{\delta\rho} c_{\alpha\delta}^{\rho} c_{\beta\rho}^{\delta}. \tag{211}$$

To prove (210) we introduce the matrix $X = \Lambda g\tilde{\Lambda} - g$. It satisfies the equation

$$\dot{X} = AX + X\tilde{A} + Ag + g\tilde{A}, \tag{212}$$

where $A = \|A_{\alpha\beta}\|$, $A_{\alpha\beta} = \sum_\nu c_{\alpha\nu}^{\beta} f^\nu$. But $Ag + g\tilde{A} \equiv 0$, since the coefficients $c_{\alpha\nu\beta} = \sum_\delta c_{\alpha\nu}^{\delta} g_{\delta\beta}$ are antisymmetrical with respect to all three indices [489]. Consequently, the unique solution to the *homogeneous* equation (212) with zero initial condition is $X \equiv 0$. Note that matrix $\Lambda(t)$ is unimodular for any semisimple algebra. Such algebras possess nondegenerate Killing–Cartan tensors [489], so the identity $\det \Lambda(t) \equiv 1$ is the immediate consequence of the identity (210).

Let us introduce matrix $Z = \|Z_{\alpha\beta}\|$, whose elements are defined as

$$Z_{\alpha\beta} = \tfrac{1}{2}\langle \hat{z}_\alpha\hat{z}_\beta + \hat{z}_\beta\hat{z}_\alpha\rangle - Z_\alpha Z_\beta \equiv Z'_{\alpha\beta} - Z''_{\alpha\beta}, \quad Z_\alpha = \langle \hat{z}_\alpha\rangle.$$

Then the universal invariants, $\mathcal{G}_m$, are given by the coefficients of the expansion

$$\mathcal{G}(\gamma;t) = \det[Z(t) - \gamma g] = \sum_{m=0}^{N} \gamma^m \mathcal{G}_m = \mathcal{G}(\gamma;0). \tag{213}$$

Using the quantities $Z'_{\alpha\beta}$ or $Z''_{\alpha\beta}$ instead of $Z_{\alpha\beta}$ in (213), one obtains similar expressions $\mathcal{G}'_m$ and $\mathcal{G}''_m$, which are also universal invariants. Since both matrices, Z and g, are symmetrical, all coefficients in the expansion (213) are different from zero, so the number of independent invariants equals the number of the algebra generators N. The equivalent set of invariants can be written as (for $\det g \neq 0$)

$$\mathcal{L}_m = \mathrm{Tr}\left(\left[Z(t)g^{-1}\right]^m\right), \qquad g^{-1} = \|g^{\alpha\beta}\|. \tag{214}$$

The simplest invariant coincides (up to a constant factor) with the mean value of the Casimir operator: $\mathcal{G}'_{N-1} \sim \mathcal{L}'_1 = \mathrm{Tr}\left(Z' g^{-1}\right) = \langle \hat{z}_\alpha \hat{z}_\beta g^{\alpha\beta} \rangle$. Another simple invariant is $\mathcal{G}_0 \sim \det\left[Z(t)\right]$. It does not depend on the structure constants.

Consider, for example, the set of operators

$$\begin{array}{ll} \hat{z}_1 = \hat{p}^2, & [\hat{z}_1\,,\,\hat{z}_2] = -4i\hbar\hat{z}_3, \\ \hat{z}_2 = \hat{x}^2, & [\hat{z}_2\,,\,\hat{z}_3] = 2i\hbar\hat{z}_2, \\ \hat{z}_3 = \frac{1}{2}\left(\hat{p}\hat{x} + \hat{x}\hat{p}\right), & [\hat{z}_3\,,\,\hat{z}_1] = 2i\hbar\hat{z}_1, \end{array} \tag{215}$$

which form the algebra isomorphic to $sl(2,R) \sim su(1,1) \sim so(1,2)$. These algebras are frequently used in quantum optics. The only nonzero coefficients of the Killing–Cartan tensor are $g_{12} = g_{21} = -2g_{33}$. Since $\det g \neq 0$, we have three invariants which depend on the *fourth-order moments* of the coordinate and momentum operators,

$$\begin{aligned} \mathcal{G}_0 &= Z_{11}Z_{22}Z_{33} + 2Z_{12}Z_{23}Z_{31} - Z_{11}Z_{23}^2 - Z_{22}Z_{31}^2 - Z_{33}Z_{12}^2, \\ \mathcal{G}_1 &= Z_{11}Z_{22} - Z_{12}^2 + 4\left(Z_{12}Z_{33} - Z_{13}Z_{23}\right), \\ \mathcal{G}_2 &= Z_{12} - Z_{33}, \end{aligned}$$

as well as invariants $\mathcal{G}'_m$ and $\mathcal{G}''_m$. Generalizations to the case $\hat{z}_1 = \hat{p}^2 + b\hat{x}^{-2}$ (the so-called "singular" or "pseudoharmonical" oscillator) were studied in [205,484].

Calculating mean values of the powers of operator $\hat{K}_{2n}(t)$ (205), we obtain another set of universal invariants, which are expressed in terms of the moments of an arbitrary order $2j$ and the second-order variances [487]:

$$\mathcal{K}_{2m}^{(j)} = \left\langle \left(\hat{\mathbf{R}}\tilde{\Sigma}^{-1}\mathcal{M}(t)\left[\Sigma^{-1}\mathcal{M}(t)\right]^{2m}\Sigma^{-1}\hat{\mathbf{R}} \right)^j \right\rangle.$$

A simple one-dimensional example is the invariant ($\hbar = 1$)

$$\begin{aligned} \mathcal{K} &= \Delta^{-1}\left\langle (\delta\hat{p})^2(\delta\hat{x})^2 + (\delta\hat{x})^2(\delta\hat{p})^2 \right\rangle \\ &+ \Delta^{-2}\left[\sigma_{pppp}\sigma_{xx}^2 + \sigma_{xxxx}\sigma_{pp}^2 + 4\sigma_{pppx}\sigma_{xx}\sigma_{xp} + 4\sigma_{xxxp}\sigma_{pp}\sigma_{xp} + 6\sigma_{ppxx}\sigma_{px}^2\right], \end{aligned}$$

where

$$\sigma_{ab\ldots c} \equiv \int W(p,q)\delta a\delta b\ldots\delta c\,\frac{dpdq}{2\pi}, \qquad \delta a \equiv a - \langle\hat{a}\rangle.$$

For other approaches to the higher-order invariants see, e.g., [484, 490–492]. Constructions similar to the universal invariants were also studied under different names ("kinematical invariants" or "generalized beam quality factors") in several papers devoted to the problem of propagation of paraxial optical beams in media with quadratic dependence of the refraction index on the transverse coordinates [493–498] and for paraxial particle beams in accelerators [499, 500]. Their existence is explained by the symplectic structure of transformations like (56). For a detailed review see [487, 501].

6 Conclusion

We have considered different methods of solving the Schrödinger and master equations for quantum systems described by means of multidimensional nonstationary quadratic Hamiltonians. We have also discussed the general properties of wave functions and density matrices (Wigner functions) of such systems. These functions are expressed in terms of solutions of certain linear ordinary differential equations for the "ε-function" (35) or the "Λ-matrix" (57). On the other hand, knowledge of the general structure of solutions enables us to solve the "inverse" problem of the so-called *"quantum control"*, when one has to find such a time dependence of parameters that the evolution of a quantum system would follow a prescribed law (or would result in a prescribed final state). In this case, the functions $\varepsilon(t)$ and $\Lambda(t)$ are actually known, and the coefficients of the Hamiltonian can be easily found from equations like (35) or (57). For the first studies in this direction see, e.g., [502–506]. Many interesting results have been obtained in [255,507–511]. For recent publications see, e.g., [512–515].

Another interesting topic, which was not discussed here due to lack of space, is the generation of nonclassical states in dielectric media with time-dependent parameters and in cavities with moving boundaries. We confine ourselves to a few references [248,255,395,516–523]. A detailed review can be found in [524].

Bibliography

[1] Feynman, R.P., Space-time approach to non-relativistic quantum mechanics. *Rev. Mod. Phys.* (1948) **20** 367–387.

[2] Dirac, P.A.M., The physical interpretation of the quantum dynamics. *Proc. Roy. Soc. London* A (1927) **113** 621–641.

[3] Kennard, E.H., Zur Quantenmechanik einfacher Bewegungstypen. *Z. Phys.* (1927) **44** 326–352.

[4] Schwinger, J. On gauge invariance and vacuum polarization. *Phys. Rev.* (1951) **82** 664–679.

[5] Dodonov, V.V., Malkin, I.A., and Man'ko, V.I., The Green function and thermodynamical properties of quadratic systems. *J. Phys.* A (1975) **8** L19–L22.

[6] Dodonov, V.V., Malkin, I.A., and Man'ko, V.I., Integrals of the motion, Green functions and coherent states of dynamical systems. *Int. J. Theor. Phys.* (1975) **14** 37–54.

[7] Dodonov, V.V., Malkin, I.A., and Man'ko, V.I., Invariants, Green functions and coherent states of dynamical systems. *Teor. Mat. Fiz.* (1975) **24** 164–176 [*Sov. Phys. – Theor. & Math. Phys.* (1975) **24** 746–754].

[8] Malkin, I.A. and Man'ko, V.I., *Dynamical Symmetries and Coherent States of Quantum Systems*. Nauka, Moscow, 1979 (in Russian).

[9] Dodonov, V.V. and Man'ko, V.I., Integrals of motion and the dynamics of nonstationary quadratic Fermi–Bose systems of the general type. In: *Quantization, Gravitation and Group Methods in Physics. Proc. Lebedev Phys. Inst., vol. 152* (A.A. Komar, ed.), pp. 145–193. Nauka, Moscow, 1983 [translated by Nova Science, Commack, 1988, as supplemental volume to vol.176, pp. 197–261].

[10] Dodonov, V.V. and Man'ko, V.I., Invariants and correlated states of nonstationary quantum systems. In: *Invariants and the Evolution of Nonstationary Quantum Sys-*

tems. *Proc. Lebedev Phys. Inst., vol. 183* (M.A. Markov, ed.), pp. 71–181. Nauka, Moscow, 1987 [translated by Nova Science, Commack, 1989, pp. 103–261].

[11] Dodonov, V.V. and Man'ko, V.I., Evolution of multidimensional systems. Magnetic properties of ideal gases of charged particles. In: *Invariants and the Evolution of Nonstationary Quantum Systems. Proc. Lebedev Phys. Inst., vol. 183* (M.A. Markov, ed.), pp. 182–286. Nauka, Moscow, 1987 [translated by Nova Science, Commack, 1989, pp. 263–414].

[12] Landovitz, L.F., Transition probabilities in a strong external field, *Phys. Rev.* A (1975) **11** 67–70.

[13] Campbell, W.B., Finkler, P., Jones, C.E., and Misheloff, M.N., Path integrals with arbitrary generators and the eigenfunction problem. *Ann. Phys.* (NY) (1976) **96** 286–302.

[14] Landovitz, L.F., Levine, A.M., and Schreiber, W.M., Time-dependent harmonic oscillators. *Phys. Rev.* A (1979) **20** 1162–1168.

[15] Landovitz, L.F., Levine, A.M., Ozizmir, E., and Schreiber, W.M., Time-dependent linear quantum systems. *J. Chem. Phys.* (1983) **78** 291–294.

[16] Bartlett, M.S. and Moyal, J.E., The exact transition probabilities of quantum-mechanical oscillator calculated by the phase-space method. *Proc. Cambr. Phil. Soc.* (1949) **45** 545–553.

[17] Feynman, R.P., Mathematical formulation of the quantum theory of electromagnetic interaction. *Phys. Rev.* (1950) **80** 440–457.

[18] Feynman, R.P., An operator calculus having applications in quantum electrodynamics. *Phys. Rev.* (1951) **84** 108–128.

[19] Glauber, R.J., Some notes on multiple-boson processes. *Phys. Rev.* (1951) **84** 395–400.

[20] Ludwig, G., Die erzwungenen Schwingungen des harmonischen Oszillators nach der Quantentheorie. *Z. Phys.* (1951) **130** 468–476.

[21] Husimi, K., Miscellanea in elementary quantum mechanics. II. *Prog. Theor. Phys.* (1953) **9** 381–402.

[22] Kerner, E.H., Note on the forced and damped oscillator in quantum mechanics. *Canad. J. Phys.* (1958) **36** 371–377.

[23] Carruthers, P. and Nieto, M., Coherent states and the forced quantum oscillator. *Amer. J. Phys.* (1965) **33** 537–544.

[24] Gilbey, D.M. and Goodman, F.O., Quantum theory of forced harmonic oscillator. *Amer. J. Phys.* (1966) **34** 143–152.

[25] Caldirola, P., Forze non concervative nella meccanica quantistica. *Nuovo Cim.* (1941) **18** 393–400.

[26] Kanai, E., On the quantization of the dissipative syatems. *Prog. Theor. Phys.* (1948) **3** 440–441.

[27] Papadopoulos, G.J., Brownian motion with quantum dynamics. *J. Phys.* A (1973) **6** 1479–1497.

[28] Hasse, R.W., On the quantum mechanical treatment of dissipative systems. *J. Math. Phys.* (1975) **16** 2005–2011.

[29] Dodonov, V.V. and Man'ko, V.I., Loss-energy states of nonstationary quantum systems. *Nuovo Cim.* B (1978) **44** 265–273.

[30] Moreira, I.C., Propagators for the Caldirola–Kanai–Schrödinger equation. *Lett. Nuovo Cim.* (1978) **23** 294–298.

[31] Dodonov, V.V., Kurmyshev, E.V., and Man'ko, V.I., Exact Green function of a damped oscillator. *Phys. Lett.* A (1979) **72** 10–12.

[32] Dodonov, V.V. and Man'ko, V.I., Coherent states and the resonance of a quantum

damped oscillator. *Phys. Rev.* A (1979) **20** 550–560.

[33] Jannussis, A.D., Brodimas, G.N., and Streclas, A.A., Propagator with friction in quantum mechanics. *Phys. Lett.* A (1979) **74** 6–10.

[34] Khandekar, D.C. and Lawande, S.V., Exact solution of a time-dependent quantal harmonic oscillator with damping and a perturbative force. *J. Math. Phys.* (1979) **20** 1870–1877.

[35] Remaud, B. and Hernández, E.S., Constants of motion and non-stationary wave functions for the damped time-dependent harmonic oscillator. *Physica* A (1980) **103** 35–54.

[36] Landovitz, L.F., Levine, A.M., and Schreiber, W.M., Transition amplitudes for time-dependent harmonic oscillators. *J. Math. Phys.* (1980) **21** 2159–2163.

[37] Colegrave, R.K. and Abdalla, M.S., Harmonic oscillator with exponentially decaying mass. *J. Phys.* A (1981) **14** 2269–2280.

[38] Remaud, B., Dorso, C., and Hernández, E.S., Coherent state propagation in open systems. *Physica* A (1982) **112** 193–213.

[39] Caldirola, P. and Lugiato, L.A., Connection between the Schrödinger equation for dissipative systems and the master equation. *Physica* A (1982) **116** 248–264.

[40] Nemes, M.C. and de Toledo Piza, A.F.R., Quantization of a phenomenological viscous force. *Phys. Rev.* A (1983) **27** 1199–1202.

[41] Cervero, J.M. and Villarroel, J., On the quantum theory of the damped harmonic oscillator. *J. Phys.* A (1984) **17** 2963–2971.

[42] De Smedt, P. and Gonzalez Lopez, A., On the asymptotic behaviour of the solutions of the Caldirola–Kanai equation. *Lett. Math. Phys.* (1986) **12** 291–300.

[43] Levine, A.M., Ozizmir, E., and Schreiber, W.M., Damped quantum mechanical oscillators with thermal fluctuations. *J. Chem. Phys.* (1987) **86** 908–913.

[44] Yeon, K.H., Um, C.I., and George, T.F., Coherent states for the damped harmonic oscillator. *Phys. Rev.* A (1987) **36** 5287–5291.

[45] Aliaga, J., Crespo, G., and Proto, A.N., Thermodynamics of squeezed states for the Kanai–Caldirola Hamiltonian. *Phys. Rev.* A (1990) **42** 4325–4335.

[46] Baskoutas, S., Jannussis, A., and Mignani, R., Time evolution of Caldirola–Kanai oscillators. *Nuovo Cim.* B (1993) **108** 953–966.

[47] Baseia, B., Bagnato, V.S., Marchiolli, M.A., and de Oliveira, M.C., Particle trapping by oscillating fields: influence of dissipation upon the squeezing effect. *Quant. Semiclass. Opt.* (1996) **8** 1147–1158.

[48] Marchiolli, M.A. and Mizrahi, S.S., Dissipative mass-accreting quantum oscillator. *J. Phys.* A (1997) **30** 2619–2635.

[49] Menon, V.J., Chanana, N., and Singh, Y., A fresh look at the BCK frictional Lagrangian. *Prog. Theor. Phys.* (1997) **98** 321–329.

[50] Messer, J., Friction in quantum mechanics. *Acta Phys. Austr.* (1979) **50** 75–91.

[51] Dekker, H., Classical and quantum mechanics of the damped harmonic oscillator. *Phys. Rep.* (1981) **80** 1–112.

[52] Bose, S.K., Dubey, U.B., and Varma, N., Study of certain aspects of anharmonic, time-dependent and damped harmonic-oscillator systems. *Fortschr. Phys.* (1989) **37** 761–818.

[53] Kleber, M., Exact solutions for time-dependent phenomena in quantum mechanics. *Phys. Rep.* (1994) **236** 331–393.

[54] Um, C.-I., Yeon, K.-H., and George, T.F., The quantum damped harmonic oscillator. *Phys. Rep.* (2002) **362** 63–192.

[55] Colegrave, R.K. and Abdalla, M.S., Harmonic oscillator with strongly pulsating mass. *J. Phys.* A (1982) **15** 1549–1559.

[56] Abdalla, M.S. and Ramjit, U.A.T., Quantum mechanics of the damped pulsating oscillator. *J. Math. Phys.* (1989) **30** 60–65.

[57] Lo, C.F., Liu, Y.T., and Li, C.B., Squeezing properties of a pulsating oscillator. *Quant. Semiclass. Opt.* (1995) **7** 843–848.

[58] Lo, C.F. and Kiang, D., Dissipative dynamics of a single particle in an ion trap. *Int. J. Mod. Phys.* B (2000) **14** 993–1006.

[59] Dodonov, V.V. and Nikonov, D.E., Application of Feynman propagators to the motion of one-dimensional wave packets. *J. Sov. Laser Research* (1991) **12** 461–464.

[60] Baskoutas, S., Jannussis, A., and Mignani, R., Dissipative tunnelling of the inverted Caldirola–Kanai oscillator. *J. Phys.* A (1994) **27** 2189–2196.

[61] Papadopoulos, G.J., Microwave-assisted tunnelling in the presence of dissipation. *J. Phys.* A (1997) **30** 5497–5510.

[62] Adamian, G.G., Antonenko, N.V., and Scheid, W., Tunneling with dissipation in open quantum systems. *Phys. Lett.* A (1998) **244** 482–488.

[63] Isar, A., Sandulescu, A., and Scheid, W., Dissipative tunneling through a parabolic potential in the Lindblad theory of open quantum systems. *Europ. Phys. J.* D (2000) **12** 3–10.

[64] Fujiwara, I., Operator calculus of quantized operator. *Prog. Theor. Phys.* (1952) **7** 433–448.

[65] Pechukas, P. and Light, J.C., On exponential form of time-displacement operators in quantum mechanics. *J. Chem. Phys.* (1966) **44** 3897–3912.

[66] *Problems in Quantum Mechanics* (D. Ter Haar, ed.), problems 20–25. Pion Limited, London, 1975 (3rd edition).

[67] Brown, L.S. and Carson, L.J., Quantum-mechanical parametric amplification. *Phys. Rev.* A (1979) **20** 2486–2497.

[68] Granovskiy, Y.I., Dimashko, Y.A., and Zhedanov, A.S., Parametric instability of linear systems in quantum theory. *Izv. VUZ Fiz.* (1980) **23** 111–121 (in Russian) [translated in *Sov. Phys. J. (USA)* (1980) **23** 164–173].

[69] Meyer, H.D., On the forced harmonic oscillator with time-dependent frequency. *Chem. Phys.* (1981) **61** 365–383.

[70] Wolf, K. B., On time-dependent quadratic Hamiltonians. *SIAM J. Appl. Math.* (1981) **40** 419–431.

[71] Bassetti, B., Montaldi, E., and Raciti, M., Some remarks on the time-dependent harmonic oscillator. *Lett. Nuovo Cim.* (1982) **33** 469–474.

[72] Urrutia, L.F. and Hernández, E., Calculation of the propagator for a time-dependent damped, forced harmonic oscillator using the Schwinger action principle. *Int. J. Theor. Phys.* (1984) **23** 1105–1127.

[73] Gazdy, B. and Micha, D.A., The linearly driven parametric oscillator: Application to collisional energy transfer. *J. Chem. Phys.* (1985) **82** 4926–4936.

[74] Brown, L.S., Squeezed states and quantum-mechanical parametric amplification. *Phys. Rev.* A (1987) **36** 2463–2466.

[75] Aliaga, J., Crespo, G., and Proto, A.N., Nonzero-temperature coherent and squeezed states for the harmonic oscillator: the time-dependent frequency case. *Phys. Rev.* A (1990) **42** 618–626.

[76] Crespo, G., Proto, A.N., Plastino, A., and Otero, D., Information-theory approach to the variable-mass harmonic oscillator. *Phys. Rev.* A (1990) **42** 3608–3617.

[77] Tsue, Y. and Fujiwara, Y., Time-dependent variational approach in terms of squeezed coherent states. *Prog. Theor. Phys.* (1991) **86** 443–467.

[78] Mizrahi, S.S., Moussa, M.H.Y., and Baseia, B., The quadratic time-dependent Hamiltonian: evolution operator, squeezing regions in phase space and trajectories.

Int. J. Mod. Phys. B (1994) **8** 1563–1576.

[79] de Toledo Piza, A.F.R., Classical equations for quantum squeezing and coherent pumping by the time-dependent quadratic Hamiltonian. *Phys. Rev.* A (1995) **51** 1612–1616.

[80] Mostafazadeh, A., Inverting a time-dependent harmonic oscillator potential by a unitary transformation and a class of exactly solvable oscillators. *Phys. Rev.* A (1997) **55** 4084–4088.

[81] Dodonov, V.V., Marchiolli, M.A., Korennoy, Y.A., Man'ko, V.I., and Moukhin, Y.A., Parametric excitation of photon-added coherent states. *Phys. Scripta* (1998) **58** 469–480.

[82] Cervero, J.M., Squeezing and dynamical symmetries. *Int. J. Theor. Phys.* (1999) **38** 2095–2109.

[83] Mostafazadeh, A., Generalized adiabatic product expansion: a nonperturbative method of solving the time-dependent Schrödinger equation. *J. Math. Phys.* (1999) **40** 3311–3326.

[84] Boschi, C.D.E., Ferrari, L., and Lewis, H.R., Reduction method for the linear quantum or classical oscillator with time-dependent frequency, damping, and driving. *Phys. Rev.* A (2000) **61** 010101.

[85] Nieto, M.M. and Truax, D.R., Schrödinger equation with time-dependent p^2 and x^2 terms. *Int. J. Mod. Phys.* A (2002) **17** 1559–1575.

[86] Ermakov, V., Differentzial'nyya uravneniya vtorago poryadka. Usloviya integriruemosti v konechnom vide [Differential equations of the second order. Conditions of integrability in a finite form]. *Universitetskiya Izvestiya (Kiev) [Bulletin of Kiev University]* (1880), vol. **20**, no. 9, section III, pp. 1–25.

[87] Milne, W.E., The numerical determination of characteristic numbers. *Phys. Rev.* (1930) **35** 863–867.

[88] Pinney, E., The nonlinear differential equation $y'' + p(x)y + cy^{-3} = 0$. *Proc. Am. Math. Soc.* (1950) **1** 681.

[89] Reid, J.L. and Ray, J.R., Ermakov systems, non-linear superposition, and solutions of non-linear equations of motion. *J. Math. Phys.* (1980) **21** 1583–1587.

[90] Leach, P.G.L., Generalized Ermakov systems. *Phys. Lett.* A (1991) **158** 102–106.

[91] Kaushal, R.S., Construction of exact invariants for time dependent classical dynamical systems. *Int. J. Theor. Phys.* (1998) **37** 1793–1856.

[92] Seymour, P.W., Leipnik, R.B., and Nicholson, A.F., Charged particle motion in a time-dependent axially symmetric magnetic field. *Austral. J. Phys.* (1965) **18** 553–565.

[93] Lewis, H.R., Class of exact invariants for classical and quantum time-dependent harmonic oscillators. *J. Math. Phys.* (1968) **9** 1976–1986.

[94] Salusti, E. and Zirilli, F., On the time-dependent harmonic oscillator. *Lett. Nuovo Cim.* (1970) **4** 999–1003.

[95] Malkin, I.A. and Man'ko, V.I., Coherent states and excitation of N-dimensional non-stationary forced oscillator. *Phys. Lett.* A (1970) **32** 243–244.

[96] Popov, V.S. and Perelomov, A.M., Parametric excitation of a quantum oscillator *Zhurn. Eksp. Teor. Fiz.* (1969) **56** 1375–1390 [*Sov. Phys. – JETP* 1969 **29** 738–745].

[97] Eliezer, C.J. and Gray, A., A note on the time-dependent harmonic oscillator. *SIAM J. Appl. Math.* (1976) **30** 463–468.

[98] Kim, S.P., A class of exactly solved time-dependent quantum harmonic oscillators. *J. Phys.* A (1994) **27** 3927–3936.

[99] Agayeva, R.G., Non-adiabatic parametric excitation of oscillator-type systems. *J. Phys.* A (1980) **13** 1685–1699.

[100] Møller, K.B. and Henriksen, N.E., On wave-packet dynamics in a decaying quadratic potential. *Phys. Scripta* (1997) **55** 542–546.

[101] Agarwal, G.S. and Kumar, S.A., Exact quantum-statistical dynamics of an oscillator with time-dependent frequency and generation of nonclassical states. *Phys. Rev. Lett.* (1991) **67** 3665–3668.

[102] Dodonov, V.V., Man'ko, V.I., and Zhivotchenko, D.V., Squeezing an harmonic oscillator with a sawtooth pulse. *J. Sov. Laser Research* (1993) **14** 127–145.

[103] Dodonov, V.V., Man'ko, O.V., and Man'ko, V.I., Time-dependent oscillator with Kronig–Penney excitation. *Phys. Lett.* A (1993) **175** 1–4.

[104] Junker, G., Results on the propagator of a periodically kicked harmonic oscillator. In: *Path Integrals in Physics* (V. Sayakanit, J.-O. Berananda, and W. Sritrakool, eds.), pp. 290–300. World Scientific, Singapore, 1994.

[105] Dodonov, V.V., Lukin, M.D., and Man'ko, V.I., Squeezing for the one-mode electromagnetic field oscillator with δ-kicked frequency. *Nuovo Cim.* B (1994) **109** 1023–1037.

[106] Dodonov, V.V., Man'ko, O.V., Man'ko, V.I., Polynkin, P.G., and Rosa, L., Delta-kicked Landau levels. *J. Phys.* A (1995) **28** 197–208.

[107] Man'ko, O.V., Damped oscillator with Kronig–Penney excitation. *Nuovo Cim.* B (1996) **111** 1111–1123.

[108] Karner, G., Man'ko, V.I., and Streit, L., Quasi-energies, loss-energies, and stochasticity. *Rep. Math. Phys.* (1991) **29** 177–193.

[109] Man'ko, V.I. and Mendes, R.V., Lyapunov exponent in quantum mechanics. A phase-space approach. *Physica* D (2000) **145** 330–348.

[110] Bunimovich, L., Jauslin, H.R., Lebowitz, J.L., Pellegrinotti, A., and Nielaba, P., Diffusive energy growth in classical and quantum driven oscillators. *J. Stat. Phys.* (1991) **62** 793–817.

[111] Dodonov, V.V., Irregular behaviour of a quantum kicked oscillator. *Phys. Lett.* A (1996) **214** 27–32.

[112] Grothaus, M., Khandekar, D.C., da Silva, J.L., and Streit, L., The Feynman integral for time-dependent anharmonic oscillators. *J. Math. Phys.* (1997) **38** 3278–3299.

[113] Newman, T.J. and Ziu, R.K.P., Three manifestations of the pulsed harmonic potential. *J. Phys.* A (1998) **31** 9621–9640.

[114] Janszky, J. and Yushin, Y.Y., Squeezing via frequency jump. *Opt. Commun.* (1986) **59** 151–154.

[115] Graham, R., Squeezing and frequency changes in harmonic oscillators. *J. Mod. Opt.* (1987) **34** 873–879.

[116] Ma, X. and Rhodes, W., Squeezing in harmonic oscillators with time-dependent frequencies. *Phys. Rev.* A (1988) **39** 1941–1947.

[117] Kiss, T., Janszky, J., and Adam, P., Time evolution of harmonic oscillators with time-dependent parameters: A step-function approximation. *Phys. Rev.* A (1994) **49** 4935–4942.

[118] Hagedorn, G.A., Loss, M., and Slawny, J., Non-stochasticity of time-dependent quadratic hamiltonians and the spectra of canonical transformations. *J. Phys.* A (1986) **19** 521–531.

[119] Abdalla, M.S. and Colegrave, R.K., Absolute squeezing of a coherent state by resonant modulation of frequency. *Phys. Lett.* A (1993) **181** 341–344.

[120] Averbukh, I., Sherman, B., and Kurizki, G., Enhanced squeezing by periodic frequency modulation under parametric-instability conditions. *Phys. Rev.* A (1994) **50** 5301–5308.

[121] Yoshimura, M., Catastrophic particle production under periodic perturbation. *Prog.*

Theor. Phys. (1995) **94** 873–898.

[122] Boschi, C.D.E. and Ferrari, L., Quantum states of an oscillator with periodic time-dependent frequency under quasiresonant condition: Unperturbed evolution, perturbative effects, and anharmonic effects. *Phys. Rev.* A (1999) **59** 3270–3279.

[123] Mollow, B.R., Quantum-mechanical harmonic oscillator under a random quadratic perturbation. *Phys. Rev.* A (1970) **2** 1477–1480.

[124] Ferrari, L., Quantum oscillator with fluctuating time-dependent frequency. *Phys. Rev.* A (1998) **57** 2347–2356.

[125] Garnier, J., Energy distribution of the quantum harmonic oscillator under random time-dependent perturbations. *Phys. Rev.* E (1999) **60** 3676–3687.

[126] Hacyan, S., The quantum sling and the Schrödinger cat. *Found. Phys. Lett.* (1996) **9** 225–233.

[127] Dremin, I.M. and Man'ko, V.I., Particles and nuclei as quantum slings. *Nuovo Cim.* A (1998) **111** 439–444.

[128] Raiford, M.T., Statistical dynamics of quantum oscillators and parametric amplification in a single mode. *Phys. Rev.* A (1970) **2** 1541–1558.

[129] Lu, E.Y.C., Quantum correlations in two-photon amplification. *Lett. Nuovo Cim.* (1972) **3** 585–589.

[130] Raiford, M.T., Degenerate parametric amplification with time-dependent pump amplitude and phase. *Phys. Rev.* A (1974) **9** 2060–2069.

[131] Abdalla, M.S., Quantum statistics of the degenerate parametric oscillator. *Physica* A (1994) **210** 461–475.

[132] Shimoda, K., Takahasi, H, and Townes, C.H., Fluctuations in amplification of quanta with application to maser amplifiers. *J. Phys. Soc. Japan* (1957) **12** 686–700.

[133] Louisell, W.H., Yariv, A., and Siegman, A.E., Quantum fluctuations and noise in parametric processes. I. *Phys. Rev.* (1961) **124** 1646–1654.

[134] Haus, H.A. and Mullen, J.A., Quantum noise in linear amplifiers. *Phys. Rev.* (1962) **128** 2407–2413.

[135] Heffner, H., The fundamental noise limit of linear amplifiers. *Proc. IRE* (1962) **50** 1604–1608.

[136] Gordon, J.P., Louisell, W.H., and Walker, L.P., Quantum fluctuations and noise in parametric processes. II. *Phys. Rev.* (1963) **129** 481–485.

[137] Takahasi, H., Information theory of quantum-mechanical channels. In: *Advances in Communication Systems. Theory and Applications, vol. 1* (A.V. Balakrishnan, ed.), pp. 227–310. Academic, New York, 1965.

[138] Mollow, B.R. and Glauber, R.J., Quantum theory of parametric amplification. I and II. *Phys. Rev.* (1967) **160** 1076–1096, 1097–1108.

[139] Mišta, L. and Peřina, J., Quantum statistics of parametric amplification. *Czechosl. J. Phys.* B (1978) **28** 392-404.

[140] Caves, C.M., Quantum limits on noise in linear amplifiers. *Phys. Rev.* D (1982) **26** 1817–1839.

[141] Friberg, S. and Mandel, L., Coherence properties of the linear photon amplifier. *Opt. Commun.* (1983) **46** 141–148.

[142] Yurke, B. and Denker, J.S., Quantum network theory. *Phys. Rev.* A (1984) **29** 1419–1437.

[143] Collett, M.J. and Gardiner, C.W., Squeezing of intracavity and traveling-wave light field produced in parametric amplification. *Phys. Rev.* A (1984) **30** 1386–1391.

[144] Loudon, R. and Shepherd, T.J., Properties of the optical quantum amplifier. *Opt. Acta* (1984) **31** 1243–1269.

[145] Yurke, B., Optical back-action-evading amplifiers. *J. Opt. Soc. Am.* B (1985) **2** 732–738.

[146] Stenholm, S., Amplification of squeezed states. *Opt. Commun.* (1986) **58** 177–180.

[147] Milburn, G.J., Steyn-Ross, M.L., and Walls, D.F., Linear amplifiers with phase-sensitive noise. *Phys. Rev.* A (1987) **35** 4443–4445.

[148] Stenholm, S., Transfer properties of quantum amplifiers. *Acta Phys. Polon.* A (1990) **78** 221–229.

[149] Anwar, J. and Zubairy, M.S., Effect of squeezing on the degenerate parametric oscillator. *Phys. Rev.* A (1992) **45** 1804–1809.

[150] Matsuo, K., Wigner distribution functions on a linear amplifier. *Phys. Rev.* A (1993) **47** 3337–3345.

[151] Senitzky, I.R., Detection of quantum noise. *Phys. Rev.* A (1993) **48** 4629–4638.

[152] Anwar, J. and Zubairy, M.S., Quantum-statistical properties of noise in a phase-sensitive linear amplifier. *Phys. Rev.* A (1994) **49** 481–484.

[153] Kim, M.S., Photon statistics in two-photon linear amplifiers. *Opt. Commun.* (1995) **114** 262-268.

[154] Haus, H.A., From classical to quantum noise. *J. Opt. Soc. Am.* B (1995) **12** 2019–2036.

[155] Ban, M., Generation of a single-mode squeezed state via a non-degenerate parametric amplifier. *Phys. Lett.* A (1997) **233** 284–290.

[156] Rekdal, P.K. and Skagerstam, B.S.K., Quantum dynamics of non-degenerate parametric amplification. *Phys. Scripta* (2000) **61** 296–306.

[157] Filip, R., Amplification of Schrödinger-cat state in a degenerate optical parametric amplifier. *J. Opt.* B (2001) **3** S1–S6.

[158] Abd Al-Kader, G.M., Linear amplifier and the coherent states superpositions. *Phys. Scripta* (2001) **63** 372–378.

[159] Ivanova, E.V., Malkin, I.A., and Man'ko, V.I., Invariants and radiation of some nonstationary systems. *Int. J. Theor. Phys.* (1977) **16** 503–515.

[160] Mehta, C.L. and Sudarshan, E.C.G., Time evolution of coherent states. *Phys. Lett* (1966) **22** 574–576.

[161] Mehta, C.L., Chand, P., Sudarshan, E.C.G., and Vedam, R., Dynamics of coherent states. *Phys. Rev.* (1967) **157** 1198–1206.

[162] Berezin, F.A., Canonical operator transformation in representation of secondary quantization. *Doklady AN SSSR* (1961) **137** 311–314 [*Sov. Phys. – Doklady* (1961) **6** 212–215].

[163] Berezin, F.A., *The Method of Second Quantization*. Academic, New York, 1966.

[164] Moshinsky, M. and Quesne, C., Linear canonical transformations and their unitary representations. *J. Math. Phys.* (1971) **12** 1772–1780.

[165] Wolf, K.B., Canonical transforms. I. Complex linear transforms. *J. Math. Phys.* (1974) **15** 1295–1301.

[166] Kramer, P., Moshinsky, M., and Seligman, T.H., Complex extensions of canonical transformations and quantum mechanics. In: *Group Theory and its Applications, vol. III* (E.M. Loebl, ed.), pp. 249–332. Academic, New York, 1975.

[167] Tikochinsky, Y., Transformation brackets for generalized Bogolyubov-boson transformations. *J. Math. Phys.* (1978) **19** 270–276.

[168] Mollow, B.R., Quantum statistics of coupled oscillator systems. *Phys. Rev.* (1965) **162** 1256–1273.

[169] Helstrom, C.V., Quasi-classical analysis of coupled oscillators. *J. Math. Phys.* (1967) **8** 37–42.

[170] Malkin, I.A., Man'ko, V.I., and Trifonov, D.A., Linear adiabatic invariants and co-

herent states. *J. Math. Phys.* (1973) **14** 576–582.

[171] Peřina, J., Peřinová, V., and Horák, R., Evolution of photon statistics of light propagating through a random medium. I. Solutions of dynamic field equations, characteristic functions and quasidistributions. *Czech. J. Phys.* B (1973) **23** 975–992.

[172] Peřina, J., Peřinová, V., Mišta, L., and Horák, R., Damped solutions for the photon statistics of radiation propagating through a random medium. I. Field equations and their solutions. *Czechosl. J. Phys.* B (1974) **24** 374–388.

[173] Leach, P.G.L., On the theory of time-dependent linear canonical transformations as applied to Hamiltonians of the harmonic oscillator type. *J. Math. Phys.* (1977) **18** 1608–1611.

[174] Dhara, A.K. and Lawande, S.V., Time-dependent invariants and the Feynman propagator. *Phys. Rev.* A (1984) **30** 560–567.

[175] Howard, S.D. and Roy, S.K., Group theoretical techniques on phase space and the calculation of quantum mechanical propagators. *J. Phys.* A (1989) **22** 4865–4876.

[176] Yeon, K.H., Walls, D.F., Um, C.I., George, T.F., and Pandey, L.N., Quantum correspondence for linear canonical transformations on general Hamiltonian systems. *Phys. Rev.* A (1998) **58** 1765–1774.

[177] Berezin, F.A., *Introduction to Algebra and Analysis with Anticommuting Elements*, Moscow State Univ. Press, Moscow, 1983.

[178] Nagamachi, S. and Nishimura, T., Linear canonical supertransformation. *J. Math. Phys.* (1993) **34** 1757–1772.

[179] Castaños, O., López-Peña, R., and Man'ko, V.I., Noether theorem and time-dependent quantum invariants. *J. Phys.* A (1994) **27** 1751–1770.

[180] Kulsrud, M., Exact quantum dynamical solutions for oscillator-like systems. *Phys. Rev.* (1956) **104** 1186–1188.

[181] Van Vleck, J.H., The correspondence principle in the statistical interpretation of quantum mechanics. *Proc. Nat. Acad. Sci. USA* (1928) **14** 178–188.

[182] Morette, C., On the definition and approximation of Feynman path integrals. *Phys. Rev.* (1951) **81** 848–852.

[183] Maslov, V.P. and Fedoriuk, M.V., *Semi-Classical Approximation in Quantum Mechanics*. Reidel, Dordrecht, 1981.

[184] Feynman, R.P. and Hibbs, A.R., *Quantum Mechanics and Path Integrals*. McGraw-Hill, New York, 1965.

[185] Schulman, L.S., *Techniques and Applications of Path Integration*. Wiley, New York, 1981.

[186] Littlejohn, R.G., The semiclassical evolution of wave packets. *Phys. Rep.* (1986) **138** 193–291.

[187] Molzahn, F.H. and Osborn, T.A., A phase-space fluctuation method for quantum dynamics. *Ann. Phys.* (NY) (1994) **230** 343–394.

[188] Papadopoulos, G.J., Gaussian path integrals. *Phys. Rev.* D (1975) **11** 2870–2875.

[189] Khandekar, D.C. and Lawande, S.V., Exact propagator for a time-dependent harmonic oscillator with and without a singular perturbation. *J. Math. Phys.* (1975) **16** 384–388.

[190] Hillery, M. and Zubairy, M.S., Path-integral approach to problems in quantum optics. *Phys. Rev.* A (1982) **26** 451–460.

[191] Cheng, B.K., Exact evaluation of the propagator for the damped harmonic oscillator. *J. Phys.* A (1984) **17** 2475–2484.

[192] Gerry, C.C., On the path integral quantization of the damped harmonic oscillators. *J. Math. Phys.* (1984) **25** 1820–1822.

[193] Junker, G. and Inomata, A., Transformation of the free propagator to the quadratic

propagator. *Phys. Lett.* A (1985) **110** 195–198.

[194] Cheng, B.K., The propagator of the time-dependent forced harmonic oscillator with time-dependent damping. *J. Math. Phys.* (1986) **27** 217–220.

[195] de Souza, C.F. and de Souza Dutra, A., The propagator for a time-dependent mass subject to a harmonic potential with a time-dependent frequency. *Phys. Lett.* A (1987) **123** 297–301.

[196] Wei, J. and Norman, E., Lie algebraic solution of linear differential equations. *J. Math. Phys.* (1963) **4** 575–577.

[197] Wilcox, R.M., Exponential operators and parameter differentiation in quantum physics. *J. Math. Phys.* (1967) **8** 962–982.

[198] Suzuki, M., Decomposition formulas of exponential operators and Lie exponentials with some applications to quantum mechanics and statistical physics. *J. Math. Phys.* (1985) **26** 601–612.

[199] Chumakov, S.M., Dodonov, V.V., and Man'ko, V.I., Correlation functions of the non-stationary quantum singular oscillator. *J. Phys.* A (1986) **19** 3229–3239.

[200] Prants, S.V., An algebraic approach to quadratic parametric processes. *J. Phys.* A (1986) **19** 3457–3462.

[201] Dattoli, G., Solimeno, S., and Torre, A., Algebraic time-ordering techniques and harmonic oscillator with time-dependent frequency. *Phys. Rev.* A (1986) **34** 2646–2653.

[202] Gerry, C.C., Ma, P.K., and Vrscay, E.R., Dynamics of SU(1,1) coherent states driven by a damped harmonic oscillator. *Phys. Rev.* A (1989) **39** 668–674.

[203] Lo, C.F., Squeezing in harmonic oscillator with time-dependent mass and frequency. *Nuovo Cim.* B (1990) **105** 497–506.

[204] Profilo, G. and Soliana, G., Group-theoretical approach to the classical and quantum oscillator with time-dependent mass and frequency. *Phys. Rev.* A (1991) **44** 2057–2065.

[205] Dodonov, V.V., Man'ko, V.I., and Zhivotchenko, D.V., Quasienergies and chaotic behaviour of periodically delta-kicked quantum singular oscillator. *Nuovo Cim.* B (1993) **108** 1349–1363.

[206] Penna, V., Compact versus noncompact quantum dynamics of time-dependent $su(1,1)$-valued Hamiltonians. *Ann. Phys.* (NY) (1996) **245** 389–407.

[207] Delgado, F.C., Mielnik, B., and Reyes, M.A., Squeezed states and Helmholtz spectra. *Phys. Lett.* A (1998) **237** 359–364.

[208] Bechler, A., Compact and noncompact dynamics of the $SU(1,1)$ coherent states driven by a coherence preserving Hamiltonian. *J. Phys.* A (2001) **34** 8081–8100.

[209] Khandekar, D.C. and Lawande, S.V., Feynman path integrals: some exact results and applications. *Phys. Rep.* (1986) **137** 115–229.

[210] Inomata, A., Kuratsuji, H., and Gerry, C.C., *Path Integrals and Coherent States of $SU(2)$ and $SU(1,1)$*. World Scientific, Singapore, 1992.

[211] Grosche, C. and Steiner, F., *Handbook of Feynman Path Integrals* (Springer Tracts in Modern Physics, vol. 145). Springer, Berlin, 1998.

[212] Fernandez, F.M., On the time evolution operator for time-dependent quadratic Hamiltonians. *J. Math. Phys.* (1989) **30** 1522–1524.

[213] Farina, C. and Segui-Santonja, A.J., Schwinger's method for a harmonic oscillator with a time-dependent frequency. *Phys. Lett.* A (1993) **184** 23–28.

[214] Weyl, H., Quantenmechanik und Gruppentheorie. *Z. Phys.* (1928) **46** 1–46.

[215] Wigner, E., On the quantum corrections for thermodynamic equilibrium. *Phys. Rev.* (1932) **40** 749–759.

[216] Dirac, P.A.M., Note on exchange phenomena in the Thomas atom. *Proc. Camb.*

Phil. Soc. (1930) **26** 376–385.

[217] Landau, L., Das Dämpfungsproblem in der Wellenmechanik. *Z. Phys.* (1927) **45** 430–441 [The damping problem in wave mechanics. In: *Collected papers of L.D. Landau* (D. Ter Haar, ed.), pp. 8–18. Gordon & Breach, New York, 1965].

[218] von Neumann, J., Wahrscheinlichkeitstheoretischer Aufbau der Quantenmechanik. *Göttin. Nachr.* (1927) 245–272. [Translated in: *Collected Works of J. von Neumann* (A.H. Taub, ed.), vol. 1, pp. 208–235. Pergamon, New York, 1961].

[219] Takabayasi, T., The formulation of quantum mechanics in terms of ensemble in phase space. *Prog. Theor. Phys.* (1954) **11** 341–373.

[220] Mori, H., Oppenheim, I., and Ross, J., Some topics in quantum statistics: the Wigner function and transport theory. In: *Studies in Statistical Mechanics* (J. de Boer and J.E. Uhlenbeck, eds.), vol. 1, pp. 217–298. North Holland, Amsterdam, 1962.

[221] Kubo, R., Wigner representation of quantum operators and its applications to electrons in a magnetic field. *J. Phys. Soc. Japan* (1964) **19** 2127–2139.

[222] Carruthers, P. and Zachariasen, F., Quantum collision theory with phase space distributions. *Rev. Mod. Phys.* (1983) **55** 245–285.

[223] Balazs, N.L. and Jennings, B.K., Wigner's function and other distribution functions in mock phase spaces. *Phys. Rep.* (1984) **104** 347–391.

[224] Englert, B.-G., On the operator bases underlying Wigner's, Kirkwood's and Glauber's phase space functions. *J. Phys.* A (1989) **22** 625–640.

[225] Takahashi, K., Distribution functions in classical and quantum mechanics. *Prog. Theor. Phys. Suppl.* (1989) **98** 109–156.

[226] Dodonov, V.V., Man'ko, V.I., and Ossipov, D.L., Gauge-invariant Weyl representation. An oscillator in an inhomogeneous magnetic field. In: *Theory of Nonstationary Quantum Oscillator. Proc. Lebedev Phys. Inst., vol. 191* (M.A. Markov, ed.), pp. 3–45. Nauka, Moscow, 1989 [translated by Nova Science, Commack, 1992, as vol. 198, pp. 1–62].

[227] Lee, H.W., Theory and application of the quantum phase-space distribution functions. *Phys. Rep.* (1995) **259** 147–211.

[228] Schleich, W.P., *Quantum Optics in Phase Space*. Wiley-VCH, Berlin, 2001.

[229] Bolivar, A.O., The Wigner representation of classical mechanics, quantization and classical limit. *Physica* A (2001) **301** 219–240.

[230] Curtright, T., Uematsu, T., and Zachos, C., Generating all Wigner functions. *J. Math. Phys.* (2001) **42** 2396–2415.

[231] Levanda, M. and Fleurov, V., A Wigner quasi-distribution function for charged particles in classical electromagnetic fields. *Ann. Phys.* (NY) (2001) **292** 199–231.

[232] Karasev, M.V. and Osborn, T.A., Symplectic areas, quantization, and dynamics in electromagnetic fields. *J. Math. Phys.* (2002) **43** 756–788.

[233] Lobo, A.C. and Nemes, M.C., A coordinate independent formulation of the Weyl–Wigner transform theory. *Physica* A (2002) **311** 111–129.

[234] Dodonov, V.V., Man'ko, O.V., and Man'ko, V.I., Quantum nonstationary oscillator: models and applications. *J. Russ. Laser Research* (1995) **16** 1–56.

[235] Krüger, J.G. and Poffyn, A., Quantum mechanics in phase space. III. Linear transformations. *Physica* A (1978) **91** 99–112.

[236] Ekert, A.K. and Knight, P.L., Canonical transformation and decay into phase-sensitive reservoirs. *Phys. Rev.* A (1990) **42** 487–493.

[237] Marinov, M.S., An alternative to the Hamilton–Jacobi approach in classical mechanics. *J. Phys.* A (1979) **12** 31–47.

[238] Akhundova, E.A., Dodonov, V.V., and Man'ko, V.I., Wigner functions of quadratic systems. *Physica* A (1982) **115** 215–231.

[239] Nieto, L.M. and Noriega, J.M., Phase-space propagators for quantum quadratic hamiltonians in one and two dimensions. *Int. J. Theor. Phys.* (1988) **27** 1043–1058.

[240] Gadella, M., Gracia-Bondía, J.M., Nieto, L.M., and Várilly, J.C., Quadratic Hamiltonians in phase-space quantum mechanics. *J. Phys.* A (1989) **22** 2709–2738.

[241] García-Calderón, G. and Moshinsky, M., Wigner distribution functions and the representation of canonical transformations in quantum mechanics. *J. Phys.* A (1980) **13** L185–L188.

[242] Bloch, F., Zur Theorie des Austauschproblems und der Remanenzerscheinung der Ferromagnetika. *Z. Phys.* (1932) **74** 295–335.

[243] Dodonov, V.V. and Man'ko, V.I., Density matrices and Wigner functions of quasiclassical quantum systems. In: *Group Theory, Gravitation and Elementary Particle Physics. Proc. Lebedev Phys. Inst., vol. 167* (A.A. Komar, ed.), pp.7–79. Nauka, Moscow, 1986 [translated by Nova Science, Commack, 1987, pp. 7–101].

[244] Simon, R., Sudarshan, E.C.G., and Mukunda, N., Gaussian–Wigner distributions in quantum mechanics and optics. *Phys. Rev.* A (1987) **36** 3868–3880.

[245] Leach, P.G.L., Invariants and wavefunctions for some time-dependent harmonic oscillator-type Hamiltonians. *J. Math. Phys.* (1977) **18** 1902–1907.

[246] Hartley, J.G. and Ray, J.R., Coherent states for the time-dependent harmonic oscillator. *Phys. Rev.* D (1982) **25** 382–386.

[247] Lo, C.F., Generating displaced and squeezed number states by a general driven time-dependent oscillator. *Phys. Rev.* A (1991) **43** 404–409.

[248] Dodonov, V.V., George, T.F., Man'ko, O.V., Um, C.I., and Yeon, K.H., Exact solutions for a mode of the electromagnetic field in resonator with time-dependent characteristics of the internal medium. *J. Sov. Laser Research* (1992) **13** 219–230.

[249] Abe, S. and Ehrhardt, R., Effects of anharmonicity on nonclassical states of the time-dependent harmonic oscillator. *Phys. Rev.* A (1993) **48** 986–994.

[250] Yeon, K.H., Lee, K.K., Um, C.I., George, T.F., and Pandey, L.N., Exact quantum theory of a time-dependent bound quadratic Hamiltonian system. *Phys. Rev.* A (1993) **48** 2716–2720.

[251] Pedrosa, I.A., Serra, G.P., and Guedes, I., Wave functions of a time-dependent harmonic oscillator with and without a singular perturbation. *Phys. Rev.* A (1997) **56** 4300–4303.

[252] Dodonov, V.V., Kurmyshev, E.V., and Man'ko, V.I., Generalized uncertainty relation and correlated coherent states. *Phys.Lett.* A (1980) **79** 150–152.

[253] Dodonov, V.V., Klimov, A.B., and Man'ko, V.I., Physical effects in correlated quantum states. In: *Squeezed and Correlated States of Quantum Systems. Proc. Lebedev Phys. Inst., vol. 200* (M.A. Markov, ed.), pp. 56–105. Nauka, Moscow, 1991 [translated by Nova Science, Commack, 1993, as vol. 205, pp. 61–107].

[254] Dodonov, V.V., Man'ko, O.V., and Man'ko, V.I., Correlated states of string and gravitational waveguide. In: *Squeezed and Correlated States of Quantum Systems. Proc. Lebedev Phys. Inst., vol. 200* (M.A. Markov, ed.), pp. 155–217. Nauka, Moscow, 1991 [translated by Nova Science, Commack, 1993, as vol. 205, pp. 163–229].

[255] Dodonov V.V. and Man'ko V.I., Correlated and squeezed coherent states of time-dependent quantum systems. In: *Modern Nonlinear Optics, Part 3. Advances in Chemical Physics Series, vol. LXXXV* (M. Evans and S. Kielich, eds.), pp. 499–530. Wiley, New York, 1994.

[256] Berry, M.V., Quantal phase factors accompanying adiabatic changes. *Proc. Roy. Soc. London* A (1984) **392** 45–57.

[257] Dodonov, V.V. and Man'ko, V.I., Adiabatic invariants, correlated states and Berry's

phase. In: *Topological Phases in Quantum Theory* (B. Markovski and S.I. Vinitsky, eds.), pp.74–83. World Scientific, Singapore, 1989,

[258] Morales, D.A., Correspondence between Berry's phase and Lewis' phase for quadratic Hamiltonians. *J. Phys.* A (1988) **21** L889–L892.

[259] Mizrahi, S.S., The geometrical phase: an approach through the use of invariants. *Phys. Lett.* A (1989) **138** 465–468.

[260] Cerveró, J.M. and Lejarreta, J.D., $SO(2,1)$-invariant systems and the Berry phase. *J. Phys.* A (1989) **22** L663–L666.

[261] Leach, P.G.L., Berry's phase and wavefunctions for time-dependent Hamilton systems. *J. Phys.* A (1990) **23** 2695–2699.

[262] Dittrich, W. and Reuter, M., Berry phase contribution to the vacuum persistence amplitude; effective action approach. *Phys, Lett.* A (1991) **155** 94–98.

[263] Benedict, M.G. and Schleich, W., On the correspondence of semiclassical and quantum phases in cyclic evolutions. *Found. Phys.* (1993) **23** 389–397.

[264] Abe, S., Adiabatic holonomy and evolution of fermionic coherent state. *Phys. Rev.* D (1989) **39** 2327–2331.

[265] *Topological Phases in Quantum Theory* (B. Markovski and S.I. Vinitsky, eds.). World Scientific, Singapore, 1989.

[266] Vinitsky, S.I., Derbov, V.L., Dubovik, V.M., Markovski, B.L., and Stepanovskii, Yu.P., Topological phases in quantum mechanics and polarization optics. *Uspekhi Fiz. Nauk* (1990) **160** 1–49 [*Sov. Phys. – Uspekhi* (1990) **33** 403–428].

[267] Anandan, J., Christian, J., and Wanelik, K., Resource letter GPP-1: Geometric phases in physics. *Am. J. Phys.* (1997) **65** 180–185.

[268] Penning, F.M., Glow discharge at low pressure between coaxial cylinders in an axial magnetic field. *Physica* (1936) **3** 873–894.

[269] Brown, L.S. and Gabrielse, G., Geonium theory: Physics of a single electron or ion in a Penning trap. *Rev. Mod. Phys.* (1986) **58** 233–311.

[270] Dehmelt, H., Less is more: Experiments with an individual atomic particle at rest in free space. *Am. J. Phys.* (1990) **58** 17–27.

[271] Paul, W., Electromagnetic traps for charged and neutral particles. *Rev. Mod. Phys.* (1990) **62** 531–540.

[272] Cook, R.J., Shankland, D.J., and Wells, A.L., Quantum theory of particle motion in a rapidly oscillating field. *Phys. Rev.* A (1985) **31** 564–567.

[273] Brown, L.S., Quantum motion in a Paul trap. *Phys. Rev. Lett.* (1991) **66** 527–529.

[274] Cirac, J.I., Garay, L.J., Blatt, R., Parkins, A.S., and Zoller, P., Laser cooling of trapped ions: The influence of micromotion. *Phys. Rev.* A (1994) **49** 421–432.

[275] Zangg, T., Meystre, P., Lenz, G., and Wilkens, M., Theory of adiabatic cooling in cavities. *Phys. Rev.* A (1994) **49** 3011–3021.

[276] Ammann, H. and Christensen, N., Delta kick cooling: a new method for cooling atoms. *Phys. Rev. Lett.* (1997) **78** 2088–2091.

[277] Peik, E., Electrodynamic trap for neutral atoms. *Eur. Phys. J.* D (1999) **6** 179–183.

[278] Gheorghe, V.N. and Werth, G., Quasienergy states of trapped ions. *Eur. Phys. J.* D (2000) **10** 197–203.

[279] Glauber, R.J., Quantum theory of particle trapping by oscillating fields. In: *Quantum Measurements in Optics. NATO ASI Series B: Physics, vol. 282* (P. Tombesi and D.F. Walls, eds.), pp. 3–14. Plenum, New York, 1992.

[280] Baseia, B. and Bagnato, V., Time-dependent Hamiltonian: Squeezing effects in particle trapping by oscillating fields. *Nuovo Cim.* B (1994) **109** 1129–1134.

[281] Schrade, G., Man'ko, V.I., Schleich, W.P., and Glauber, R.J., Wigner functions in the Paul trap. *Quant. Semiclass. Opt.* (1995) **7** 307–325.

[282] Baseia, B. Vyas, R., and Bagnato, V., Particle trapping by oscillating fields: connecting squeezing with cooling. *Mod. Phys. Lett.* B (1996) **10** 661–669.

[283] Bardroff, P.J., Leichtle, C., Schrade, G., and Schleich, W.P., Endoscopy in the Paul trap: Measurement of the vibratory quantum state of a single ion. *Phys. Rev. Lett.* (1996) **77** 2198–2201.

[284] Hacyan, S., Squeezed states and uncertainty relations in rotating frames and Penning trap. *Phys. Rev.* A (1996) **53** 4481–4487.

[285] Castaños, O., Jáuregui, R., López-Peña, R., Recamier, J., and Manko, V.I., Schrödinger-cat states in Paul traps *Phys. Rev.* A (1997) **55** 1208–1216.

[286] Schrade, G., Bardroff, P.J., Glauber, R.J., Leichtle, C., Yakovlev, V, and Schleich, W.P., Endoscopy in the Paul trap: The influence of the micromotion. *Appl. Phys.* B (1997) **64** 181–191.

[287] Castaños, O., Hacyan, S., López-Peña, R., and Manko, V.I., Schrödinger cat states in a Penning trap. *J. Phys.* A (1998) **31** 1227–1237.

[288] Combescure, M., Crystallization of trapped ions: a quantum approach. *Ann. Phys.* (NY) (1990) **204** 113–123.

[289] Dodonov, V.V., Man'ko, V.I., and Rosa, L., Quantum singular oscillator as a model of a two-ion trap: An amplification of transition probabilities due to small time variations of the binding potential. *Phys. Rev.* A (1998) **57** 2851–2858.

[290] Bogolyubov, N.N. and Tyablikov, S.V., An approximate method of finding the lowest energy levels of electrons in a metal. *Zhurn. Eksp. Teor. Fiz.* (1949) **19** 256–268 (in Russian).

[291] Titulaer, U.M., Ergodic features of harmonic-oscillator systems. I. *Physica* (1973) **70** 257–275.

[292] Tsallis, C., Diagonalization methods for the general bilinear Hamiltonian of an assembly of bosons. *J. Math. Phys.* (1978) **19** 277–286.

[293] Colpa, J.H., Diagonalization of quadratic boson Hamiltonians. *Physica* A (1978) **93** 327–343.

[294] Broadbridge, P. and Hurst, C.A., Canonical forms for quadratic Hamiltonians. *Physica* A (1981) **108** 39–62.

[295] Williamson, J., On the algebraic problem concerning the normal forms of linear dynamical systems. *Amer. J. Math.* (1936) **58** 141–163.

[296] Laub, A.J. and Meyer, K., Canonical forms for symplectic and Hamiltonian matrices. *Celest. Mech.* (1974) **9** 213–238.

[297] Arnold, V.I., *Mathematical Methods of Classical Mechanics*. Springer, Berlin, 1978.

[298] Seleznyova, A.N., Unitary transformations for the time-dependent quantum oscillator. *Phys. Rev.* A (1995) **51** 950–959.

[299] Nieto, M.M. and Truax, D.R., Time-dependent Schrödinger equations having isomorphic symmetry algebras. I. Classes of interrelated equations. *J. Math. Phys.* (2000) **41** 2741–2752.

[300] Tucker, J. and Walls, D.F., Quantum theory of parametric frequency conversion. *Ann. Phys.* (NY) (1969) **52** 1–15.

[301] Lu, E.Y.C., Quantum theory of nonlinear optical processes with time-dependent pump amplitude and phase–frequency conversion. *Phys. Rev.* A (1973) **8** 1053–1061.

[302] Walls, D.F., Quantum statistics of the cyclotron resonance infrared detector. *J. Phys.* A (1975) **8** 751–758.

[303] Estes, L.E., Keil, T.H., and Narducci, L.M., Quantum-mechanical description of two coupled harmonic oscillators. *Phys. Rev.* (1968) **175** 286–299.

[304] Abdalla, M.S., Isotropic time-dependent coupled oscillators. *Phys. Rev.* A (1987) **35** 4160–4166.

[305] Sandulescu, A., Scutaru, H., and Scheid, W., Open quantum system of two coupled harmonic oscillators for application in deep inelastic heavy ion collisions. *J. Phys.* A (1987) **20** 2121–2131.

[306] Yeon, K.H., Um, C.I., Kahng, W.H., and George, T.F., Propagators for driven coupled harmonic oscillators. *Phys. Rev.* A (1988) **38** 6224–6230.

[307] Abdalla, M.S., Anisotropic time-dependent coupled oscillators. *Phys. Rev.* A (1990) **41** 3775–3781.

[308] Kim, Y.S. and Man'ko, V.I., Time-dependent mode coupling and generation of two-mode squeezed states. *Phys. Lett.* A (1991) **157** 226–228.

[309] Celeghini, E., Rasetti, M., and Vitiello, G., Quantum dissipation. *Ann. Phys.* (NY) (1992) **215** 156–170.

[310] Lo, C.F., Time evolution of two coupled general driven time-dependent oscillators. *Nuovo Cim.* B (1995) **110** 1015–1024.

[311] Cervero, J.M. and Lejarreta, J.D., Generalized two-mode harmonic oscillator: $SO(3,2)$ dynamical group and squeezed states. *J. Phys.* A (1996) **29** 7545–7560.

[312] Feranchuk, I.D. and Tolstik, A.L., Operator method for coupled anharmonic oscillators. *J. Phys.* A (1999) **32** 2115–2128.

[313] de Castro, A.S.M. and Dodonov V.V., Quantum coupled oscillators versus forced oscillator. *J. Opt.* B (2001) **3** 228–237.

[314] Bakasov, A.A., Bakasova, N.V., Bashkirov, E.K., and Chmielowski, V., Impossibility of steady squeezing for two-mode linear system without self-action. *J. Phys.* I (France) (1991) **1** 1217–1227.

[315] Peřinová, V., Lukš, A., Křepelka, J., Sibilia, C., and Bertolotti, M., Quantum statistics of light in a lossless linear coupler. *J. Mod. Opt.* (1991) **38** 2429–2457.

[316] Abdalla, M.S., Statistical properties of the time evolution operator for two coupled oscillators. *J. Mod. Opt.* (1993) **40** 1369–1385.

[317] Man'ko, O.V. and Yeh, L., Correlated squeezed states of two coupled oscillators with delta-kicked frequencies. *Phys. Lett.* A (1994) **189** 268–276.

[318] Baseia, B., Dantas, C.A.M., and Bagnato, V., On the quantum noise reduction of a laser-cooled ion coupled to a source oscillator. *Mod. Phys. Lett.* B (1994) **8** 1833–1845.

[319] Abdalla, M.S., Quantum treatment of the time-dependent coupled oscillators. *J. Phys.* A (1996) **29** 1997-2012.

[320] Castaños, O., López-Peña, R., and Manko, V.I., Schrödinger cat states of a non-stationary generalized oscillator. *J. Phys.* A (1996) **29** 2091–2109.

[321] Hacyan, S., Evolution and entanglement of Fock states. *Rev. Mex. Fís.* (1997) **43** 519–526.

[322] Bykov, V.P., Variance of a two-mode field under parametric excitation conditions. *Kvant. Elektron.* (1997) **24** 973–977 [*Quantum Electronics* (1997) **27** 944–948].

[323] Kalmykov, S.Y. and Veisman, M.E., Quantum-statistical properties of two coupled modes of electromagnetic field. *Phys. Rev.* A (1998) **57** 3943–3951.

[324] Karpati, A., Adam, P., Janszky, J., Bertolotti, M., and Sibilia, C., Nonclassical light in complex optical systems. *J. Opt.* B (2000) **2** 133–139.

[325] Ji, J.Y. and Peak, D., The vacuum excitation and squeezing properties of two quantum oscillators with delta-kicked interactions. *J. Phys.* A (2001) **34** 3429–3435.

[326] Senitzky, I.R., Effects of zero-point energy in elementary-system interactions. *Phys. Rev.* A (1998) **57** 40–47.

[327] Janszky, J., Sibilia, C., Bertolotti, M., and Yushin, Y., Non-classical light in a linear

coupler. *J. Mod. Opt.* (1988) **35** 1757–1765.

[328] Chmielowski, W. and Chizhov, A.V., Correlation properties of Bose systems of polariton type. *Teor. Mat. Fiz.* (1991) **86** 285–293 [*Theor. & Math. Phys.* (1991) **86** 196–202].

[329] Bonato, C.A. and Baseia, B., Transference of squeezing in coupled oscillators. *Int. J. Theor. Phys.* (1994) **33** 1445–1460.

[330] Gomes, A.R., Baseia, B., and Marques, G.C., Mutual transfer of squeezing effect in coupled two-photon lasers. *Mod. Phys. Lett.* (1995) **9** 999–1015.

[331] Rodrigues, H., Portes Jr., D., Duarte, S.B., and Baseia, B., Transferring squeezing and statistics in coupled circuits. *Physica* A (2002) **311** 188–198.

[332] de Castro, A.S.M. and Dodonov V.V., Squeezing exchange and entanglement between resonancely coupled modes. *J. Russ. Laser Research* (2002) **23** 93–121.

[333] Fu, J., Gao, X.C., Xu, J.B., and Zou, X.B., Exchange of nonclassical properties between two interacting modes of light and mutual conversion of the Fock and coherent states. *Canad. J. Phys.* (1999) **77** 211–220.

[334] Parkins, A.S. and Kimble, H.J., Quantum state transfer between motion and light. *J. Optics* B (1999) **1** 496–504.

[335] de Oliveira, M.C., Mizrahi, S.S., and Dodonov, V.V., Information transfer in the course of a quantum interaction. *J. Optics* B (1999) **1** 610–617.

[336] de Castro, A.S.M., Dodonov V.V., and Mizrahi, S.S., Quantum state exchange between coupled modes. *J. Opt.* B (2002) **4** S191–S199.

[337] Fan, H.-Y., Squeezing in the triatomic linear molecule model revealed by virtue of IWOP technique. *J. Phys.* A (1993) **26** 151–158.

[338] Abdalla, M.S., Ahmed, M.M.A., and Al-Homidan, S., Quantum statistics of three modes coupled oscillators. *J. Phys.* A (1998) **31** 3117–3139.

[339] El-Orany, F.A.A., Peřina, J., and Abdalla, M.S., Statistical properties of three quantized interacting oscillators. *Phys. Scripta* (2001) **63** 128–140.

[340] Mazur, P. and Montroll, E., Poincaré cycles, ergodicity, and irreversibility in assemblies of coupled harmonic oscillators. *J. Math. Phys.* (1960) **1** 70–84.

[341] Ullersma, P., An exactly solvable model for Brownian motion. III. Motion of a heavy mass in a linear chain *Physica* (1966) **32** 74–89.

[342] Rzążewski, K. and Zakowicz, W., On interaction of harmonic oscillators with radiation field. *Nuovo Cim.* B (1971) **1** 111–122.

[343] Papadopoulos, G.J., Exact nonequilibrium density matrix for a particle in a harmonic chain. *Physica* (1974) **74** 529–545.

[344] Razavy, M., Quantum-mechanical irreversible motion of an infinite chain. *Canad. J. Phys.* (1979) **57** 1731–1737.

[345] Lindenberg, K. and West, B.J., Statistical properties of quantum systems: the linear oscillator. *Phys. Rev.* A (1984) **30** 568–582.

[346] Dodonov, V.V., George, T.F., Man'ko, O.V., Um, C.I., and Yeon, K.H., Propagators for quantum oscillator chains. *J. Sov. Laser Research* (1991) **12** 385–394.

[347] Dodonov, V.V., Man'ko, O.V., and Man'ko, V.I., Correlated states of a quantum oscillator and of a quantum chain of oscillators with a delta-bump in frequency. *J. Sov. Laser Research* (1992) **13** 196–214.

[348] Dodonov, V.V., Man'ko, O.V., and Man'ko, V.I., Nonstationary parametric chain of oscillators. In: *Theory of the Interaction of Multilevel Systems with Quantized Fields. Proc. Lebedev Phys. Inst., vol. 208* (V.I.Man'ko and M.A.Markov, eds.), pp. 179–206. Nauka, Moscow, 1992 [translated by Nova Science, Commack, 1996, as vol. 209, pp. 117–161].

[349] Schwinger, J., Brownian motion of a quantum oscillator. *J. Math. Phys.* (1961) **2**

407–432.

[350] Rau, J., Relaxation phenomena in spin and harmonic oscillator systems. *Phys. Rev.* (1963) **129** 1880–1888.

[351] Louisell, W.H. and Walker, L.P., Density operator theory of harmonic oscillator relaxation. *Phys. Rev.* (1965) **137**B 204–211.

[352] Ford, W.G., Kac, M., and Mazur, P., Statistical mechanics of assemblies of coupled oscillators. *J. Math. Phys.* (1965) **6** 504–515.

[353] Ullersma, P., An exactly solvable model for Brownian motion. II. Derivation of the Fokker–Planck equation and the master equation. *Physica* (1966) **32** 56–73.

[354] Agarwal, G.S., Entropy, the Wigner distribution function, and the approach to equilibrium of a system of coupled harmonic oscillators. *Phys. Rev.* A (1971) **3** 828–831.

[355] Agarwal, G.S., Brownian motion of a quantum oscillator. *Phys. Rev.* A (1971) **4** 739–747.

[356] Braun, E. and Godoy, S.V., Quantum statistical effects of the motion of an oscillator interacting with a radiation field. *Physica* A (1977) **86** 337–354.

[357] Han, D., Kim, Y.S., and Noz, M.E., Illustrative example of Feynman's rest of the universe. *Amer. J. Phys.* (1999) **67** 61–66.

[358] Dodonov, V.V., Klimov, A.B., and Man'ko, V.I., Quantum multidimensional systems with quadratic Hamiltonians. Evolution of distinguished subsystems. In: *Theory of the Interaction of Multilevel Systems with Quantized Fields. Proc. Lebedev Phys. Inst., vol. 208* (V.I.Man'ko and M.A.Markov, eds.), pp. 105–178. Nauka, Moscow, 1992 [translated by Nova Science, Commack, 1996, as vol. 209, pp. 1–115].

[359] *Bateman Manuscript Project, Higher Transcendental Functions* (A. Erdélyi, ed.). McGraw-Hill, New York, 1953.

[360] Dodonov, V.V., Malkin, I.A., and Man'ko, V.I., Coherent states of a charged particle in a time-dependent uniform electromagnetic field of a plane current. *Physica* (1972) **59** 241–256.

[361] Chernikov, N.A., System with Hamiltonian of time dependent quadratic form in x and p. *Zh. Eksp. Teor. Fiz.* (1967) **53** 1006–1017 [*Sov. Phys. – JETP* (1968) **26** 603–608].

[362] Holz, A., N-dimensional anisotropic oscillator in a uniform time-dependent electromagnetic field. *Lett. Nuovo Cim.* (1970) **4** 1319–1323.

[363] Dodonov, V.V., Man'ko, O.V., and Man'ko, V.I., Multidimensional Hermite polynomials and photon distribution for polymode mixed light. *Phys. Rev.* A (1994) **50** 813–817.

[364] Dattoli, G., Torre, A., Lorenzutta, S., and Maino, G., Coupled harmonic oscillators, generalized harmonic-oscillator eigenstates and coherent states. *Nuovo Cim.* B (1996) **111** 811–823.

[365] Kok, P. and Braunstein, S.L., Multi-dimensional Hermite polynomials in quantum optics. *J. Phys.* A (2001) **34** 6185–6195.

[366] Dodonov, V.V. and Man'ko, V.I., New relations for two-dimensional Hermite polynomials. *J. Math. Phys.* (1994) **35** 4277–4294.

[367] Dodonov, V.V., Asymptotic formulae for two-variable Hermite polynomials. *J. Phys.* A (1994) **27** 6191–6203.

[368] Dattoli, G. and Torre, A., Phase space formalism: the generalized harmonic-oscillator functions. *Nuovo Cim.* B (1995) **110** 1197–1212.

[369] Wünsche, A., General Hermite and Laguerre two-dimensional polynomials. *J. Phys.* A (2000) **33** 1603–1629.

[370] Courant, E.D. and Snyder, H.S., Theory of the alternating-gradient synchrotron.

Ann. Phys. (NY) (1958) **3** 1–48.

[371] Lewis, H.R., Jr., Classical and quantum systems with time-dependent harmonic-oscillator-type Hamiltonians. *Phys. Rev. Lett* (1967) **18** 510–512.

[372] Lewis, H.R., Jr. and Riesenfeld, W.B., An exact quantum theory of the time-dependent harmonic oscillator and of a charged particle in a time-dependent electromagnetic field. *J. Math. Phys.* (1969) **10** 1458–1473.

[373] Leach, P.G.L., Quadratic Hamiltonians, quadratic invariants and the symmetry group $SU(n)$. *J. Math. Phys.* (1978) **19** 446–451.

[374] Abe, S. and Ehrhardt, R., Method of invariant in nonstationary field theory. *Z. Phys.* C (1993) **57** 471–474.

[375] Abe, S., Invariants for time-dependent fermion system. *Phys. Lett.* A (1993) **181** 359–365.

[376] Monteoliva, D.B., Mirbach, B., and Korsch, H.J., Global and local dynamical invariants and quasienergy state of time-periodic Hamiltonians. *Phys. Rev.* A (1998) **57** 746–752.

[377] Andrews, M., Invariant operators for quadratic Hamiltonians. *Amer. J. Phys.* (1999) **67** 336–343.

[378] Bergman, E.E. and Holz, A., Exact solutions of an n-dimensional anisotropic oscillator in a uniform magnetic field. *Nuovo Cim.* B (1972) **7** 265–276.

[379] Hall, R.L. and Schwezinger, B., The complete exact solution to the translation-invariant N-body harmonic oscillator problem. *J. Math. Phys.* (1979) **20** 2481–2483.

[380] Moshinsky, M. and Winternitz, P., Quadratic Hamiltonians in phase space and their eigenstates. *J. Math. Phys.* (1980) **21** 1667–1682.

[381] Groenewold, H.J., On the principles of elementary quantum mechanics. *Physica* (1946) **12** 405–460.

[382] Krüger, J.G. and Poffyn, A., Quantum mechanics in phase space. II. Eigenfunctions of the Liouville operator. *Physica* A (1977) **87** 132–144.

[383] Akhundova, E.A., Dodonov, V.V., and Man'ko, V.I., Eigenfunctions of quadratic Hamiltonians in Wigner's representation. *Teor. Mat. Fiz.* (1984) **60** 413–422 [*Theor. & Math. Phys.* (1984) **60** 907–913].

[384] Budanov, V.G., Methods of Weyl representation of the phase space and canonical transformations. 1. *Teor. Mat. Fiz.* (1984) **61** 347–363 [*Theor. & Math. Phys.* (1984) **61** 1183–1195].

[385] Dodonov, V.V. and Man'ko, V.I., Phase space eigenfunctions of multidimensional quadratic Hamiltonians. *Physica* A (1986) **137** 306–316.

[386] Balazs, N.L. and Jennings, B.K., Unitary transformations, Weyl's association and the role of canonical transformations. *Physica* A (1983) **121** 576–586.

[387] Dodonov, V.V., Man'ko, V.I., and Shakhmistova, O.V., Wigner functions of a particle in a time-dependent uniform field. *Phys. Lett.* A (1984) **102** 295–297.

[388] Berry, M.V. and Balazs, N.L., Nonspreading wave packets. *Amer. J. Phys.* (1979) **47** 264–267.

[389] Yan, Y.J. and Mukamel, S., Semiclassical dynamics in Liouville space: Application to molecular electronic spectroscopy. *J. Chem. Phys.* (1988) **88** 5735–5748.

[390] Gantmakher, F.R., *The Theory of Matrices*. Nauka, Moscow, 1966.

[391] Peřinová, V., Křepelka, J., Peřina, J., Lukš, A., and Szlachetka, P., Entropy of optical fields. *Opt. Acta* (1986) **33** 15–32.

[392] Lukš, A. and Peřinová, V., Entropy of shifted Gaussian states. *Czechosl. J. Phys.* (1989) **39** 392–407.

[393] Holevo, A.S., Sohma, M., and Hirota, O., Capacity of quantum Gaussian channels.

Phys. Rev. A (1999) **59** 1820–1828.

[394] Dodonov, V.V., Man'ko, O.V., and Man'ko, V.I., Photon distribution for one-mode mixed light with a generic Gaussian Wigner function. *Phys. Rev.* A (1994) **49** 2993–3001.

[395] Dodonov, A.V. and Dodonov, V.V., Nonstationary Casimir effect in cavities with two resonantly coupled modes. *Phys. Lett.* A (2001) **289** 291–300.

[396] Dodonov, V.V. and Man'ko, V.I., Wigner functions of a damped quantum oscillator. In: *Group Theoretical Methods in Physics. Proc. Second Int. Seminar, Zvenigorod, 1982* (M.A. Markov, V.I. Man'ko and A.E. Shabad, eds.), vol. 2, pp. 109–122. Nauka, Moscow, 1983 (in Russian) [vol. 1, pp. 705–717. Harwood Academic, London, 1985 (in English)].

[397] Dodonov, V.V. and Man'ko, O.V., Quantum damped oscillator in a magnetic field. *Physica* A (1985) **130** 353–366.

[398] Dodonov, V.V. and Man'ko, V.I., Evolution equations for the density matrices of linear open systems. In: *Classical and Quantum Effects in Electrodynamics. Proc. Lebedev Phys. Inst., vol. 176* (A.A.Komar, ed.), pp. 53–60. Nova Science, Commack, 1988.

[399] Lindblad, G., Generators of quantum dynamical semigroups. *Commun. Math. Phys.* (1976) **48** 119–130.

[400] Bausch, R., Bewegungsgesetze nicht abgeschlossener Quantensysteme. *Z. Phys.* (1966) **193** 246–265.

[401] Belavin, A.A., Zel'dovich, B.Y., Perelomov, A.M., and Popov, V.S., Relaxation of quantum systems with equidistant spectra. *Zhurn. Eksp. Teor. Fiz.* (1969) **56** 264–274 [*Sov. Phys. – JETP* 1969 **29** 145–150].

[402] Gorini, V., Kossakowski, A., and Sudarshan, E.C.G., Completely positive dynamical semigroups of N-level systems. *J. Math. Phys.* (1976) **17** 821–825.

[403] Jordan, T.F. and Sudarshan, E.C.G., Dynamical mappings of density operators in quantum mechanics. *J. Math. Phys.* (1961) **2** 772–775.

[404] Jordan, T.F., Pinsky, M.A., and Sudarshan, E.C.G., Dynamical mappings of density operators in quantum mechanics. II. Time dependent mappings. *J. Math. Phys.* (1962) **3** 848–852.

[405] Kraus, K., General state changes in quantum theory. *Ann. Phys.* (NY) (1971) **64** 311–335.

[406] Davies, E.B., *Quantum Theory of Open Systems*. Academic, London, 1976.

[407] Spohn, H., Kinetic equations from Hamiltonian dynamics: Markovian limits. *Rev. Mod. Phys.* (1980) **52** 569–615.

[408] Alicki, R. and Lendi, K., *Quantum Dynamical Semigroups and Applications*. Springer, Berlin, 1987.

[409] Lax, M., Quantum noise. IV. Quantum theory of noise sources. *Phys. Rev.* (1966) **145** 110–129.

[410] Lax, M., Fluctuations and Cooperative Phenomena in Classical and Quantum Physics. In: *1966 Brandeis Summer Lectures Series, Statistical Physics, vol. II* (M. Chretien, ed.), pp. 270–478. Gordon & Breach, New York, 1968.

[411] Haken, H., Cooperative phenomena in systems far from thermal equilibrium. *Rev. Mod. Phys.* (1975) **47** 67–121.

[412] Weidlich, W. and Haake, F., Coherence properties of the statistical operator in a laser model. *Z. Phys.* (1965) **185** 30–47.

[413] Scully, M.O. and Lamb, W.E., Jr., Quantum theory of an optical maser. I. General theory. *Phys. Rev.* (1967) **159** 208–226.

[414] Louisell, W.H. and Marburger, J.H., Solutions of the damped oscillator Fokker–

Planck equation. *IEEE J. Quant. Electron.* (1967) **3** 348–358.

[415] Bonifacio, R. and Haake, F., Quantum mechanical master equation and Fokker–Planck equation for the damped harmonic oscillator. *Z. Phys.* (1967) **200** 526–540.

[416] Glauber, R.J., Coherence and quantum detection. In: *Quantum Optics. Proceedings of the International School of Physics "Enrico Fermi", Course XLII, Varenna, 1967* (R.J. Glauber, ed.), pp. 15–56. Academic, New York, 1969.

[417] Li, K.-H., Physics of open systems. *Phys. Rep.* (1986) **134** 1–85.

[418] Grabert, H., Schramm, P., and Ingold, G.-L., Quantum Brownian motion: the functional integral approach. *Phys. Rep.* (1988) **168** 115–207.

[419] Shanta, P., Chaturvedi, S., Srinivasan, V., and Mancini, F., Time-dependent Bogoliubov transformations and the damped harmonic oscillator. *Mod. Phys. Lett.* A (1993) **8** 1999–2009.

[420] Barchielli, A., Continual measurements for quantum open systems. *Nuovo Cim.* B (1983) **74** 113–137.

[421] Dekker, H. and Valsakumar, M.C., A fundamental constraint on quantum mechanical diffusion coefficients. *Phys. Lett.* A (1984) **104** 67–71.

[422] Sandulescu, A. and Scutaru, H., Open quantum systems and the damping of collective modes in deep inelastic collisions. *Ann. Phys.* (NY) (1987) **173** 277–317.

[423] Isar, A., Sandulescu, A., Scutaru, H., Stefanescu, E., and Scheid, W., Open quantum systems. *Int. J. Mod. Phys.* E (1994) **3** 635–714.

[424] Vacchini, B., Completely positive quantum dissipation. *Phys. Rev. Lett.* (2000) **84** 1374–1377.

[425] Glauber, R.J., Coherent and incoherent states of the radiation field. *Phys. Rev.* (1963) **131** 2766–2788.

[426] Drummond, P.D. and Gardiner, C.W., Generalised P-representations in quantum optics. *J. Phys.* A (1980) **13** 2353–2368.

[427] Bausch, R. and Stahl, A., On description of noise in quantum systems. *Z. Phys.* (1967) **204** 32–46.

[428] Gardiner, C.W. and Collett, M.J., Input and output in damped quantum systems: Quantum stochastic differential equations and the master equation. *Phys. Rev.* A (1985) **31** 3761–3774.

[429] Tombesi, P. and Mecozzi, A., Generation of macroscopically distinguishable quantum states and detection by the squeezed-vacuum technique. *J. Opt. Soc. Am.* B (1987) **4** 1700–1709.

[430] Kennedy, T.A.B. and Walls, D.F., Squeezed quantum fluctuations and macroscopic quantum coherence. *Phys. Rev.* A (1988) **37** 152–157.

[431] Marte, M.A.M., Ritsch, H., and Walls, D.F., Squeezed-reservoir lasers. *Phys. Rev.* A (1988) **38** 3577–3588.

[432] Tombesi, P., Parametric oscillator in a squeezed bath. *Phys. Lett.* A (1988) **132** 241–243.

[433] Ginzel, C., Banacloche, J.G., and Schenzle, A., Statistical properties of a laser with a squeezed reservoir. *Acta Phys. Polon.* A (1990) **78** 123–139.

[434] Hu, B.L. and Matacz, A., Quantum Brownian motion in a bath of parametric oscillators: A model for system-field interactions. *Phys. Rev.* D (1994) **49** 6612–6635.

[435] Zoubi, H. and Ben-Aryeh, Y., The evolution of harmonic oscillator Wigner functions described by the use of group representations. *Quant. Semiclass. Opt.* (1998) **10** 447–458.

[436] Dung, H.T., Joshi, A., and Knöll, L., Field damping in a squeezed thermal reservoir. *J. Mod. Opt.* (1998) **45** 1067–1083.

[437] van der Plank, R.W.F. and Suttorp, L.G., Generalized master equation for systems

in nonideal cavities with squeezed baths. *Eur. Phys. J.* D (1998) **3** 183–193.

[438] Jakob, M., Abranyos, Y., and Bergou, J.A., Quantum measurement apparatus with a squeezed reservoir: Control of decoherence and nonlocality in phase space. *Phys. Rev.* A (2001) **64** 062102.

[439] Dupertuis, M.-A. and Stenholm, S., Rigged-reservoir response. I. General theory. *J. Opt. Soc. Am.* B (1987) **4** 1094–1101.

[440] Dupertuis, M.-A., Barnett, S.M., and Stenholm, S., Rigged-reservoir response. II. Effects of a squeezed vacuum. *J. Opt. Soc. Am.* B (1987) **4** 1102–1108.

[441] Leonhardt, U., Quantum statistics of a two-mode $SU(1,1)$ interferometer. *Phys. Rev.* A (1994) **49** 1231–1242.

[442] Kim, M.S. and Bužek, V., Photon statistics of superposition states in phase-sensitive reservoirs. *Phys. Rev.* A (1993) **47** 610–619.

[443] Dung, H.T. and Knöll, L., Density matrix for photons in a phase-sensitive reservoir. *J. Mod. Opt.* (1999) **46** 859–874.

[444] Golubev, Y.M. and Sokolov, I.V., Photon antibunching in a coherent light source and suppression of the photorecording noise. *Zhurn. Eksp. Teor. Fiz.* (1984) **87** 408–416 [*Sov. Phys. – JETP* (1984) **60** 234–238].

[445] Bergou, J., Davidovich, L., Orszag, M., Benkert, C., Hillery, M., and Scully, M.O., Influence of the pumping statistics in lasers and masers. *Opt. Commun.* (1989) **72** 82–86.

[446] Milburn, G.J., Intrinsic decoherence in quantum mechanics. *Phys. Rev.* A (1991) **44** 5401–5406.

[447] Moya-Cessa, H., Bužek, V., Kim, M.S., and Knight, P.L., Intrinsic decoherence in the atom-field interaction. *Phys. Rev.* A (1993) **48** 3900–3905.

[448] Bužek, V. and Konôpka, M., Dynamics of open systems governed by the Milburn equation. *Phys. Rev.* A (1998) **58** 1735–1739.

[449] Xu, J.-B., Zou, X.-B., and Yu, J.-H., Influence of intrinsic decoherence on nonclassical properties of the two-mode Raman coupled model. *Eur. Phys. J.* D (2000) **10** 295–300.

[450] Bonifacio, R., Time as a statistical variable and intrinsic decoherence. *Nuovo Cim.* B (1999) **114** 473–488.

[451] Mancini, S., Vitali, D., Tombesi, P., and Bonifacio, R., Preserving quantum coherence via random modulation. *J. Opt.* B (2002) **4** S300–S306.

[452] Dodonov, V.V. and Mizrahi, S.S., Stationary states in saturated two-photon processes and generation of phase-averaged mixtures of even and odd quantum states. *Acta Phys. Slovaca* (1998) **48** 349–360.

[453] Alicki, R., Path integrals and stationary phase approximation for quantum dynamical semigroup. Quadratic systems. *J. Math. Phys.* (1982) **23** 1370–1376.

[454] Dodonov, V.V. and Man'ko, V.I., Integrals of motion of pure and mixed quantum systems. *Physica* A (1978) **94** 403–412.

[455] Zel'dovich, B.Y., Perelomov, A.M., and Popov, V.S., Relaxation of quantum oscillator. *Zhurn. Eksp. Teor. Fiz.* (1968) **55** 589–605 [*Sov. Phys. – JETP* (1969) **28** 308–316].

[456] Arnoldus, H.F., Density matrix for photons in a cavity. *J. Opt. Soc. Am.* B (1996) **13** 1099–1106.

[457] Schell, A. and Barakat, R., Approach to equilibrium of single mode radiation in a cavity. *J. Phys.* A (1973) **6** 826–836.

[458] Peřina, J., Peřinová, V., and Mišta, L., Damped solutions for the photon statistics of radiation propagating through a random medium. II. Characteristic functions, quasi-distributions and photon counting statistics. *Czechosl. J. Phys.* B (1974) **24**

482–505.

[459] Rockower, E.B., Abraham, N.B., and Smith, S.R., Evolution of quantum statistics of light. *Phys. Rev.* A (1978) **17** 1100–1112.

[460] Agarwal, G.S., Master equation in phase-space formulation of quantum optics. *Phys. Rev.* (1969) **178** 2025–2035.

[461] Jakeman, E., Statistics of binomial number fluctuations. *J. Phys.* A (1990) **23** 2815–2825.

[462] Caldeira, A.O. and Leggett, A.J., Influence of damping on quantum interference: An exactly soluble model. *Phys. Rev.* A (1985) **31** 1059–1066.

[463] Walls, D.F. and Milburn, G.J., Effect of dissipation on quantum coherence. *Phys. Rev.* A (1985) **31** 2403–2408.

[464] Savage, C.M. and Walls, D.F., Damping of quantum coherence: The master-equation approach. *Phys. Rev.* A (1985) **32** 2316–2323.

[465] Zurek, W.H., Preferred states, predictability, classicality and the environment-induced decoherence. *Prog. Theor. Phys.* (1993) **89** 281–312.

[466] Agarwal, G.S., Master equation methods in quantum optics. In: *Progress in Optics, vol. XI* (E. Wolf, ed.), pp. 1–76. North Holland, Amsterdam, 1973.

[467] Ingarden, R.S. and Kossakowski, A., On the connection of nonequilibrium information thermodynamics with non-Hamiltonian quantum mechanics of open systems. *Ann. Phys.* (NY) (1975) **89** 451–485.

[468] Braunstein, S.L., Damping of quantum superpositions. *Phys. Rev.* A (1992) **45** 6803–6810.

[469] Kim, M.S. and Bužek, V., Schrödinger-cat states at finite temperature: Influence of a finite temperature heat bath on quantum interferences. *Phys. Rev.* A (1992) **46** 4239–4251.

[470] Gallis, M.R., Emergence of classicality via decoherence described by Lindblad operators. *Phys. Rev.* A (1996) **53** 655–660.

[471] Saito, H. and Hyuga, H., Relaxation of Schrödinger cat states and displaced thermal states in a density operator representation. *J. Phys. Soc. Japan* (1996) **65** 1648–1654.

[472] Karrlein, R. and Grabert, H., Exact time evolution and master equations for the damped harmonic oscillator. *Phys. Rev.* A (1997) **55** 153–164.

[473] Kohen, D., Marston, C.C., and Tannor, D.J., Phase space approach to theories of quantum dissipation. *J. Chem. Phys.* (1997) **107** 5236–5253.

[474] Fischer, W., Leschke, H., and Müller, P., On the averaged quantum dynamics by white-noise Hamiltonians with and without dissipation. *Ann. Phys.* (Berlin) (1998) **7** 59–100.

[475] Isar, A., Uncertainty, entropy and decoherence of the damped harmonic oscillator in the Lindblad theory of open quantum systems. *Fortschr. Phys.* (1999) **47** 855–879.

[476] Isar, A., Sandulescu, A., and Scheid, W., Purity and decoherence in the theory of a damped harmonic oscillator. *Phys. Rev.* E (1999) **60** 6371–6381.

[477] Dodonov, V.V., Mizrahi, S.S., and de Souza Silva. A.L., Decoherence and thermalization dynamics of a quantum oscillator. *J. Opt.* B (2000) **2** 271–281.

[478] Arnoldus, H.F., Temporal correlations between photon detections from damped single-mode radiation. *Opt. Commun.* (2000) **182** 381–391.

[479] Marian, P. and Marian, T.A., Environment-induced nonclassical behaviour. *Europ. Phys. J.* D (2000) **11** 257–265.

[480] Turchette, Q.A., Myatt, C.J., King, B.E., Sackett, C.A., Kielpinski, D., Itano, W.M., Monroe, C., and Wineland, D.J., Decoherence and decay of motional quantum states of a trapped atom coupled to engineered reservoirs. *Phys. Rev.* A (2000) **62** 053807.

[481] Davidovich, L., Brune, M., Raimond, J.M., and Haroche, S., Mesoscopic quantum coherences in cavity QED: Preparation and decoherence monitoring schemes. *Phys. Rev.* A (1996) **53** 1295–1309.

[482] *Modern Studies of Basis Quantum Concepts and Phenomena, Proceedings of Nobel Symposium 104, Gimo, Sweden, 1997* (E.B. Karlsson and E. Brändas, eds.), *Phys. Scripta* (1998) **T76** 1–232.

[483] Plenio, M.B. and Knight, P.L., The quantum-jump approach to dissipative dynamics in quantum optics. *Rev. Mod. Phys.* (1998) **70** 101–144.

[484] Dodonov, V.V. and Man'ko, V.I., Universal invariants of quantum systems and generalized uncertainty relations. In: *Group Theoretical Methods in Physics. Proc. Second Int. Seminar, Zvenigorod, 1982* (M.A. Markov, V.I. Man'ko, and A.E. Shabad, eds.), vol. 2, pp. 11–33. Nauka, Moscow, 1983 (in Russian) [vol. 1, pp. 591–612. Harwood Academic, London, 1985 (in English)].

[485] Hernández, E.S. and Remaud, B., Quantal fluctuations and invariant operators for a general time-dependent harmonic oscillator. *Phys. Lett.* A (1980) **75** 269–272.

[486] Turner, R.E. and Snider, R.F., A phase space moment method for classical and quantum dynamics. *Canad. J. Phys.* (1981) **59** 457–470.

[487] Dodonov, V.V., Universal integrals of motion and universal invariants of quantum systems. *J. Phys.* A (2000) **33** 7721–7738.

[488] Robertson, H.P., An indeterminacy relation for several observables and its classical interpretation. *Phys. Rev.* (1934) **46** 794–801.

[489] Barut, A.O. and Rączka, R., *Theory of Group Representations and Applications.* World Scientific, Singapore, 1986.

[490] Dodonov, V.V. and Man'ko, O.V., Universal invariants of paraxial optical beams. In: *Group Theoretical Methods in Physics, Proc. Third Seminar, Yurmala, 1985* (V.V. Dodonov, M.A. Markov, and V.I. Man'ko, eds.), vol. 2, pp. 523–530. VNU Science Press, Utrecht, 1986.

[491] Dragt, A.J., Neri, F., and Rangarajan, G., General moment invariants for linear Hamiltonian systems. *Phys. Rev.* A (1992) **45** 2572–2585.

[492] Dragoman, D., Higher-order moments of the Wigner distribution function in first-order optical systems. *J. Opt. Soc. Am.* A (1994) **11** 2643–2646.

[493] Simon, R., Mukunda, N., and Sudarshan, E.C.G., Partially coherent beams and a generalized $ABCD$-law. *Opt. Commun.* (1988) **65** 322–328.

[494] Bastiaans, M.J., Second-order moments of the Wigner distribution function in first-order optical systems. *Optik* (1991) **88** 163–168.

[495] Kauderer, M., First-order sources in first-order systems: second-order correlations. *Appl. Opt.* (1991) **30** 1025–1035.

[496] Serna, J., Martínez-Herrero, R., and Mejías, P.M., Parametric characterization of general partially coherent beams propagating through $ABCD$ optical systems. *J. Opt. Soc. Am.* A (1991) **8** 1094–1098.

[497] Dattoli, G., Mari, C., Richetta, M., and Torre, A., On the generalized Twiss parameters and Courant–Snyder invariant in classical and quantum optics. *Nuovo Cim.* B (1992) **107** 269–287.

[498] Bastiaans, M.J., $ABCD$ law for partially coherent Gaussian light, propagating through first-order optical systems. *Opt. Quant. Electron.* (1992) **24** S1011–S1019.

[499] Holm, D.D., Lysenko, W.P., and Scovel, J.C., Moment invariants for the Vlasov equation. *J. Math. Phys.* (1990) **31** 1610–1615.

[500] Neri, F. and Rangarajan, G., Kinematic moment invariants for linear Hamiltonian systems. *Phys. Rev. Lett.* (1990) **64** 1073–1075.

[501] Dodonov, V.V. and Man'ko, O.V., Universal invariants of quantum-mechanical and

optical systems. *J. Opt. Soc. Am.* A (2000) **17** 2403–2410.

[502] Butkovskii, A.G. and Samoilenko, Y.I., Control of quantum systems. *Automation and Remote Control* (1979) **40** 485–502, 629–645.

[503] Butkovskii, A.G. and Pustyl'nikova, E.I., Controlling the coherent states of a quantum oscillator. *Automation and Remote Control* (1982) **43** 1393–1398.

[504] Rubin, M.H., On the control of quantum statistical systems. *J. Stat. Phys.* (1982) **28** 177–188.

[505] Belavkin, V.P., Theory of the control of observable quantum systems. *Automation and Remote Control* (1983) **44** 178–188.

[506] Butkovskii, A.G. and Pustyl'nikova, E.I., Control of coherent states of quantum systems with a quadratic Hamiltonian. *Automation and Remote Control* (1984) **45** 1000–1008.

[507] Mielnik, B., Evolution loops. *J. Math. Phys.* (1986) **27** 2290–2306.

[508] Butkovskii, A.G. and Samoilenko, Y.I., *Control of Quantum Mechanical Processes and Systems*. Kluwer, Dordrecht, 1990.

[509] Krause, J.L., Whitnell, R.M. , Wilson, K.R., Yan, Y.J., and Mukamel, S., Optical control of molecular dynamics: molecular cannons, reflectrons, and wave-packet focusers. *J. Chem. Phys.* (1993) **99** 6562–6578.

[510] Fernández, D.J. and Mielnik, B., Controlling quantum motion. *J. Math. Phys.* (1994) **35** 2083–2104.

[511] Messina, M. and Wilson, K.R., A semiclassical implementation of quantum control using Gaussian wave packet dynamics. *Chem. Phys. Lett.* (1995) **241** 502–510.

[512] Cao, J.S., Messina, M., and Wilson, K.R., Quantum control of dissipative systems: Exact solutions. *J. Chem. Phys.* (1997) **106** 5239–5248.

[513] Brown, F.L.H. and Silbey, R.J., Quantum control for arbitrary linear and quadratic potentials *Chem. Phys. Lett.* (1998) **292** 357–368.

[514] Fan, H.Y. and Wünsche, A., Design of squeezing. *J. Opt.* B (2000) **2** 464–469.

[515] Ramakrishna, V., Flores, K.L., Rabitz, H., and Ober, R.J., Quantum control by decompositions of $SU(2)$. *Phys. Rev.* A (2000) **62** 053409.

[516] Dodonov, V.V., Klimov, A.B., and Man'ko, V.I., Generation of squeezed states in a resonator with a moving wall. *Phys. Lett.* A (1990) **149** 225–228.

[517] Lobashov, A.A. and Mostepanenko, V.M., Quantum effects associated with parametric generation of light and the theory of squeezed states. *Teor. Mat. Fiz.* (1991) **88** 340–357 [*Theor. & Math. Phys.* (1991) **88** 913–925].

[518] Dodonov, V.V., Klimov, A.B., and Nikonov, D.E., Quantum phenomena in nonstationary media. *Phys. Rev.* A (1993) **47** 4422–4429.

[519] Artoni, M., Bulatov, A., and Birman, J., Zero-point noise in a nonstationary dielectric cavity. *Phys. Rev.* A (1996) **53** 1031–1035.

[520] Cirone, M., Rzążewski, K., and Mostowski, J. Photon generation by time-dependent dielectric: A soluble model. *Phys. Rev.* A (1997) **55** 62–66.

[521] Artoni, M., Bulatov, A., and Seery, B.D., Nonclassical phase of the electromagnetic field in a nonstationary dielectric. *Phys. Rev.* A (1998) **58** 3345–3348.

[522] Dodonov, V.V. and Andreata, M.A., Squeezing and photon distribution in a vibrating cavity. *J. Phys.* A (1999) **32** 6711–6726.

[523] Mendonça, J.T., Guerreiro, A., and Martins, A.M., Quantum theory of time refraction. *Phys. Rev.* A (2000) **62** 033805.

[524] Dodonov, V.V., Nonstationary Casimir Effect and analytical solutions for quantum fields in cavities with moving boundaries. In: *Contemporary Optics and Electrodynamics, part 1. Advances in Chemical Physics, vol. 119, part 1* (M. Evans, ed.), pp. 309–394. Wiley, New York, 2001.

Chapter 4

Even and odd coherent states and tomographic representation of quantum mechanics and quantum optics

V. I. Man'ko

1 Introduction

One of the aims of this chapter is to give a brief review of properties of even and odd coherent states [1] as the simplest representatives of nonclassical states which have important applications in modern quantum optics. Another goal is to discuss the properties of a new "tomographic representation" of quantum mechanics and classical statistical mechanics. In the quantum domain, it is also called the "probability representation of quantum mechanics", because in this representation the quantum state is described by the conventional positive probability distribution function. We show that the most classical state of quantum mechanics, i.e., coherent state, can also be constructed in the "quantum-like" description of states in classical statistical mechanics.

On the other hand, such nonclassical states as even and odd coherent states are admissible only in the quantum domain. The formal solutions to the evolution equation of classical statistical mechanics, which can be expressed in terms of even and odd coherent states, do not describe real states in classical mechanics, since they result in a negative-valued density in the classical phase space. One can make a remark that Feynman suggested using the notion of negative probability in the quantum domain [2]. To deal with negative probabilities in the classical domain, one also needs to change essentially our understanding of the notion of "chance" in the classical approach and to consider this notion like a temperature in C-scale which can be measured by negative values (see also [3]).

In quantum mechanics, the superposition principle has a key role and the phenomenon of quantum interference of two complex probability amplitudes gives the main qualitative difference of classical and quantum pictures [4]. Within the framework of quantum mechanics, the states known as the closest to classical ones are the coherent states of the harmonic oscillator [5]. In [5], the name "coherent state" was introduced for a state described by the wave function which is a Gaussian packet with equal and minimal admissible dimensionless dispersions of position and momentum. A superposition of these states and its nonclassical properties have been demonstrated in [6]. The simplest representatives of the coherent state superpositions are even and odd coherent states introduced in [1] (see also [7]). In this work, the name "even and odd coherent states" has been given to even and odd superpositions of two Gaussian packets describing coherent states. In [8], these states were discussed as a subclass of some generic set of nonclassical states.

A scheme of generation of even and odd coherent states of a trapped ion has been proposed in [9]. This scheme gives the possibility of studying quantum interference phenomena with essentially higher stability than the realization of the even and odd coherent states in quantum optics [10]. The importance of the even and odd coherent states of the electromagnetic field is also related to the possibilities of reducing the noise influence on the signal in the process of quantum-state signal transmission used in optical communications [11]. Also, the even and odd coherent states might be used as alternatives to squeezed states of light to improve the sensitivity of interferometric gravitational wave detectors [12]. For large amplitudes of the partners of the superposition of two coherent states, these states of light and their slight modifications were interpreted as "Schrödinger cat states" in [13] where their generation due to propagation of initially coherent light in Kerr medium was suggested. One can use generalized correlated states [14] as the partners of the superposition to take into account the influence of mode quadrature correlations on the nonclassical properties of light.

Various types of the "Schrödinger cat states" and different possibilities of their generation were discussed, e.g., in [15–29]. A review of the properties of Schrödinger cat states and other "quantum macroscopic superpositions" was given in [30] (for some modifications and analogues see, e.g., [31]). Due to the simplicity of even and odd coherent states, it is worth studying their nonclassical properties in detail

Also it is interesting to compare the properties of the most classical states, such as coherent states, with the properties of very nonclassical "Schrödinger cat states" within the framework of the unified description in which classical and quantum fluctuations are associated with the same object — tomographic probability distribution. Fluctuations of physical observables and noises in physical processes play an important role in both, the classical and quantum domains. However, there is an essential difference in the nature of classical and quantum fluctuations. The latter cannot be annihilated by any means due to the quantum uncertainty principle (uncertainty relation) [32–34]. In contradistinction to quantum noise, classical fluctuations, in principle, can be reduced to zero (by increasing the accuracy of measurement and by decreasing the temperature).

Traditionally, classical and quantum fluctuations are described in the framework of different mathematical formalisms. Classical systems are described, in the most general case, in terms of nonnegative probability-distribution functions, which obey the classical Liouville equation (we confine ourselves here to closed systems). The quantum fluctuations are associated with the description of pure quantum states in terms of a complex wave function $\psi(x, t)$ satisfying the Schrödinger equation [35] or, in the generic case, in terms of a complex density matrix [36–38] $\rho(x, x', t) = \rho^*(x', x, t)$ satisfying certain restrictions, which ensure the normalization and positive definiteness of the statistical operator.

However, it was shown recently that there exists a possibility of describing quantum systems using only nonnegative "classical" probabilities [39–41]. One of the aims of this chapter is to describe the current status of this new "tomographic" approach to quantum mechanics.

2 Single-mode even and odd coherent states

Usually, the coherent state $|\alpha\rangle$ is defined as [5]

$$|\alpha\rangle = D(\alpha)|0\rangle\,, \tag{1}$$

where the displacement operator $D(\alpha)$ is

$$D(\alpha) = \exp[\alpha a^\dagger - \alpha^* a]. \tag{2}$$

Here α is a complex number, $a, a^\dagger$ are the annihilation and creation operators for the bosonic field mode, and $|0\rangle$ is the vacuum state, i.e., $a|0\rangle = 0$, $\langle 0|0\rangle=1$. The decomposition of the coherent state in terms of number states $|n\rangle$ reads

$$|\alpha\rangle = \exp\left(-|\alpha|^2/2\right)\sum_n \frac{\alpha^n}{\sqrt{n!}}|n\rangle\,. \tag{3}$$

Therefore the photon distribution function for coherent states has the Poissonian form [5]

$$P(n) = \frac{|\alpha|^{2n}}{n!}\,\exp\left(-|\alpha|^2\right). \tag{4}$$

The coherent state $|\alpha\rangle$ is also the eigenstate of the photon annihilation operator with eigenvalue α,

$$a|\alpha\rangle = \alpha|\alpha\rangle\,. \tag{5}$$

The coherent states are minimum uncertainty states. In the coordinate representation, the wave function of the coherent state reads (in dimensionless units)

$$\psi_\alpha(x) = \pi^{-1/4}\exp\left(-\frac{|\alpha|^2}{2} - \frac{x^2}{2} + \sqrt{2}\alpha x - \frac{\alpha^2}{2}\right). \tag{6}$$

In [1,7] it was proposed to consider superpositions of two coherent states $|\alpha\rangle$ and $|-\alpha\rangle$. One of the possible superpositions describes the even coherent state

$$|\alpha_+\rangle = N_+\left(|\alpha\rangle + |-\alpha\rangle\right), \qquad N_+ = \frac{\exp\left(|\alpha|^2/2\right)}{2\sqrt{\cosh|\alpha|^2}}. \tag{7}$$

The odd coherent state is defined as

$$|\alpha_-\rangle = N_-\left(|\alpha\rangle - |-\alpha\rangle\right), \qquad N_- = \frac{\exp\left(|\alpha|^2/2\right)}{2\sqrt{\sinh|\alpha|^2}}. \tag{8}$$

Even/odd coherent states can be generated from a vacuum in the following way

$$|\alpha_\pm\rangle = D(\alpha_\pm)|0\rangle, \tag{9}$$

where even/odd displacement operators read [1,7]

$$D(\alpha_+) = \cosh(\alpha a^\dagger - \alpha^* a), \quad D(\alpha_-) = \sinh(\alpha a^\dagger - \alpha^* a).$$

Both even and odd coherent states are normalized eigenstates of operator a^2

$$a^2|\alpha_\pm\rangle = \alpha^2|\alpha_\pm\rangle, \tag{10}$$

where α is an arbitrary complex number. It is worth mentioning that by applying operator a to an even coherent state $|\alpha_+\rangle$, one obtains an odd coherent state with the same label α, but with a different normalization constant:

$$a|\alpha_+\rangle = \alpha\sqrt{\tanh|\alpha|^2}\,|\alpha_-\rangle. \tag{11}$$

Similarly,

$$a|\alpha_-\rangle = \alpha\sqrt{\coth|\alpha|^2}\,|\alpha_+\rangle. \tag{12}$$

The decomposition of the even and odd coherent states in terms of number states can be obtained by using equation (3)

$$|\alpha_+\rangle = N_+ e^{-|\alpha|^2/2}\sum_n \frac{1+(-1)^n}{\sqrt{n!}}\alpha^n|n\rangle\,, \tag{13}$$

$$|\alpha_-\rangle = N_- e^{-|\alpha|^2/2}\sum_n \frac{1-(-1)^n}{\sqrt{n!}}\alpha^n|n\rangle\,. \tag{14}$$

The even coherent state can only be expressed in terms of even number states and the odd coherent state can only be expressed in terms of odd number states. Consequently, the probability of finding an odd number of photons in the even coherent states equals zero, as does the probability of finding an even number of photons in the odd coherent states. Thus the probability distribution functions for these states strongly oscillate:

$$P_{(+)}(n) = \begin{cases} \dfrac{|\alpha|^{4k}}{\cosh|\alpha|^2(2k)!} & \text{for n=2k} \\ 0 & \text{for n=2k+1} \end{cases},$$

$$P_{(-)}(n) = \begin{cases} 0 & \text{for n=2k} \\ \dfrac{|\alpha|^{2(2k+1)}}{\sinh|\alpha|^2(2k+1)!} & \text{for n=2k+1} \end{cases}.$$

Expectation values of the first-order moments for the annihilation and creation operators in the even and odd coherent states are equal to zero

$$\langle\alpha_\pm|a|\alpha_\pm\rangle = 0\,, \tag{15}$$

due to Eqs. (11), (12), and the orthogonality property $\langle\alpha_\pm|\alpha_\mp\rangle = 0$. The expectation values of the second-order moments are

$$\begin{aligned}
\langle\alpha_\pm|a^2|\alpha_\pm\rangle &= \alpha^2, \\
\langle\alpha_+|a^\dagger a|\alpha_+\rangle &= |\alpha|^2 \tanh|\alpha|^2, \\
\langle\alpha_-|a^\dagger a|\alpha_-\rangle &= |\alpha|^2 \coth|\alpha|^2.
\end{aligned} \tag{16}$$

Defining the quadratures of the electromagnetic field mode as

$$X_1 = \left(a + a^\dagger\right)/2, \quad X_2 = \left(a - a^\dagger\right)/(2i),$$

one obtains the following expressions for their variances in the even coherent state:

$$4\Delta X_1^2 = 2|\alpha|^2 \tanh|\alpha|^2 + 2|\alpha|^2 \cos 2\theta + 1, \tag{17}$$

$$4\Delta X_2^2 = 2|\alpha|^2 \tanh|\alpha|^2 - 2|\alpha|^2 \cos 2\theta + 1. \tag{18}$$

We have introduced the modulus and phase of the complex coherent state amplitude as $\alpha{=}|\alpha|e^{i\theta}$. Eqs. (17) and (18) show that for $\theta{=}\pi/2$ the first quadrature has some amount of squeezing for small values of $|\alpha|$, while for θ=0 the second quadrature is squeezed. There is a possibility of squeezing alternatively in both the quadratures depending upon the phase of the complex amplitude α. For odd coherent states we get the same expressions as in (17) and (18), but $\tanh|\alpha|^2$ should be replaced by $\coth|\alpha|^2$. Also, odd coherent states do not exhibit the property of second-order squeezing.

The variances of the photon number operator $n = a^\dagger a$ for the even and odd coherent states are given by ($\sigma_n \equiv \langle n^2\rangle - \langle n\rangle^2$)

$$\begin{aligned}
\sigma_{n_+} &= \langle\alpha_+|(a^\dagger a)^2|\alpha_+\rangle - \langle\alpha_+|a^\dagger a|\alpha_+\rangle^2, \\
&= |\alpha|^4 + |\alpha|^2 \tanh|\alpha|^2 - |\alpha|^4 \tanh{}^2|\alpha|^2,
\end{aligned} \tag{19}$$

$$\begin{aligned}
\sigma_{n_-} &= \langle\alpha_-|(a^\dagger a)^2|\alpha_-\rangle - \langle\alpha_-|a^\dagger a|\alpha_-\rangle^2, \\
&= |\alpha|^4 + |\alpha|^2 \coth|\alpha|^2 - |\alpha|^4 \coth{}^2|\alpha|^2.
\end{aligned} \tag{20}$$

For large values of $|\alpha|$, they become practically equal: $\sigma_{n_\pm} \approx |\alpha|^2$.

3 Multimode even and odd coherent states

The multimode even and odd coherent states were introduced in [42], and their properties related to the parametric excitation of a multimode oscillator were discussed in [43]. Following these studies, we define the multimode even and odd coherent states (MEOCS) as

$$|\mathbf{A}_\pm\rangle = N_\pm\left(|\mathbf{A}\rangle \pm |-\mathbf{A}\rangle\right), \tag{21}$$

where the multimode coherent state $|\mathbf{A}\rangle$ is

$$|\mathbf{A}\rangle = |\alpha_1, \alpha_2, \ldots, \alpha_n\rangle = D(\mathbf{A})|\mathbf{0}\rangle, \tag{22}$$

and the multimode coherent state is created from multimode vacuum state $|\mathbf{0}\rangle$ by the multimode displacement operator $D(\mathbf{A})$ [5]. The definition of MEOCS is the obvious generalization of the single-mode even and odd coherent state given in [1,7]. The normalization constants are given by

$$N_+ = \frac{\exp\left(|\mathbf{A}|^2/2\right)}{2\sqrt{\cosh(|\mathbf{A}|^2)}}, \quad N_- = \frac{\exp\left(|\mathbf{A}|^2/2\right)}{2\sqrt{\sinh(|\mathbf{A}|^2)}}, \tag{23}$$

where $\mathbf{A} = (\alpha_1, \alpha_2, \ldots, \alpha_n)$ is a complex vector, whose modulus equals

$$|\mathbf{A}|^2 = |\alpha_1|^2 + |\alpha_2|^2 + \cdots + |\alpha_n|^2 = \sum_{m=1}^{n} |\alpha_m|^2. \tag{24}$$

MEOCS can be decomposed into multimode number states as

$$|\mathbf{A}_\pm\rangle = N_\pm e^{-\frac{1}{2}|\mathbf{A}|^2} \sum_n \frac{\alpha_1^{n_1}\cdots\alpha_n^{n_n}}{\sqrt{n_1!\cdots n_n!}}(1 \pm (-1)^{n_1+n_2+\cdots+n_n})|\mathbf{n}\rangle, \tag{25}$$

where $|\mathbf{n}\rangle = |n_1, n_2, \ldots, n_n\rangle$ is the multimode number state.

The following important relations hold for MEOCS:

$$a_i|\mathbf{A}_+\rangle = \alpha_i\sqrt{\tanh(|\mathbf{A}|^2)}\,|\mathbf{A}_-\rangle, \quad a_i|\mathbf{A}_-\rangle = \alpha_i\sqrt{\coth(|\mathbf{A}|^2)}\,|\mathbf{A}_+\rangle. \tag{26}$$

From the above equations it can be easily verified that expectation values of the annihilation and creation operators equal zero for the multimode even and odd coherent states. The second-order statistical moments of the creation/annihilation operators are given by the following expressions:

$$\begin{aligned}
\langle\mathbf{A}_\pm|a_i a_k|\mathbf{A}_\pm\rangle &= \alpha_i\alpha_k, \\
\langle\mathbf{A}_+|a_i^\dagger a_k|\mathbf{A}_+\rangle &= \alpha_i^*\alpha_k \tanh(|\mathbf{A}|^2), \\
\langle\mathbf{A}_-|a_i^\dagger a_k|\mathbf{A}_-\rangle &= \alpha_i^*\alpha_k \coth(|\mathbf{A}|^2).
\end{aligned}$$

The photon number fluctuations are described by means of the (co)variances

$$\sigma_{ik}^\pm = \langle\mathbf{A}_\pm|n_i n_k|\mathbf{A}_\pm\rangle - \langle\mathbf{A}_\pm|n_i|\mathbf{A}_\pm\rangle\langle\mathbf{A}_\pm|n_k|\mathbf{A}_\pm\rangle\,,$$

where $n_i = a_i^\dagger a_i$. The following expressions hold for these specific fourth-order statistical moments:

$$\begin{aligned}
\sigma_{ik}^+ &= |\alpha_i|^2\tanh(|\mathbf{A}|^2)\,\delta_{ik} + |\alpha_i\alpha_k|^2\mathrm{sech}^2(|\mathbf{A}|^2), \\
\sigma_{ik}^- &= |\alpha_i|^2\coth(|\mathbf{A}|^2)\,\delta_{ik} - |\alpha_i\alpha_k|^2\mathrm{cosech}^2(|\mathbf{A}|^2).
\end{aligned}$$

Comparing the variance of the photon number σ_{jj} with its mean value in the same mode $\langle n_j\rangle$, we see that for the even states the photon statistics is always super-Poissonian and for the odd states it is always sub-Poissonian.

The inequality $\sigma_{ik}^\pm \neq 0$ for $i \neq k$ shows the existence of the statistical dependence between different modes in MEOCS, which can be characterized by the correlation coefficients

$$R_{jk}^\pm = \frac{\sigma_{jk}^\pm}{\sqrt{\sigma_{jj}^\pm\sigma_{kk}^\pm}} = \frac{\pm|\alpha_j\alpha_k|}{\sqrt{\left[|\alpha_j|^2 \pm \frac{1}{2}\sinh(2|\mathbf{A}|^2)\right]\left[|\alpha_k|^2 \pm \frac{1}{2}\sinh(2|\mathbf{A}|^2)\right]}}.$$

This coefficient, being exponentially small for $|\mathbf{A}| \gg 1$, becomes essential in the case of small values of $|\mathbf{A}|$. For example, in the two-mode case we have, for $|\alpha_1|^2 + |\alpha_2|^2 \ll 1$, the following interesting relations:

$$R_{12}^{-} \approx -1, \quad R_{12}^{+} \approx \frac{|\alpha_1 \alpha_2|}{\sqrt{(|\alpha_1|^2 + 2|\alpha_2|^2)\,(|\alpha_2|^2 + 2|\alpha_1|^2)}}.$$

We observe almost 100% anticorrelation (which does not depend on the concrete values of the amplitudes $|\alpha_1|$ and $|\alpha_2|$) in the case of *odd two-mode coherent states*. In the case of *even* states, the maximum value $R_{12}^{+} = 1/3$ is achieved for $|\alpha_1| = |\alpha_2|$, whereas $R_{12}^{+} \approx |\alpha_1|/(|\alpha_2|\sqrt{2}) \ll 1$ if $|\alpha_1| \ll |\alpha_2|$.

The probabilities of finding $\mathbf{n} = (n_1, n_2, \ldots, n_n)$ photons (more precisely, n_1 photons in the first mode, n_2 photons in the second mode, and so on) in MEOCS are as follows

$$\begin{aligned} P_+(\mathbf{n}) &= \frac{|\alpha_1|^{2n_1}|\alpha_2|^{2n_2}\cdots|\alpha_n|^{2n_n}}{n_1!n_2!\cdots n_n!\cosh(|\mathbf{A}|^2)}, \quad n_1 + n_2 + \cdots + n_n = 2k, \\ P_-(\mathbf{n}) &= \frac{|\alpha_1|^{2n_1}|\alpha_2|^{2n_2}\cdots|\alpha_n|^{2n_n}}{n_1!n_2!\cdots n_n!\sinh(|\mathbf{A}|^2)}, \quad n_1 + n_2 + \cdots + n_n = 2k+1. \end{aligned}$$

These probabilities are equal to zero if $\sum n_j$ is an odd number for the even states, or if this sum is an even number in the case of odd states. The probability of finding $2k$ photons in all modes (i.e., for all possible combinations of integers $n_1, n_2, \ldots$, satisfying the restriction $\sum n_j = 2k$) of the even coherent state equals

$$P_+(2k) = \frac{|\mathbf{A}|^{4k}}{(2k)!\cosh(|\mathbf{A}|^2)}. \tag{27}$$

Similarly, the probability of finding 2k+1 photons in the odd coherent state equals

$$P_-(2k+1) = \frac{|\mathbf{A}|^{2(2k+1)}}{(2k+1)!\sinh(|\mathbf{A}|^2)}. \tag{28}$$

The statistical operator of MEOCS reads $\rho_\pm = |\mathbf{A}_\pm\rangle\langle\mathbf{A}_\pm|$. Therefore the Husimi–Kano Q-function [44,45] can be calculated as

$$\begin{aligned} Q_+(\mathbf{B},\mathbf{B}^*) &= \langle\mathbf{B}|\rho_+|\mathbf{B}\rangle = 4\,N_+^2 \exp\left(-|\mathbf{A}|^2 - |\mathbf{B}|^2\right)|\cosh(\mathbf{A}\mathbf{B}^*)|^2, \\ Q_-(\mathbf{B},\mathbf{B}^*) &= \langle\mathbf{B}|\rho_-|\mathbf{B}\rangle = 4\,N_-^2 \exp\left(-|\mathbf{A}|^2 - |\mathbf{B}|^2\right)|\sinh(\mathbf{A}\mathbf{B}^*)|^2, \end{aligned}$$

where $|\mathbf{B}\rangle = |\beta_1, \beta_2, \ldots, \beta_n\rangle$ is another multimode coherent state.

The Wigner function of the diadic operator $|\mathbf{A}\rangle\langle\mathbf{B}|$, constructed from the multimode coherent states, was found in [46]

$$W_{\mathbf{A},\mathbf{B}}(\mathbf{q},\mathbf{p}) = 2^N \exp\left[2\,(\mathbf{A}\mathbf{Z}^* + \mathbf{Z}\mathbf{B}^* - \mathbf{Z}\mathbf{Z}^*) - \mathbf{A}\mathbf{B}^* - \tfrac{1}{2}|\mathbf{A}|^2 - \tfrac{1}{2}|\mathbf{B}|^2\right],$$

where $\mathbf{Z} = (\mathbf{q} + i\mathbf{p})\,/\sqrt{2}$,

$$\mathbf{A}\mathbf{Z}^* = \alpha_1 Z_1^* + \alpha_2 Z_2^* + \cdots \alpha_n Z_n^*, \quad \mathbf{Z}\mathbf{Z}^* = Z_1 Z_1^* + Z_2 Z_2^* + \cdots + Z_n Z_n^*.$$

The Wigner function of MEOCS is a superposition of functions $W_{\mathbf{A},\mathbf{B}}(\mathbf{q},\mathbf{p})$ with all possible different combinations of vector labels $\mathbf{A}$ and $-\mathbf{A}$:

$$\begin{aligned} W_{\mathbf{A}}^{(\pm)}(\mathbf{q},\mathbf{p}) &= |N_{\pm}|^2\Big[W_{(\mathbf{A},\mathbf{A})}(\mathbf{q},\mathbf{p}) \pm W_{(\mathbf{A},-\mathbf{A})}(\mathbf{q},\mathbf{p}) \\ &\qquad \pm W_{(-\mathbf{A},\mathbf{A})}(\mathbf{q},\mathbf{p}) + W_{(-\mathbf{A},-\mathbf{A})}(\mathbf{q},\mathbf{p})\Big]. \end{aligned}$$

The evolution of one-mode and multimode even and odd coherent states due to a parametric excitation, action of an external force, or an influence of environment was studied in detail, e.g., in [16,43,47].

Entangled coherent states

Any multimode coherent state is a product of independent coherent states of each mode, thus its photon distribution function is the product of independent Poissonian distribution functions. But in the case of multimode even and odd coherent states we cannot factorize the photon distribution functions due to the presence of non-factorable terms $\cosh|\mathbf{A}|^2$ or $\sinh|\mathbf{A}|^2$ in the right-hand sides of equations (27) and (28). This fact indicates the existence of "statistical entanglement" between different modes in MEOCS.

The family of "entangled coherent states" has much in common with multimode even/odd coherent states, and under certain conditions representatives of one family can be considered as particular cases of another. Considering the tensor product of two Hilbert spaces, one can construct the following most general entangled superposition of four different coherent states:

$$|\psi\rangle = \mathcal{N}\left(|\alpha\rangle_1|\beta\rangle_2 + re^{i\phi}|\gamma\rangle_1|\delta\rangle_2\right), \quad r \geq 0, \tag{29}$$

$$\mathcal{N}^{-2} = 1 + r^2 + 2r\,\mathrm{Re}s, \quad s = e^{i\phi}\langle\alpha|\gamma\rangle\langle\beta|\delta\rangle,$$

$$|s|^2 = \exp(-|\mu|^2 - |\nu|^2), \quad \mu = \alpha - \gamma, \quad \nu = \beta - \delta.$$

Such superpositions and their different specific combinations were studied by many authors, beginning with Tombesi and Mecozzi [48], who considered the following state:

$$|\psi\rangle = \frac{1}{2}\left(e^{-i\pi/4}\left[|\beta\rangle_1 - |-\beta\rangle_1\right]|0\rangle_2 + |0\rangle_1\left[|-i\beta\rangle_2 + |i\beta\rangle_2\right]\right).$$

Sanders [49] introduced the "entangled coherent states", which can be written as

$$|\psi\rangle_S = \mathcal{N}_S\left(|\alpha\rangle_1|\beta\rangle_2 + i|-i\beta\rangle_1|i\alpha\rangle_2\right), \tag{30}$$

whereas Chai [50] considered the states (29) with $r = 1$, $\gamma = \alpha^*$ and $\delta = \beta^*$, as well as the states $|\psi\rangle \sim |\alpha\rangle_1|\alpha\rangle_2 \pm |-\alpha\rangle_1|-\alpha\rangle_2$, which are nothing but the special ("diagonal") case of the even/odd two-mode coherent states.

The cases of $r = 1$, $\beta = \alpha$, $\gamma = \delta = -\alpha^*$ or $\beta = \gamma = -\alpha^*$, $\delta = \alpha$ were analyzed in [51], whereas the case of r arbitrary, but $\gamma = \beta$ and $\delta = \alpha$ was discussed in [52]. The generic case was considered recently in [53,54]. The distinguished role of the *odd* "diagonal" two-mode coherent states in many processes (such as teleportation) was demonstrated recently in [54–56]. For other properties and various applications of even/odd and entangled coherent states see, e.g., [57–65].

4 Quantum mechanics in tomographic picture

During the last 70 years, several different attempts have been launched in order to construct a bridge between the classical and quantum pictures [66–69]. For example, Feynman introduced the path integral method [70]. Wigner [71] proposed a *real* quasiprobability distribution function

$$W(q,p,t) = \int \rho\,(q+u/2\,,q-u/2\,,t)\; e^{-ipu}\,du\,. \tag{31}$$

We use the normalization

$$\int W(q,p,t)\,dq\,dp/(2\pi) = 1\,, \tag{32}$$

so that the "quantum purity parameter" can be expressed as

$$\mu_0 \equiv \mathrm{Tr}\,\widehat{\rho}^2(t) = \int W^2(q,p,t)\,dq\,dp/(2\pi)\,. \tag{33}$$

This parameter is an invariant of the *unitary* evolution. One has $\mu_0 = 1$ for pure states with $\rho_\psi(x,x',t) = \psi(x,t)\psi^*(x',t)$, otherwise $0 < \mu_0 < 1$.

The Wigner function assumes negative values in some regions of the phase space for almost all "nonclassical quantum states", excepting the cases when it has the Gaussian form. For this reason, $W(q,p)$ is not a "true" probability density, so the name "quasiprobability distribution" is associated with it.

4.1 Tomographic-probability distributions

Let us consider, in the one-dimensional case, an operator $\hat{X}$ which is a linear combination of position $\hat{q}$ and momentum $\hat{p}$ operators [72, 73],

$$\hat{X} = \mu\hat{q} + \nu\hat{p}\,, \tag{34}$$

with real parameters μ and ν. Evidently, operator $\hat{X}$ is Hermitian, thus it is related to a measurable observable. The nonnegative probability (marginal) density for the observable (34) is given by

$$w(X,\mu,\nu) = \langle X|\hat{\rho}|X\rangle\,, \tag{35}$$

where $\hat{\rho}$ is the statistical operator, while the eigenstate $|X\rangle$ of operator (34) can be written as

$$|X\rangle = \int dq\,\langle q|X\rangle\,|q\rangle\,, \tag{36}$$

with $|q\rangle$ the position eigenket. The wave function $\langle q|X\rangle$ can be easily calculated by using the equality

$$\langle q|\hat{X}|x\rangle = \langle q|\mu\hat{q}+\nu\hat{p}|X\rangle\,, \tag{37}$$

and then transforming it in a partial differential equation

$$X\,\langle q|X\rangle = \mu q\,\langle q|X\rangle - i\nu\frac{\partial}{\partial q}\langle q|X\rangle\,. \tag{38}$$

The solution is

$$\langle q|X\rangle = (2\pi|\nu|)^{-1/2}\exp\left[i\frac{X}{\nu}q - \frac{i}{2}\frac{\mu}{\nu}q^2\right]. \tag{39}$$

As soon as $\mu \to 1$ and $\nu \to 0$, the wave function (39) tends to $\delta(q-X)$.

Equation (35) can be formally rewritten as

$$w(X,\mu,\nu) = \mathrm{Tr}\left\{\hat{\rho}\,\hat{\Pi}_X(\mu,\nu)\right\}, \tag{40}$$

where the transformed projector is given by

$$\hat{\Pi}_X(\mu,\nu) = \hat{U}(\mu,\nu)\hat{\Pi}_x\,\hat{U}^{-1}(\mu,\nu), \quad \hat{\Pi}_x = |x\rangle\langle x|. \tag{41}$$

The unitary operator $\hat{U}(\mu,\nu)$ is related to the symplectic group representation [74]

$$\hat{U}(\mu,\nu) = \exp\left[\frac{i\phi}{2}\left(\hat{p}^2+\hat{q}^2\right)\right]\exp\left[\frac{i\lambda}{2}\left(\hat{q}\hat{p}+\hat{p}\hat{q}\right)\right], \tag{42}$$

where the rotation and scaling parameters ϕ and λ are expressed in terms of μ and ν as follows

$$\mu = e^{\lambda}\cos\phi, \quad \nu = e^{-\lambda}\sin\phi,$$

$$\phi = \frac{1}{2}\arcsin(2\mu\nu), \quad e^{-2\lambda} = \frac{1+\sqrt{1-4\mu^2\nu^2}}{2\mu^2}.$$

Explicitly, the measurable marginal probability (40) can be expressed in terms of the density matrix elements $\rho(y+\nu k, y) = \langle y+\nu k|\hat{\rho}|y\rangle$ (in the representation of the density matrix over the position eigenkets) as [72]

$$w(X,\mu,\nu) = \int dy\,dk\,\exp\left[-ikX + \frac{i\mu\nu k^2}{2} + iky\mu\right]\rho(y+\nu k, y). \tag{43}$$

It can be written also as the Fourier transform of the characteristic function

$$w(X,\mu,\nu) = \int dk\,e^{-ikX}\langle e^{ik\hat{X}}\rangle, \tag{44}$$

and in terms of the Wigner function [39–41] (we use dimensionless units)

$$w(X,\mu,\nu,t) = \int \delta(X-\mu q-\nu p)W(q,p,t)\,\frac{dq\,dp}{2\pi}. \tag{45}$$

The marginal distribution satisfies the homogeneity properties

$$w(x,\mu,\kappa\nu) = \frac{1}{\kappa}w(x/\kappa,\mu/\kappa,\nu), \quad w(x,\kappa\mu,\nu) = \frac{1}{\kappa}w(x/\kappa,\mu,\nu/\kappa).$$

It is normalized as

$$\int w(X,\mu,\nu,t)\;dX = 1. \tag{46}$$

It is remarkable that function $w(X,\mu,\nu,t)$ is *nonnegative* for any values of real auxiliary variables μ and ν. This is clearly seen from equation (35), and also from the simple formula, which holds in the case of pure states,

$$w(X,\mu,\nu) = \frac{1}{2\pi|\nu|}\left|\int \Psi(y)\exp\left(\frac{i\mu y^2}{2\nu} - \frac{iyX}{\nu}\right) dy\right|^2. \tag{47}$$

Consequently, function $w(X,\mu,\nu,t)$ can be considered as a true (classical) probability distribution [39–41]. Its physical meaning is as follows: it is the probability distribution of the generalized position observable $X(\mu,\nu) = \mu q + \nu p$, measured in different reference frames in the phase space.

Example of Gaussian distributions

A simple example is the single-mode Gaussian state, which is determined completely by the first-order mean values $\langle p\rangle$ and $\langle q\rangle$, by variances σ_{pp} and σ_{qq}, and by covariance σ_{pq}. In this case the Wigner function reads

$$W(q,p) = \frac{1}{\sqrt{\det\sigma}}\exp\left[-\frac{1}{2}(\widetilde{p},\widetilde{q})\,\sigma^{-1}\begin{pmatrix}\widetilde{p}\\ \widetilde{q}\end{pmatrix}\right], \tag{48}$$

where

$$\widetilde{p} = p - \langle p\rangle, \quad \widetilde{q} = q - \langle q\rangle, \quad \sigma = \begin{pmatrix}\sigma_{pp} & \sigma_{pq}\\ \sigma_{pq} & \sigma_{qq}\end{pmatrix}.$$

The purity parameter of the state (48) equals $\mu_0 = (4\det\sigma)^{-1/2}$. The tomographic probability in the Gaussian state has the form

$$w_{\mathrm{G}}(X,\mu,\nu,t) = \frac{1}{\sqrt{2\pi\sigma_X(\mu,\nu,t)}}\exp\left\{-\frac{[X-\bar{X}(\mu,\nu,t)]^2}{2\sigma_X(\mu,\nu,t)}\right\}, \tag{49}$$

where

$$\sigma_X(\mu,\nu,t) = \mu^2\sigma_{qq}(t) + \nu^2\sigma_{pp}(t) + 2\mu\nu\,\sigma_{pq}(t), \tag{50}$$

$$\bar{X}(\mu,\nu,t) = \mu\langle q(t)\rangle + \nu\langle p(t)\rangle. \tag{51}$$

Even and odd coherent states in the tomographic representation

Using the relation between the wave function and the tomogram one can obtain the marginal distribution for the even and odd coherent states

$$w_\pm(X,\mu,\nu) = \frac{N_\pm^2}{\sqrt{\pi(\mu^2+\nu^2)}}\exp\left[-\frac{1}{2}(\alpha+\alpha^*)^2 - \frac{X^2}{\mu^2+\nu^2}\right.$$
$$\left.+\nu\left(\frac{\alpha^2}{\nu-i\mu}+\frac{\alpha^{*2}}{\nu+i\mu}\right)\right]\left|\exp\left(\frac{i\sqrt{2}\alpha X}{i\mu-\nu}\right)\pm\exp\left(-\frac{i\sqrt{2}\alpha X}{i\mu-\nu}\right)\right|^2. \tag{52}$$

It is the image of the nonclassical "Schrödinger cat" state in the probability representation of quantum mechanics. For the harmonic oscillator, the evolution of the tomogram is described by the same formula with the replacement α by αe^{it}.

Symplectic tomography

If one knows function $w\,(X,\mu,\nu,t)$, then the Wigner function can be reconstructed by means of the linear integral transformation [39–41]

$$W(q,p,t) = \int w\,(X,\mu,\nu,t) \exp\left[-i\,(\mu q + \nu p - X)\right] \frac{d\mu\, d\nu\, dX}{2\pi}\,. \tag{53}$$

The relation (43) can be inverted [73] as

$$\hat{\rho} = \int dX\, d\mu\, d\nu\; w(X,\mu,\nu)\, \hat{\mathcal{K}}(X,\mu,\nu)\,, \tag{54}$$

where the kernel operator takes the form

$$\hat{\mathcal{K}}(X,\mu,\nu) = (2\pi)^{-1} \exp\left[-iX + i\mu\nu/2\right] \exp\left(i\mu\hat{q}\right) \exp\left(i\nu\hat{p}\right)\,. \tag{55}$$

The scheme under discussion was called "symplectic tomography" [73], since the linear canonical transformation (34) belongs to the symplectic group $Sp(2,R)$. The description of quantum systems in terms of the density matrices or Wigner functions is completely equivalent to the description in terms of "tomographic probabilities". The advantage of the latter ones consists in their nonnegativity and in a possibility of direct measurement in experiments. A disadvantage is the appearance of extra parameters μ,ν. This is the price one must pay in order to use nonnegative "classical probabilities" in quantum mechanics.

Reconstruction of the wave function

Knowing the statistical operator of the pure quantum state $\hat{\rho}_\psi = |\psi\rangle\langle\psi|$, one can restore the Hilbert-space vector $|\psi\rangle$ (up to a phase factor) through a *linear* transformation [75]

$$|\psi\rangle_r = \int dq\, |q\rangle \frac{\langle q|\hat{\rho}_\psi|r\rangle}{\sqrt{\langle r|\hat{\rho}_\psi|r\rangle}}\,, \tag{56}$$

where r may be an arbitrary point (provided $\langle r|\hat{\rho}_\psi|r\rangle \neq 0$). Using the inverse transformation to (31) ($\hbar = 1$)

$$\rho(q,q') = \frac{1}{2\pi} \int W\left(\frac{q+q'}{2}\,,p\right) e^{ip(q-q')}\, dp\,, \tag{57}$$

one can write the wave function $\psi_r(q)$ explicitly in terms of the Wigner function $W_\psi(q,p)$ in the form

$$\psi_r(q) = \int W_\psi\left(\frac{q+r}{2}\,,p\right) e^{ip(q-r)}\, dp \left[2\pi \int W_\psi(r,p)\, dp\right]^{-1/2}\,. \tag{58}$$

A similar reconstruction formula in terms of the tomographic probability reads [75] (here we choose the special case $r=0$)

$$\psi_0(q) = \frac{\int w_\psi(Y,\mu,q) \exp\left[i\left(Y - \frac{1}{2}\mu q\right)\right]\, d\mu\, dY}{\left[2\pi \int w_\psi(Y,\mu,0)\, e^{iY}\, d\mu\, dY\right]^{1/2}}\,. \tag{59}$$

One should take into account that, for pure states, one has the equality

$$\frac{1}{2\pi}\int w_\psi(X,\mu,\nu)\,w_\psi(Y,-\mu,-\nu)\,e^{i(X+Y)}\,d\mu\,d\nu\,dX\,dY = 1\,, \tag{60}$$

which is another form of the purity condition $\mathrm{Tr}\,\widehat{\rho}^2 = 1$ (for the analysis of the purity conditions in terms of phase-space distributions see, e.g., [76–78]). Methods of reconstruction of the *time-dependent phase* of the wave function (satisfying the Schrödinger equation) from the Wigner function were considered in [76, 79]. Reviews of other methods of detection and quantum state reconstruction in quantum optics can be found in [80, 81].

4.2 Evolution of tomographic probabilities

The quantum Liouville–von Neumann equation for the statistical operator

$$\dot{\widehat{\rho}} + i\left[\widehat{H},\widehat{\rho}\right] = 0 \tag{61}$$

or its counterpart for the density matrix $\rho(x,x',t)$

$$\frac{\partial\rho}{\partial t} = \frac{i}{2m}\left(\frac{\partial^2\rho}{\partial x^2} - \frac{\partial^2\rho}{\partial x'^2}\right) - i\left[V(x) - V(x')\right]\rho(x,x',t) \tag{62}$$

results in the following evolution equation (in dimensionless units) for the tomographic probability of a closed quantum system:

$$\begin{aligned}\dot{w} - \mu\frac{\partial}{\partial\nu}w - i&\left[V\left(-\left(\frac{\partial}{\partial X}\right)^{-1}\frac{\partial}{\partial\mu} - i\frac{\nu}{2}\frac{\partial}{\partial X}\right)\right.\\ &\left.-V\left(-\left(\frac{\partial}{\partial X}\right)^{-1}\frac{\partial}{\partial\mu} + i\frac{\nu}{2}\frac{\partial}{\partial X}\right)\right] w = 0\,.\end{aligned} \tag{63}$$

Actually, this is an integro-differential equation, due to the presence of the operator $(\partial/\partial X)^{-1}$. Introducing the operator

$$\hat{Q} = -\left(\frac{\partial}{\partial X}\right)^{-1}\frac{\partial}{\partial\mu}, \tag{64}$$

one can rewrite equation (63) in the equivalent form

$$\begin{aligned}&\dot{w} - \mu\frac{\partial w}{\partial\nu} - \frac{\partial V}{\partial q}\left(\hat{Q}\right)\nu\frac{\partial}{\partial X}w\\ &= 2\sum_{n=1}^{\infty}(-1)^n\frac{V^{2n+1}\left(\hat{Q}\right)}{(2n+1)!}\left(\frac{\nu}{2}\frac{\partial}{\partial X}\right)^{2n+1} w\,,\end{aligned} \tag{65}$$

where $V^k(x)$ is the k-th derivative of the function $V(x)$. Equations (63) and (65) can be compared with the quantum evolution equation for the Wigner quasidistribution $W(q,p,t)$ [71, 82]

$$\frac{\partial W}{\partial t} = -\frac{p}{m}\frac{\partial W}{\partial q} + \sum_{n=0}^{\infty}\frac{(\hbar/2i)^{2n}}{(2n+1)!}\frac{\partial^{2n+1}V}{\partial q^{2n+1}}\frac{\partial^{2n+1}W}{\partial p^{2n+1}}, \tag{66}$$

which also can be written in the integro-differential form

$$\frac{\partial W}{\partial t} = -\frac{p}{m}\frac{\partial W}{\partial q} + \int_{-\infty}^{\infty} W(q, p+\xi, t) J(q,\xi)\, d\xi\,, \tag{67}$$

where

$$J(q,\xi) = \frac{i}{\pi\hbar^2}\int_{-\infty}^{\infty} dy\, [V(q+y) - V(q-y)] \exp[-2iy\xi/\hbar]\,.$$

4.3 Tomographic propagators

Let $G(x, y, t)$ be the propagator of the Schrödinger equation (considered in detail in Chapter 3), which describes the evolution of the wave function according to the relation

$$\psi(x,t) = \int G(x,y,t)\psi(y,0)\, dy\,. \tag{68}$$

Then (in the case of the unitary evolution) the propagator $K(x, x', y, y', t)$ of equation (62) for the density matrix is given by the product

$$K(x,x',y,y',t) = G(x,y,t)G^*(x',y',t)\,. \tag{69}$$

The evolution equation for the tomographic probabilities (63) is linear; consequently, its solutions can be expressed by means of the "tomographic propagator" $\Pi(y, \mu, \nu, y', \mu', \nu', t)$ as follows:

$$w\,(X,\mu,\nu,t) = \int \Pi\,(X,\mu,\nu,X',\mu',\nu',t)\, w\,(X',\mu',\nu',0)\, dX'\, d\mu'\, d\nu'\,.$$

The tomographic propagator Π is related to the Schrödinger propagator G by means of the transformation

$$\Pi(x,\mu,\nu,x',\mu',\nu',t) = \int k^2 \exp\left[ik\left(x' - x - \frac{k\mu'\nu'}{2} - \mu' z + \mu a\right)\right] \times G\left(a + \frac{k\nu}{2}, z + k\nu', t\right) G^*\left(a - \frac{k\nu}{2}, z, t\right) \frac{dk\, dz\, da}{4\pi^2}\,. \tag{70}$$

In turn, the density matrix propagator $K(x, x', z, z', t)$ can be expressed in terms of $\Pi(y, \mu, x - x', y', \mu', \nu', t)$:

$$K(x,x',z,z',t) = \int \frac{d\mu\, d\mu'\, dy\, dy'\, d\nu'}{(2\pi)^2|\nu'|}\Pi(y,\mu,x-x',y',\mu',\nu',t) \times \exp\left\{i\left(y - \mu\frac{x+x'}{2}\right) - i\frac{z-z'}{\nu'}y' + i\frac{z^2 - z'^2}{2\nu'}\mu'\right\}\,. \tag{71}$$

In the special case of *quadratic Hamiltonians* the right-hand side of the evolution equation (65) disappears, and the last term in the left-hand side becomes proportional to $\nu\,\partial/\partial\mu$. Consequently, the evolution equation turns into the partial differential equation of the first order, whose propagator is reduced to the product of delta-functions (similar to the propagator for the Wigner functions of closed quadratic systems, considered in Chapter 3)

$$\Pi\left(X,\mu,\nu,X',\mu',\nu',t\right)=\delta\left(X-X'+\mathcal{N}\Lambda^{-1}\Delta\right)\delta\left(\mathcal{N}'-\mathcal{N}\Lambda^{-1}\right), \qquad (72)$$

with the vectors $\mathcal{N}=(\nu,\mu)$ and $\mathcal{N}'=(\nu',\mu')$. Here matrix Λ and vector Δ are exactly the same matrix and vector that define linear integrals of motion considered in Chapter 3. Formula (72) holds for multimode systems as well.

For the harmonic oscillator, there exist stationary solutions to the evolution equation (63) in the form

$$\begin{aligned} w_n(X,\mu,\nu) &= \left[\pi\left(\mu^2+\nu^2\right)\right]^{-1/2}2^{-n}(n!)^{-1} \\ &\quad\times\exp\left(-\frac{X^2}{\mu^2+\nu^2}\right)H_n^2\left(\frac{X}{\sqrt{\mu^2+\nu^2}}\right), \end{aligned} \qquad (73)$$

where $H_n(x)$ is the Hermite polynomial.

For more details and concrete examples related to the "probability representation of quantum mechanics" see, e.g., [83–93].

4.4 Classical statistical mechanics in the "quantum" and tomographic representations

The most general state of a classical system is described by a nonnegative distribution function $f(q,p,t)$, normalized as $\int f(p,q,t)\,dq\,dp=1$. Considering $f(q,p,t)$ as an analogue of the Wigner function, one can introduce a complex "density matrix" (cf. [66, 78]; we use dimensionless variables)

$$R(x,x',t)=R^*(x',x,t)=\int f\left(\frac{x+x'}{2},p,t\right)e^{ip(x-x')}\,dp\,, \qquad (74)$$

and a real nonnegative tomographic-probability function [72]

$$w\left(x,\mu,\nu,t\right)=\int f\left(q,p,t\right)\delta(x-\mu q-\nu p)\,dq\,dp\,. \qquad (75)$$

Functions (74) and (75) completely determine the classical distribution function, in view of the relationships

$$f\left(q,p,t\right)=\frac{1}{2\pi}\int R\left(q+\frac{u}{2},q-\frac{u}{2},t\right)e^{-ipu}\,du \qquad (76)$$

and

$$f\left(q,p,t\right)=\frac{1}{4\pi^2}\int w\left(x,\mu,\nu,t\right)e^{i(x-\mu q-\nu p)}\,dx\,d\mu\,d\nu\,. \qquad (77)$$

There exists a specific class of classical distributions, whose "density matrices" can be factorized as $R_\psi(x,x',t)=\psi(x,t)\psi^*(x',t)$. For these distributions, equation (76) takes the form

$$f\left(q,p,t\right)=\frac{1}{2\pi}\int\psi\left(q+\frac{u}{2},t\right)\psi^*\left(q-\frac{u}{2},t\right)e^{-ipu}\,du\,. \qquad (78)$$

The admissible functions $\psi(x,t)$ are those which give a nonnegative probability distribution $f(q,p,t)$ in (78). One of the nesessary conditions of the existence of the representation (78) is the equality

$$\mu_0 = 2\pi \int f^2\,(q,p,t)\;dp\,dq = 1\,. \tag{79}$$

In the simplest case of the Hamiltonian $H = p^2/2m + V(q)$, the probability distribution $f(q,p,t)$ satisfies the Liouville equation

$$\frac{\partial f}{\partial t} + \frac{p}{m}\frac{\partial f}{\partial q} - \frac{\partial f}{\partial p}\frac{\partial V}{\partial q} = 0\,. \tag{80}$$

Given an arbitrary initial distribution function $f_0(q,p)$, the solution to equation (80) reads

$$f(q,p,t) = f_0\left(q_0(q,p,t),\, p_0(q,p,t)\right), \tag{81}$$

where $q_0(q,p,t)$ and $p_0(q,p,t)$ are the integrals of motion which have the initial values $q_0(q,p,0) = q$ and $p_0(q,p,0) = p$. Since the canonical transformation of position and momentum preserves the phase volume, and in view of Eq. (81), the value of the "purity parameter" μ_0 in Eq. (79) is preserved in time for any potential $V(q)$ in (80). Equation (80) can be rewritten in terms of the complex "density matrix" $R(x,x',t)$ as

$$\frac{\partial R}{\partial t} = \frac{i}{2}\left(\frac{\partial^2 R}{\partial x^2} - \frac{\partial^2 R}{\partial x'^2}\right) - i(x-x')V'\left(\frac{x+x'}{2}\right)R(x,x',t)\,. \tag{82}$$

An equivalent equation for the "tomographic probability" distribution $w(x,\mu,\nu,t)$ reads

$$\dot{w} - \mu\frac{\partial}{\partial \nu}w - \frac{\partial V}{\partial q}\left(\hat{Q}\right)\nu\frac{\partial}{\partial X}w = 0\,, \tag{83}$$

where the argument of the function $\partial V/\partial q$ is replaced by the operator (64).

The quantum-like complex wave function can be introduced as a mathematical ansatz to look for the solution of equation (82) in the factorized form. For the Gaussian probability density, such as (48), the factorization can be performed, provided the variance matrix possesses the property $\det\sigma = 1/4$. Then the density-like matrix can be represented as a product of wave functions of squeezed and correlated states. If $\sigma_{pp} = \sigma_{qq} = 1/2$ and $\langle q\rangle = \sqrt{2}\,\mathrm{Re}\,\alpha$, $\langle p\rangle = \sqrt{2}\,\mathrm{Im}\,\alpha$, the complex density-like matrix of the *classical* Gaussian state can be written in the form $\mathcal{R}_\alpha(x,x') = \Psi_\alpha(x)\Psi^*_\alpha(x')$, where the function $\Psi_\alpha(x)$ is given by Eq. (6).

4.5 Discussion

We have shown that states of classical and quantum systems can be described by means of the same "tomographic probability" distribution $w(X,\mu,\nu,t)$. This function is nothing but the nonnegative and normalized probability distribution of position X measured in different reference frames in the phase space; the reference frames being labelled by real parameters μ amd ν [39, 41]. If one knows the tomographic probability, the standard probability distribution $f(q,p,t)$ in the phase space can be reconstructed for classical systems and the Wigner quasidistribution $W(q,p,t)$ can be reconstructed for quantum systems. Classical and quantum tomographic-probability distributions obey different evolution equations, (83) and (65), respectively. However, these equations coincide in the case of quadratic Hamiltonians.

The difference between quantum and classical tomographic probabilities is connected not only with different dynamics, but also with different initial conditions imposed for solving the evolution equations. Some tomographic-probability distributions are admissible only in the quantum domain, some are admissible only in the classical domain. The tomographic probabilities, which are admissible only in the classical domain, have the fluctuations σ_X (50) of the variable X which, for some reference-frame parameters μ and ν, violate the uncertainty relation. The tomographic probabilities, which are admissible only in quantum domain, provide the Wigner function which takes negative values in some points of the phase space. A common subset of the "classical" and "quantum" probability distributions is the family of Gaussian solutions (49) of Eqs. (63) and (83). In the classical domain, these tomographic probabilities are admissible for arbitrary values of the initial dispersion matrix. In the quantum domain, they are admissible only for the initial dispersion matrix, which satisfies the Schrödinger–Robertson uncertainty relation $\sigma_{pp}\sigma_{qq} - \sigma_{pq}^2 \geq 1/4$ (in dimensionless units).

Examples of the tomographic approach to problems of classical statistical mechanics can be found, e.g., in [84, 89]. The tomographic-probability description was successfully applied in quantum mechanics of spinless particles [75, 87, 88, 92, 94], and this approach was extended to the systems with spin [95–102]. The "wave function" factorization $R_\psi(x,x',t) = \psi(x,t)\psi^*(x',t)$ was used for the description of classical electronic beams within the framework of the thermal wave model in [103–106]. Another field of application is the theory of analytic signals [107, 108]. For instance, an invertable map of the tomographic probability onto the wave function of a pure state was found in an explicit form in [109]. For applications to image processing see [110–112].

Acknowledgements

The author would like to acknowledge the Russian Foundation for Basic Research for the partial support under Project No. 99-02-17753.

Bibliography

[1] Dodonov, V.V., Malkin, I.A., and Man'ko, V.I., Even and odd coherent states and excitations of a singular oscillator. *Physica* (1974) **72** 597–615.

[2] Feynman, R.P., Negative probabilities. In: *Quantum Implications, Essays in Honour of David Bohm* (B.J. Hiley and F.D. Peats, eds.), pp. 235–248. Routledge, London, 1987.

[3] Mückenheim, W., A review of extended probabilities. *Phys. Rep.* (1986) **133** 337–401.

[4] Dirac, P.A.M., *The Principles of Quantum Mechanics*, 4th ed., Oxford University Press, Oxford, 1987.

[5] Glauber, R.J., Coherent and incoherent states of the radiation field. *Phys. Rev.* (1963) **131** 2766–2788.

[6] Cahill, K.E. and Glauber, R.J., Density operators and quasiprobability distributions. *Phys. Rev.* (1969) **177** 1882–1902.

[7] Malkin, I.A. and Man'ko, V.I., *Dynamical Symmetries and Coherent States of Quantum Systems*. Nauka, Moscow, 1979 (in Russian).

[8] Nieto, M.M. and Truax, D.R., Squeezed states for general systems. *Phys. Rev. Lett.* (1993) **71** 2843–2846.

[9] de Matos Filho, R.L. and Vogel, W., Even and odd coherent states of the motion of a trapped ion. *Phys. Rev. Lett.* (1996) **76** 608–611.

[10] Haroche, S., Mesoscopic coherences in cavity QED. *Nuovo Cim.* B (1995) **110** 545–556.

[11] Sasaki, M. and Hirota, O., Two examples of measurement processes illustrating Helstrom's optimum decision bound. *Phys. Lett.* A (1996) **210** 21–25.

[12] Ansari, N.A., Di Fiore, L., Man'ko, M.A., Man'ko, V.I., Solimeno, S., and Zaccaria, F., Quantum limits in interferometric gravitational wave antennas in the presence of even and odd coherent states. *Phys. Rev.* A (1994) **49** 2151–2156.

[13] Yurke, B. and Stoler, D., Generating quantum mechanical superpositions of macroscopically distinguishable states via amplitude dispersion. *Phys. Rev. Lett.* (1986) **57** 13–16.

[14] Sudarshan, E.C.G., Chiu, C.B., and Bhamathi, G., Generalized uncertainty relations and characteristic invariants for the multimode states. *Phys. Rev.* A (1995) **52** 43–54.

[15] Brune, M., Haroche, S., Raimond, J.M., Davidovich, L., and Zagury, N., Manipulation of photons in a cavity by dispersive atom-field coupling. Quantum-nondemolition measurements and generation of "Schrödinger cat" states. *Phys. Rev.* A (1992) **45** 5193–5214.

[16] Bužek, V., Vidiella-Barranco, A., and Knight, P.L., Superpositions of coherent states – squeezing and dissipation. *Phys. Rev. A* (1992) **45** 6570–6585.

[17] Reid, M.D. and Krippner, L., Macroscopic quantum superposition states in nondegenerate parametric oscillation. *Phys. Rev.* A (1993) **47** 552–555.

[18] Gerry, C.C. and Hach III, E.E., Generation of even and odd coherent states in a competitive two-photon process. *Phys. Lett.* A (1993) **174** 185–189.

[19] Varada, G.V. and Agarwal, G.S., Quantum-statistical properties of a particle in a double-harmonic-oscillator potential: Generation of Schrödinger-cat states. *Phys. Rev.* A (1993) **48** 4062–4067.

[20] Chumakov, S.M., Klimov, A.B., and Sanchez-Mondragon, J.J., General properties of quantum optical systems in a strong field limit. *Phys. Rev.* A (1994) **49** 4972–4978.

[21] Gerry, C.C., Complementarity and quantum erasure with dispersive atom–field interactions. *Phys. Rev.* A (1996) **53** 1179–1182.

[22] Gerry, C.C., Generation of four-photon coherent states in dispersive cavity QED. *Phys. Rev.* A (1996) **53** 3818–3821.

[23] de Matos Filho, R.L. and Vogel, W., Even and odd coherent states of the motion of a trapped ion. *Phys. Rev. Lett.* (1996) **76** 608–611.

[24] Agarwal, G.S., Puri, R.R., and Singh, R.P., Atomic Schrödinger cat states. *Phys. Rev.* A (1997) **56** 2249–2254.

[25] Gerry, C.C. and Grobe, R., Generation and properties of collective atomic Schrödinger-cat states. *Phys. Rev.* A (1997) **56** 2390–2396.

[26] Vitali, D. and Tombesi, P., Generation and detection of linear superpositions of classically distinguishable states of a radiation mode. *Int. J. Mod. Phys.* B (1997) **11** 2119–2140.

[27] Delgado, A., Klimov, A.B., Retamal, J.C., and Saavedra, C. Macroscopic field superpositions from collective interactions. *Phys. Rev.* A (1998) **58** 655–662.

[28] Moya-Cessa, H., Wallentowitz, S., and Vogel, W., Quantum-state engineering of a trapped ion by coherent-state superpositions. *Phys. Rev.* A (1999) **59** 2920-2925.

[29] De Martini, F., Fortunato, M., Tombesi, P., and Vitali, D., Generating entangled superpositions of macroscopically distinguishable states within a parametric oscillator. *Phys. Rev.* A (1999) **60** 1636–1651.

[30] Bužek, V. and Knight, P.L., Quantum interference, superposition states of light, and nonclassical effects. In: *Progress in Optics, vol. XXXIV* (E. Wolf, ed.), pp. 1–158. North Holland, Amsterdam, 1995.

[31] Spiridonov, V., Universal superpositions of coherent states and self–similar potentials. *Phys. Rev.* A (1995) **52** 1909–1935.

[32] Heisenberg, W., Über den anschaulichen Inhalt der quantentheoretischen Kinematik und Mechanik. *Z. Phys.* (1927) **43** 172–198.

[33] Schrödinger, E., Zum Heisenbergschen Unschärfeprinzip. *Ber. Kgl. Akad. Wiss. Berlin* (1930) **24** 296–303.

[34] Robertson, H.P., A general formulation of the uncertainty principle and its classical interpretation. *Phys.Rev.* (1930) **35** 667.

[35] Schrödinger, E., Quantisierung als Eigenwertproblem. II. *Ann. Phys.* (Leipzig) (1926) **79** 489–527. [Quantization as a problem of proper values. II. In: Schrödinger, E., *Collected papers on wave mechanics*, pp. 13–40. Chelsey, New York, 1978].

[36] Landau, L., Das Dämpfungsproblem in der Wellenmechanik. *Z. Phys.* (1927) **45** 430–441 [The damping problem in wave mechanics. In: *Collected papers of L.D. Landau* (D. Ter Haar, ed.), pp. 8–18. Gordon & Breach, New York, 1965].

[37] von Neumann, J., *Mathematische Grundlagen der Quantenmechanik*. Springer, Berlin, 1932.

[38] Landau, L.D. and Lifshits, E.M., *Quantum Mechanics* (2nd edition), Pergamon, Oxford, 1965.

[39] Mancini, S., Manko, V.I., and Tombesi, P., Symplectic tomography as classical approach to quantum systems. *Phys. Lett.* A (1996) **213** 1–6.

[40] Manko, V.I., Classical formulation of quantum mechanics. *J. Russ. Laser Research* (1996) **17** 579–584.

[41] Mancini, S., Manko, V.I., and Tombesi, P., Classical–like description of quantum dynamics by means of symplectic tomography. *Found. Phys.* **27** (1997) 801–824.

[42] Ansari, N.A. and Man'ko, V.I., Photon statistics of multimode even and odd coherent light. *Phys. Rev.* A (1994) **50** 1942–1945.

[43] Dodonov, V.V., Man'ko, V.I., and Nikonov, D.E., Even and odd coherent states for multimode parametric systems. *Phys. Rev.* A (1995) **51** 3328–3336.

[44] Husimi, K., Some formal properties of the density matrix. *Proc. Phys. Math. Soc. Japan* (1940) **22** 264–314.

[45] Kano, Y., A new phase space distribution function in statistical theory of electromagnetic field. *J. Math. Phys.* (1965) **6** 1913–1915.

[46] Dodonov, V.V. and Man'ko, V.I., *Invariants and the Evolution of Nonstationary Quantum Systems. Proc. Lebedev Phys. Inst., vol. 183* (M.A. Markov, ed.). Nauka, Moscow, 1987 [translated by Nova Science, Commack, 1989].

[47] Dodonov, V.V., Mizrahi, S.S., and de Souza Silva, A.L., Decoherence and thermal-

ization dynamics of a quantum oscillator. *J. Opt.* B (2000) **2** 271–281.

[48] Tombesi, P. and Mecozzi, A., Generation of macroscopically distinguishable quantum states and detection by the squeezed-vacuum technique. *J. Opt. Soc. Am.* B (1987) **4** 1700-1709.

[49] Sanders, B.C., Entangled coherent states. *Phys. Rev.* A (1992) **45** 6811–6815.

[50] Chai, C.L., Two-mode nonclassical state via superpositions of two-mode coherent states. *Phys. Rev.* A (1992) **46** 7187–7191.

[51] Jeong, H., Kim, M.S., and Lee, J., Quantum-information processing for a coherent superposition state via a mixed entangled coherent channel. *Phys. Rev.* A (2001) **64** 052308.

[52] Rice, D.A. and Sanders, B.C., Complementarity and entangled coherent states. *Quant. Semiclass. Opt.* (1998) **10** L41–L47.

[53] Fu, H., Wang, X, and Solomon, A.I., Maximal entanglement of nonorthogonal states: classification. *Phys. Lett.* A (2001) **291** 73–76.

[54] Dodonov, V.V., Castro, A.S.M., and Mizrahi, S.S., Covariance entanglement measure for two-mode continuous variable systems. *Phys. Lett.* A (2002) **296** 73–81.

[55] van Enk, S.J. and Hirota, O., Entangled coherent states: Teleportation and decoherence. *Phys. Rev.* A (2001) **64** 022313.

[56] Wang, X.G. and Sanders, B.C., Multipartite entangled coherent states. *Phys. Rev.* A (2002) **65** 012303.

[57] Huang, H. and Agarwal, G.S., General linear transformations and entangled states. *Phys. Rev.* A (1994) **49** 52–60.

[58] Sanders, B.C., Lee, K.S., and Kim, M.S., Optical homodyne measurements and entangled coherent states. *Phys. Rev.* A (1995) **52** 735–741.

[59] Gerry, C.C., Generation of Schrödinger cats and entangled coherent states in the motion of a trapped ion by a dispersive interaction. *Phys. Rev.* A (1997) **55** 2478–2481.

[60] Gilchrist, A., Deuar, P., and Reid, M.D., Contradiction of quantum mechanics with local hidden variables for quadrature phase measurements on pair–coherent states and squeezed macroscopic superpositions of coherent states. *Phys. Rev.* A (1999) **60** 4259–4271.

[61] Sanders, B.C. and Rice, D.A., Nonclassical fields and the nonlinear interferometer. *Phys. Rev.* A (2000) **61** 013805.

[62] Recamier, J., Castaños, O., Jáuregui, R., and Frank, A., Entanglement and generation of superpositions of atomic coherent states. *Phys. Rev.* A (2000) **61** 063808.

[63] Massini, M., Fortunato, M., Mancini, S., and Tombesi, P., Synthesis and characterization of entangled mesoscopic superpositions for a trapped electron. *Phys. Rev.* A (2000) **62** 041401.

[64] Wang, X.G., Sanders, B.C., and Pan, S.H., Entangled coherent states for systems with $SU(2)$ and $SU(1,1)$ symmetries. *J. Phys.* A (2000) **33** 7451–7467.

[65] Luis, A., Equivalence between macroscopic quantum superpositions and maximally entangled states: Application to phase-shift detection. *Phys. Rev.* A (2001) **64** 054102.

[66] Shirokov, Y.M., Combined algebra for quantum and classical mechanics. *Teor. Mat. Fiz.* (1976) **28** 308–319 [*Theor. & Math. Phys.* (1976) **28** 806–813].

[67] Turner, R.E. and Snider, R.F., Superoperator methods of calculating cross-sections: III. Unified description of classical and quantal scattering. *Can. J. Phys.* (1980) **58** 1171–1182.

[68] Bell, J.S., *Speakable and Unspeakable in Quantum Mechanics*, Cambridge University Press, Cambridge, 1987.

[69] Wang, L. and O'Connell, R.F., Quantum mechanics without wave functions. *Found. Phys.* (1988) **18** 1023–1033.

[70] Feynman, R.P., Space-time approach to non-relativistic quantum mechanics. *Rev. Mod. Phys.* (1948) **20** 367–387.

[71] Wigner, E., On the quantum corrections for thermodynamic equilibrium. *Phys. Rev.* (1932) **40** 749–759.

[72] Mancini, S., Man'ko, V.I., and Tombesi, P., Wigner function and probability distribution for shifted and squeezed quadratures. *Quant. Semiclass. Opt.* (1995) **7** 615–623.

[73] D'Ariano, G.M., Mancini, S., Man'ko, V.I., and Tombesi, P., Reconstructing the density operator by using generalized field quadratures. *Quant. Semiclass. Opt.* (1996) **8** 1017–1027.

[74] Mancini, S., Man'ko, V.I., and Tombesi, P., Different realizations of the tomographic principle in quantum state measurement. *J. Mod. Opt.* (1997) **44** 2281–2292.

[75] Man'ko, V.I., Marmo, G., Sudarshan, E.C.G., and Zaccaria, F., On the relation between Schrödinger and von Neumann equations. *J. Russ. Laser Research* (1999) **20** 421–437.

[76] Takabayasi, T., The formulation of quantum mechanics in terms of ensemble in phase space. *Prog. Theor. Phys.* (1954) **11** 341–373.

[77] Tatarskii, V.I., The Wigner representation of quantum mechanics. *Uspekhi Fiz. Nauk* (1983) **139** 587–619 [Sov. Phys. Usp. (1983) **26** 311–327].

[78] Muga, J.G. and Snider, R.F., Violation of the pure-state condition by the classically evolved Wigner function. *Europhys. Lett.* (1992) **19** 569–573.

[79] Leavens, C.R. and Mayato, R.S., On constructing the wave function of a quantum particle from its Wigner phase-space distribution function. *Phys. Lett.* A (2001) **280** 163–172.

[80] Bužek, V., Adam, G., and Drobný, G., Reconstruction of Wigner functions on different observation levels. *Ann. Phys.* (N.Y.) (1996) **245** 37–97.

[81] Welsch, D.-G., Vogel, W., and Opatrný, T., Homodyne detection and quantum state reconstruction. In: *Progress in Optics, vol. XXXIX* (E. Wolf, ed.), pp. 63–211. North Holland, Amsterdam, 1999.

[82] Moyal, J.E., Quantum mechanics as a statistical theory. *Proc. Camb. Phil. Soc.* (1949) **45** 99–124.

[83] Manko, O.V., Symplectic tomography of nonclassical states of a trapped ion. *J. Russ. Laser Research* (1996) **17** 439–448.

[84] Manko, O. and Manko, V.I., Quantum states in probability representation and tomography. *J. Russ. Laser Research* (1997) **18** 407–444.

[85] Man'ko, V.I. and Safonov, S.S., The damped quantum oscillator and a classical representation of quantum mechanics. *Teor. Mat. Fiz.* (1997) **112** 467–478 [*Theor. & Math. Phys.* (1997) **112** 1172–1181].

[86] Man'ko, V.I. and Safonov, S.S., Quantum damped oscillator in probability representation. *J. Russ. Laser Research* (1997) **18** 537–560.

[87] Man'ko, V.I., Rosa, L., and Vitale, P., Time-dependent invariants and Green functions in the probability representation of quantum mechanics. *Phys. Rev* A (1998) **58** 3291–3303.

[88] Man'ko, V.I., Rosa, L., and Vitale, P., Probability representation in quantum field theory. *Phys. Lett.* B (1998) **439** 328–336.

[89] Manko, O. and Manko, V.I., "Classical" propagator and path integral in the probability representation of quantum mechanics. *J. Russ. Laser Research* (1999) **20**

67–76.

[90] Man'ko, O.V., Classical propagators of quadratic quantum systems. *Teor. Mat. Fiz.* (1999) **121** 285–296 [*Theor. & Math. Phys.* (1999) **121** 1496–1505].

[91] Man'ko, O.V., Classical propagator for an ion in a Penning trap and for Raman scattering process. *Izv. Ross. Akad. Nauk, Ser. Fiz.* (1999) **63** 1095–1100.

[92] Man'ko. V., Moshinsky, M., and Sharma, A., Diffraction in time in terms of Wigner distributions and tomographic probabilities. *Phys. Rev* A (1999) **59** 1809–1815.

[93] Manko, O.V., Classical propagator for quadratic quantum systems. Example of a trapped ion. *Fortschr. Phys.* (2000) **48** 643–647.

[94] Man'ko, V.I., Marmo, G., Sudarshan, E.C.G., and Zaccaria, F., Inner composition law of pure states as a purification of impure states. *Phys. Lett.* A (2000) **273** 31–36.

[95] Dodonov, V.V. and Man'ko, V.I., Positive distribution description for spin states. *Phys. Lett.* A (1997) **229** 335–339.

[96] Manko, V.I. and Manko, O.V., Spin state tomography. *J. Exp. Theor. Phys.* (1997) **85** 430–434.

[97] Man'ko, V.I. and Safonov, S.S., Tomography of quantum states of a symmetric rotor. *Phys. Atom. Nucl.* (1997) **61** 585–591.

[98] Andreev, V.A., Man'ko, V.I., and Safonov, S.S., Spin states and probability distribution functions. *J. Russ. Laser Research* (1998) **19** 340–368.

[99] Andreev, V.A. and Man'ko, V.I., Tomography of two-particle spin states *J. Exp. Theor. Phys.* (1998) **87** 239–245.

[100] Klimov, A.B. and Man'ko, V.I., Symplectic tomography of the Jaynes–Cummings model. *J. Russ. Laser Research* (2000) **21** 205–213.

[101] Terra Cunha, M.O., Man'ko, V.I., and Scully, M.O., Quasiprobability and probability distributions for spin-$^1/_2$ states. *Found. Phys. Lett.* (2001) **14** 103–117.

[102] Klimov, A.B., Man'ko, O.V., Man'ko, V.I., Smirnov, Y.F., and Tolstoy, V.N., Tomographic representation of spin and quark states. *J. Phys.* A (2002) **35** 6101–6123.

[103] Fedele, R. and Man'ko, V.I., Role of semiclassical description in the quantumlike theory of light rays. *Phys. Rev* E (1999) **60** 6042–6050.

[104] Fedele, R., Man'ko, M.A., and Man'ko, V.I., Wave–optics applications in charged–particle–beam transport. *J. Russ. Laser Research* (2000) **21** 1–33.

[105] Fedele, R., Man'ko, M.A., and Man'ko, V.I., Charged–particle–beam propagator in wave electron optics: phase space and tomographic pictures. *J. Opt. Soc. Am.* A (2000) **17** 2506–2512.

[106] De Nicola, S., Fedele, R., and Man'ko, V.I., Classical and quantum–like approaches to charged–particle fluids in a quadrupole. *Phys. Scripta* (2002) **65** 345–349.

[107] Man'ko, M.A., Man'ko, V.I., and Mendes, R.V., Tomograms and other transforms: a unified view. *J. Phys.* A (2001) **34** 8321–8332.

[108] Man'ko, M.A., Tomograms, wavelets, and quasidistributions in the geometric picture. *J. Russ. Laser Research* (2001) **22** 505–533.

[109] Man'ko, V.I. and Mendes, R.V., Noncommutative time–frequency tomography. *Phys. Lett.* A (1999) **263** 53–61.

[110] Man'ko, M.A., Fractional Fourier transform in information processing, tomography of optical signal, and Green function of harmonic oscillator. *J. Russ. Laser Research* (1999) **20** 226–238.

[111] Man'ko, M.A., Fractional Fourier analysis and quantum propagators. In: *Quantum Theory and Symmetries* (H.-D. Doebner, J.-D. Hennig, W. Lücke, and V.K. Dobrev, eds.), pp. 226–231. World Scientific, Singapore, 2000.

[112] Man'ko, M.A., Quasidistributions, tomography, and fractional Fourier transform in signal analysis. *J. Russ. Laser Research* (2000) **21** 411–437.

Chapter 5

The binomial states of light

Antonio Vidiella Barranco

1 Introduction

The quantization of the eletromagnetic field energy was originally proposed by Einstein [1]. This probably constituted the first application of the quantum of energy introduced by Planck a couple of years before. That heuristic quantization of the field led to the explanation of the photoelectric effect, and originated the quantum theory of light. Einstein proposed that the radiation field (of frequency $\nu = \omega/2\pi$) was itself organized as a form of "needles" (*nadelstrahlung*) which would carry a definite amount of energy, $E = \hbar\omega$, where $2\pi\hbar = h$ is Planck's constant. About ten years later, in paper [2] he pointed out that each quantum of radiation must also possess the linear momentum $\vec{p} = \hbar\vec{k}$, where $\vec{k}$ is the field wave vector. Those entities of well-defined energy, or energy quanta, are nowadays called photons (after Lewis [3]). Despite of the success of Einstein's theory, the physics community was reluctant in accepting the concept of photon. The proper treatment of the electromagnetic field within quantum theory, as well as its first application was accomplished by Dirac [4], following the development of quantum theory as a whole, according to which dynamical quantities are associated to operators. The free field (canonical) quantization is accomplished analogously to that of the single harmonic oscillator, based on a pair of non-hermitian operators $\hat{a}$ and $\hat{a}^\dagger$. The Hamiltonian of a single mode of the field of frequency ω is expressed in a very compact form in terms of those operators as [1]

$$\hat{H} = \hbar\omega(\hat{a}^\dagger\hat{a} + \tfrac{1}{2}). \tag{1}$$

The spectrum of $\hat{H}$ is nondegenerate and discrete: $E_n = \hbar\omega(n + 1/2)$, with $n = 0, 1, 2\ldots$, and eigenstates $|n\rangle$ designated as number or Fock states

$$\hat{a}^\dagger\hat{a}|n\rangle = n|n\rangle. \tag{2}$$

The energy separation between two consecutive Fock states is precisely equal to $\hbar\omega$, the energy of one photon, and the Fock states are thus states with a precise

[1] It will be employed the angular frequency ω instead of $\nu = \omega/2\pi$.

number of photons. The operators $\hat{a}^\dagger$ and $\hat{a}$ are such that

$$\hat{a}^\dagger|n\rangle = \sqrt{n+1}|n+1\rangle; \quad \hat{a}|n\rangle = \sqrt{n}|n-1\rangle, \tag{3}$$

i.e., they can be interpreted as photon creation and annihilation operators. Therefore photons are elementary excitations of the radiation field within the quantum theory of radiation. The Fock states $|n\rangle$ arise naturally within the quantum formalism, although their controlled generation as well as a nondemolition detection has only been accomplished recently within high-Q microwave cavities [5]. Light available in nature is essentially thermal light, consisting of an *incoherent* mixture of Fock states, or a statistical mixture, rather than pure Fock states, although atoms undergoing transitions are typical sources of optical (or microwave) photons. The density operator of a single-mode (frequency ω) field in thermal equilibrium at a temperature T can be written as

$$\hat{\rho}_T = \sum_{n=0}^{\infty} P_n^T |n\rangle\langle n|; \qquad P_n^T = \left(\frac{\overline{n}}{1+\overline{n}}\right)^n, \tag{4}$$

or a mixture of Fock states weighted by a geometrical (Bose) distribution with mean photon number

$$\overline{n} = \frac{1}{e^{\hbar\omega/kT} - 1}, \tag{5}$$

k being Boltzmann's constant. The first artificially engineered light, or laser light [6–10], may also be viewed as a mixture of Fock states. After averaging out phases, we have the state for the laser approximately given by an expression analogous to that in equation (4), but with:

$$P_n^L = \frac{\overline{n}^n e^{-\overline{n}}}{n!}, \tag{6}$$

which is a Poisson distribution having a mean photon number

$$\overline{n} = \frac{\beta/C}{1 - \beta/C}, \tag{7}$$

where C is the loss rate in the laser, and β a parameter that depends on the pump rate, the number of atoms in the active medium, atomic dipole moments, the field frequency and decay constants of the levels involved [11]. Due to fact that light is commonly generated by uncontrolled sources (many atoms incoherently emitting many photons), it basically consists of mixtures of Fock states rather than pure Fock states. Nevertheless, there may be defined other states with uncertain number of photons, although pure. One of the most important are the coherent (quasi-classical) states [12, 13]. Those states, introduced in quantum physics by Schrödinger [12], were "revived" by Glauber [13] in the context of quantum optics, precisely because they are the closest pure states to laser light. The coherent states are eigenstates of the annihilation operator $\hat{a}$, or

$$\hat{a}|\alpha\rangle = \alpha|\alpha\rangle, \tag{8}$$

being the amplitude α a complex number. They can be represented as an expansion in the Fock state basis as

$$|\alpha\rangle = \sum_{n=0}^{\infty} \frac{\sqrt{\overline{n}}^n e^{-\overline{n}/2}}{\sqrt{n!}} e^{i\varphi n}|n\rangle; \qquad \overline{n} = |\alpha|^2, \tag{9}$$

and exhibit a Poissonian statistics, $P_n^c = |\alpha|^{2n}e^{-|\alpha|^2}/n!$, such as that of laser light. Distributions are therefore quite important in quantum optics. Both the Poisson distribution (coherent state) and the Kronecker delta $P_n^F = \delta_{nm}$ (Fock state) may be regarded as special cases of the more "basic" binomial distribution [14]. This means that two of the most important quantum states of the field can be expressed using a single photon number distribution function. The study of pure states associated to that distribution, named *binomial states*, is going to be the subject of this chapter. The name "binomial states" was first introduced in quantum optics by Stoler, Saleh and Teich in 1985 [15]. These states are interesting themselves not only because of their interpolating properties, but also due to the interesting statistical properties they might show. The single-mode binomial states of the quantized electromagnetic field can be defined [15] in terms of the Fock state basis as

$$|p, M; \theta_n\rangle = \sum_{n=0}^{M} e^{i\theta_n} B_n^M \, |n\rangle, \tag{10}$$

where

$$B_n^M = \left[\frac{M!}{n!\,(M-n)!} p^n (1-p)^{M-n}\right]^{1/2}, \tag{11}$$

p and M being real parameters and θ_n relative phases. Their photon number distribution

$$P_m^B = |\langle m|p, M; \theta_n\rangle|^2 = \frac{M!}{n!\,(M-n)!} p^n (1-p)^{M-n}$$

is simply a binomial distribution, i.e., there is a probability P_m^B of occurrence of m photons (each one with probability p), having M independent ways of doing it. It is interesting to note that there is a maximum permissible number of photons, M, in a field prepared in a binomial state, i.e., $P_m^M = 0$ for $m > M$. This fact will be discussed in more detail, especially regarding generation schemes. For simplicity I consider here the case of real expansion coefficients, or $\theta_n = 0$ for any n. [2]

From equations (10) and (11), we can see that given any (finite) M, if $p = 0$, $|p, M\rangle$ is reduced to the vacuum state $|0\rangle$. Moreover, for $p = 1$, we obtain the Fock state containing M photons, $|n\rangle = |M\rangle$. Now in the limit $p \to 0$ and $M \to \infty$, but with $pM = \alpha^2$ constant, the binomial distribution turns into a Poisson distribution, and $|p, M\rangle$ becomes a coherent state $|\alpha\rangle$ having a real amplitude α. Binomial states belong therefore to the class of *intermediate states*, and it would be interesting to investigate their properties as we vary the values of p and M.

[2] Henceforth the binomial states will be denoted as $|p, M; \theta_n = 0\rangle \equiv |p, M\rangle$.

The binomial states obey the following eigenvalue equation [16]

$$\hat{B}|p, M\rangle = \sqrt{p}M|p, M\rangle, \tag{12}$$

$$\hat{B} = \sqrt{p}\hat{a}^\dagger\hat{a} + \sqrt{1-p}\sqrt{M\hat{I} - \hat{a}^\dagger\hat{a}}\ \hat{a} \tag{13}$$

$\hat{I}$ being the identity operator. The application of the ladder operator formalism leads to the conclusion that binomial states are in fact special SU(2) coherent states [16, 17]. Of course equation (13) reduces to the eigenvalue equations (2) and (8) in the appropriate limits.

This chapter is organized as follows: in section 2, the general properties of binomial states will be presented, focusing especially on their nonclassical features. In section 3, we will discuss the basic behaviour of fields prepared in binomial states while interacting with matter within a simple model of quantum optical resonance. The loss of coherence of binomial states confined in a lossy cavity is the subject of section 4. In section 5, we will present different proposals for the generation of quantum states, particularly the binomial states. In section 6, some other intermediate states related to binomial states will be briefly introduced. Section 7 concludes the chapter.

2 Nonclassical properties

Quantum states of light can be classified according to their statistical properties. They are usually compared to a reference state, namely, the coherent state. The reason for that is the proximity of coherent states to classical stable waves. If we consider an optical, single-mode cavity of volume V, the electric field of the mode can be expressed in terms of the creation and annihilation operators as (at $t = 0$, for instance)

$$\hat{E}(z) = \sqrt{\frac{\hbar\omega}{\epsilon_0 V}}\,(\hat{a} + \hat{a}^\dagger)\,\sin kz, \tag{14}$$

We easily see that for a Fock state, $\langle\hat{E}\rangle = \langle n|\hat{E}|n\rangle = 0$, for any number of photons n. This does not have any classical correspondence, because a classical stable wave has $\overline{E} \neq 0$ [11]. Nevertheless, for a coherent state $|\alpha\rangle$

$$\langle\alpha|\hat{E}|\alpha\rangle = \sqrt{\frac{\hbar\omega}{\epsilon_0 V}}\,\mathrm{Re}\,\alpha, \tag{15}$$

which means that it can be viewed as a "quasiclassical" state [12]. From completely different arguments, it is possible to reach the same conclusion, as I will discuss in section 4. Thus, our reference state will be the coherent state, and other states of the field will be here named *nonclassical states*. Examples of important nonclassical states of the field are the Fock states, the squeezed states [18, 19], photon-added states [20, 21], the Schrödinger cat states [22], and the binomial states [15]. The interpolating properties of the binomial states will allow us to tune amongst different states, which, in their turn, will exhibit diverse nonclassical properties. Here I will present the most relevant statistical properties regarding the binomial states, such as noise reduction in convenient variables, e.g., reduction in quadrature noise (squeezing) as well as in photon number noise (sub-Poissonian character).

2.1 Statistical properties

The quadrature operators, here denoted as $\hat{X}$ and $\hat{Y}$, are defined as:

$$\hat{X} = \left(\hat{a} + \hat{a}^\dagger\right)/2, \qquad \hat{Y} = \left(\hat{a} - \hat{a}^\dagger\right)/(2i). \tag{16}$$

The quadrature operators do not commute, i.e., $[\hat{X}, \hat{Y}] = i/2$, and as a consequence their variances obey the uncertainty relation:

$$\langle \Delta\hat{X}^2 \rangle \langle \Delta\hat{Y}^2 \rangle \geq 1/16. \tag{17}$$

Second-order (or ordinary) squeezing occurs [18, 19] if any of the quadratures present reduction in their second-order moments below the coherent state level, i.e., either $\langle \Delta\hat{X}^2 \rangle < 1/4$ or $\langle \Delta\hat{Y}^2 \rangle < 1/4$. For a reference (coherent) state, $\langle \Delta\hat{X}^2 \rangle = \langle \Delta\hat{Y}^2 \rangle = 1/4$. The equality in equation (17) stands for minimum uncertainty states. It is convenient to define a second-order squeezing index as:

$$S_x^{(2)} = 4\langle \Delta\hat{X}^2 \rangle - 1. \tag{18}$$

If $S_x < 0$ ($S_y < 0$), there will be (ordinary or second-order) squeezing in the $\hat{X}$ ($\hat{Y}$) quadrature. Fock and coherent states do not exhibit quadrature squeezing.

For the binomial state $|p, M\rangle$ the squeezing indexes are given by [23]

$$S_x^{(2)} = 2pM + 2p[M(M-1)]^{1/2} \sum_{n=0}^{M-2} B_n^M B_n^{M-2} - 4pM \Big(\sum_{n=0}^{M-1} B_n^M B_n^{M-1} \Big)^2,$$

$$S_y^{(2)} = 2pM - 2p[M(M-1)]^{1/2} \sum_{n=0}^{M-2} B_n^M B_n^{M-2}.$$

Another quantity useful for characterizing nonclassical behaviour is Mandel's Q parameter, [3] defined as:

$$Q = \frac{\langle \Delta\hat{n}^2 \rangle - \langle \hat{n} \rangle}{\langle \hat{n} \rangle}. \tag{19}$$

If $Q = 0$, as in the case of a coherent state, the photon statistics is Poissonian. If $Q > 0$ ($Q < 0$), we have a super(sub)-Poissonian field, respectively. A signature of nonclassical behaviour, for instance, would be a distribution narrower than a Poissonian one, i.e., when $Q < 0$. For the binomial states, we have $\langle \hat{n} \rangle = pM$ and $\langle \Delta\hat{n}^2 \rangle = p(1-p)M$, so that

$$Q = \frac{p(1-p)M - pM}{pM} = -p. \tag{20}$$

[3]Mandel's parameter is related to the second-order field correlation function, $g^{(2)}(0)$ as $Q = \langle \hat{n} \rangle [g^{(2)}(0) - 1]$.

This means that binomial states are basically sub-Poissonian, for any value of p, and independently of M, except in the coherent state limit (also including the vacuum state). The extreme case of sub-Poissonian field is the Fock state one, which has $Q = -1$.

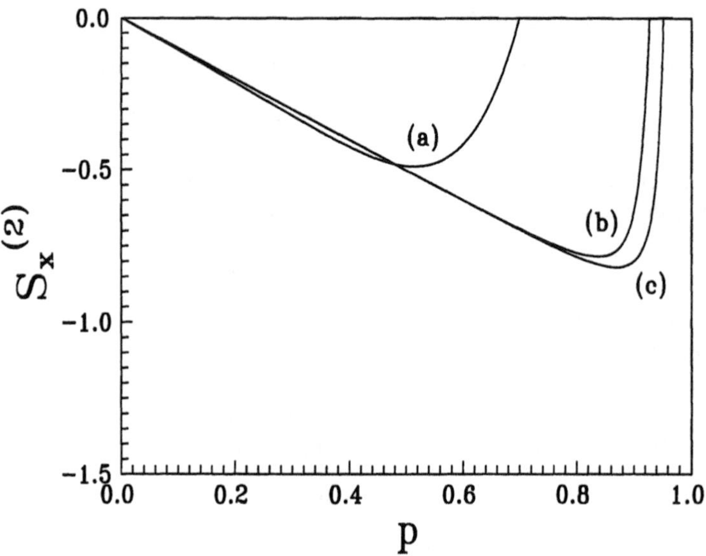

Figure 1: Squeezing index of the $\hat{X}$ quadrature as a function of p for a binomial state having: (a) $M = 5$, (b) $M = 50$, (c) $M = 100$.

In Figure 1, there is a plot of $S_x^{(2)}$ as a function of p for different values of M, showing that the binomial state is squeezed within a considerable range of values of p, with a maximum of squeezing (minimum of S_x) that depends on M. There is obviously no squeezing when $p = 0$ (vacuum) and when $p \to 1$ (Fock state). The binomial states are not minimum uncertainty states, except in the coherent state limit ($M \to \infty$), which also includes the vacuum state as a special case.

2.2 Phase properties

In order to complement the characterization of the binomial states, I would like now to discuss their phase properties using the Pegg–Barnett Hermitian phase operator [24–26]. The Pegg–Barnett formalism is based on the "construction" of states of well-defined phase, or phase states $|\theta\rangle$, in a finite-dimensional Hilbert space (of dimension $d + 1$):

$$|\theta_m\rangle = \frac{1}{(d+1)^{1/2}} \sum_{n=0}^{d} \exp(in\theta_m)|n\rangle. \tag{21}$$

The particular phase states in equation (21) form an orthonormal set provided

$$\theta_m = \theta_0 + 2\pi m/(d+1), \qquad m = 0, 1, ..., d, \tag{22}$$

where θ_0 is an arbitrary reference phase. After calculating the relevant expectation values, the limit $d \to \infty$ has to be taken in order to be consistent with the quantum mechanical formalism, in which an infinite state space is used.

A Hermitian phase operator $\hat{\Phi}_\theta$ can be constructed using the phase states themselves, or

$$\hat{\Phi}_\theta = \sum_{m=0}^{d} \theta_m |\theta_m\rangle\langle\theta_m|. \tag{23}$$

It is also possible to define a continuous phase probability distribution for a general field state described by $\hat{\rho}$ as:

$$dP(\theta) = \langle\theta_m|\hat{\rho}|\theta_m\rangle d\theta. \tag{24}$$

It is not difficult to see that for a binomial state the phase probability distribution can be written as:

$$P(\theta) = \frac{1}{2\pi}\left\{1 + 2\sum_{n>n'} B_n^M B_{n'}^M \cos[(n-n')\theta]\right\}, \tag{25}$$

where $\theta_0 = 0$ has been chosen. Now we may analyze what happens with the phase probability distribution as we vary p. A plot of $P(\theta)$ as a function of p and θ is shown in Figure 2, after a numerical evaluation of equation (25), with $M = 1$. We notice that for both the vacuum case $p = 0$ and the Fock state case $p = 1$, the

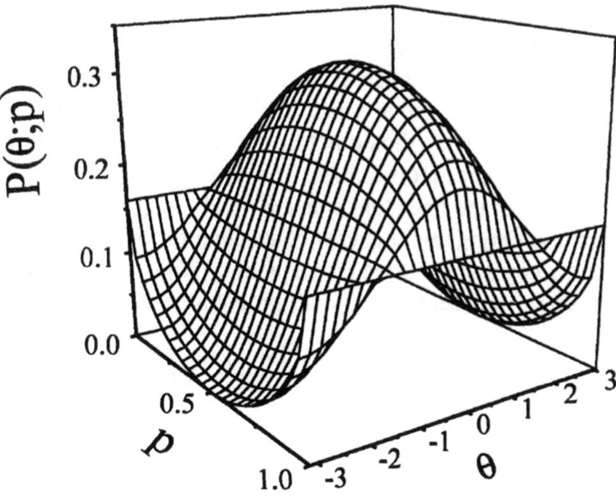

Figure 2: Phase probability distribution of a binomial state having $M = 1$ as a function of p and θ.

phase is uniformly distributed, with probability equal to $1/2\pi$. For intermediate values of p, a peak centred at $\theta = 0$ starts being shaped, meaning that the phase becomes better defined around its mean value.

2.3 Quasiprobabilities

The formulation of quantum mechanics in phase space [27] has shown itself particularly useful in quantum optics. The quasiprobability distributions in the coherent state basis have been providing not only an alternative way of calculating quantum mechanical expectation values, but also useful pictures and explanations of nonclassical features of light fields [28–30] such as squeezing, for instance. Moreover, quasiprobabilities are of central importance in the field of quantum state reconstruction [31–33]. In fact, what is obtained in reconstruction schemes are the quasiprobability distributions themselves rather than the density operator (or the state vector).

If a field is prepared in a quantum state described by $\hat{\rho}$, we can define an infinite number of (s-parametrized) quasiprobability distributions in phase space as [34]

$$P(\beta; s) = \frac{1}{\pi^2} \int d^2\xi \, C(\xi; s) \, \exp(\beta\xi^* - \beta^*\xi), \tag{26}$$

where the quantum characteristic function is

$$C(\xi; s) = \mathrm{Tr}[\hat{D}(\xi)\hat{\rho}] \exp(s|\xi|^2/2). \tag{27}$$

Here $\beta = x + iy$, with (x, y) being c-numbers, corresponding to the quadratures $(\hat{X}, \hat{Y})$, respectively, and

$$\hat{D}(\xi) = \exp(\xi\hat{a}^\dagger - \xi^*\hat{a})$$

is Glauber's displacement operator [13]. For particular values of s we obtain the well-known distributions, e.g., for $s = 0$ we have the Wigner function, and for $s = -1$ the (Husimi) Q function. All distributions are plagued with some kind of problem, such as negativity, and this is generally regarded as a signature of nonclassical effects. They should not be considered true probability distributions, hence the name quasiprobabilities. The Wigner function, associated to the symmetric ordering of the operators $\hat{a}$ and $\hat{a}^\dagger$, and the Q-function, associated to their anti-normal ordering, are the most convenient for our purposes. Instead of the phase space integration method (see equation (26), I am going to use the (s-parametrized) series representation of quasiprobabilities, given by [35]

$$P(\beta; s) = \frac{2}{\pi} \sum_{k=0}^{\infty} (-1)^k \frac{(1+s)^k}{(1-s)^{k+1}} \langle \beta, k|\hat{\rho}|\beta, k\rangle, \tag{28}$$

where $|\beta, k\rangle = \hat{D}(\beta)|k\rangle$ are the displaced Fock states. The expression above is suitable for straightforward numerical evaluation. If we take $s = -1$, for instance, equation (28) becomes the familiar expression for the Q-function

$$Q(\beta) = \frac{1}{\pi} \langle \beta|\hat{\rho}|\beta\rangle. \tag{29}$$

By taking $s = 0$ in equation (28), we obtain a series representation for the Wigner function

$$W(\beta) = \frac{2}{\pi} \sum_{k=0}^{\infty} (-1)^k \langle \beta, k|\hat{\rho}|\beta, k\rangle. \tag{30}$$

Now we simply insert equation (10) into (30), which yields

$$W(\beta) = \frac{2}{\pi} \sum_{k=0}^{\infty} \left| \sum_{n=0}^{M} \left[\frac{M!}{(M-n)!} p^n (1-p)^{M-n} \right]^{1/2} \chi_{nk}(\beta) \right|^2 . \tag{31}$$

In the expression above, the matrix elements $\chi_{nk}(\beta) = \langle n|\hat{D}|k\rangle$ are

$$\chi_{nk}(\beta) = \begin{cases} \sqrt{\frac{k!}{n!}} \exp(-|\beta|^2/2) \beta^{n-k} \mathcal{L}_k^{n-k}(|\beta|^2) & \text{if } n \geq k \\ \sqrt{\frac{n!}{k!}} \exp(-|\beta|^2/2) (\beta^*)^{k-n} \mathcal{L}_n^{k-n}(|\beta|^2) & \text{if } n \leq k \end{cases} , \tag{32}$$

where $\mathcal{L}_n^{\alpha}(|\beta|^2)$ are the generalized Laguerre polynomials.

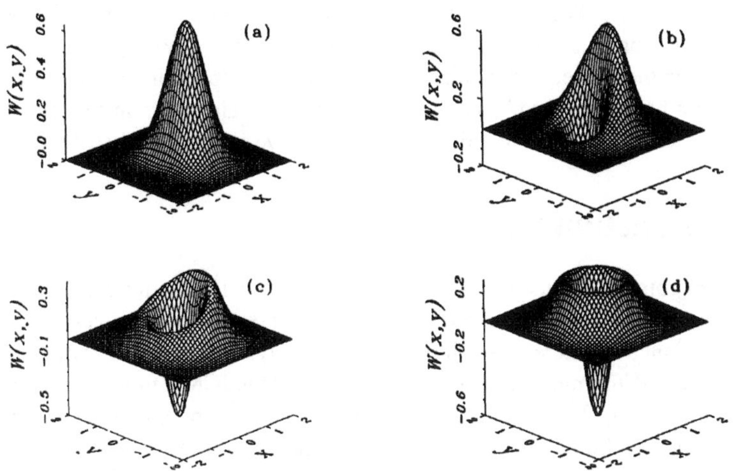

Figure 3: Wigner functions of binomial states having $M = 1$ for different values of p: (a) $p = 0$, (b) $p = 0.5$, (c) $p = 0.9$, and (d) $p = 1$.

I shall remark that the Wigner function is equivalent to other representations of quantum states, e.g., the density operator $\hat{\rho}$. Now I would like to discuss changes in the Wigner function as we vary the parameters p and M. We may analyze graphically the transition from the vacuum state $|0\rangle$ ($p = 0$) to the Fock state $|M\rangle$ ($p = 1$) containing M photons, for instance. In Figure 3 there are plots of the Wigner function of a binomial state with $M = 1$ for different values of p. When $p = 0$ (Figure 3a), we note the well-known Gaussian characteristic of the vacuum state. By increasing the probability of having individual photons to $p = 0.5$, the state acquires a positive "coherent" amplitude, which corresponds to a gain of energy in the field ($\langle\hat{n}\rangle = pM$), as can be seen in Figure 3b. By further increasing the probability p to $p = 0.9$, we clearly see a deformation in the Wigner function around the origin (Figure 3c), and finally, when $p = 1$, the Wigner function representing the one-photon Fock state $|M = 1\rangle$ is formed (Figure 3d). The phenomenon of squeezing can be "followed" from the phase space point of view. We note that as p is increased, at some stage a deformation in the Wigner function starts, in order to close the "ring", symmetrically posed around the origin, which characterizes the Fock state ($p = 1$). That deformation shows the decrease in noise (squeezing) in the $\hat{X}$ quadrature (along the x direction) at the expense of an increase in noise in the $\hat{Y}$ quadrature (along the y direction).

The negativity of the Wigner function is also connected to nonclassical effects. In general it is not straightforward to relate a particular effect to that negativity. For instance, the Wigner function of a squeezed state is non-negative. We may conjecture that the growth of the negative part of the Wigner function as p is increased is related to the linear increase of the sub-Poissonian character of the binomial states ($Q = -p$). In fact the Wigner function becomes more and more negative almost linearly as p is increased.

The phase space representation discussed here can be connected to the phase probability distribution shown in the preceding subsection [4] (Figure 2). As the Wigner function is shifted from the origin (p going from 0 to 1), there is a better definition in the phase, up to a minimum value for its dispersion. Then, as the Fock state starts being "formed" (p close to 1), there is again a loss in that sharpness, leading, eventually, to the situation in which we have a randomly distributed phase.

3 Interaction with matter

Now I am going to discuss the interaction of quantized light with matter. In general this constitutes a complicated problem, although it is possible to investigate many features of such an interaction within a simple, analytically solvable model, known as the Jaynes–Cummings model (JCM) [36,37]. In the JCM a mode of the field (in a "lossless" cavity [5] of volume V), is considered to be dipole-coupled to an atom (electric atomic dipole $\hat{\mathbf{D}}$). Despite the large number of energy states available in the atom, a monochromatic field [6] will basically select a pair of them (denoted as $|e\rangle$ and $|g\rangle$) having a difference in energy of $E_e - E_g = \hbar\omega_0 \approx \hbar\omega$. The Hamiltonian describing such a system after neglecting rapidly oscillating terms (the rotating wave approximation) will be

$$\hat{H} = \tfrac{1}{2}\hbar\omega_0\sigma_z + \hbar\omega\,\hat{a}^\dagger\hat{a} + \hbar g(\sigma_+\hat{a} + \hat{a}^\dagger\sigma_-), \tag{33}$$

where $\sigma_z = |e\rangle\langle e| - |g\rangle\langle g|$, $\sigma_+ = |e\rangle\langle g|$, $\sigma_- = |g\rangle\langle e|$ are the atomic operators,

$$g = e\,[\omega/(\epsilon_0\hbar V)]^{1/2}\,\hat{\epsilon}\cdot\langle g|\hat{\mathbf{D}}|e\rangle \tag{34}$$

is the atom-field coupling constant, and e is the electron charge. Hamiltonian (33) admits exact diagonalization, so that the time evolution is easily obtained. The atom-field density operator can be written in the atomic basis as [38,39]

$$\hat{\rho}_{af}(t) = \begin{pmatrix} \hat{A}\hat{\rho}^f(0)\hat{A}^\dagger & \hat{A}\hat{\rho}^f(0)\hat{B}^\dagger \\ \hat{B}\hat{\rho}^f(0)\hat{A}^\dagger & \hat{B}\hat{\rho}^f(0)\hat{B}^\dagger \end{pmatrix}, \tag{35}$$

[4] Even though a qualitative connection can be established between amplitude/phase associated to quasiprobabilities and mean photon number/Pegg–Barnett phase, their relation is not yet fully understood.

[5] The influence of losses on the cavity field will be discussed in section 4.

[6] Electric field polarized in the $\hat{\epsilon}$ direction and frequency ω.

where

$$\hat{A} = \cos[gt(\hat{a}\hat{a}^\dagger)^{1/2}], \qquad \hat{B} = -i\hat{a}^\dagger \sin[gt(\hat{a}\hat{a}^\dagger)^{1/2}](\hat{a}\hat{a}^\dagger)^{-1/2}.$$

That is the solution for an atom initially non-correlated with the field and prepared in the upper state, or $\hat{\rho}_{af}(0) = |e\rangle\langle e|\hat{\rho}_f(0)$.

3.1 Field evolution

Several aspects of the interaction of quantized fields with matter have already been discussed in the literature [40–42], and particularities relative to each subsystem (either the atom or the field), such as the atomic dynamics or the field "purity", are generally investigated. We are therefore more interested in the evolution of the reduced density operators which are obtained from the total density operator in equation (35) through a partial trace operation $\hat{\rho}_{a(f)}(t) = \mathrm{Tr}_{f(a)}[\hat{\rho}(t)]$. A way of determining the field state purity is by means of its von Neumann entropy [43]

$$S = -\mathrm{Tr}\left\{\hat{\rho}_f(t)\ln[\hat{\rho}_f(t)]\right\}. \tag{36}$$

For any pure state $S = 0$, and $S > 0$ for a statistical mixture of pure states. We may calculate the field entropy as the atom–field interaction takes place. In the case of having an initial pure state $\hat{\rho}_{af}(0) = |\Psi\rangle\langle\Psi|$, the atomic entropy equals the field entropy, according to the Araki–Lieb theorem [44]. It is sometimes more convenient to calculate the atomic entropy, given that it is then reduced to the task of diagonalizing a 2×2 matrix. The atomic (reduced) density operator from equation (35) is

$$\hat{\rho}_a(t) = \mathrm{Tr}_f[\hat{\rho}_{af}(t)] = \begin{pmatrix} \lambda_{11} & \lambda_{12} \\ \lambda_{12}^* & \lambda_{22} \end{pmatrix}, \quad \lambda_{ij} = \sum_{n=0}^{\infty}\langle n|\rho_{ij}|n\rangle. \tag{37}$$

The atomic (field) entropy will be

$$S = -\eta_+\ln(\eta_+) - \eta_-\ln(\eta_-), \tag{38}$$

with

$$\eta_\pm = \frac{1}{2}\left\{1 \pm \left[(\lambda_{11} - \lambda_{22})^2 + 4|\lambda_{12}|^2\right]^{1/2}\right\}.$$

The case in which the field is initially prepared in a coherent state [39] is normally addressed. The field becomes almost pure only at specific times, when the subsystems atom and field are non-correlated [39, 45, 46]. The time evolution of the field purity is strongly dependent on the field statistics, and for a binomial state we might expect different behaviours of the purity as we vary p and M. In particular, it would be interesting to see the case for which the field is highly sub-Poissonian (p close to 1) but still having a relatively small phase spread. [7]

[7] I shall remark that for a Fock state ($p = 1$), the phase is not defined, which corresponds, in classical terms, to a randomly distributed phase.

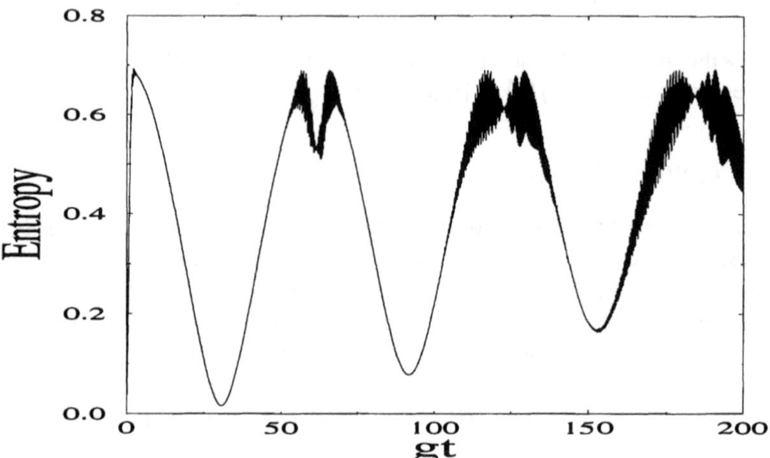

Figure 4: Field entropy as a function of time, for the field initially prepared in a coherent state having $|\alpha|^2 = \overline{n} = 95$.

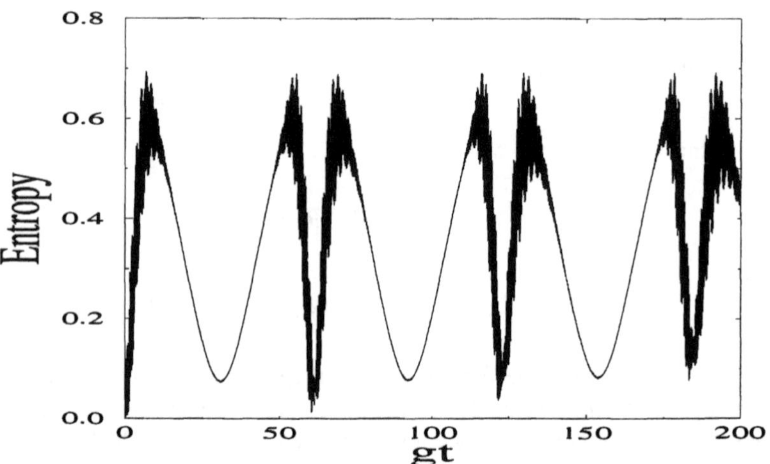

Figure 5: Field entropy as a function of time, for the field initially prepared in a binomial state having $M = 100$ and $p = 0.95$ ($\overline{n} = 95$).

For an initial field prepared in a binomial state $|p, M\rangle$, the matrix elements $\lambda_{ij} = \lambda^*_{ji}$ are

$$\lambda_{11} = \sum_{n=0}^{M} |B_n^M|^2 \cos^2[\lambda t(n+1)^{1/2}], \quad \lambda_{22} = \sum_{n=0}^{M-1} |B_n^M|^2 \sin^2[\lambda t(n+1)^{1/2}],$$

$$\lambda_{12} = (pM)^{1/2} \sum_{n=0}^{M-1} B_n^M B_n^{M-1} \cos[\lambda t(n+2)^{1/2}] \sin^2[\lambda t(n+1)^{1/2}].$$

In Figure 4 I show the field entropy as a function of the scaled time gt for a field initially prepared in a coherent state having $\alpha^2 = 95$. We note the field becoming more pure at certain times, as well as a progressive degradation in the field purity (increase in entropy), as time goes on. For comparison, I have plotted in Figure 4 the field entropy as a function of the scaled time gt, for an initial coherent state, and in Figure 5 the entropy for an initial binomial state having $p = 0.95$ and $M = 100$. We note that for the binomial state, the field becomes (periodically) almost pure at certain times, being basically a statistical mixture ($S > 0$) most of the time. Nevertheless the minimum entropy is almost the same at every cicle, contrarily to what happens for an initial coherent state (see Figure 4). This is due to the highly sub-Poissonian character of the field, which has a Mandel parameter rather close to unity ($Q = -0.95$). In fact, for an initial Fock state at $t = 0$, the field entropy is periodic [47].

In order to complement our discussion on the field dynamics, we may also follow the field evolution by using the phase space representation previously introduced in section 2. It is convenient in this case to employ the Q-function instead of the Wigner function. In the case of an initial binomial state, the time-dependent Q-function will read [42]

$$Q(\beta;t) = \frac{e^{-|\beta|^2}}{\pi}\left(\left|\sum_{n=0}^{M} \frac{(\beta^*)^n}{\sqrt{n!}} B_n^M \cos[\lambda t(n+1)^{1/2}]\right|^2 + \left|\sum_{n=0}^{M-1} \frac{(\beta^*)^{n+1}}{\sqrt{(n+1)!}} B_n^M \sin[\lambda t(n+1)^{1/2}]\right|^2\right). \tag{39}$$

In Figure 6 there are contour plots of the Q-function in equation (39) for three different times. At $t = t_R/2$ (where $t_R = 2\pi|\alpha|/g$ is the "revival time": see below) we note that the Q-function has split into two branches, each one having a shape similar to the original state. This corresponds to the superposition of two binomial states [42]. At $t = t_R$, the recombination of the two branches is so remarkable that the resulting Q-function is almost indistinguishable from the one at $t = 0$, apart from a reflection relatively to the origin. It is then reasonable to suppose that a pure state having a binomial photon number distribution is generated at $t = t_R$. This is in agreement with the result that at this time the field is almost in a pure state (see Figure 5).

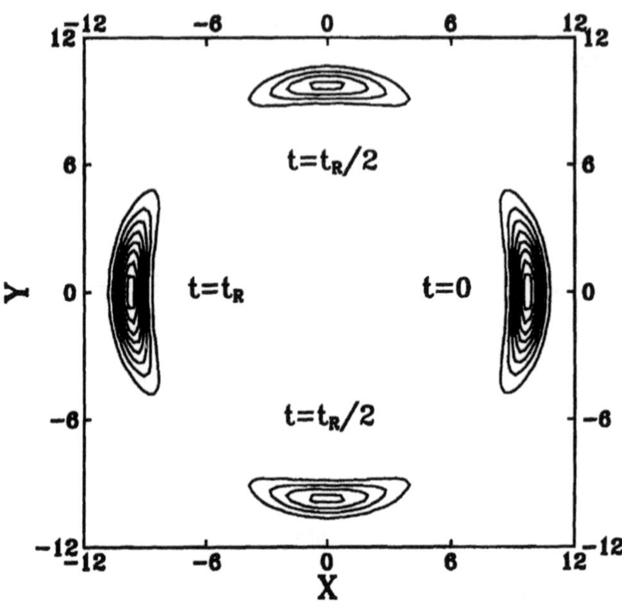

Figure 6: Contour plot of the Q-function relative to the cavity field undergoing a Jaynes–Cummings type dynamics at three different times: (a) $t = 0$, (b) $t = t_R/2$, and (c) $t = t_R$.

3.2 Atomic evolution

We may also calculate the evolution of the atomic population, for instance, the probability of occupation of level $|e\rangle$ minus the probability of occupation of level $|g\rangle$, represented by the population inversion, which for an initial state $|e\rangle|\varphi\rangle_f$ is given by

$$P_e - P_g = \langle\sigma_z\rangle = \sum_{n=0}^{\infty} P_n \cos(2gt\sqrt{n+1}). \tag{40}$$

This gives us not only information on how the atom is excited and de-excited by the field, but also information about the properties of the field itself.

In Figure 7 there is a plot of the atomic inversion as a function of gt, for an initial coherent state with $|\alpha|^2 = 95$, and in Figure 8 a plot for an initial binomial state having $p = 0.95$ and $M = 100$. We note how the structure of oscillations changes as the field statistics is modified. For an initial coherent state the well-known collapses and revivals [48] of the atomic inversion occur, which are beats due to the fact that there is a spread in the photon number distribution of the initial field. A "revival" of the oscillations becomes apparent at a "revival time" $t_R = 2\pi|\alpha|/g$, in which a re-phasing of the more representative terms in the sum in equation (40) happens. Note that there is never a return to the initial state (corresponding to $\langle\sigma_z\rangle = 1$). However, the field becomes almost pure at half of the revival time instead (see Figure 4), that is, when the field and atom are non-correlated [45, 46]. In the case of the binomial state considered here, the basic

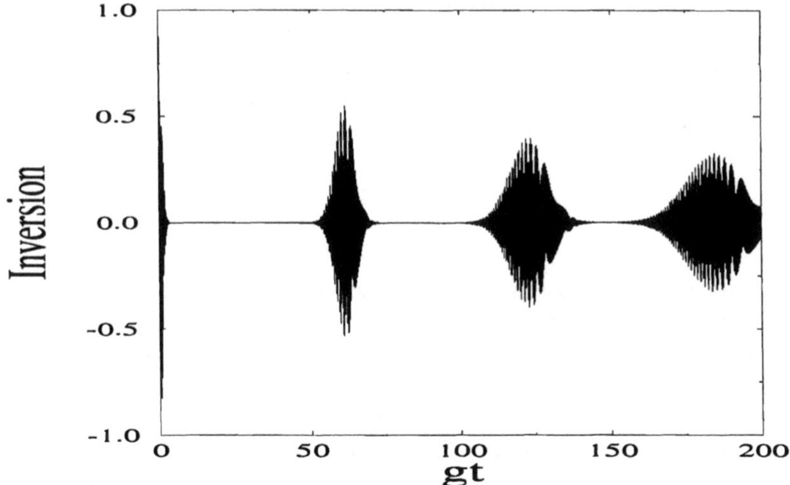

Figure 7: Atomic inversion as a function of time for a field initially prepared in a coherent state having $|\alpha|^2 = 95$.

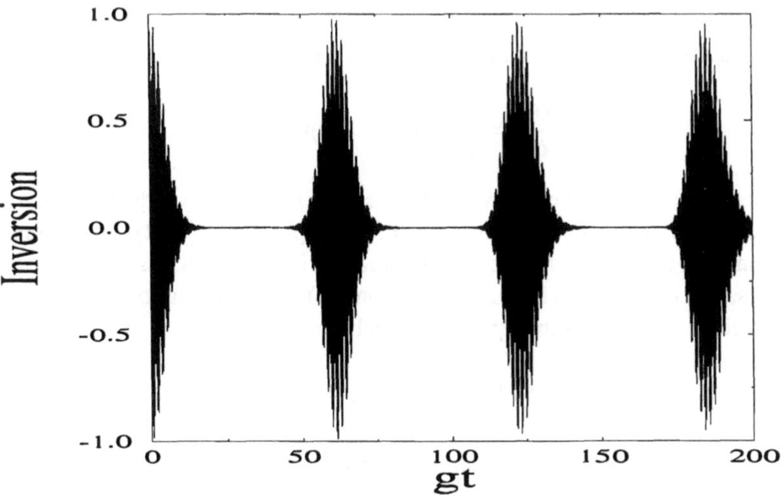

Figure 8: Atomic inversion as a function of time for a field initially prepared in a binomial state having $M = 100$ and $p = 0.95$.

structure of collapses and revivals is maintained, but the oscillations get closer to the extreme values of ± 1, as shown in Figure 8. This is because the initial field is highly sub-Poissonian, so that the atom has a tendency to perform Rabi cycles just as in the case of having the field initially in a Fock state. In this case the revival time is $t_R = 2\pi(pM)^{1/2}/g$, and the field also almost returns to a pure state at the revival time ($t = t_R/2$) itself, differently from the coherent state case (compare Figures 4 and 5).

4 Decay

Now I am going to discuss the evolution of a field mode confined in a lossy cavity. This is an interesting problem, narrowly connected to the much-debated issue of loss of coherence in quantum mechanics. It is already known that different mechanisms of loss will affect different states in distinct ways. Normally we adopt a model which represents either the action of an environment or some other (intrinsic) mechanism [49], and which results in the elimination of the non-diagonal elements of the density operator in a given basis (decoherence). For instance, the coherent states of the field are the only states that remain pure after being prepared and left to decay inside a lossy cavity (at zero temperature) [50]. This feature reassures us that coherent states are indeed quasi-classical states [12]. Here I am going to consider the simple case of a single mode confined inside a lossy cavity and prepared in a binomial state. The decay of binomial states can be analyzed under the light of their interpolating properties. This will of course yield very different patterns for the field purity, depending on the state parameters p and M, as time goes on. The cavity field decay is modelled through the phenomenological coupling of the field to an "external" reservoir, here taken at zero temperature. This results in the standard master equation for the field density operator $\hat{\rho}$ [51]

$$\frac{\partial \hat{\rho}}{\partial t} = \frac{\gamma}{2}\left(2\hat{a}\hat{\rho}\hat{a}^\dagger - \hat{a}^\dagger\hat{a}\hat{\rho} - \hat{\rho}\hat{a}^\dagger\hat{a}\right), \tag{41}$$

where $\gamma = \omega/Q$ is the decay constant, and Q is the cavity's quality factor. The decay of the field energy is given by $\langle\hat{n}(t)\rangle = \langle\hat{n}(0)\rangle \exp(-\gamma t)$. That represents a nonunitary evolution for the density operator, even though it is normalized at all times, or $\mathrm{Tr}\hat{\rho}(t) = 1$. The master equation (41) admits a solution via superoperators. If we define the superoperators $\hat{J}$ and $\hat{L}$ such that

$$\hat{J}\hat{\rho}(t) = \gamma\hat{a}\hat{\rho}(t)\hat{a}^\dagger, \qquad \hat{L}\hat{\rho}(t) = -\frac{\gamma}{2}\left[\hat{a}^\dagger\hat{a}\hat{\rho}(t) + \hat{\rho}(t)\hat{a}^\dagger\hat{a}\right], \tag{42}$$

the master equation assumes a simpler form

$$\partial\hat{\rho}/\partial t = (\hat{J} + \hat{L})\hat{\rho}(t). \tag{43}$$

This has the formal solution

$$\hat{\rho}(t) = \exp[(\hat{J} + \hat{L})t]\hat{\rho}(0). \tag{44}$$

Although the superoperators $\hat{J}$ and $\hat{L}$ do not commute, their commutator is basically $\hat{J}$ itself ($[\hat{L}, \hat{J}] = -\gamma \hat{J}$), so that the exponential can be disentangled [52] either as

$$\hat{\rho}(t) = \exp(\hat{L}t) \exp\left[\hat{J}\left(1 - e^{-\gamma t}\right)/\gamma\right] \hat{\rho}(0), \tag{45}$$

or as

$$\hat{\rho}(t) = \exp\left[\hat{J}\left(e^{\gamma t} - 1\right)/\gamma\right] \exp(\hat{L}t)\hat{\rho}(0). \tag{46}$$

If we employ the solution as in equation (45) for a field initially prepared in a coherent state, or $\hat{\rho}(0) = |\alpha\rangle\langle\alpha|$, after some algebra we obtain

$$\hat{\rho}(t) = |\alpha e^{-\gamma t/2}\rangle\langle\alpha e^{-\gamma t/2}|. \tag{47}$$

This means that under the evolution dictated by the master equation (41) the coherent state remains not only pure but also coherent although with an exponentially decaying amplitude so that it eventually reaches the vacuum state, or $\hat{\rho}(\infty) = |0\rangle\langle 0|$. As I have already mentioned, it may be shown that the coherent state is the only pure state of the field having such a property [50]. If instead the field is prepared in a Fock state, or $\hat{\rho}(0) = |M\rangle\langle M|$ we may use the other solution [equation (46)] to calculate its time evolution. The result is

$$\hat{\rho}(t) = \sum_{l=0}^{M} \frac{M!}{l!(M-l)!}(1 - e^{-\gamma t})^l (e^{-\gamma t})^{M-l} |M - l\rangle\langle M - l|, \tag{48}$$

which is a statistical mixture of Fock states with a binomial weight $p = 1 - e^{-\gamma t}$. In other words, dissipation destroys the coherence of a Fock state, contrarily to what happens to a coherent state. This is also an example of a binomial distribution arising in a physical process in quantum optics, the Fock state becoming a mixture of Fock states (with binomial weight) even at short times. [8]

We have concluded that the "extreme" states, the coherent and the Fock states, are affected very differently when exposed to dissipation. To see how a general binomial state behaves while dissipating, we use the general solution of equation (41) in the Fock basis, for the field initially prepared in a binomial state,

$$\hat{\rho}(t) = \sum_{k,j=1}^{M} B_k^M B_j^M \exp[-\frac{\gamma t}{2}(k + j)]$$
$$\times \sum_{m=0}^{min(k,j)} \frac{1}{m!} \left[\frac{k!j!}{(k-m)!(j-m)!}\right]^{1/2} (e^{\gamma t} - 1)^m |k - m\rangle\langle j - m|.$$

In order to investigate the purity of the field as this is dissipated in the cavity, we may calculate either the field's von Neumann entropy (see equation (36)), or simply the *linear entropy*

$$\varsigma(t) = 1 - \mathrm{Tr}\left[\hat{\rho}(t)^2\right]. \tag{49}$$

[8] One has to be careful, though, because the master equation (41) is not valid for arbitrarily short times.

The linear entropy ζ has computational advantages as well as a behaviour very similar to von Neumann's entropy.

In Figure 9 we have plots of the linear entropy ζ as a function of time for different values of p. At $t = 0$, the field is in a pure state, or $\zeta = 0$. Parameter ζ becomes larger as time goes on, as the field gets closer to a statistical mixture. This in general occurs in a time much shorter than the characteristic time of the system, or $\tau \ll 1/\gamma$, as we may note in Figure 9. The linear entropy curve reaches a maximum value, before starting to decrease towards zero as the field dissipates to the vacuum (final) state. We note a high sensitivity of the field purity to the parameter p, for a given value of M. For instance, the maximum of the linear entropy curve is reduced by almost a half even for a modest variation of p of 10%, as seen in Figure 9. This might be attributed to the fact that this specific binomial state is practically a minimum uncertainty state, with respect to the quadrature variances, for a wide range of values of p [53]. In Figure 10 the product of the quadrature variances $\langle\Delta\hat{X}\rangle\langle\Delta\hat{Y}\rangle$ is plotted as a function of p, for different values of M, and we note that the binomial state is basically a minimum uncertainty state in the range $0 \leq p < 0.7$. For values closer to $p = 1$, the binomial state gets closer to a Fock state $|M\rangle$ (containing 20 photons in the example given here), being no longer a minimum uncertainty state, which means a considerable increase in the quadrature variances product.

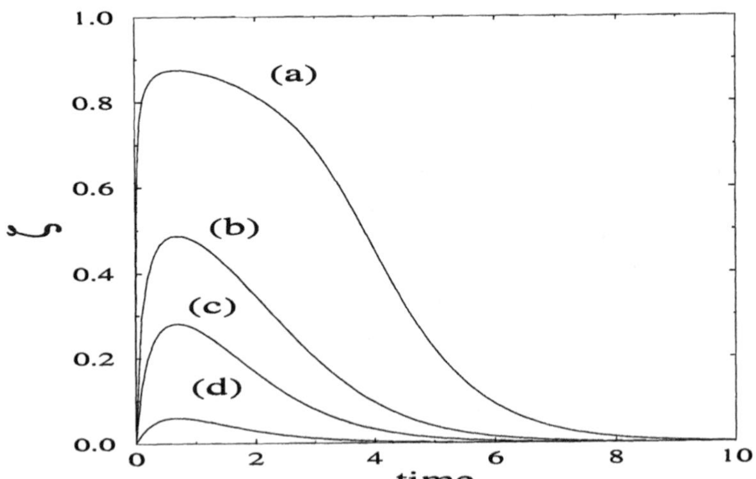

Figure 9: Field linear entropy ζ as a function of time for a binomial cavity field ($M = 20$) in a dissipative cavity for different values of p: (a) $p = 1.0$, (b) $p = 0.9$, (c) $p = 0.8$, (d) $p = 0.5$.

I would like to point out that other interesting aspects of the influence of losses on the structure of binomial states have also been addressed in the literature. For instance, the behaviour of the field prepared in various states subjected to one-photon and two-photon losses, as well as the competition between both processes, have been already discussed, e.g., in [54,55], where the following more general

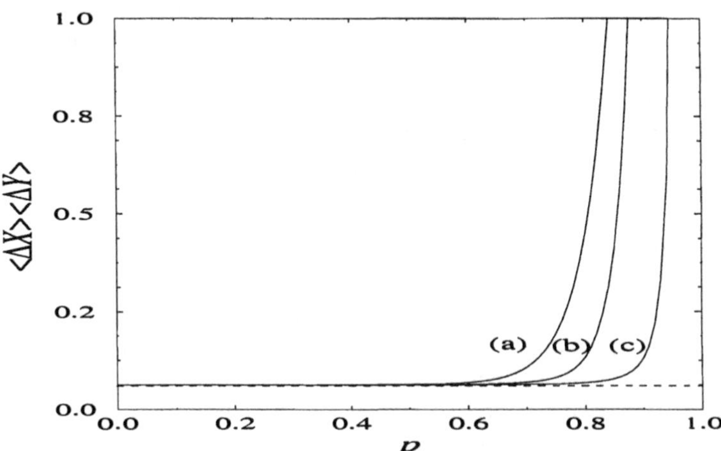

Figure 10: Product of quadrature variances $\langle\Delta\hat{X}\rangle\langle\Delta\hat{Y}\rangle$ as a function of p for different values of M: (a) $M = 10$, (b) $M = 20$, and (c) $M = 100$.

master equation was considered:

$$\frac{\partial\hat{\rho}}{\partial t} = \frac{1}{2}\sum_{n=1}^{2} D_n \left(2\hat{a}^n\hat{\rho}\hat{a}^{\dagger n} - \hat{a}^{\dagger n}\hat{a}^n\hat{\rho} - \hat{\rho}\hat{a}^{\dagger n}\hat{a}^n\right), \tag{50}$$

D_n being positive decay constants. This equation describes the coupling to two independent reservoirs, characterized by one- and two-photon transitions.

The equation for the photon number distribution of the cavity field (the diagonal elements $p_n = \langle n|\hat{\rho}|n\rangle$) can be solved exactly by using the method of the generating function [54] and yields

$$p_m(t) = \sum_{n=m}^{\infty} A_n\, B_{nm}^{(\nu)} e^{-\lambda_n t},$$

where $\lambda_n = D_1 n + D_2 n(n-1)$, $\nu = D_1/D_2$, and

$$B_{nm}^{(\nu)} = \frac{(-1)^{n-m}(n+\nu-1)_m(m+\nu)_{n-m}}{2^n m!(n-m)!} F(m-n, 1-n; m+\nu; -1),$$

$F(a,b;c;z)$ being Gauss' hypergeometric function. For the binomial state, the function A_n reads

$$A_n^{(bin)} = (2p)^n \frac{(M-n+1)_n}{(M+\nu)_n}\frac{2n-1+\nu}{n-1+\nu} F(n-M, n; 1-M-\nu; 1-2p).$$

Analytical expressions for p_m for other states found in [54] allowed us to study the evolution of Mandel's Q parameter and to calculate the mean transition time from the initial state to the equilibrium state of the field as a function of the ratio between the decay constants $\nu = D_1/D_2$. The conclusion was that the transition occurs in a finite time even for highly excited initial states, if the decay constant associated to two-photon losses, D_2, is nonzero. It is also worth mentioning that even weak two-photon losses may drastically change the stationary state of the field [55]. The effects of the environment are normally destructive, although some states of the field are more robust to dissipation than others.

5 Generation

The discussions presented in the preceding sections are based on the assumption that somehow a binomial state had been previously generated inside an optical cavity. The generation of nonclassical states is a central issue in quantum optics, yet attempts to generate pure states of the field have been elusive for a long time. The difficulties are mainly due to the fact that extreme control is required, and even the tiniest portion of noise is sufficient to destroy a pure state, as we have seen in the last section, where we supposed that the field was already generated and had been left to interact with its environment. As an example of successful generation (as well as detection) of specific nonclassical states of light we may cite the Fock state generation [5] and the Schrödinger type state (superposition of two coherent states) [56]. In the paper where the binomial states were first introduced [15], a way to generate them was suggested, using a mixture of N_2 and CO_2 molecules. As the N_2 molecules undergo transitions in M vibrational levels, the probability p of emission of individual photons will depend on the pressure of the CO_2 gas, which controls the relaxation mechanism. Collisional de-excitation becomes dominant as the pressure is increased. In this case no photons are produced, and this would correspond to the situation in which $p \to 0$. By decreasing the CO_2 pressure, radiative transitions become more likely, and in principle p could be increased up to $p = 1$. In another example of generation, using a free-electron laser [57,58], this tuning depends on the laser parameters. In this case we have a time-dependent probability $p(t)$ of emission of individual photons, which is basically given by $p(t) = \sin^2 \Omega t$, where $\Omega = \left[\Lambda^2 + |\vec{k}|^2 p_0^2/m^2\right]^{1/2}$, $\Lambda = e^2/(2m\omega\epsilon_0 V_W)$, and V_W is the wiggler volume. The field has a wave vector $\vec{k}$, and the electron an initial momentum p_0 and mass m. This would allow a fine-tuning of p, simply by choosing a convenient interaction time. The extreme values of p, for instance, can be obtained at $t = \pi/2\Omega$ ($p = 1$), and at $t = \pi/\Omega$ ($p = 0$). Although these two schemes would in principle allow the production of a state having a binomial photon number distribution, there is no guarantee that it will be a pure state. Therefore we need a more proper treatment, based on a Hamiltonian.

More generally, there have been proposed, although not yet experimentally implemented, various schemes aiming at the generation of arbitrary pure states of the field, generically known as *quantum state engineering*. Here I will discuss a few of them. Quantum state generation schemes can be divided into two classes: nonunitary processes involving the collapse of the wave function, and unitary processes. In the former methods, usually conveniently prepared atoms provide the necessary energy and coherence for the building up of the (single mode) intracavity field. The atoms successively cross the cavity itself, and can be measured or not, constituting what it is known as a *micromaser*. Unitary engineering processes are perhaps not so widely discussed, but they constitute alternatives to micromaser-based schemes. Of course any experiment of quantum state engineering would demand extraordinary control. In what follows I will discuss different propositions for the generation of binomial states.

5.1 State generation via micromaser

A scheme based on Fock state superpositions

One of the first propositions for a quantum state engineering scheme was based on the concept of micromaser [59]. The basic idea of such a method is the injection of M (two-level) atoms prepared in superpositions of energy eigenstates into a cavity whose main mode is cooled down to the vacuum state $|0\rangle$. As the atoms leave the cavity, they are measured so that they collapse in one of their energy eigenstates $|e\rangle$ or $|g\rangle$ (see Figure 11).

We would like to generate a specific field state of the type

$$|\Psi\rangle = \sum_{n=o}^{M} A_n^M |n\rangle, \tag{51}$$

and we have to find out how the atoms should be prepared in order to do so. [9]

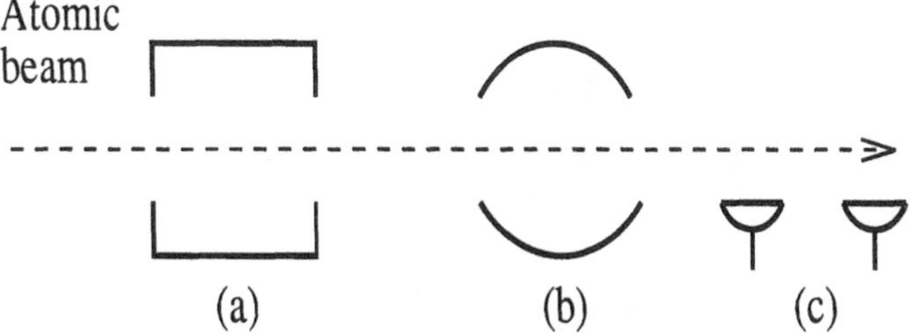

Figure 11: Schematic diagram showing a typical micromaser-based experiment; an atomic beam crosses a Ramsey zone (a), where the atoms are prepared in suitable superposition states. Then the atoms cross the high-Q cavity (b), where they interact with the field mode, being eventually measured by the detectors in (c).

A condition to be fulfilled is that *all the M atoms* must be detected in the lower energy state $|g\rangle$, otherwise the quantum state generation will not be accomplished. The necessity for such a *conditional measurement* is one of the shortcomings of the method, because the probability of having all M atoms in the ground state after detection, decreases considerably as M gets larger. An adaptation of that scheme for binomial and reciprocal-binomial states has already been presented in the literature [60]. Before entering the high-Q cavity, the atoms, provenient

[9]Note that there is in such a field a maximum number of photons, M.

from an oven, cross a Ramsey (microwave) zone (see Figure 11), where they are prepared in a superposition state of the type

$$|\phi\rangle = c_g^{(k)}|g\rangle_k + c_e^{(k)}|e\rangle_k, \tag{52}$$

the index k labelling the k-th atom. The method consists in determining the atomic coefficients c_g and c_e that would lead to the "target" field state in equation (51) after the successive interactions and detections of M atoms. The basic atom–field dynamics is dictated by the (on-resonance) Jaynes–Cummings Hamiltonian (33). Solving Schrödinger's equation, [10] we find that if the state of the field before the passage of the k-th atom is

$$|\varphi^{(k-1)}\rangle = \sum_n \varphi_n^{(k-1)}|n\rangle,$$

the joint atom–field state after the atom has left the cavity will be [11]

$$\begin{aligned}|\Phi^{af}\rangle = \sum_n \varphi_n^{(k-1)}\Big[&C_n^{(k)}|n\rangle|e\rangle - iS_n^{(k)}|n+1\rangle|g\rangle \\ &+\frac{c_g}{c_e}\left(C_{n-1}^{(k)}|n\rangle|g\rangle - iS_{n-1}^{(k)}|n-1\rangle|e\rangle\right)\Big],\end{aligned}$$

where

$$C_n^{(k)} = \cos(g\tau_k\sqrt{n+1}), \qquad S_n^{(k)} = \sin(g\tau_k\sqrt{n+1}),$$

and τ_k is the interaction time of the k-th atom with the field. [12] If M atoms are detected in the ground state $|g\rangle$ after having passed the cavity, the field and atomic coefficients are connected by the recurrence relations [60]

$$\begin{aligned} A_n^M &= (1-\delta_{n,0})A_{n-1}^{M-1}c_e^{(M)}\sin(g\tau_M\sqrt{n}) \\ &\quad +(1-\delta_{n,M})A_n^{M-1}c_g^{(M)}\cos(g\tau_M\sqrt{n}), \end{aligned} \tag{53}$$

with $A_0^0 = 1$ and τ_M is the interaction time relative to the Mth atom. Now if we want to prepare a binomial state having $M = 2$, for instance,

$$A_n^M = \left[\frac{M!}{n!\,(M-n)!}p^n(1-p)^{M-n}\right]^{1/2},$$

we obtain the following relations for the atomic coefficients [60],

$$\frac{c_e^{(1)}}{c_g^{(1)}} = \frac{[p/2(1-p)]^{1/2}}{\sin(\pi/\sqrt{2})}, \qquad \frac{c_e^{(2)}}{c_g^{(2)}} = [2p/(1-p)]^{1/2},$$

for interaction times $\tau_1 = \tau_2 = \pi/2g$. Of course the atoms should have very low dispersion velocities as they enter the high-Q cavity, and the normalization condition should be fulfilled, or $|c_e^{(i)}|^2 + |c_g^{(i)}|^2 = 1$, so that we obtain the atomic coefficients as a function of p, or

$$c_g^{(2)} = (1+p)^{-1/2}, \qquad c_e^{(2)} = \left[2p/(1-p^2)\right]^{1/2}. \tag{54}$$

[10] With the field initially in the vacuum state $|0\rangle$ and the atom in the superposition state (52).

[11] Because we are dealing with pure states it is not necessary to employ the more general density operator solution already presented in equation (35).

[12] Note that we must have $c_e \neq 0$.

A scheme based on coherent state superpositions

Any pure state may be expressed as a continuous superposition of coherent states

$$|\Psi\rangle = \int d^2\alpha\, G(\alpha, \alpha^*)|\alpha\rangle, \tag{55}$$

where $d^2\alpha = d(\mathrm{Re}\,\alpha)d(\mathrm{Im}\,\alpha)$. Due to the coherent states' overcompleteness, there is an infinite number of expansion functions G, although they might be chosen in such a way that they are nonzero in convenient sub-spaces, e.g., straight lines and circles [13] [61]. Despite the fact that coherent states form a continuous basis, it is possible to construct, within a good approximation, general quantum states as *discrete superpositions of coherent states* [61]. Here I am going to discuss the case of superpositions in a circle, through which it is possible to generate the binomial states. Consider the discrete coherent state superposition

$$|\psi_N\rangle = \sum_{k=1}^{N} F_k |Re^{i\phi_k}\rangle, \tag{56}$$

where the coherent states $|Re^{i\phi_k}\rangle$ are equally distributed in a circle of radius R in phase space, with

$$\phi_k = \phi_0 + \tfrac{1}{2}(2k - N - 1)\Delta\phi; \qquad k = 1, \cdots, N, \tag{57}$$

and $\Delta\phi$ being the angle between two adjacent coherent states. The coefficients F_k are proportional to the one-dimensional continuous distribution $F_R(\phi_k)$. We may also define the *misfit parameter*

$$\epsilon \equiv 1 - |\langle\Psi|\psi_N\rangle| = 1 - \left|\sum_{n=0}^{\infty} A_n^* a_n^{(N)}\right|, \tag{58}$$

$$|\psi_N\rangle = \sum_{n=0}^{\infty} a_n^{(N)}|n\rangle, \qquad |\Psi\rangle = \sum_{n=0}^{\infty} A_n|n\rangle,$$

which" measures" the deviation of the approximate state $|\psi_N\rangle$ in equation (56) from the "actual" target state $|\Psi\rangle$.

The larger the number of coherent states in the discrete superposition, the more accurate will be the approximation (smaller ϵ). It is worth noting that it is possible to obtain a rather good approximation even for a small number of component states, e.g., around $N = 20$ [61]. Superpositions of coherent states as in equation (56) can be generated by injecting conveniently prepared two-level atoms within cavities which are appreciably detuned from the cavity field mode [56, 61].

The Wigner function corresponding to the state $|\psi_N\rangle$ is given by

$$W_N(\beta) = \frac{2}{\pi}e^{-R^2-2|\beta|^2} \sum_{k,l=1}^{N} F_l^* F_k \exp\left[2R\left(\beta e^{-i\phi_l} + \beta^* e^{i\phi_k}\right) - R^2 e^{i(\phi_k-\phi_l)}\right],$$

and can be readily obtained from the expressions above.

[13]Those are representations in phase space.

5.2 Unitary evolution schemes

We have been discussing schemes based on nonunitary evolution, but yielding pure states of the field. It would be interesting to broaden the possibilities by considering other methods. In this section I will discuss basically two methods involving Hamiltonians which might allow quantum state generation through some kind of unitary evolution.

A time-dependent Hamiltonian

Several atoms are in general necessary for the implementation of the "superposition" methods discussed above. It is possible, in principle, to control a cavity field using a *single atom* by choosing an appropriate interaction [62]. The central ingredient of the proposal is the time-dependent Hamiltonian

$$\hat{H} = \frac{1}{2}\omega_0\sigma_z + \omega_c\hat{a}^\dagger\hat{a} + [r(t)e^{-i\omega_L t} + g(t)\hat{a}]\sigma_+ + [r^*(t)e^{i\omega_L t} + g^*(t)\hat{a}^\dagger]\sigma_- ,$$

where ω_c is the cavity field frequency. The atom can make transitions through two channels: the channel of coupling strength $r(t)$, a classical (driving) field of frequency ω_L having adjustable amplitude and the coupling to the quantized field with coupling constant $g(t)$. In practice this could be accomplished by using a three-level atom excited via two Raman channels [62], as depicted in Figure 12.

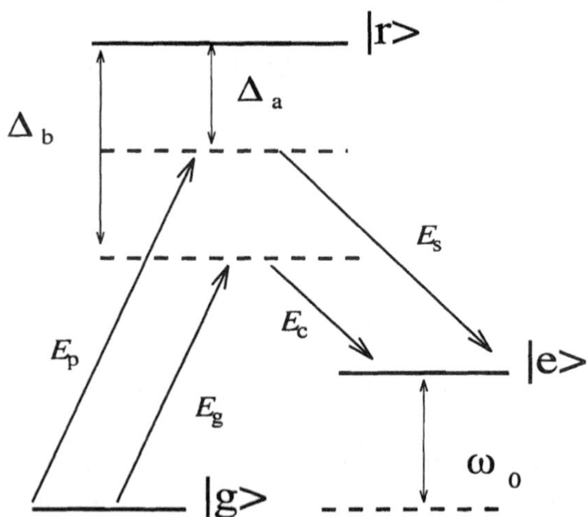

Figure 12: Raman scheme for the quantum state engineering in a proposed method involving a time-dependent Hamiltonian, as discussed in reference [50].

Two classical external fields $\vec{E}_p(t)$ and $\vec{E}_s(t)$ have frequencies ω_p and ω_s respectively, such that their difference $\omega_p - \omega_s = \omega_0$ equals the difference in energy between the levels $|e\rangle$ and $|g\rangle$. Another channel contains a classical external field $E_g(t)$ (frequency ω_g), as well as the quantized field $\vec{E}_c$ of frequency ω_c. The couplings can then be written as

$$r(t) = -[\vec{d}_{rg} \cdot \vec{E}_p(t)][\vec{d}_{er} \cdot \vec{E}_s^*(t)]/\Delta_a, \tag{59}$$

$$g(t) = -[\vec{d}_{rg} \cdot \vec{E}_g(t)][\vec{d}_{er} \cdot \vec{E}_c^*(t)]/\Delta_b, \tag{60}$$

where $\vec{d}_{ij} = \langle i|\vec{d}|j\rangle$ is the atomic dipole matrix element, and Δ_i is the detuning between the laser and the level $|r\rangle$. By switching on and off the classical fields, it is possible to manipulate the couplings in order to obtain the necessary time-dependence for a feasible quantum state engineering scheme. Choosing the initial atom–field state as $|\Psi(0)\rangle = |g\rangle|0\rangle$, we may look for an evolution operator that will lead us, after a time τ, to the "target" state

$$|\Psi(\tau)\rangle = \sum_{n=0}^{M} \mathcal{A}_n^M |n\rangle|g\rangle, \tag{61}$$

i.e., with the atom and field subsystems non-correlated (disentangled), so that the field is sure to remain in a pure state. We may divide the duration of the experiment τ into $2M$ sub-intervals, with every step taking time $\tau' = \tau/2M$. Then the evolution operator can be represented as

$$\hat{U}(\tau) = \hat{Q}_M \hat{C}_M \hat{Q}_{M-1} \hat{C}_{M-1} \cdots \hat{Q}_2 \hat{C}_2 \hat{Q}_1 \hat{C}_1,$$

where

$$\hat{Q}_j = \begin{pmatrix} \cos|g_j|\tau'(\hat{a}\hat{a}^\dagger)^{1/2} & -i\hat{a}e^{i\phi_j}\dfrac{\sin[|g_j|\tau'(\hat{a}^\dagger\hat{a})^{1/2}}{(\hat{a}^\dagger\hat{a})^{1/2}} \\ -i\hat{a}^\dagger e^{-i\phi_j}\dfrac{\sin|g_j|\tau'(\hat{a}\hat{a}^\dagger)^{1/2}}{(\hat{a}\hat{a}^\dagger)^{1/2}} & \cos|g_j|\tau'(\hat{a}^\dagger\hat{a})^{1/2} \end{pmatrix},$$

$$\hat{C}_j = \begin{pmatrix} \cos|r_j|\tau' & -i\hat{a}e^{i\theta_j}\sin|r_j|\tau' \\ -i\hat{a}^\dagger e^{-i\theta_j}\sin|r_j|\tau' & \cos|r_j|\tau' \end{pmatrix},$$

and $r_j = |r_j|e^{\theta_j}$, $g_j = |g_j|e^{\phi_j}$. From the conditions

$$|\Psi(\tau)\rangle = \hat{U}(\tau)|g\rangle|0\rangle, \qquad |g\rangle|0\rangle = \hat{U}(\tau)^\dagger|\Psi(\tau)\rangle,$$

it is possible to write down equations for the sets of values $\{r_j\}$ and $\{c_j\}$

$$\alpha_j \cos|g_j|\sqrt{j}\tau' + i\beta_j e^{-i\phi_j}\sin|g_j|\sqrt{j}\tau' = 0, \tag{62}$$

$$\mu_j \cos|r_j|\tau' + i\nu_j e^{i\theta_j}\sin|r_j|\sqrt{j}\tau' = 0, \tag{63}$$

with

$$\alpha_j = \langle j, g|F_{j+1}\rangle, \qquad \beta_j = \langle j-1, e|F_{j+1}\rangle,$$
$$\mu_j = \langle j-1, e|Q_J^\dagger|F_{j+1}\rangle, \qquad \nu_j = \langle j-1, j|Q_J^\dagger|F_{j+1}\rangle,$$
$$|F_J\rangle = \hat{C}_j^\dagger \hat{Q}_j^\dagger \hat{C}_{j+1}^\dagger \hat{Q}_{j+1}^\dagger \cdots \hat{C}_M \hat{Q}_M|\Psi(\tau)\rangle, \qquad |F_{M+1}\rangle = |\Psi(\tau)\rangle.$$

Equations (62) and (63) can be solved numerically, in order to determine the values of $\{r_j\}$ and $\{c_j\}$. In this way it is possible to find the values for the couplings that would allow the generation of any prescribed state, including the binomial state. The couplings can be controlled via the Raman transitions, according to equations

(59) and (60). I would like to point out that there is no need to have equal time intervals (τ'), and that the scheme should work in such a way that each decay channel is effective at a time. As it is usual in this kind of experiment, there are several possible sources of noise that may hamper the process, which demands a careful design of the experiment. However, discussions on the influence of noise on these generation schemes are not normally found in the literature, despite of the importance of the issue in quantum state generation.

A tailor-made nonlinear Hamiltonian

The schemes discussed above rely upon the proper transfer of coherence from individual atoms to a field confined within a cavity. Now I am going to discuss a somewhat different approach, based on nonlinear interactions [63]. That scheme is based on the fact that a Fock state can be generated through unitary transformations [64]. A generalization of this method leads to a quantum state engineering method, which would make possible, in principle, the generation of arbitrary states. I assume that the "target" state has the form $|\Psi\rangle = \sum_{n=0}^{M} A_n|n\rangle$, i.e., it contains no more than M photons. In order to build up a suitable Hamiltonian, we first consider specific interaction times $t_m = (m + 1/2)\pi$, such that we may write

$$|\Psi\rangle = e^{-i\hat{H}t_m/\hbar}|0\rangle = \hat{H}|0\rangle, \tag{64}$$

which is true provided that we have $\hat{H}|\Psi\rangle = |0\rangle$ as well. A general admissible form for $\hat{H}$ is

$$\hat{H} = \sum_{n=0}^{M} A_n \left(|0\rangle\langle n| + |n\rangle\langle 0|\right) - A_0 \sum_{k=0}^{M} |k\rangle\langle k| + \hat{P}\hat{F}\hat{P}, \tag{65}$$

where

$$\hat{P} = \hat{I} - \sum_{k=0}^{M} |k\rangle\langle k|, \tag{66}$$

and $\hat{F}$ is any Hermitian operator. I am considering the coefficients A_n real for simplicity. In order to eliminate higher-order powers of $\hat{a}$ and $\hat{a}^\dagger$, the following decomposition for the operator $\hat{F}$ can be assumed:

$$\hat{F}(\hat{a}, \hat{a}^\dagger) = \hat{f}_0(\hat{a}^\dagger\hat{a}) + \sum_{m=1}^{M} \left[\hat{f}_m(\hat{a}^\dagger\hat{a})\frac{\hat{a}^m}{\sqrt{m!}} + \frac{\hat{a}^{\dagger m}}{\sqrt{m!}}\hat{f}_m^*(\hat{a}^\dagger\hat{a})\right]. \tag{67}$$

Substituting formulae (67) and (66) into equation (65), we obtain the conditions that should be imposed on the auxiliary functions $f_l(k) \equiv \langle k|\hat{f}_l(\hat{a}^\dagger\hat{a})|k\rangle$ in order to cancel out most of the terms in $\hat{H}$ [63]:

$$f_0(0) = A_0; \qquad f_0(k) = -A_0; \qquad f_m(0) = A_m; \qquad f_m(k) = 0, \tag{68}$$

for $k, m = 1, 2, \ldots M$. These conditions are not enough to determine the operators $\hat{f}_m(\hat{a}^\dagger, \hat{a})$, but they can be used as a guide for doing that. We may, for instance, write the operators $\hat{f}$ as series of powers of $\hat{a}^\dagger\hat{a}$

$$\hat{f}_0(\hat{a}^\dagger\hat{a}) = A_0\left[2\left(1 - \hat{a}^\dagger\hat{a}\right)\mathcal{F}(\hat{a}^\dagger\hat{a}) - 1\right], \tag{69}$$

$$\hat{f}_m(\hat{a}^\dagger \hat{a}) = A_m \, \mathcal{F}(\hat{a}^\dagger \hat{a}); \qquad m \neq 0, \tag{70}$$

where

$$\mathcal{F}(\hat{a}^\dagger \hat{a}) = \sum_{l=0}^{M} D_l \, (\hat{a}^\dagger \hat{a})^l. \tag{71}$$

Due to the condition $f_m(0) = A_m \, \mathcal{F}(\hat{a}^\dagger \hat{a})|0\rangle = |0\rangle$ should hold, which leads to $D_0 = 1$. We may also use the condition $f_m(k) = 0$ [see equation (68)] to determine the remaining coefficients D_l. The successive application of the function $\mathcal{F}(\hat{a}^\dagger \hat{a})$ onto M Fock states $|1\rangle, |2\rangle, \ldots, |M\rangle$, leads to the following M coupled linear equations for the coefficients D_l:

$$\begin{aligned}
1 \; &+ \; D_1 + D_2 + \ldots + D_M = 0 \\
1 \; &+ \; 2D_1 + 4D_2 + \ldots + 2^M D_M = 0 \\
&\vdots \\
1 \; &+ \; MD_1 + M^2 D_2 + \ldots + M^M D_M = 0.
\end{aligned} \tag{72}$$

The solution of this set of equations always exists, which completely determines the functions $\hat{f}_m$. Therefore there is a specific set of numerical coefficients D_l for every value of M, which are of the form

$$D_1 = \alpha_1/M!, \quad D_2 = \alpha_2/M!, \ldots \quad D_M = \alpha_M/M!,$$

with $|\alpha_1| < |\alpha_2| < \ldots < |\alpha_M|$. In particular, $\alpha_M = (-1)^M/M!$. This result is convenient for us because the powers of $\hat{a}^\dagger \hat{a}$ will be multiplied by increasingly smaller coefficients, i.e., the relative importance of the higher-order terms will be diminished.

The final form of our generating Hamiltonian will then be

$$\hat{H}_{|\Psi\rangle} = f_0 \left(\hat{a}^\dagger \hat{a}\right) + \sum_{m=1}^{M} \frac{A_m}{\sqrt{m!}} \left[\mathcal{F}(\hat{a}^\dagger \hat{a}) \, \hat{a}^m + \hat{a}^{\dagger m} \, \mathcal{F}(\hat{a}^\dagger \hat{a})\right]. \tag{73}$$

As an example I am going to show how a simple state, namely the superposition of the vacuum state with a one-photon state, or qubit

$$|\phi\rangle_{01} = (1-p)^{1/2}|0\rangle + p^{1/2}|1\rangle, \tag{74}$$

can be generated within such a scheme. In this case

$$\mathcal{F}(\hat{a}^\dagger \hat{a}) = 1 - \hat{a}^\dagger \hat{a}, \tag{75}$$

and the corresponding Hamiltonian is

$$\hat{H} = (1-p)^{1/2} \left[1 - 4\hat{a}^\dagger \hat{a} + 2 \left(\hat{a}^\dagger \hat{a}\right)^2\right] + p^{1/2} \left[(1 - \hat{a}^\dagger \hat{a})\hat{a} + \hat{a}^\dagger (1 - \hat{a}^\dagger \hat{a})\right]. \tag{76}$$

Our generating Hamiltonian looks like a typical Hamiltonian found in nonlinear optics. This can be helpful while seeking an effective interaction that could be experimentally implemented. If we pump with a classical field (polarized along the y direction) $E_c = Ee^{-i\Omega t} + E^* e^{i\Omega t}$ a nonlinear medium characterized by linear and nonlinear susceptibilities $\chi^{(1)}$ and $\chi^{(3)}$ respectively, the coupling of the signal (polarized along the x direction), the pump and the output fields will be described by the following Hamiltonian [64]:

$$\begin{aligned}\hat{H}_{NL} = \chi^{(1)}_{xx}\hat{a}^\dagger\hat{a} + \chi^{(1)}_{yy}|E|^2 + +\chi^{(3)}_{xxxx}\hat{a}^{\dagger 2}\hat{a}^2 + \chi^{(3)}_{xyxy}\hat{a}^\dagger\hat{a}|E|^2 \\ +\chi^{(3)}_{yyyy}|E|^2|E|^2 + (\chi^{(1)}_{xy}\hat{a}^\dagger E + \chi^{(3)}_{xxyy}\hat{a}^{\dagger 2}E^2 \\ +\chi^{(3)}_{xxxy}\hat{a}^{\dagger 2}\hat{a}E + \chi^{(3)}_{xyyy}\hat{a}^\dagger|E|^2 E + \text{h.c.}).\end{aligned} \tag{77}$$

Both the pump and the signal are travelling waves propagating along the z axis (see Figure 13). The expression for our generator [in equation (76)] has the same

Figure 13: Diagram depicting a quantum state engineering scheme based on nonlinear interactions. A classical pump field supplies energy for the generation of nonclassical states $|\phi\rangle$, from the vacuum, $|0\rangle$, assisted by a nonlinear medium (NLM).

form as the Hamiltonian in equation (77), which could allow, at least in principle, the preparation of a nonlinear medium in such a way that we would end up with the Hamiltonian in equation (76). The linear and nonlinear susceptibilities have to assume specific values in order to make possible the correspondence between both Hamiltonians. For instance, there are no terms in equation (76) proportional to $\hat{a}^{\dagger 2}$, which means that we need $\chi^{(3)}_{xxyy} = 0$. By comparing terms proportional to $\hat{a}^\dagger$, $\hat{a}^{\dagger 2}\hat{a}$ and to $\hat{a}^\dagger\hat{a}$, we obtain

$$\chi^{(1)}_{xy}(E_0)E + \chi^{(3)}_{xyyy}|E|^2 E = -\chi^{(3)}_{xxxy}E = p^{1/2}, \tag{78}$$

$$\chi^{(1)}_{xx}(E_0) + \chi^{(3)}_{xyxy}|E|^2 = -\chi^{(3)}_{xxxx} = -2(1-p)^{1/2}. \tag{79}$$

These relations among the different susceptibilities are of the same form as the ones found in reference [64]. However, in our case we are generating a state which is the one-photon state coherently superposed to the vacuum, a fact which is embodied in the "tuning" parameter p. The susceptibilities $\chi^{(3)}_{ijkl}$ are of course a feature of the chosen crystal and the first-order susceptibilities $\chi^{(1)}_{ij}$ can be changed by the application of a static field E_0. We can control the generation procedure by adjusting the values of the pump field E and the static field E_0 in order to satisfy the relations (78) and (79). From equation (78) we see that the probability of having one photon in the field, p, is proportional to the pump field amplitude E, as one would expect. We note that the conditions in equations (78) and (79) connect the quantum superposition principle, represented by the coherent superposition of the vacuum with the one-photon state, with "macroscopic features" such as nonlinear susceptibilities in a crystal.

Despite its being possible, through this method, to construct Hamiltonians for the generation of virtually any state of the field, its experimental implementation may not be an easy task, especially for larger values of M. This is because it requires higher-order powers of $\hat{a}^\dagger\hat{a}$. For the generation of a different state, just as an example, consider the superposition of the vacuum state with a state containing, for instance, two photons

$$|\phi\rangle_{02} = A_0|0\rangle + A_2|2\rangle. \tag{80}$$

The generating Hamiltonian will read

$$\hat{H} = A_0\left[1 - 5(\hat{a}^\dagger\hat{a}) + 4(\hat{a}^\dagger\hat{a})^2 - (\hat{a}^\dagger\hat{a})^3\right] + \frac{A_2}{\sqrt{2}}\left[\mathcal{F}(\hat{a}^\dagger\hat{a})\hat{a}^2 + \hat{a}^{\dagger 2}\mathcal{F}(\hat{a}^\dagger\hat{a})\right],$$

$$\mathcal{F}(\hat{a}^\dagger\hat{a}) = 1 - \frac{3}{2}\hat{a}^\dagger\hat{a} + \frac{1}{2}(\hat{a}^\dagger\hat{a})^2.$$

The presence of the terms of the type $\hat{a}^{\dagger 2}\left(\hat{a}^\dagger\hat{a}\right)^2$ means that a fourth-order nonlinear susceptibility, $\chi^{(4)}$, should take part in the generation process of the state $|\phi\rangle_{02}$ in equation (80).

The methods presented here are ideal propositions in the sense that the destructive effects of the environment have not been taken into account. In section 4 we have already discussed the influence of the environment on a cavity field, and we have seen how the field purity is affected by losses. This gives us an insight into the destructive effects of the environment which might also occur during the quantum state field generation process.

6 Other intermediate states

In this chapter I have been primarily concerned with various aspects involving the binomial states of the quantized field, having as a central point the interpolating properties of the states themselves. Other intermediate, and more general interpolating states can be defined, and several theoretical studies of them can be found in the literature. A natural way of introducing other intermediate states might be by using generalized distributions which have as a special case the binomial distribution, for instance.

6.1 Negative binomial states

The negative binomial states constitute a class of interpolating states similar to binomial states [65,66]. They can be expressed in terms of the Fock state basis as

$$|p, M\rangle^- = \sum_{n=0}^{\infty} B_n^-(p, M)|n\rangle, \tag{81}$$

where M may be an arbitrary positive number (not necessarily integer), and

$$B_n^-(p, M) = \left[\binom{-M}{n} (-p^2)^n (1-p^2)^M \right]^{1/2} \quad (82)$$

is the "square root" of the negative binomial distribution [14]. In the limit of $p \to 0$, the state $|p, M\rangle^-$ reduces to the vacuum state. If $p \to 0$, $M \to \infty$ but with $p^2 M = \alpha^2$ finite, the negative binomial state reduces to the coherent state $|\alpha\rangle$. Nevertheless, differently to the binomial states, in the limit of $p \to 1$, they become the phase states

$$|\theta\rangle = \frac{1}{\sqrt{2}} \sum_{n=0}^{\infty} \exp(in\theta)|n\rangle, \quad (83)$$

instead of the Fock state $|M\rangle$. The phase states above are considered to be unphysical (not normalized), although they satisfy the completeness relation

$$\int_{-\pi}^{\pi} d\theta |\theta\rangle\langle\theta| = 1.$$

Moreover, they constitute the ground for the definition of an Hermitian phase operator [24] [see for instance equation (21)]. Therefore, the negative binomial states are *intermediate phase-coherent states*. I would like to point out that Fock states have an undefined phase; because the binomial state is an *intermediate Fock-coherent state*, it can be considered a kind of complementary to the negative binomial state.

The mean photon number and Mandel's Q parameter in a negative binomial state are

$$\langle\hat{n}\rangle = \frac{Mp^2}{1-p^2}; \qquad Q = \frac{p^2}{1-p^2} > 0, \qquad \langle\hat{n}\rangle = MQ, \quad (84)$$

i.e., the negative binomial states are basically super-Poissonian [14] (note that Mandel's factor is independent of the parameter M). They also exhibit quadrature squeezing [65].

6.2 Hypergeometric states

The hypergeometric states are a generalization of the binomial states, and can be written in terms of a (three-parameter) hypergeometric distribution [67]

$$|L, p, M\rangle = \sum_{n=0}^{M} H_n^M(p, L)|n\rangle, \quad (85)$$

$$H_n^M(p, L) = \left[\binom{Lp}{n} \binom{L(1-p)}{M-n} \right]^{1/2} \binom{L}{M}^{-1/2}, \quad (86)$$

[14] Apart from the vacuum and coherent state cases which are Poissonian.

and L being a real number satisfying $L \geq \{M/p, M/(1-p)\}$. Note that Lp is not generally an integer. In the limit of $L \to \infty$, the hypergeometirc state becomes the binomial state, or $|L, p, M\rangle \to |p, M\rangle$. The mean photon number in the hypergeometric state does not depend on L, and indeed it is the same as for the corresponding binomial state, $\langle \hat{n} \rangle = pM$. However, the photon number fluctuations depend on L. Mandel's parameter

$$Q = (1-p)\frac{L-M}{L-1} - 1 \tag{87}$$

is in general negative, meaning that the hypergeometric state is basically sub-Poissonian. It is worth mentioning that squeezing is particularly sensitive to the parameter L, especially when it gets small [67], i.e., far from the binomial state limit.

6.3 Superpositions of binomial states

Superpositions of binomial states are also nonclassical states [42, 68]. Particular superpositions of binomial states might interpolate between the Fock and the Schrödinger cat-type states [30]. Such states can be written as [42]

$$|\psi_s\rangle = \mathcal{N} \sum_{n=0}^{M} \left[1 + (-1)^n e^{i\phi}\right] B_n^M |n\rangle, \tag{88}$$

$$|\mathcal{N}|^{-2} = 2\left(1 + \cos\phi \sum_{n=0}^{M} (-1)^n |B_n^M|^2\right).$$

They may be generated from binomial states, either in nonlinear media or in interactions with single two-level atoms [42].

After fixing a particular value for the relative phase ϕ, e.g., $(\phi = \pi)$, we obtain for $M \neq 0$, in the limit of $p \to 1$:

$$|\psi_s\rangle_{p\to 1} = \begin{cases} |M\rangle & \text{for } M \text{ odd} \\ |M-1\rangle & \text{for } M \text{ even} \end{cases}. \tag{89}$$

We note that for $M = 0$ we have as a result the vacuum state $|0\rangle$. Thus in the Fock state limit ($p \to 1$), the resulting state will depend on whether M is even or odd. We also have that, in the limit of $p \to 0$, the superposition of binomial states above becomes the one-photon state, or $|\Psi_s\rangle_{p\to 0} = |1\rangle$, irrespective of M being odd or even, instead of the vacuum state $|0\rangle$, as occurs with ordinary binomial states in that particular limit. This characterizes a "state selection" process, in which combinations of values of the phase ϕ and M lead to different Fock states. A particularly curious situation happens for $\phi = \pi$ and $M = 2$. In this case, the resulting superposition state will be the one-photon state $|1\rangle$ for *any value of p*.

The statistical properties of superpositions of binomial states substantially differ from those of the binomial states and are highly sensitive to the relative phase ϕ. For instance, Mandel's Q parameter reads

$$Q = 2\mathcal{N}^2 p^2 M(M-1)\left[1 + \cos\phi \sum_{n=0}^{M-2} (-1)^n |B_n^{M-2}|^2\right] \Big/ \langle \hat{n} \rangle - \langle \hat{n} \rangle,$$

with

$$\langle \hat{n} \rangle = \langle \hat{a}^\dagger \hat{a} \rangle = 2\mathcal{N}^2 \left[pM + \cos\phi \sum_{n=0}^{M} (-1)^n n |B_n^M|^2 \right].$$

We may fix a value for ϕ, and then vary p from $p = 0$ to $p = 1$. We note that $Q \to -1$ as $p \to 1$. This is consistent with the fact that our superposition of binomial states approaches, in that limit, a Fock state (either $|M\rangle$ or $|M-1\rangle$). However, $Q \to 1$ as $p \to 0$, for $\phi = 0$ and $Q \to -1$ for $\phi = \pi$. The latter case is understandable since for $p \to 0$ with $\phi = \pi$, the superposition state becomes the one-photon state $|1\rangle$, which has, of course, $Q = -1$. If $\phi = 0$, though, the ratio $\langle \Delta \hat{n}^2 \rangle / \langle \hat{n} \rangle \to 2$, in such a way that $Q \to 1$. Moreover, for $\phi = \pi/2$, $Q \to 0$ as $p \to 0$. This might seem strange, because in the limit of $p \to 0$, for both $\phi = 0$ and $\phi = \pi/2$, the resulting state is the vacuum state $|0\rangle$ anyway. One has to be careful, though, in the way limits are taken. Namely, for a coherent state $|\alpha\rangle$, in which $\langle \Delta \hat{n}^2 \rangle / \langle \hat{n} \rangle = 1$, $Q = 0$ for any α. But it also happens that the vacuum state is a particular case of a Fock state $|n\rangle$, for $n = 0$. Because the Fock state basis is a discrete one, limits cannot be carried out in the same way as in the coherent state case, and this means that Mandel's Q parameter would be undefined for the vacuum state. In the binomial state case, as well as for a superposition of binomial states, however, it is possible to continuously "tune" them from any Fock state $|M\rangle$ to the vacuum state, and the result will depend on specific features of the states themselves (phases, etc.). These superpositions of binomial states exhibit squeezing in the $\hat{Y}$ quadrature [42] rather than in the $\hat{X}$ as it stands for the binomial states. This happens because superposing two states placed along the x axis will result in a noisier $\hat{X}$ quadrature, and hence squeezing may only exist in the $\hat{Y}$ quadrature.

7 Conclusion

I have presented a thorough discussion of the binomial states of the quantized field. They are characterized by two parameters, the probability of "occurrence" of a photon, p, and the maximum number of photons, M, having as special cases the coherent and Fock states, two of the most important states of the quantized field. I have investigated the way the states' nonclassical properties are modified by tuning p and M. Properties such as squeezing may substantially change, and we are able to understand those variations by considering the shape of quasiprobabilities in phase space. The interesting fact is that squeezing occurs just due to the deformation needed to "form" the Fock state, which is, in its turn, non-squeezed (they have Wigner functions circularly symmetric relative to the origin). Quasiprobabilities are indeed very interesting tools for investigating field behaviour while it interacts with matter. It is possible to "follow" the field evolution and find at which times the field reorganizes itself as a form of a pure state, for instance. If the initial field is highly sub-Poissonian, but has a small phase spread, the field will become almost pure periodically, although the atomic inversion retains its character of collapses and revivals. The connection between the field statistics of the initial field and its properties as time goes on is also made clear in the case of the (free) field confined in a lossy cavity. I have discussed how coherence is lost as the field parameters p and M are varied. Depending on the initial field, the quality of loss of coherence changes significantly, and I have discussed the connection between field properties, such as the quadrature noise, and the degree of maximum field purity, for instance. Generation schemes are essential to any discussion involving quantum states of light. Here I have presented a few of them, which can be extended for the generation of virtually any quantum state of light. Both nonunitary (micromaser-based) and unitary (nonlinear interactions) processes, accompanied by experimental suggestions, have been discussed. For completeness, I have also included a section in which are presented other intermediate states, which are more general, and somehow related to the binomial states themselves. For the most recent publications see, e.g., [69–73].

Acknowledgements

I would like to thank Dr. J.A. Roversi for a critical reading of the manuscript. This work was partially supported by Conselho Nacional de Desenvolvimento Científico e Tecnológico (CNPq), Brazil.

Bibliography

[1] Einstein, A., Über einen die Erzeugung und Verwandlung des Lichtes betreffenden heuristischen Gesichtspunkt. *Ann. Phys.* (Leipzig) (1905) **17** 132–148 [On a heuristic point of view about the creation and conversion of light. English translation in: D. Ter Haar, *The Old Quantum Theory*, pp. 91–107. Pergamon, New York, 1967].

[2] Einstein, A., Zur Quantentheorie der Strahlung. *Mitt. Phys. Ges. Zürich* (1916) **18** 47–62 [On the quantum theory of radiation. English translation in: D. Ter Haar, *The Old Quantum Theory*, pp. 167–183. Pergamon, New York, 1967].

[3] Lewis, G.N., Thermodynamic consequences of the conservation of photons. *Nature* (1926) **118** 874–875.

[4] Dirac, P.A.M., The quantum theory of the emission and absorption of radiation. *Proc. Roy. Soc. Lond.* A (1927) **114** 243–265.

[5] Nogues, G., Rauschenbeutel, A., Osnaghi, S., Brune, M., Raimond, J.M., and Haroche S., Seeing a single photon without destroying it. *Nature* (1999) **400** 239–242.

[6] Gordon, J.P., Zeiger, H.J., and Townes, C.H., Molecular microwave oscillator and new hyperfine structure in the microwave spectrum of $NH3$. *Phys. Rev.* (1954) **95** 282–284.

[7] Gordon, J.P., Zeiger, H.J., and Townes, C.H., Maser — new type of microwave, amplifier, frequency standard, and spectrometer. *Phys. Rev.* (1955) **99** 1264–1274.

[8] Basov, N.G. and Prokhorov, A.M., Application of molecular beams to the radio spectroscopic study of the rotation spectra of molecules. *Zh. Exp. Teor. Fiz.* (1954) **27** 431–438 (in Russian).

[9] Basov, N.G. and Prokhorov, A.M., On possible methods of producing active molecules for a molecular generator. *Zh. Exp. Teor. Fiz.* (1955) **28** 249 [*Sov. Phys. JETP* (1955) **1** 184].

[10] Maiman, T.H., Stimulated optical radiation in ruby. *Nature* (1960) **187** 493–494.

[11] Loudon R. *The Quantum Theory of Light*, Clarendon, Oxford, 1973.

[12] Schrödinger E., Der stetige Übergang von der Mikro- zur Makromechanik. *Naturswissenschaften* **14** (1926) 664–666.

[13] Glauber R., Coherent and incoherent states of radiation field. *Phys. Rev.* (1963) **131** 2766–2788.

[14] Feller, W., *An Introduction to Probability Theory and its Applications*. Wiley, New York, 1971.

[15] Stoler, D., Saleh, B.E.A., and Teich, M.C., Binomial states of the quantized radiation field. *Optica Acta* (1985) **32** 345–355.

[16] Fu, H.C. and Sasaki, R., Generalized binomial states: Ladder operator approach. *J. Math. Phys.* (1996) **29** 5637–5644.

[17] Fan, H.Y. and Jing S.C., Connection of a type of q-deformed binomial state with q-spin coherent states. *Phys. Rev.* A (1994) **50** 1909–1912.

[18] Stoler, D., Equivalence classes of minimum uncertainty packets. *Phys. Rev.* D (1970) **1** 3217–3219.

[19] Yuen, H., Two-photon coherent states of radiation field. *Phys. Rev.* A (1976) **13** 2226–2243.

[20] Dakna, M., Knöll, L., and Welsch, D.G., Photon added state preparation via conditional measurement on a beam splitter. *Opt. Commun.* (1998) **145** 309–321.

[21] Dodonov, V.V., Marchiolli, M.A., Korennoy, Y.A., Man'ko, V.I., and Moukhin Y.A., Dynamical squeezing of photon-added coherent states. *Phys. Rev.* A (1998) **58** 4087–4094.

[22] Dodonov, V.V., Malkin, I.A., and Man'ko, V.I., Even and odd coherent states and excitations of a singular oscillator. *Physica* (1974) **72** 597–615.

[23] Vidiella-Barranco, A. and Roversi, J.A., Statistical and phase properties of the binomial states of the electromagnetic field. *Phys. Rev.* A (1994) **50** 5233–5241.

[24] Pegg, D.T. and Barnett, S.M., Unitary phase operator in quantum mechanics. *Europhys. Lett.* (1988) **6** 483–487.

[25] Pegg, D.T. and Barnett, S.M., Quantum optical phase. *J. Mod. Opt.* (1994) **44** 225–264.

[26] Pegg, D.T. and Barnett, S.M., Phase measurement by projection synthesis. *Phys. Rev. Lett.* (1996) **76** 4148–4150.

[27] Hillery, M., O'Connell, R.F., Scully, M.O., and Wigner, E.P., Distribution functions in physics: Fundamentals. *Phys. Rep.* (1984) **106** 121–167.

[28] Schleich, W.P., Walls, D.F., and Wheeler, J.A., Area of overlap and interference in phase-space versus Wigner pseudoprobabilities. *Phys. Rev.* A (1988) **38** 1177–1186.

[29] Eiselt, J. and Risken, H., Quasiprobability distributions for the Jaynes–Cummings model with cavity damping. *Phys. Rev.* A (1991) **43** 346–360.

[30] Bužek, V., Vidiella-Barranco, A., and Knight, P.L., Superpositions of coherent states: Squeezing and dissipation. *Phys. Rev.* A (1992) **45** 6570–6585.

[31] Vogel, K. and Risken, H., Determination of quasiprobability distributions in terms of probability-distributions for the rotated quadrature phase. *Phys. Rev.* A (1989) **40** 2847–2849.

[32] Leonhardt, U., *Measuring the Quantum State of Light*, CUP, Cambridge, 1997.

[33] Moya-Cessa, H., Roversi, J.A., Dutra, S.M., and Vidiella-Barranco, A., Recovering coherence from decoherence: a method of quantum state reconstruction. *Phys. Rev.* A (1999) **60** 4029–4033.

[34] Cahill, K.E. and Glauber, R.J., Density operators and quasiprobability distributions. Phys. Rev. (1969) **177** 1882–1902.

[35] Moya-Cessa, H. and Knight, P.L., Series representation of quantum field quasiprobabilities. *Phys. Rev.* A (1993) **48** 2479–2481.

[36] Jaynes, E.T. and Cummings, F.W., Comparison of quantum and semiclassical radiation theories with application to the beam maser. *IEEE* (1963) **51** 89–108.

[37] Shore, B.W. and Knight, P.L., The Jaynes–Cummings model. *J. Mod. Opt.* (1993) **40** 1195–1238.

[38] Stenholm, S., Quantum theory of electromagnetic fields interacting with atoms and molecules. *Phys. Rep.* (1973) **6** 1–121.

[39] Phoenix, S.J.D., and Knight, P.L., Fluctuations and entropy in models of quantum optical resonance. *Ann. Phys.* (1988) **186** 381–407.

[40] Joshi, A. and Puri, R.R., Effects of the binomial field distribution on collapse and revival phenomena in the Jaynes–Cummings model. *J. Mod. Opt.* (1987) **34** 1421–1431.

[41] Goggin, M.E., Sharma, M.P., and Gavrielides, A., Effects of the binomial field distribution on collapse and revival phenomena in three-level atoms. *J. Mod. Opt.* (1990) **37** 99–108.

[42] Vidiella-Barranco, A. and Roversi, J.A., Quantum superpositions of binomial states of light. *J. Mod. Opt.* (1995) **42** 2475–2493.

[43] Von Neumann J., *Mathematical Foundations of Quantum Mechanics*, Princeton Univ. Press, Princeton, 1955.

[44] Araki, H. and Lieb, E.H., Entropy inequalities. *Commun. Math. Phys.* (1970) **18** 160–170.

[45] Gea-Banacloche, J., Collapse and revival of the state vector in the Jaynes–Cummings model: an example of state preparation by a quantum apparatus. *Phys. Rev. Lett.* (1990) **65** 3385–3388.

[46] Gea-Banacloche, J., Atom-state and field-state evolution in the Jaynes–Cummings model for large initial fields. *Phys. Rev.* A (1991) **44** 5913–5931.

[47] Moya-Cessa, H. and Vidiella-Barranco, A., Interaction of squeezed light with two-level atoms. *J. Mod. Opt.* (1992) **39** 2481–2499.

[48] Eberly, J.H., Narozhny, N.B., and Sanchez-Mondragon, J.J., Periodic spontaneous collapse and revival in a simple quantum model. *Phys. Rev. Lett.* (1980) **44** 1323–1326.

[49] *Decoherence and the Appearance of a Classical World in Quantum Theory* (D. Giulini, E. Joos, C. Kiefer, J. Kupsch, I.-O. Stamatescu, and H.D. Zeh, eds.). Springer, Berlin, 1996.

[50] Dutra, S.M., Decoherence as the process of generation of coherent states. *J. Mod. Opt.* (1998) **45** 759–764.

[51] Louisell, W.H., *Quantum Statistical Properties of Radiation.* Wiley, New York, 1973.

[52] Radmore, P.M. and Barnett, S.M., *Methods in Theoretical Quantum Optics*, Clarendon, Oxford, 1997.

[53] Vidiella-Barranco, A. and Roversi, J.A., Binomial states, statistical mixtures and dissipation. In: *Proceedings of the "XIX Encontro Nacional de Física da Matéria Condensada", Águas de Lindóia, Brazil, 2–6 Sept. 1996* (J.W. Tabosa, ed.), pp. 33–36.

[54] Dodonov, V.V. and Mizrahi, S.S., Competition between one- and two-photon absorption processes. *J. Phys.* A (1997) **30** 2915–2935.

[55] Dodonov, V.V. and Mizrahi, S.S., Exact stationary photon distributions due to competition between one- and two-photon processes. *J. of Phys. A* (1997) **30** 5657–5667.

[56] Brune, M., Haroche, S. Raimond, J.M., Davidovich, L., and Zagury, N., Manipulation of photons in a cavity by dispersive atom–field coupling: quantum nondemolition measurements and generation of Schrödinger cat states. *Phys. Rev.* A (1992) **45** 5193–5214.

[57] Lee, C.T., Photon antibunching in a free-electron laser. *Phys. Rev A* (1985) **31** 1213–

1215.

[58] Dattoli, G., Gallardo, J., and Torre, A., Binomial states of the quantized radiation field — comment. *J. Opt. Soc. Am. B* (1987) **4** 185–187.

[59] Vogel, K., Akulin, V.M. and Schleich, W.P., Quantum state engineering of the radiation field. *Phys. Rev. Lett.* (1993) **71** 1816–1819.

[60] Moussa, M.H.Y. and Baseia, B., Generation of the reciprocal-binomial state. *Phys Lett. A* (1998) **238** 223–226.

[61] Szabo, S., Adam, P., Janszky, J., and Domokos, P., Construction of quantum states o the radiation field by discrete coherent-state superpositions. *Phys. Rev.* A (1996) **5** 2698–2710.

[62] Law, C.K. and Eberly, J.H., Arbitrary control of a quantum electromagnetic field *Phys. Rev. Lett.* (1996) **76** 1055–1058.

[63] Vidiella-Barranco, A. and Roversi, J.A., Quantum state engineering via unitary trans formations. *Phys. Rev.* A (1998) **58** 3349–3352.

[64] Kilin, S.Y. and Horoshko, D.B., Fock state generation by the methods of nonlinea optics. *Phys. Rev. Lett.* (1995) **74** 5206–5207.

[65] Fu, H.C. and Sasaki, R., Negative binomial states of quantized radiation fields. *J Phys. Soc. Jpn.* (1997) **66** 1989–1994.

[66] Wang, X.G. and Fu, H.C., Negative binomial states of quantized radiation field an their excitations are nonlinear coherent states. *Mod. Phys. Lett. B* (1999) **13** 617–623

[67] Fu, H.C. and Sasaki, R., Hypergeometric states and their nonclassical properties. *J Math. Phys.* (1997) **38** 2154–2166.

[68] Abdalla, M.S., Mahran, M.H., and Obada, A.-S.F., Statistical properties of the eve binomial state. *J. Mod. Opt.* (1994) **41** 1889–1902.

[69] Fan, H.Y., Liu, N.L., and Li, H., Two types of correlated binomial – negative binomia states and their nonclassical properties. *Mod. Phys. Lett. B* (1999) **13** 1047–1054.

[70] Wang, X.G., Phase properties of hypergeometric states and negative hypergeometric states. *J. Opt.* B (2000) **2** 29–32.

[71] Wang, X.G. and Fu, H.C., Excited binomial states and excited negative binomia states of the radiation field and some of their statistical properties. *Int. J. Theor. Phys* (2000) **39** 1437–1444.

[72] El-Orany, F.A.A., Abdalla, M.S., Obada, A.-S.F., and Abd Al-Kader, G.M., Influence of squeezing operator on the quantum properties of various binomial states. *Int. J Mod. Phys.* B (2001) **15** 75–100.

[73] Liao, J., Wang, X.G., Wu, L.A., and Pan, S.H., Nonclassical properties and hole burning in the real and imaginary binomial states. *Int. J. Mod. Phys.* B (2001) **15** 2115–2123.

Chapter 6

Nonclassical states of light propagating in Kerr media

Ryszard Tanaś

1 Introduction

A century ago Planck discovered that it was possible to explain properties of black-body radiation by introducing discrete packets of energy, which we now call photons. It was the beginning of the quantum era. Nonclassical properties of optical fields have been the subject of intense studies for more than a decade now. Phenomena such as photon antibunching and squeezing, which have no classical analogues, are well known. To observe them it is essential to transform nonlinearly an optical field in one of a great variety of nonlinear optical processes. Among the nonlinear processes that can serve this purpose there is one that we wish to discuss here. This is the optical Kerr effect, or to be more precise the effect of self-phase modulation of the optical field propagating in a nonlinear, isotropic medium. If the intensity of light propagating through the nonlinear medium is sufficiently high, the refractive index of the medium depends on the intensity causing a nonlinear change of phase of the propagating field. In a classical optical Kerr configuration there are two beams: a strong, linearly polarized beam that makes the isotropic medium birefringent, and a weak, probe beam that detects the birefringence of the medium. An alternative variant of the optical Kerr effect is the propagation of a single, strong beam with elliptical polarization that serves the double purpose of both inducing and detecting the birefringence of the medium. As a result, one observes the rotation of the polarization ellipse of the propagating beam, an effect first observed by Maker *et al.* [1]. If the field is circularly polarized, a one-mode description of the field is possible, and we encounter a pure form of self-phase modulation, i.e., the effect in which the field modulates its own phase. This one-mode case is particularly interesting because it can be reduced to an anharmonic oscillator problem, which is probably the simplest problem to tackle when dealing with nonlinear systems.

Quantum description of the field propagating in a Kerr medium, which will be the subject of our concern in this chapter, reveals a number of interesting features that we are going to address here. Nonclassical effects associated with the anharmonic oscillator model have been discussed in many papers [2-24], and more generally, effects associated with light propagation in Kerr media have been studied in [25-44].

In this chapter we review a number of quantum features of optical fields associated with propagation of intense light through nonlinear, isotropic media. We begin with the classical description of field propagation, introducing the nonlinear polarization of the medium which enters as the source term into the approximate field equations obtained from the Maxwell equations in the slowly varying amplitude approximation. This establishes the classical background for further quantum considerations. Next, we quantize the field and construct the effective Hamiltonian from which we get equations of motion for the quantum fields. We discuss a number of quantum effects such as photon antibunching, squeezing, the formation of Schrödinger cats and kittens, changes in field polarization due to the quantum nature of the field, as well as the quantum description of the field phase. The characteristic feature of quantum evolution — the periodicity — is strongly affected by dissipation. We give exact analytical formulae describing the quantum evolution of the field, including dissipation. We have collected results illustrating various aspects of quantum evolution and we believe that this review, although far from complete, will be a useful source of information on the subject.

2 Kerr media: classical background

Propagation of strong laser light through a nonlinear medium makes the isotropic medium birefringent; for example, an elliptical polarization of the light is rotating as the beam traverses the medium, an effect observed by Maker *et al.* [1] in the pioneering years of nonlinear optics. The refractive index of the medium depends on the intensity of light, the effect usually referred to as the *optical Kerr effect*, and the medium exhibiting this effect is called a *Kerr medium*. There is no need for field quantization to describe the birefringence induced by strong light. It can be explained with classical fields. However, we are interested here in the nonclassical properties of light fields propagating in a nonlinear medium, but before we start the quantum description let us briefly summarize the classical results.

The third-order polarization induced in the medium by a monochromatic light field of frequency ω can be written in the form [45,46]

$$P_i^{(+)} = \sum_{jkl} \chi_{ijkl}(-\omega, -\omega, \omega, \omega) E_j^{(-)} E_k^{(+)} E_l^{(+)} , \tag{1}$$

where $\chi_{ijkl}(-\omega, -\omega, \omega, \omega)$ is the third-order nonlinear susceptibility tensor of the medium and the electromagnetic field is decomposed into the positive- and negative-frequency parts

$$E_i(z,t) = E_i^{(+)} \mathrm{e}^{-i(\omega t - kz)} + E_i^{(-)} \mathrm{e}^{i(\omega t - kz)} , \tag{2}$$

with $k = n(\omega)\omega/c$ and $n(\omega)$ being the linear refractive index of the medium. We assume that the field propagates in direction z of the laboratory coordinate frame. The nonlinear susceptibility tensor $\chi_{ijkl} = \chi_{ijkl}(-\omega,-\omega,\omega,\omega)$ of an isotropic medium has the form (we suppress the ω dependence in order to shorten the notation)

$$\chi_{ijkl} = \chi_{xxyy}\,\delta_{ij}\delta_{kl} + \chi_{xyxy}\,\delta_{ik}\delta_{jl} + \chi_{xyyx}\,\delta_{il}\delta_{jk}\,. \tag{3}$$

Taking into account the permutation symmetry of the tensor χ with respect to the first and second pairs of indices, we have additionally $\chi_{xyxy} = \chi_{xyyx}$. This allows us to write (1) in vector form with only two nonlinearity parameters

$$\mathbf{P}^{(+)} = \chi_{xxyy}\,\mathbf{E}^{(-)}(\mathbf{E}^{(+)}\cdot\mathbf{E}^{(+)}) + 2\chi_{xyxy}\,(\mathbf{E}^{(-)}\cdot\mathbf{E}^{(+)})\mathbf{E}^{(+)}\,, \tag{4}$$

where the positive-frequency part of the electric field amplitude can be written in two alternative forms

$$\mathbf{E}^{(+)} = E_x^{(+)}\hat{\mathbf{e}}_x + E_y^{(+)}\hat{\mathbf{e}}_y = E_+^{(+)}\hat{\mathbf{e}}_+ + E_-^{(+)}\hat{\mathbf{e}}_-\,. \tag{5}$$

In (5), $\hat{\mathbf{e}}_x$ and $\hat{\mathbf{e}}_y$ are the unit vectors, $E_x^{(+)}$ and $E_y^{(+)}$ are the components of the field amplitudes in a Cartesian basis; alternatively $\hat{\mathbf{e}}_+$ and $\hat{\mathbf{e}}_-$ are the unit vectors, $E_+^{(+)}$ and $E_-^{(+)}$ are the amplitude components in a circular basis. The positive- and negative-frequency parts of the field amplitudes $\mathbf{E}^{(-)}$ and $\mathbf{E}^{(+)}$ are, for classical fields, complex conjugate to each other. For quantum fields they become Hermitian conjugate operators. The relations between the two bases are given by

$$\hat{\mathbf{e}}_\pm = \frac{1}{\sqrt{2}}\left(\hat{\mathbf{e}}_x \pm i\hat{\mathbf{e}}_y\right)\,, \quad E_\pm^{(+)} = \frac{1}{\sqrt{2}}\left(E_x^{(+)} \mp iE_y^{(+)}\right)\,, \tag{6}$$

and additionally the following relations hold true

$$\begin{aligned} \left|E_x^{(+)}\right|^2 + \left|E_y^{(+)}\right|^2 &= \left|E_+^{(+)}\right|^2 + \left|E_-^{(+)}\right|^2\,, \\ 2\,E_+^{(+)}E_-^{(+)} &= E_x^{(+)2} + E_y^{(+)2}\,. \end{aligned} \tag{7}$$

From the Maxwell equations, in the slowly varying amplitude approximation, the amplitude of the field propagating through the medium obeys the equation [46]

$$\frac{\partial \mathbf{E}^{(+)}}{\partial z} = i\,\frac{2\pi\omega^2}{k\,c^2}\mathbf{P}^{(+)}\,, \tag{8}$$

which on inserting (4) leads to the system of coupled equations

$$\begin{aligned} \frac{\partial E_x^{(+)}}{\partial z} &= i\,\frac{2\pi\omega^2}{k\,c^2}\Big[\chi_{xxyy}\,E_x^{(-)}\left(E_x^{(+)2} + E_y^{(+)2}\right) \\ &\qquad + 2\chi_{xyxy}\left(\left|E_x^{(+)}\right|^2 + \left|E_y^{(+)}\right|^2\right)E_x^{(+)}\Big]\,, \\ \frac{\partial E_y^{(+)}}{\partial z} &= i\,\frac{2\pi\omega^2}{k\,c^2}\Big[\chi_{xxyy}\,E_y^{(-)}\left(E_x^{(+)2} + E_y^{(+)2}\right) \\ &\qquad + 2\chi_{xyxy}\left(\left|E_x^{(+)}\right|^2 + \left|E_y^{(+)}\right|^2\right)E_y^{(+)}\Big]\,. \end{aligned} \tag{9}$$

For isotropic media the circular basis is more natural, and employing (6) and (7) gives us the equations

$$\begin{aligned}\frac{\partial E_{\pm}^{(+)}}{\partial z} &= i\,\frac{2\pi\omega^2}{k\,c^2}\left[2\chi_{xxyy}\left|E_{\mp}^{(+)}\right|^2\right.\\&\left.+\,2\chi_{xyxy}\left(\left|E_{+}^{(+)}\right|^2+\left|E_{-}^{(+)}\right|^2\right)\right]E_{\pm}^{(+)},\end{aligned}\tag{10}$$

from which it immediately follows that

$$\frac{\partial}{\partial z}\left|E_{\pm}^{(+)}\right|^2=0\,,\tag{11}$$

i.e., intensities of both circular components are conserved during the propagation. Of course, the conservation of the intensities of the two circular components is only valid for media without absorption (real χ). This conservation makes the solution of (10) a trivial task, and we get the simple exponential solution [47]

$$E_{\pm}^{(+)}(z)=\exp(iz\Phi_{\pm})E_{\pm}^{(+)}(0)\tag{12}$$

with

$$\begin{aligned}\Phi_{\pm} &= \frac{2\pi\omega^2}{k\,c^2}\left[2\chi_{xyxy}\left|E_{\pm}^{(+)}\right|^2+2(\chi_{xxyy}+\chi_{xyxy})\left|E_{\mp}^{(+)}\right|^2\right]\\&= \frac{2\pi\omega^2}{k\,c^2}\left[(\chi_{xxyy}+2\chi_{xyxy})\left(\left|E_{+}^{(+)}\right|^2+\left|E_{-}^{(+)}\right|^2\right)\right.\\&\left.\mp\,\chi_{xxyy}\left(\left|E_{+}^{(+)}\right|^2-\left|E_{-}^{(+)}\right|^2\right)\right].\end{aligned}\tag{13}$$

Knowing the solution (12), in the circular basis, it is straightforward to perform the inverse transformation to the Cartesian basis and write down the solutions for the Cartesian components of the field amplitudes

$$E_x^{(+)}(z)=\frac{1}{2}\left[\left(\mathrm{e}^{iz\Phi_+}+\mathrm{e}^{iz\Phi_-}\right)E_x^{(+)}(0)-i\left(\mathrm{e}^{iz\Phi_+}-\mathrm{e}^{iz\Phi_-}\right)E_y^{(+)}(0)\right],$$

$$E_y^{(+)}(z)=\frac{1}{2}\left[i\left(\mathrm{e}^{iz\Phi_+}-\mathrm{e}^{iz\Phi_-}\right)E_x^{(+)}(0)+\left(\mathrm{e}^{iz\Phi_+}+\mathrm{e}^{iz\Phi_-}\right)E_y^{(+)}(0)\right].$$

The solution (12), together with (13), shows that the nonlinear interaction in a Kerr medium appears as an intensity-dependent phase of the field (self-phase modulation or intensity-dependent refractive index). Since the two circular components of the field accumulate different phases along the path z in the medium, the polarization of the field alters. For elliptically polarized light the polarization ellipse undergoes self-induced rotation [1].

The polarization of light propagating in the medium can be conveniently expressed by the Stokes parameters which, in terms of the field amplitudes, can be written in the form [48]

$$
\begin{aligned}
s_0 &= \left|E_x^{(+)}\right|^2 + \left|E_y^{(+)}\right|^2 = \left|E_+^{(+)}\right|^2 + \left|E_-^{(+)}\right|^2 , \\
s_1 &= E_x^{(-)}E_x^{(+)} - E_y^{(-)}E_y^{(+)} = E_+^{(-)}E_-^{(+)} + E_-^{(-)}E_+^{(+)} \\
&= s_0 \cos 2\eta \cos 2\theta , \\
s_2 &= E_x^{(-)}E_y^{(+)} + E_y^{(-)}E_x^{(+)} = -i\left(E_+^{(-)}E_-^{(+)} - E_-^{(-)}E_+^{(+)}\right) \\
&= s_0 \cos 2\eta \sin 2\theta , \\
s_3 &= -i\left(E_x^{(-)}E_y^{(+)} - E_y^{(-)}E_x^{(+)}\right) = \left|E_+^{(+)}\right|^2 - \left|E_-^{(+)}\right|^2 \\
&= s_0 \sin 2\eta ,
\end{aligned}
\tag{14}
$$

where $0 \leq \theta \leq \pi$ defines the azimuth of the polarization ellipse, i.e., the angle between the major axis of the polarization ellipse and the x axis of the Cartesian coordinate frame, and $-\pi/4 \leq \eta \leq \pi/4$ defines the ellipticity parameter; $\tan\eta$ is the ratio of the minor axis and the major axis of the polarization ellipse and the sign defines its helicity (handedness).

One can also define the degree of polarization

$$
\mathcal{P} = \sqrt{s_1^2 + s_2^2 + s_3^2}\,/s_0 . \tag{15}
$$

For completely polarized light $\mathcal{P} = 1$, for unpolarized light $\mathcal{P} = 0$, and in between the light is partially polarized with the degree $\mathcal{P}$.

Classically, as is evident from (13), the change in phase depends on two Stokes parameters s_0 (total intensity) and s_3. For linearly polarized light $\eta = s_3 = 0$ and the overall change in phase during the propagation depends only on the total intensity s_0, and there is no change in the phase difference between the two circular components, which means that the linearly polarized light should preserve its polarization. Another important and interesting case is that of circular polarization ($\eta = -\pi/4, \pi/4$). In this case the problem reduces to one-mode propagation, which is equivalent to a simple anharmonic oscillator model. It is also easy to check that the Stokes parameter s_3 preserves its value, i.e., the ellipticity of the polarization ellipse does not alter. Moreover, the degree of polarization $\mathcal{P}$ remains unity if it was unity initially. Some of these obvious classical "truths" appear not to be "truths" when the field becomes a quantum field. We shall try to clarify this situation.

3 Quantum fields

A rigorous quantum treatment of macroscopic fields in nonlinear, dispersive dielectrics is not at all a trivial task, and one can meet serious problems of a rather fundamental nature [49–53] when doing it the wrong way. It often happens that when the Hamiltonian corresponds to the correct classical energy, the equations of motion generated by this Hamiltonian are not correct, and, on the contrary, it is possible to get correct equations of motion from the Hamiltonian that do not correspond to the correct energy. The source of the problems is the fact that it is the displacement field $\mathbf{D}$ which is the canonical momentum to the vector potential $\mathbf{A}$, and thus the displacement field modes should be quantized rather than the electric field $\mathbf{E}$ modes, as is usually done. The modes in this case are the collective matter-field modes instead of pure field modes and their excitations have different physical interpretation than photons in a vacuum field. However, in quantum optics, usually a very simple quantization scheme [54,55] is used in which classical field amplitudes are replaced by appropriate operators in the Hilbert space, and the effective Hamiltonian is constructed in such a way that it leads to the Heisenberg equations of motion for the field operators which reproduce the classical equations when the field operator character is neglected. We shall follow such a simple scheme of quantization here.

To take into account the quantum character of the field propagating in the Kerr medium we express the electric field amplitude (5) in terms of the annihilation operators for the corresponding modes of the field

$$\mathbf{E}^{(+)} = i\sqrt{\frac{2\pi\hbar\omega}{n_\omega^2 V}}\left(a_x\hat{\mathbf{e}}_x + a_y\hat{\mathbf{e}}_y\right) = i\sqrt{\frac{2\pi\hbar\omega}{n_\omega^2 V}}\left(a_+\hat{\mathbf{e}}_+ + a_-\hat{\mathbf{e}}_-\right), \tag{16}$$

where n_ω is the linear refractive index of the medium, V is the volume of quantization, and a_x, a_y (a_+, a_-) are the annihilation operators for the two orthogonal modes in the Cartesian (circular) coordinate frame satisfying the bosonic commutation relations

$$\begin{aligned} [a_x, a_x^\dagger] &= [a_y, a_y^\dagger] = [a_+, a_+^\dagger] = [a_-, a_-^\dagger] = 1\,, \\ [a_x, a_y] &= [a_x, a_y^\dagger] = [a_+, a_-] = [a_+, a_-^\dagger] = 0\,. \end{aligned} \tag{17}$$

Similarly to classical relations (6) and (7), we have the following relations between the two bases

$$\hat{\mathbf{e}}_\pm = \frac{1}{\sqrt{2}}\left(\hat{\mathbf{e}}_x \pm i\hat{\mathbf{e}}_y\right), \quad a_\pm = \frac{1}{\sqrt{2}}\left(a_x \mp i a_y\right), \tag{18}$$

$$a_x^\dagger a_x + a_y^\dagger a_y = a_+^\dagger a_+ + a_-^\dagger a_-\,, \quad 2\,a_+ a_- = a_x^2 + a_y^2\,. \tag{19}$$

To write the quantum equations of motion for the field operators we need the appropriate Hamiltonian. It is easy to check that the interaction Hamiltonian that will correctly describe the field evolution in a Kerr medium can be written in two equivalent forms

$$H = \frac{\hbar\kappa}{2}\left[:(a_x^\dagger a_x + a_y^\dagger a_y)^2: + \left(d - \frac{1}{2}\right)(a_x^{\dagger 2} + a_y^{\dagger 2})(a_x^2 + a_y^2)\right], \tag{20}$$

$$H = \frac{\hbar\kappa}{2}\left[a_+^{\dagger 2} a_+^2 + a_-^{\dagger 2} a_-^2 + 4d\, a_+^\dagger a_-^\dagger a_- a_+\right], \tag{21}$$

where :: is used to denote the normal ordering of the operators, the nonlinear coupling constant κ is related to the nonlinear susceptibility of the medium

$$\kappa = \frac{V}{\hbar}\left[\frac{2\pi\hbar\omega}{n^2(\omega)V}\right]^2 2\chi_{xyxy}\,, \tag{22}$$

and the asymmetry parameter d, given by

$$2d = 1 + \chi_{xxyy}/\chi_{xyxy} \,, \tag{23}$$

describes the asymmetry of the nonlinear properties of the medium. If the nonlinear susceptibility tensor is symmetric with respect to all its indices, then $d = 1$, but generally $d \neq 1$. For atoms with a degenerate one-photon transition Ritze [27] has obtained the following results

$$d = \begin{cases} (2J-1)(2J+3)/[2(2J^2+2J+1)] \,, & J \leftrightarrow J \,, \\ (2J^2+3)/[2(6J^2-1)] \,, & J \leftrightarrow J-1 \end{cases} \tag{24}$$

The coupling between the modes depends crucially on this asymmetry parameter.

From the form (21) of the interaction Hamiltonian it is obvious that both $a_+^\dagger a_+$ and $a_-^\dagger a_-$ commute with the Hamiltonian, i.e., the number of photons in each of the circular polarization modes is a constant of motion. However, as is evident from the form (20) of the Hamiltonian, the numbers of photons $a_x^\dagger a_x$ and $a_y^\dagger a_y$ for the two Cartesian components of the field do not commute with the Hamiltonian, i.e., they are not constants of motion. This fact has very important consequences: the linear polarization of the field, contrary to the classical field, is not preserved when the quantum field propagates in the isotropic, nonlinear Kerr medium. Here we see the advantage of the circular basis over the Cartesian basis in describing quantum properties of light propagating in the Kerr medium.

The equations of motion for the annihilation operators, which correspond to classical equations of motion for the slowly varying amplitudes, can be obtained from the Hamiltonian (21) as the Heisenberg equations

$$\frac{da_\pm}{dt} = \frac{1}{i\hbar}[a_\pm, H] = -i\kappa(a_\pm^\dagger a_\pm + 2d a_\mp^\dagger a_\mp)\, a_\pm \,. \tag{25}$$

Comparing (5) and (16) and replacing time t by $-n(\omega)z/c$, it is easy to check that the quantum equations (25) are equivalent to the classical equations (10), but now the field amplitudes are the operator quantities. The replacement of t by $-n(\omega)z/c$ means the transition from the problem of the field in a cavity, when the time evolution of the field amplitudes is studied, to the problem of travelling waves, when the field amplitudes at a distance z in the medium are looked for. In the case of the propagation problem we deal with localized field operators, but we require the same commutation relations (17) for both the cavity field operators and the localized operators. This simplistic approach is not always applicable, but it is sufficient for our purposes here.

Since the numbers of photons in the two circular modes are both constants of motion, the solutions to equations (25) are, as for classical fields, given by the exponentials

$$a_\pm(\tau) = \exp\left\{ i\tau \left[a_\pm^\dagger(0) a_\pm(0) + 2d\, a_\mp^\dagger(0) a_\mp(0) \right] \right\} a_\pm(0) \,, \tag{26}$$

where $\tau = \kappa n(\omega)z/c$ (or $\tau = -\kappa t$, in a cavity problem). Despite the similarity of solution (26) to the classical solution (12), there is one fundamental difference between the two: solution (26) is the operator solution and measurable quantities can be extracted from it only after taking the expectation values of the operator solutions in the initial state of the field. The dimensionless parameter τ in (26) can be treated as either the elapsed time or the distance in the medium, depending on the problem considered. For simplicity we shall refer to it as "time" later on.

Using solution (26) and relations (18), we obtain the operator solutions for the Cartesian components of the field in the form

$$a_x(\tau) = \frac{1}{\sqrt{2}}\left[a_+(\tau) + a_-(\tau)\right], \quad a_y(\tau) = \frac{i}{\sqrt{2}}\left[a_+(\tau) - a_-(\tau)\right]. \tag{27}$$

It is also convenient to define the Hermitian Stokes operators [56]

$$\begin{aligned} S_0 &= a_x^\dagger a_x + a_y^\dagger a_y = a_+^\dagger a_+ + a_-^\dagger a_- \,, \\ S_1 &= a_x^\dagger a_x - a_y^\dagger a_y = a_+^\dagger a_- + a_-^\dagger a_+ \,, \\ S_2 &= a_x^\dagger a_y + a_y^\dagger a_x = -i(a_+^\dagger a_- - a_-^\dagger a_+) \,, \\ S_3 &= -i(a_x^\dagger a_y - a_y^\dagger a_x) = a_+^\dagger a_+ - a_-^\dagger a_- \end{aligned} \tag{28}$$

with the commutation relations

$$[S_j, S_k] = 2i\epsilon_{jkl} S_l \,, \quad [S_j, S_0] = 0 \,, \quad (j,k,l = 1,2,3) \,. \tag{29}$$

The noncommutability of the Stokes operators precludes the simultaneous measurement of the physical quantities represented by them. Apart from a factor of 2, the operators S_j $(j = 1, 2, 3)$ coincide with the components of the angular momentum operator, while S_0 represents the total number operator. Moreover, we have

$$S_1^2 + S_2^2 + S_3^2 = S_0(S_0 + 2) \,. \tag{30}$$

Expectation values of operators (28) give the Stokes parameters, i.e.,

$$s_j = \langle S_j \rangle \,, \quad (j = 0, ..., 3) \,, \tag{31}$$

which correspond to the classical Stokes parameters (14).

Looking at the form of solutions (26) it is tempting to introduce the operator equivalent of the classical phase $\Phi_\pm$, as given by (13); this, however, would mean the decorrelation of the exponential from the operators $a_\pm(0)$ and would thus lead to completely wrong results whenever the quantum properties of the field play an important role. This will become clear later on.

4 One-mode field: anharmonic oscillator model

The main goal of this chapter is to present the nonclassical properties of the field propagating in the isotropic, nonlinear Kerr medium. When there is no dissipation in the medium, the quantum properties of the field are defined by the solutions (26), or (27), and the initial state of the field. Applying the operator solution (26) to the initial state of the field we find the state of the field at time (or distance) τ. The state evolves in τ and resulting quantum properties of the field depend on τ. We shall discuss such properties for certain initial states of the field. Before proceeding any further, we note that when the field entering the medium is circularly polarized, say with the polarization vector $\hat{\mathbf{e}}_+$, then the state of the field in the orthogonal mode with the polarization vector $\hat{\mathbf{e}}_-$ is in the vacuum state, *i.e*, $a_-|0\rangle = 0$. This reduces the problem to the one-mode problem, and we can omit the operators $a_-^\dagger$ and a_- in the exponential (26) arriving at the widely discussed problem the anharmonic oscillator. It is important to emphasize here that such reduction is only possible for circularly polarized light, but not for linearly polarized light, because, as already mentioned, the linear polarization of the quantum field is not preserved during the evolution.

The simplest case of light propagation in a Kerr medium is thus a single mode of circularly polarized light, for which the operator solution (26) simplifies to

$$a(\tau) = \exp[i\tau a^\dagger(0)a(0)]\, a(0)\,, \tag{32}$$

where we have dropped the mode index for simplicity. The solution (32) can be written in a different form using the evolution operator,

$$a(\tau) = U_K^\dagger(\tau)a(0)U_K(\tau)\,, \tag{33}$$

$$U_K(\tau) = \exp\left[i\frac{\tau}{2}a^{\dagger 2}(0)a^2(0)\right] = \exp\left[i\frac{\tau}{2}\hat{n}(\hat{n}-1)\right]\,, \tag{34}$$

where $\hat{n} = a^\dagger(0)a(0)$ is the photon number operator. Assuming that the initial state of the field is $|\psi_0\rangle$, the expectation value of the field operator is given by

$$\langle a(\tau)\rangle = \langle\psi_0|a(\tau)|\psi_0\rangle = \langle\psi_0|U_K^\dagger(\tau)a(0)U_K(\tau)|\psi_0\rangle = \langle\psi(\tau)|a(0)|\psi(\tau)\rangle\,,$$

where the state of the field at time τ is given by

$$|\psi(\tau)\rangle = U_K(\tau)|\psi_0\rangle = \exp\left[i\frac{\tau}{2}{a^\dagger}^2(0)a^2(0)\right]|\psi_0\rangle = \exp\left[i\frac{\tau}{2}\hat{n}(\hat{n}-1)\right]|\psi_0\rangle\,.$$

If the initial state of the field is a coherent state $|\alpha_0\rangle$ then the state after time τ takes the form

$$\begin{aligned}|\psi_K\rangle &= U_K(\tau)|\alpha_0\rangle = \exp\left[i\frac{\tau}{2}\hat{n}(\hat{n}-1)\right]|\alpha_0\rangle \\ &= \exp(-|\alpha_0|^2/2)\sum_{n=0}^{\infty}\frac{\alpha_0^n}{\sqrt{n!}}\exp\left[i\frac{\tau}{2}n(n-1)\right]|n\rangle\,. \end{aligned}\tag{35}$$

States of the form (35) have very interesting nonclassical features and they are usually referred to as the *Kerr states*. The expectation value in the Kerr state (35) of the annihilation operator takes the form $[a = a(0)]$

$$\begin{aligned}\langle\psi_K|a|\psi_K\rangle &= \langle\alpha_0|a(\tau)|\alpha_0\rangle = \langle\alpha_0|\,e^{i\tau a^\dagger a}a\,|\alpha_0\rangle \\ &= \alpha_0\exp[|\alpha_0|^2(e^{i\tau}-1)]\,,\end{aligned}\tag{36}$$

and for the square of the annihilation operator we get

$$\langle\psi_K|a^2|\psi_K\rangle = \langle\alpha_0|a^2(\tau)|\alpha_0\rangle = \alpha_0^2\exp[i\tau + |\alpha_0|^2(e^{2i\tau}-1)]\,. \tag{37}$$

The mean number of photons is equal to

$$\langle\psi_K|a^\dagger a|\psi_K\rangle = \langle\alpha_0|a^\dagger(\tau)a(\tau)|\alpha_0\rangle = |\alpha_0|^2 \tag{38}$$

and is the same as for the initial coherent state. The expectation values calculated above are very useful and will be exploited later.

4.1 Squeezing

By inspection of the solution (32), we find $a^\dagger(\tau)a(\tau) = a^\dagger(0)a(0) = a^\dagger a$, which means that the photon number operator is a constant of motion, and thus any function of the number operator is also a constant of motion, which implies that the photon statistics does not alter during the evolution. So, an for initially coherent state with Poissonian photon statistics, the statistics remains Poissonian all the time. The nonlinear change in phase present in (32), or (35), does not affect the photon statistics of the field. The situation is completely different when quantum fluctuations of the field itself are concerned. Let us introduce the Hermitian operator

$$X_\theta = a\mathrm{e}^{-i\theta} + a^\dagger \mathrm{e}^{i\theta}\,, \tag{39}$$

which for $\theta = 0$ corresponds to the in-phase quadrature component of the field and for $\theta = \pi/2$ to the out-of-phase component. We will also use the notation

$$X_{\theta=0} = X_1\,, \quad X_{\theta=\pi/2} = X_2 \tag{40}$$

to describe the quadrature components, the commutator of which is equal to

$$[X_1, X_2] = 2i\,. \tag{41}$$

Quantum fluctuations of the quadrature components of the field are measured by their variances. Generally, the variance of operator (39) is given by

$$\begin{aligned} V_\theta &= \langle(\Delta X_\theta)^2\rangle = \langle X_\theta^2\rangle - \langle X_\theta\rangle^2 \\ &= 2\mathrm{Re}\left[\langle(\Delta a)^2\rangle \mathrm{e}^{-2i\theta}\right] + 2\left[\langle a^\dagger a\rangle - |\langle a\rangle|^2\right] + 1\,, \end{aligned} \tag{42}$$

where

$$\langle(\Delta a)^2\rangle = \langle a^2\rangle - \langle a\rangle^2\,. \tag{43}$$

For a coherent state (or vacuum), quantum fluctuations are isotropic and the variance (42) is equal to unity independently of θ.

The state of the field is said to be *squeezed state* [57] if for some θ the variance (42) becomes smaller than unity, and perfect squeezing is obtained when $V_\theta = 0$. On introducing the normally ordered variance [58] of the operator (39), which means neglecting the unity in formula (42) that comes from the boson commutation rules, the condition for squeezing can be reformulated: the state of the field is squeezed if the normally ordered variance $\langle : (\Delta X_\theta)^2 : \rangle$ becomes negative for some θ. Using (36)-(38), we get from (42) (after dropping the unity) the following expressions [2] for the normally ordered variances of the two quadrature omponents

$$\begin{aligned} \left.\begin{matrix}\langle: (\Delta X_1(\tau))^2 :\rangle \\ \langle: (\Delta X_2(\tau))^2 :\rangle\end{matrix}\right\} &= 2|\alpha_0|^2 \left\{1 - \exp[2|\alpha_0|^2(\cos\tau - 1)]\right\} \\ &\pm 2\text{Re}\left\{\alpha_0^2 \exp[i\tau + |\alpha_0|^2(e^{2i\tau} - 1)] - \alpha_0^2 \exp[2|\alpha_0|^2(e^{i\tau} - 1)]\right\}. \end{aligned} \quad (44)$$

Negative values of one of the variances in (44) mean squeezing in the corresponding quadrature of the field. The two quadratures are plotted, assuming α_0 real and $\tau = 1 \times 10^{-6}$, in Figure 1, showing oscillatory behaviour with regions of considerable squeezing. This means that the Kerr states (35) are nonclassical and exhibit

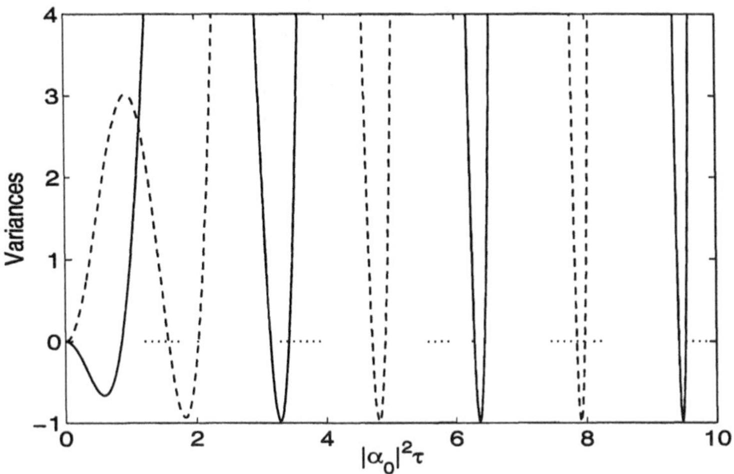

Figure 1: Normally ordered variances $\langle: (\Delta X_1)^2 :\rangle$ (solid line) and $\langle: (\Delta X_2)^2 :\rangle$ (dashed line) versus $|\alpha_0|^2\tau$.

a high degree of squeezing for appropriately chosen evolution time τ. It is clear, however, that the Kerr states are not the minimum uncertainty states. The choice $\theta = 0$, or $\theta = \pi/2$, is not always the best for maximizing the squeezing obtainable in the anharmonic oscillator model. Differentiating V_θ with respect to θ leads to the angles θ_+ and θ_-, corresponding to the maximum and minimum variances, given by the relation

$$\exp(2i\theta_\pm) = \pm\sqrt{\langle(\Delta a)^2\rangle / \langle(\Delta a^\dagger)^2\rangle}, \quad (45)$$

which leads to the extremal variances in the form [59]

$$V_\pm = \langle(\Delta X_\pm)^2\rangle = \pm 2\sqrt{\langle(\Delta a)^2\rangle\langle(\Delta a^\dagger)^2\rangle} + 2\left[\langle a^\dagger a\rangle - |\langle a\rangle|^2\right] + 1,$$

where $X_\pm = X_{\theta=\theta_\pm}$. This immediately gives us the condition for *principal squeezing* introduced by Lukš *et al.* [60]

$$\langle\Delta a^\dagger \Delta a\rangle - |\langle(\Delta a)^2\rangle| < 0. \quad (46)$$

Loudon [59] has shown that the variance (42) can be rewritten in the alternative form

$$V_\theta = \langle(\Delta X_-)^2\rangle \cos^2(\theta - \theta_-) + \langle(\Delta X_+)^2\rangle \sin^2(\theta - \theta_-)\,, \tag{47}$$

which geometrically represents Booth's elliptical lemniscate in polar coordinates. It is demonstrated in Figure 2 for the first minimum of $\langle \Delta X_1^2(\tau)\rangle$ which appears for $|\alpha_0|^2\tau = 0.59$. The circle of unit radius marks the level of vacuum fluctuations. Whenever the lemniscate is inside the circle the field is squeezed. For Kerr states (35) one gets from (45)

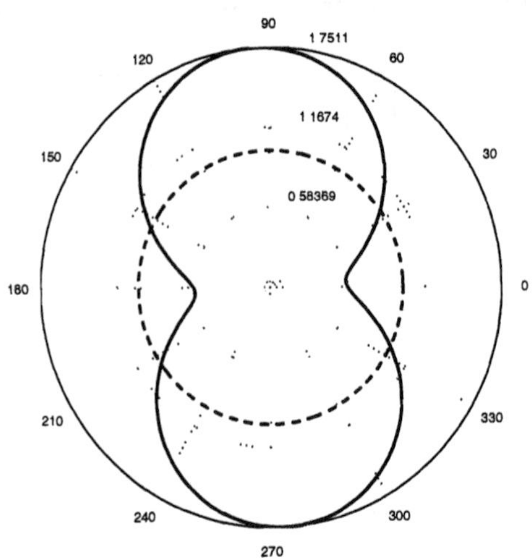

Figure 2: Plot of $\sqrt{V_\theta(\tau)}$ as a function of θ for $|\alpha_0|^2\tau = 0.59$; the dashed circle marks the level of vacuum fluctuations.

$$\exp(2i\theta_\pm) = \pm\exp(2i\varphi_0 + i\tau + |\alpha_0|^2 \sin 2\tau)\,, \tag{48}$$

which gives

$$\theta_+ = \varphi_0 + \tfrac{1}{2}(\tau + |\alpha_0|^2 \sin 2\tau)\,, \quad \theta_- = \theta_+ + \pi/2\,, \tag{49}$$

where φ_0 is the phase of the initial coherent state, $\alpha_0 = |\alpha_0|e^{i\varphi_0}$. This means that the phases $\theta_\pm = \theta_\pm(\tau)$ for which the variance approaches its extremal values depend on τ, *i.e*, they evolve in the course of the evolution of the oscillator. For given τ we can tune the squeezing by changing the phase φ_0 or the intensity $|\alpha_0|^2$ of the beam.

Since the third-order nonlinearity of the Kerr medium is usually very small, the realistic values of τ are also very small ($\tau = 1 \times 10^{-6}$ is a rather optimistic estimation [29]), and one can expect noticeable quantum effects for a large number of photons $|\alpha_0|^2 \gg 1$, such as to have $|\alpha_0|^2\tau$ of the order of unity. This makes it possible to introduce a new variable $x = |\alpha_0|^2\tau$ and derive a quite simple expression for the variance $V(\theta)$ [24] ($\gamma = \varphi_0 - \theta$)

$$V_\theta(x) = 4x\sin(\gamma + x)\,[\sin(\gamma + x) - x\cos(\gamma + x)] + 1\,. \tag{50}$$

Similar formulae can be derived even for the case when higher-order nonlinearities are included [37].

The variances are the second-order moments of the field distribution and their graphical representations, such as an ellipse for the principal squeezing or Booth's lemniscate, for V_θ exhibit twofold rotational symmetry which reflects the fact that the variance is a quadratic function of the field operators. In this respect the graphical representation of squeezing obtained for Kerr states does not differ from the representation of ideal squeezed states, which are minimum uncertainty states. There is, however, another possibility: to represent graphically quantum states by plotting their quasiprobability distributions such as Q or Wigner functions. As an example, we shall show plots of the Q-function which for the Kerr states takes the form

$$Q(\alpha,\alpha^*,\tau) = |\langle\alpha|\psi(\tau)\rangle|^2 = \exp(-|\alpha|^2 - |\alpha_0|^2)|\Sigma(\tau)|^2\,, \qquad (51)$$

where

$$\Sigma(\tau) = \sum_{n=0}^{\infty} \frac{(\alpha^*\alpha_0)^n}{n!} \exp[i\frac{\tau}{2}n(n-1)]\,. \qquad (52)$$

In Figure 3 we show the contour plots of the function $Q(\alpha,\alpha^*,\tau)$ calculated according to (51). The plots reveal the famous "crescent" or "banana" shapes known for the Kerr states. The Q-function contours do not exhibit the twofold symmetry known for the ideal squeezed states. The Kerr states are thus quite different from the ideal squeezed states.

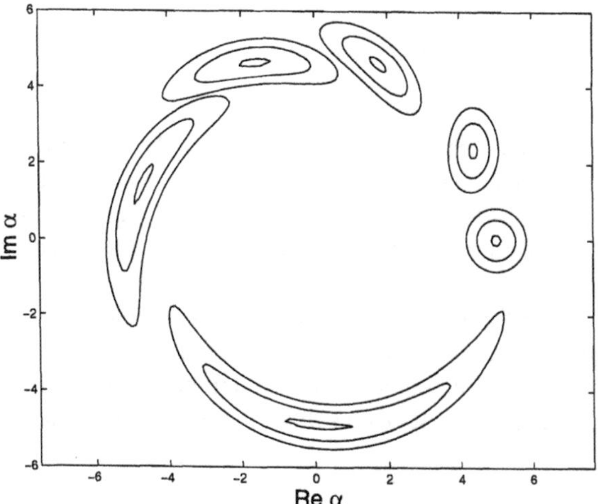

Figure 3: Contour plots of the function $Q(\alpha,\alpha^*,\tau)$ for $\alpha_0 = 5$ and various values of τ ($\tau = 0,\ 0.02,\ 0.05,\ 0.08,\ 0.12,\ 0.2$ counterclockwise starting from the circle contour). Contours are plotted at 0.98, 0.75 and 0.5 of the maximum.

4.2 Photon statistics

As we have shown in the previous section, for the one-mode field with the circular polarization being initially in a coherent state, the initial Poissonian statistics of the field is preserved during the evolution of the field in a Kerr medium. However, Kitagawa and Yamamoto [6] have shown that when the nonlinear Kerr medium is placed into one arm of the Mach–Zehnder interferometer and the reflectivities of the interferometer mirrors are chosen appropriately, the photon number fluctuations can be reduced by interference with a coherent reference field. We shall sketch here Kitagawa and Yamamoto's main results. In Figure 4 we show the

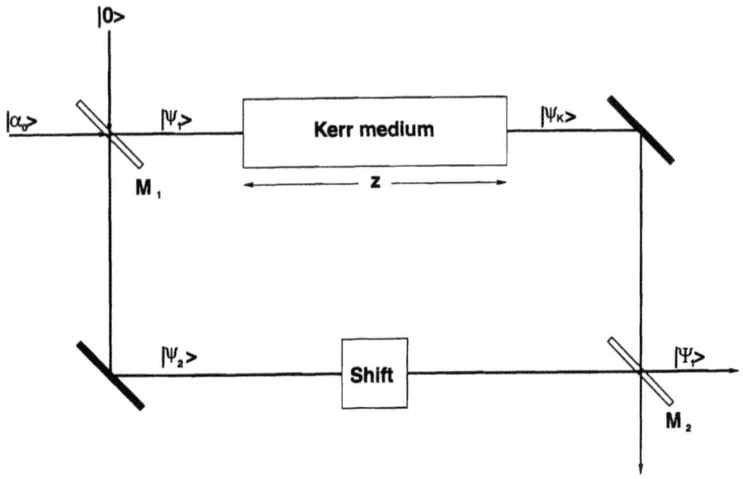

Figure 4: Nonlinear Mach–Zehnder interferometer with an optical Kerr medium.

scheme of the Mach–Zehnder interferometer with the nonlinear Kerr medium. A coherent-state input signal is divided into two parts by the first beam splitter M_1. One part $|\psi_1\rangle$ undergoes nonlinear evolution in the Kerr medium, and the other part $|\psi_2\rangle$ is not disturbed and is combined with the Kerr-medium output signal $|\psi_K(\tau)\rangle$ at the beam splitter M_2, whose reflectivity is close to unity. The output $|\Psi_1\rangle$ from the interferometer can be shown to have reduced photon-number fluctuations at the expense of increased phase fluctuations. When the input field is a coherent state $|\alpha_0\rangle$ in one input to the interferometer and the vacuum state $|0\rangle$ in the other input, after the first beam splitter there are two coherent states

$$|\psi_1\rangle = |\alpha_1\rangle, \quad |\psi_2\rangle = |\alpha_2\rangle, \qquad \alpha_1 = \sqrt{1-R_1}\,\alpha_0, \quad \alpha_2 = \sqrt{R_1}\,\alpha_0,$$

where R_1 is the reflectivity of M_1. The state $|\psi_1\rangle$ becomes the Kerr state $|\psi(\tau)\rangle$ after passage through the Kerr medium, while the state $|\psi_2\rangle$ acquires a phase shift ϕ and becomes the coherent state $|\alpha_2 e^{-i\phi}\rangle$. For high reflectivity of M_2, it can be shown that the output state $|\Psi_1\rangle$ becomes [6]

$$|\Psi_1\rangle = D(\xi)U_K(\tau)|\alpha_1\rangle\,, \tag{53}$$

where

$$D(\xi) = \mathrm{e}^{\xi a^\dagger - \xi^* a} \tag{54}$$

is the unitary displacement operator,

$$\xi = \sqrt{1-R_2}\,\alpha_2 \mathrm{e}^{-i\phi} = \sqrt{(1-R_2)R_1}\,\alpha_0 \mathrm{e}^{-i\phi} \tag{55}$$

with R_2 being the reflectivity of M_2, and $U_K(\tau)$ is given by (34). We assume here α_0 as real. Calculating the mean number of photons in the output state Ψ_1 gives the formula

$$\begin{aligned}
\langle \hat{n} \rangle &= \langle \Psi_1 | a^\dagger a | \Psi_1 \rangle = \langle \alpha_1 | U_K^\dagger(\tau) D^\dagger(\xi) a^\dagger a D(\xi) U_K(\tau) | \alpha_1 \rangle \\
&= \langle \alpha_1 | U_K^\dagger(\tau)(a^\dagger + \xi^*)(a+\xi) U_K(\tau) | \alpha_1 \rangle \\
&= \langle \alpha_1 | a^\dagger a | \alpha_1 \rangle + |\xi|^2 + 2\mathrm{Re}\left[\xi^* \langle \alpha_1 | \mathrm{e}^{i\tau a^\dagger a} a | \alpha_1 \rangle\right] \\
&= |\alpha_1|^2 + |\xi|^2 + 2\mathrm{Re}\left\{\alpha_1 \xi^* \exp(|\alpha_1|^2(\mathrm{e}^{i\tau}-1))]\right\} \\
&= |\alpha_1|^2 + |\xi|^2 + 2|\alpha_1||\xi|\mathrm{e}^{-\beta/2}\cos\vartheta\,,
\end{aligned} \tag{56}$$

where

$$\beta = 4|\alpha_1|^2 \sin^2\frac{\tau}{2}, \qquad \vartheta = \phi + |\alpha_1|^2 \sin\tau, \qquad \alpha_1 \xi^* = |\alpha_1||\xi|\mathrm{e}^{i\phi}.$$

In (56) we have used the property of the displacement operator

$$D^\dagger(\xi) a D(\xi) = a + \xi\,, \tag{57}$$

and relations (33) and (36). In a similar manner one can calculate the variance of the number of photons in the outgoing beam,

$$\begin{aligned}
\langle \hat{n}^2 \rangle - \langle \hat{n} \rangle^2 &= \langle \hat{n} \rangle + 2|\alpha_1|^2|\xi|^2 + 4|\alpha_1|^3|\xi|\mathrm{e}^{-\beta/2}\left[\cos(\tau+\vartheta) - \cos\vartheta\right] \\
&\quad + 2|\alpha_1|^2|\xi|^2\left[\mathrm{e}^{-2\beta'}\cos(\tau + 2\vartheta - \beta\sin\tau) - 2\mathrm{e}^{-\beta}\cos^2\vartheta\right],
\end{aligned} \tag{58}$$

where $\beta' = |\alpha_1|^2 \sin^2\tau$. Adjusting the phase difference between the two arms of the interferometer by choosing ϕ in such a way that $\vartheta = \pi/2$, formula (58) simplifies to that obtained by Kitagawa and Yamamoto [6]

$$\begin{aligned}
\langle \hat{n}^2 \rangle - \langle \hat{n} \rangle^2 &= \langle \hat{n} \rangle - 4|\alpha_1|^3|\xi|\mathrm{e}^{-\beta/2}\sin\tau \\
&\quad + 2|\alpha_1|^2|\xi|^2\left[1 - \mathrm{e}^{-2\beta'}\cos(\tau - \beta\sin\tau)\right].
\end{aligned} \tag{59}$$

If the variance (59) is smaller than the mean number of photons the photon statistics is sub-Poissonian. There are two commonly used parameters to measure the departure of the photon statistics from the Poissonian statistics of coherent fields. One of them is the normalized second-order (intensity) correlation function defined as

$$g^{(2)} = \frac{\langle a^{+2} a^2 \rangle}{\langle a^\dagger a \rangle^2} = \frac{\langle \hat{n}(\hat{n}-1) \rangle}{\langle \hat{n} \rangle^2}, \tag{60}$$

and the other is the q parameter introduced by Mandel [58] and defined as

$$q = \frac{\langle \hat{n}^2 \rangle - \langle \hat{n} \rangle^2}{\langle \hat{n} \rangle} - 1 = \left[g^{(2)} - 1 \right] \langle \hat{n} \rangle \,. \tag{61}$$

Whenever the value of the q parameter (or equivalently $g^{(2)} - 1$) becomes negative the photon statistics is sub-Poissonian, and the limit $q = -1$ is reached for number states with the zero photon number variance. On inserting (56) (with $\vartheta = \pi/2$) and (59) into (61), one obtains a relatively simple formula for the Mandel q parameter

$$q = \frac{|\alpha_1|^2}{1+\xi_*^2} \left\{ 2\xi_*^2 \left[1 - \mathrm{e}^{-2\beta'} \cos(\tau - \beta \sin \tau) \right] - 4\xi_* \mathrm{e}^{-\beta/2} \sin \tau \right\} . \tag{62}$$

The choice of $\xi_* = |\xi|/|\alpha_1|$ in (62) can be optimized to get the minimal value of

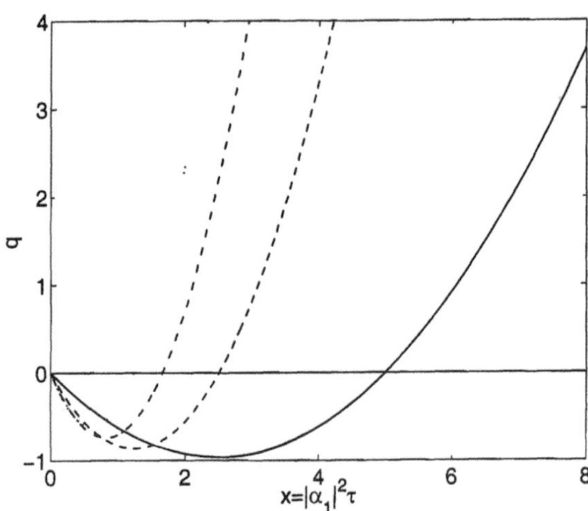

Figure 5: The Mandel q parameter versus $x = |\alpha_1|^2 \tau$ for $\tau = 1 \times 10^{-6}$ and ξ_*=0.2 (solid), 0.4 (dashed), 0.6 (dashed-dotted), 0.8 (dotted).

q for given $|\alpha_1|$ and τ. Similarly to the quadrature variance (50), for small values of $\tau \ll 1$ and a large mean number of photons $|\alpha_1|^2 \gg 1$, we can introduce the variable $x = |\alpha_1|^2 \tau$ and find that

$$q \approx -\frac{4\xi_* x(1-\xi_* x)}{1+\xi_*^2} , \quad \Longrightarrow \quad q|_{2\xi_* x = 1} \approx -\frac{1}{1+\xi_*^2} , \tag{63}$$

which shows that, for such a choice of parameters, q is negative and the photon statistics of the outgoing beam is sub-Poissonian. In fact, Kitagawa and Yamamoto [6] have shown that the photon-number variance can be reduced in this way to a value smaller than $\langle \hat{n} \rangle^{1/3}$. This means considerable reduction of photon-number fluctuations at the expense of increased fluctuations of the phase. The problem of the phase will be discussed later on.

A more general formula for q can be obtained from (56) and (58). In Figure 5 we have plotted the q parameter versus $x = |\alpha_1|^2\tau$, for $\tau = 1\times 10^{-6}$ and different values of ξ_*. The negative values of q seen in the figure are the signature of sub-Poissonian photon statistics, i.e., narrowing of the photon-number distribution.

The states (53) are obtained by applying the displacement operator (54) to the Kerr states (35), and thus the name *displaced Kerr states* has been attached to them. Quantum properties of displaced Kerr states have been studied by Wilson-Gordon *et al.* [39] and by Peřinova *et al.* [61]

4.3 Schrödinger cats and kittens

Kerr states (35) belong to a class of generalized coherent states introduced by Titulaer and Glauber [62] and discussed by Stoler [63]. They differ from coherent states by extra n-dependent phase factors in their decomposition into Fock states. Białynicka-Birula [64] has shown that, under appropriate periodic conditions, generalized coherent states go over into discrete superpositions of coherent states. Later, Yurke and Stoler [5], and Tombesi and Mecozzi [8] discussed a possibility of generating quantum mechanical superpositions of macroscopically distinguishable states in the anharmonic oscillator model. Miranowicz *et al.* [20] have shown that the superpositions of even and odd number of components are possible in the model and have found analytical formulae for such superpositions.

The Kerr state (35) can be rewritten in the form

$$|\psi_K(\tau)\rangle = \sum_{n=0}^{\infty} b_n \exp\left\{i\left[n\varphi_0 + \frac{\tau}{2}n(n-1)\right]\right\}, \tag{64}$$

where

$$b_n = \exp(-|\alpha_0|^2/2)\frac{|\alpha_0|^n}{\sqrt{n!}}, \qquad \alpha_0 = |\alpha_0|\exp(i\varphi_0). \tag{65}$$

Since $n(n-1)$ is an even number, $|\psi_K(\tau+T)\rangle = |\psi_K(\tau)\rangle$ for $T = 2\pi$. Moreover, we have

$$\exp\left[i\frac{\tau}{2}(n+2N)(n+2N-1)\right] = \exp\left[i\frac{\tau}{2}n(n-1)\right]\exp[i\tau N(2N+2n-1)],$$

which means that for

$$\tau = 2\pi\frac{M}{N} = T\frac{M}{N} \tag{66}$$

the exponential becomes periodic with the period $2N$. We assume that M and N are coprime integers. When τ is taken as a fraction of the period, according to (66), then the state (64) becomes a superposition of coherent states [64]

$$|\psi_K(\tau = TM/N)\rangle = \sum_{k=0}^{2N-1} c_k|e^{i\varphi_k}\alpha_0\rangle, \tag{67}$$

where $|\alpha_0\rangle$ is the initial coherent state. The phases φ_k are given by

$$\varphi_k = k\frac{\pi}{N}, \qquad k = 0, 1, ..., 2N-1, \tag{68}$$

and the coefficients c_k are given by the set of $2N$ equations

$$\sum_{k=0}^{2N-1} c_k \exp(in\varphi_k) = \exp\left[i\pi \frac{M}{N} n(n-1)\right], \tag{69}$$

for $n = 0, 1, \ldots, 2N-1$. Equations (69) can be rewritten as

$$\sum_{k=0}^{2N-1} c_k \exp\left\{i\frac{\pi}{N}[nk - Mn(n-1)]\right\} = 1, \tag{70}$$

which, after a summation over n and a minor rearrangement, gives

$$\sum_{k=0}^{2N-1} \frac{1}{2N} \sum_{n=0}^{2N-1} \exp\left\{i\frac{\pi}{N}[nk - Mn(n-1)]\right\} = 1. \tag{71}$$

In view of the condition $\sum_{k=0}^{2N-1} c_k c_k^* = 1$ we immediately obtain

$$c_k = \frac{1}{2N} \sum_{n=0}^{2N-1} \exp\left\{-i\frac{\pi}{N}[nk - Mn(n-1)]\right\}. \tag{72}$$

Equation (72) gives the coefficients c_k of the superposition (67) for arbitrary M and N. Because of the symmetry of the system, only one half of the coefficients c_k are different from zero and the superposition (67) has only N components although the summation contains $2N$ terms. Anticipating this, we have extended the summation twice in order to preserve N for the number of components. Thus the denominator of the fraction M/N in equation (66) determines the number of components that appear in the superposition (67), which will contain the components with either even or odd indices only. Examples of such states can be found in [20]. Coefficients (72) can be rewritten in a different form,

$$c_k = \frac{1 + (-1)^{k-M(N-1)}}{2N} \sum_{n=0}^{N-1} \exp\left[-i\frac{\pi}{N}[nk - Mn(n-1)]\right], \tag{73}$$

which explicitly shows that all c_k for which $k - M(N-1)$ is an odd number are equal to zero. That is, for $M(N-1)$ odd (even), only the coefficients with odd (even) k survive. The coefficients of the superposition have their modules equal to $1/\sqrt{N}$, and they can be written as

$$c_k = \frac{1}{\sqrt{N}} \exp(i\gamma_k), \tag{74}$$

where the phases γ_k can be formally found from the relation

$$\gamma_k = -i \ln(\sqrt{N}\, c_k) \tag{75}$$

with c_k given by (73). The trigonometric sums (72) can be summed exactly in some special cases. It is not difficult, however, to calculate c_k numerically according to (73) and then to find γ_k from (75). In particular, for $M = 1$, $N = 2$, we find from (73) that the coefficients of the superposition are

$$c_0 = 0\,, \quad c_1 = \frac{1}{\sqrt{2}}\,\mathrm{e}^{-i\pi/4}\,, \quad c_2 = 0\,, \quad c_3 = \frac{1}{\sqrt{2}}\,\mathrm{e}^{i\pi/4}\,, \tag{76}$$

and the superposition state is given by

$$|\psi_K(\tau = \pi)\rangle = \frac{1}{\sqrt{2}}\left(\mathrm{e}^{-i\pi/4}|i\alpha_0\rangle + \mathrm{e}^{i\pi/4}|-i\alpha_0\rangle\right), \tag{77}$$

which is, apart from the shift in phase, the state obtained by Yurke and Stoler [5]. If $|\alpha_0|$ is large, the state (77) is a superposition of two macroscopically distinguishable quantum states, or a *Schrödinger cat* state. For $M = 1$, $N = 3$, we have the superposition of three macroscopically distinguishable quantum states [20]

$$|\psi_K(\tau = 2\pi/3)\rangle = \frac{1}{\sqrt{3}}\left(\mathrm{e}^{i\pi/6}|\alpha_0\rangle - i|\mathrm{e}^{i\pi/3}\alpha_0\rangle + \mathrm{e}^{i\pi/6}|\mathrm{e}^{i2\pi/3}\alpha_0\rangle\right).$$

Superpositions with more than two states are often called *Schrödinger kittens.*

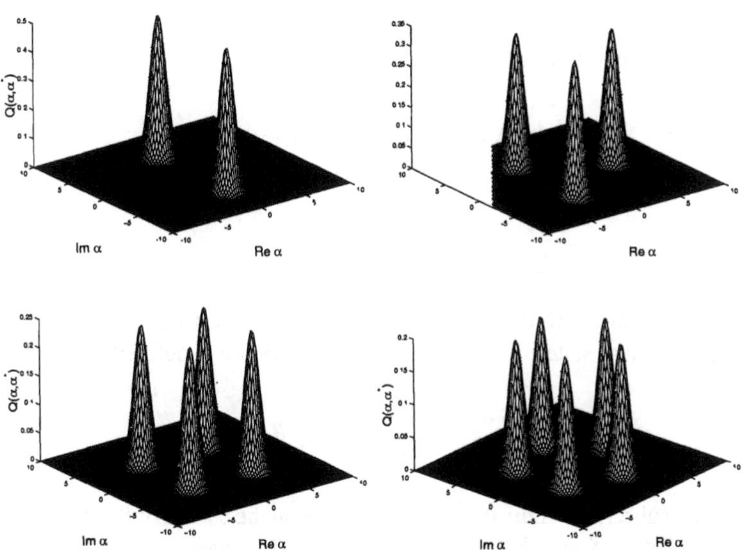

Figure 6: Plots of the Q-function for the superposition states with $\alpha_0 = 5$, $M = 1$, and $N = 2, 3, 4, 5$.

To visualize the superposition states it is convenient to plot their Q-function defined by (51). Examples are shown in Figure 6. The peaks of the quasi-probability distribution representing particular states of the superposition are located around a circle of radius $|\alpha_0|$ and they are well separated when $|\alpha_0|$ is large and the number of states entering the superposition is limited. How many well-separated states can be placed around the circle of radius $|\alpha_0|$ has been discussed in [20]. The pictures shown in Figure 6 are obtained from the same formula that gives the crescent (or banana) shape shown in Figure 3. The only difference is the choice of evolution time τ. If τ is such that the number N of components in the superposition is very large, the peaks of component states overlap and the Q-function has the shape with contours shown in Figure 3. The superpositions of coherent states similar to those discussed here, called even and odd coherent states, have been studied by Dodonov *et al.* [65,66].

4.4 Quantum phase properties

The nonclassical properties of the Kerr states come from the self-phase modulation, or self-squeezing, discussed earlier. These are just nonlinear changes of the phase that lead to squeezing without any change in photon statistics. It is thus important to look closer at the quantum phase properties of the Kerr states. To describe quantum phase properties of the Kerr states we shall use the Hermitian phase formalism introduced by Pegg and Barnett [67,68] (more on quantum phase can be found in [69, 70]). The formalism is based on introducing a finite $(s+1)$-dimensional space spanned by the number states $|0\rangle, |1\rangle, ..., |s\rangle$. The Hermitian phase operator operates on this finite space, and after all necessary expectation values have been calculated, the value of s is allowed to tend to infinity. A complete orthonormal basis of $s+1$ states is defined as

$$|\theta_m\rangle \equiv \frac{1}{\sqrt{s+1}} \sum_{n=0}^{s} \exp(in\theta_m)|n\rangle \,, \tag{78}$$

where

$$\theta_m \equiv \theta_0 + \frac{2\pi m}{s+1} \,, \quad m = 0, 1, \ldots, s \,. \tag{79}$$

The value of θ_0 is arbitrary and defines a particular basis set of $s+1$ mutually orthogonal phase states. The Hermitian phase operator is defined as

$$\hat{\phi}_\theta \equiv \sum_{m=0}^{s} \theta_m |\theta_m\rangle\langle\theta_m| \,, \tag{80}$$

where the subscript θ indicates the dependence on the choice of θ_0. The states (78) are eigenstates of the phase operator (80) with the eigenvalues θ_m restricted to lie within a phase window between θ_0 and $\theta_0 + 2\pi$. The unitary phase operator $\exp(i\hat{\phi}_\theta)$ is defined as the exponential function of the Hermitian operator $\hat{\phi}_\theta$. This operator acting on the eigenstate $|\theta_m\rangle$ gives the eigenvalue $\exp(i\theta_m)$, and can be written as [67,68]

$$\exp(i\hat{\phi}_\theta) \equiv \sum_{n=0}^{s-1} |n\rangle\langle n+1| + \exp[i(s+1)\theta_0]|s\rangle\langle 0| \,. \tag{81}$$

It is the last term in (81) that ensures the unitarity of this operator. The first sum reproduces the Susskind–Glogower [71,72] phase operator in the limit $s \to \infty$.

The expectation value of the phase operator (80) in a pure state $|\psi\rangle$ is given by

$$\langle\psi|\hat{\phi}_\theta|\psi\rangle = \sum_{m=0}^{s} \theta_m |\langle\theta_m|\psi\rangle|^2 , \tag{82}$$

where $|\langle\theta_m|\psi\rangle|^2$ gives a probability of being found in the phase state $|\theta_m\rangle$. The density of phase states is $(s+1)/2\pi$, so in the continuum limit as s tends to infinity, we can write equation (82) as

$$\langle\psi|\hat{\phi}_\theta|\psi\rangle = \int_{\theta_0}^{\theta_0+2\pi} \theta P(\theta)\, \mathrm{d}\theta , \tag{83}$$

where the continuum phase distribution $P(\theta)$ is introduced by

$$P(\theta) = \lim_{s\to\infty} \frac{s+1}{2\pi} |\langle\theta_m|\psi\rangle|^2 \tag{84}$$

with θ_m being replaced by the continuous phase variable θ. As the phase distribution function $P(\theta)$ is known, all the quantum-mechanical phase expectation values can be calculated with this function in a classical-like manner. The choice of the value θ_0 defines the 2π range window of the phase values. After taking into account (78), we can write (84) as

$$P(\theta) = \lim_{s\to\infty} \frac{1}{2\pi} \sum_{n=0}^{s}\sum_{k=0}^{s} \mathrm{e}^{-i(n-k)\theta_m} \langle n|\psi\rangle\langle\psi|k\rangle . \tag{85}$$

For the Kerr state $|\psi\rangle = |\psi_K(\tau)\rangle$ given by (64), we symmetrize the phase distribution with respect to the phase φ_0 by taking

$$\theta_0 = \varphi_0 - \frac{\pi s}{s+1} \tag{86}$$

and introducing a new phase label

$$\mu = m - s/2 , \tag{87}$$

which runs in integer steps from $-s/2$ to $s/2$. Then the phase distribution becomes symmetric in μ, and we get [73]

$$\begin{aligned} P(\theta) &= \lim_{s\to\infty} \frac{1}{2\pi} \sum_{n=0}^{s}\sum_{k=0}^{s} \exp\left\{-i\left[(n-k)\theta_\mu - \frac{\tau}{2}[n(n-1) - k(k-1)]\right]\right\} b_n b_k \\ &= \frac{1}{2\pi} \sum_{n=0}^{\infty}\sum_{k=0}^{\infty} \exp\left\{-i\left[(n-k)\theta - \frac{\tau}{2}[n(n-1) - k(k-1)]\right]\right\} b_n b_k . \end{aligned} \tag{88}$$

Now, integrals over θ are taken in the symmetric range between $-\pi$ and π. The phase distribution $P(\theta)$ is normalized such that $\int_{-\pi}^{\pi} P(\theta)\mathrm{d}\theta = 1$.

If the field is described by the density operator ρ instead of the pure state $|\psi\rangle$, equation (82) takes a more general form

$$\langle\hat{\phi}_\theta\rangle = \mathrm{Tr}(\rho\hat{\phi}_\theta) = \sum_{m=0}^{s} \langle\theta_m|\rho|\theta_m\rangle , \tag{89}$$

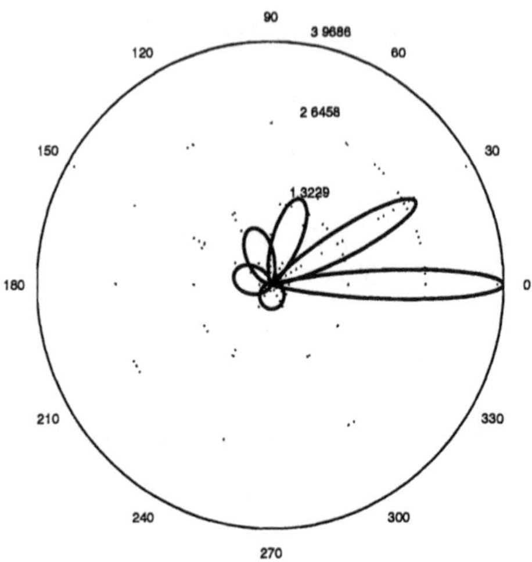

Figure 7: Phase distributions $P(\theta)$ for $\tau = 0.02,\ 0.05,\ 0.08,\ 0.12,\ 0.2$ counting counter-clockwise from the peak at $\theta = 0$ (compare Figure 3).

and the phase distribution $P(\theta)$ is given, instead of equation (88), by

$$P(\theta) = \frac{1}{2\pi} \sum_{n=0}^{\infty} \sum_{k=0}^{\infty} \mathrm{e}^{-i(n-m)\theta} \rho_{nk} \,. \tag{90}$$

This more general form is needed when the dissipation is present in the system. For a not too large mean number of photons $|\alpha_0|^2$, the amplitudes b_n go to zero very rapidly as n increases, and the phase distribution (88) can be easily evaluated numerically. All the expectation values of the phase operator can be calculated by taking the integrals over the continuous variable θ with the probability distribution $P(\theta)$ given by (88). The phase distribution $P(\theta)$ carries all the statistical information about the phase of the Kerr states. It is thus interesting to see the plots of $P(\theta)$ for different evolution times τ. We show a few examples of such plots in Figure 7. Polar plots of $P(\theta)$ show nicely the broadening of the phase distribution during the evolution and they also show clearly the appearance of the Schrödinger cats and kittens discussed in the previous section. This is illustrated in Figure 8.

If the mean number of photons is large, the approximate method [73] can be applied to find the phase distribution. In this case the Poisson photon number distribution is well approximated by a continuous Gaussian distribution

$$P(n) = \exp(-|\alpha_0|^2) \frac{|\alpha_0|^{2n}}{n!} \approx (2\pi|\alpha_0|^2)^{-1/2} \exp\left[-\frac{(|\alpha_0|^2 - n)^2}{2|\alpha_0|^2}\right] .$$

Inserting the square root of $P(n)$ for b_n and performing integration over n in (88) one gets the Gaussian distribution for the phase

$$P(\theta) = \frac{1}{(2\pi\sigma^2)^{1/2}} \exp\left[-\frac{(\theta - \bar{\theta})^2}{2\sigma^2}\right] , \tag{91}$$

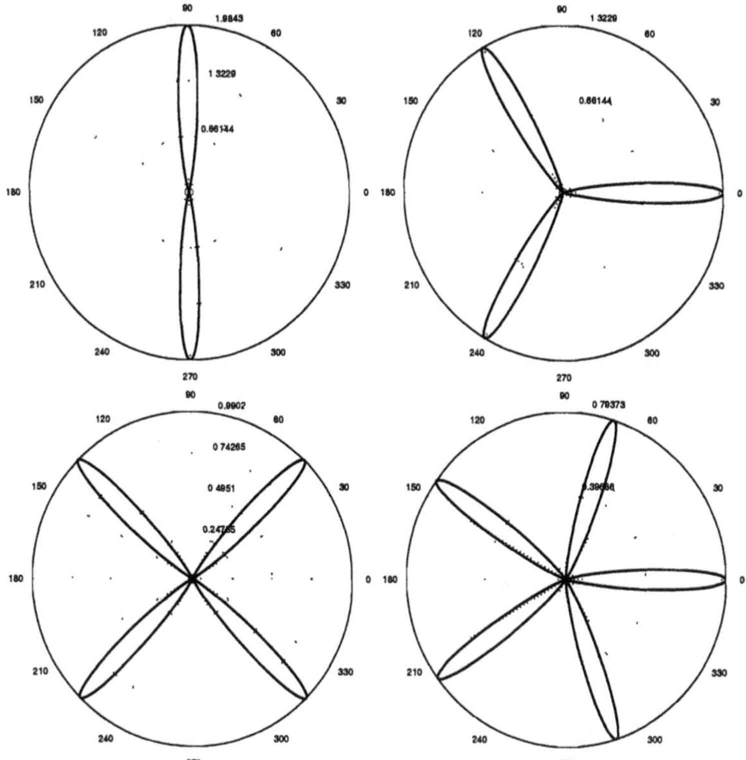

Figure 8: Plots of $P(\theta)$ for the superpositions of coherent states for $\alpha_0 = 5$ and $\tau = 2\pi/2,\ 2\pi/3,\ 2\pi/4,\ 2\pi/5$ (compare Figure 6).

where

$$\bar{\theta} = \varphi_0 + \tau\left(|\alpha_0|^2 - \tfrac{1}{2}\right), \quad \sigma^2 = |\alpha_0|^2\tau^2 + \tfrac{1}{4}. \tag{92}$$

From (92) we see that the mean phase is shifted by $\tau|\alpha_0|^2$ with respect to the initial phase φ_0 (we neglect $\frac{1}{2}$), which is consistent with the operator solution (32) when the operators are replaced by classical field amplitudes ($a(0) \to \alpha_0$). We see that the dispersion of the Gaussian distribution increases with τ. Since the photon distribution remains Poissonian with the variance $\langle(\Delta\hat{n})^2\rangle = \langle\hat{n}\rangle$, we can immediately write the phase–photon number uncertainty relation

$$\langle(\Delta\hat{\phi}_\theta)^2\rangle\langle(\Delta\hat{n})^2\rangle = \tfrac{1}{4} + |\alpha_0|^4\tau^2 \tag{93}$$

which means fast expansion of the uncertainty product during the evolution. One should, however, keep in mind that (91) is approximate and it is valid only for not too broad Gaussians. Generally, the exact formula (88) should be used to calculate the mean phase and the variance.

The mean value of the phase for the Kerr state calculated using (88) gives [73]

$$\langle\psi_K(\tau)|\hat{\phi}_\theta|\psi_K(\tau)\rangle = \varphi_0 + \frac{1}{2\pi}\int_{-\pi}^{\pi}\theta P(\theta)\,\mathrm{d}\theta$$
$$= \varphi_0 - 2\sum_{n>k} b_n b_k \frac{(-1)^{n-k}}{n-k}\sin\left\{\frac{\tau}{2}[n(n-1)-k(k-1)]\right\}, \qquad (94)$$

and for the variance of $\hat{\phi}_\theta$ [73]

$$\langle\psi_K(\tau)|(\Delta\hat{\phi}_\theta)^2|\psi_K(\tau)\rangle = \frac{\pi^2}{3}$$
$$+4\sum_{n>k} b_n b_k \frac{(-1)^{n-k}}{(n-k)^2}\cos\left\{\frac{\tau}{2}[n(n-1)-k(k-1)]\right\}$$
$$-\left\{2\sum_{n>k} b_n b_k \frac{(-1)^{n-k}}{n-k}\sin\left\{\frac{\tau}{2}[n(n-1)-k(k-1)]\right\}\right\}^2. \qquad (95)$$

Both the mean phase and the phase variance are periodic functions of τ.

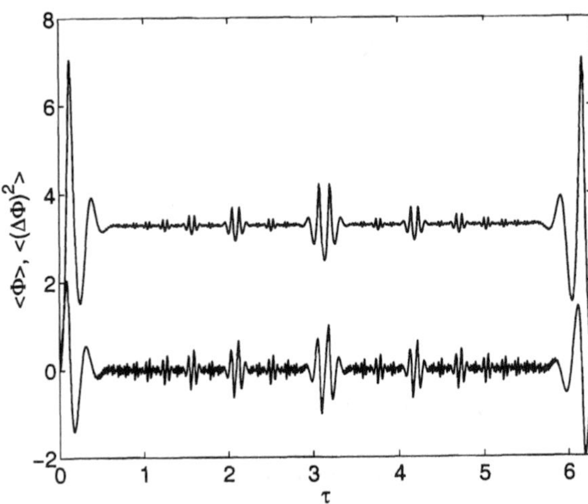

Figure 9: Mean phase and phase variance versus τ for $\alpha_0 = 5$.

In Figure 9 we show an example of the evolution of the phase and the variance for the Kerr state with the mean number of photons $\alpha_0 = 5$. The phase variance rapidly increases at the beginning of the evolution and next shows oscillations around the value $\pi^2/3$, which is the value for the uniformly distributed phase, as for example, in the case of a single number state. The values above $\pi^2/3$ come from the fact that as the mean phase is shifted during the evolution, the phase window symmetrized with respect to φ_0 becomes not symmetric with respect to the new value of the phase. The minima in the variance indicate the positions of the superpositions of coherent states. The positions of the superposition states are much more visible from the evolution of the Wehrl entropy for the system [74].

4.5 Dissipation

So far, we have discussed an ideal situation when there is no dissipation in the system, i.e., the field propagating in the Kerr medium is not absorbed and the total intensity is conserved. In real situations this is usually not the case, and one has to include damping into the model. The dissipative quantum and classical Liouville dynamics of the anharmonic oscillator has been studied by Milburn and Holmes [4] and Milburn [3]. The anharmonic oscillator model admits exact analytical solution even in the presence of dissipation, as was first shown by Milburn and Holmes [75] for the Q-function with initially coherent-state distribution. This approach has been generalized by Peřinova and Lukš [13, 21], Daniel and Milburn [16], and Milburn *et al.* [15]. Phase properties of the damped anharmonic oscillator have been studied by Gantsog and Tanaś [76]. Some generalizations of the model can be found in [77–80].

To describe the system with dissipation it is necessary to include the coupling of the nonlinear oscillator to a reservoir, which is usually represented by an infinite set of harmonic oscillators. The evolution of the anharmonic oscillator can no longer be described by the Kerr states (35), which are pure quantum states, but the density matrix must be used. By use of the standard techniques of the quantum theory of damping, the following master equation, in the Markov approximation and in the interaction picture, is obtained for the density operator of the field propagating in the Kerr medium

$$\frac{\partial\rho}{\partial t} = -i\frac{\kappa}{2}\left[{a^{\dagger}}^{2}a^{2},\rho\right] + \frac{\gamma}{2}\left([a\rho,a^{\dagger}] + [a,\rho a^{\dagger}]\right) + \gamma\bar{n}\left[[a,\rho],a^{\dagger}\right], \qquad (96)$$

where γ is the damping constant, $\bar{n} = [\exp(\hbar\omega/kT) - 1]^{-1}$ is the mean number of thermal photons for the temperature T of the reservoir. The exact solution to the master equation (96) is possible for both the "quiet" [4, 13] ($\bar{n} = 0$) and "noisy" [16, 21] ($\bar{n} \neq 0$) reservoirs. Here we quote only the solution for $\bar{n} = 0$, which is simpler and given by

$$\rho_{nm}(\tau) = \exp\left[-i\frac{\tau}{2}(n-m)\right] f_{n-m}^{(n+m)/2}(\tau) \times \sum_{l=0}^{\infty}\frac{1}{l!}\sqrt{\frac{(n+l)!(m+l)!}{n!m!}}\left[\frac{\lambda[1-f_{n-m}(\tau)]}{\lambda+i(n-m)}\right]^{l}\rho_{n+l,m+l}(0), \qquad (97)$$

where we have used the notation $\lambda = \gamma/|\kappa| > 0$, $\tau = -\kappa t > 0$, and

$$f_{n-m}(\tau) = \exp\left\{i\tau\left[(n-m)+i\lambda\right]\right\}. \qquad (98)$$

For $\lambda = 0$ the solution (97) becomes

$$\rho_{nm}(\tau) = \exp\left[i\frac{\tau}{2}[n(n-1) - m(m-1)]\right]\rho_{nm}(0), \qquad (99)$$

and describes the dynamics of the lossless anharmonic oscillator. From (99) it is clear that the diagonal matrix elements of the field density matrix do not change if there is no damping. This means that the photon statistics does not change, as discussed before. The nondiagonal elements are related to the nonlinear change of the field phase, and this change is responsible for quantum effects discussed earlier.

For the initial coherent state $|\alpha_0\rangle$ we have

$$\rho_{n+l,m+l}(0) = b_{n+l}b_{m+l}\exp[i(n-m)\varphi_0]\,, \tag{100}$$

where b_n is given by (65). On inserting (100) into (97) we get

$$\begin{aligned}\rho_{nm}(\tau) &= b_n b_m \exp\left[i(n-m)\left(\varphi_0 - \frac{\tau}{2}\right)\right] f_{n-m}^{(n+m)/2}(\tau) \\ &\quad\times \exp\left[|\alpha_0|^2\lambda\,\frac{1-f_{n-m}(\tau)}{\lambda - i(n-m)}\right],\end{aligned} \tag{101}$$

where $f_{n-m}(\tau)$ is given by (98). The solution (101) allows for calculations of the quantum mechanical mean values of the field operators. For example, the mean value of the field operator is equal to [4]

$$\begin{aligned}\langle a\rangle &= \mathrm{Tr}\{a\rho(\tau)\} = \sum_{n=0}^{\infty}\sqrt{n}\,\rho_{n,n-1}(\tau) \\ &= \alpha_0\exp\left\{-\frac{1}{2}\lambda\tau + |\alpha_0|^2\,\frac{\mathrm{e}^{i\tau(1+i\lambda)}-1}{1+i\lambda}\right\},\end{aligned} \tag{102}$$

which for $\lambda = 0$ goes over into (36). Because of damping the periodicity of the evolution present in (36) is lost in (102). The quantum recurrences will thus be lost when the damping plays an important role in the propagation process. All the quantum properties discussed earlier will be rapidly degraded by the dissipation.

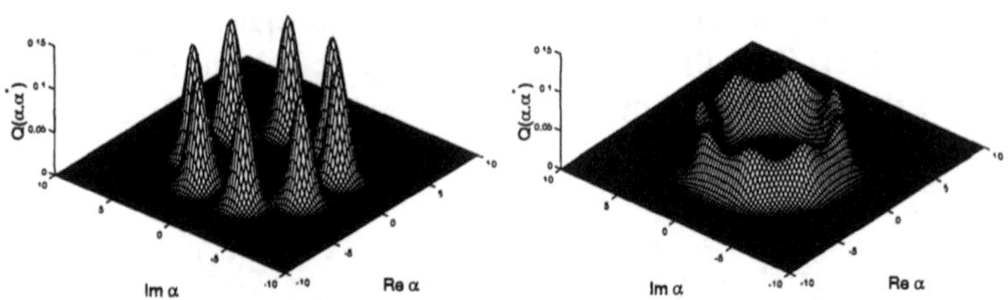

Figure 10: The Q function for $\alpha_0 = 5$, $\tau = \pi/4$ (superposition state with 8 components), and $\lambda = 0$ (left); $\lambda = 0.1$ (right).

The quasiprobability function Q can be calculated with (101) according to the formula

$$Q(\alpha,\alpha^*;\tau) = \exp(-|\alpha|^2)\sum_{n,m}\frac{\alpha^{*n}}{\sqrt{n!}}\frac{\alpha^m}{\sqrt{m!}}\rho_{nm}\,, \tag{103}$$

and the phase distribution function can be obtained by inserting (101) into (90). In Figure 10 we show examples of the Q-function for $\alpha_0 = 5$ and $\tau = \pi/4$, which corresponds to the superposition of coherent states with 8 components. The figure shows how the quantum interference leading to the superposition states is washed away by the presence of damping in the system. The left figure shows the situation without damping and the 8 peaks of the superposition are clearly distinguishable. The right figure exhibits the same Q-function but when the damping parameter $\lambda = 0.1$ is nonzero. The figure shows how the quantum coherences are suppressed by the dissipation.

5 Two-mode fields

The one-mode description of the propagation of a quantum field in Kerr media, discussed so far, is only possible if the field is circularly polarized. In the case of elliptical polarization of the incoming field, a two-mode description is necessary. Now, we proceed to study the quantum properties of the two-mode case of fields propagating in a Kerr medium. We shall concentrate on the effects that require two modes and are not observed for the one mode of a circularly polarized field.

5.1 Squeezing and Schrödinger cat states

As in the one-mode case, it is easy to calculate, from the operator solutions (26) for initially coherent state $|\alpha_+, \alpha_-\rangle \equiv |\alpha_+\rangle|\alpha_-\rangle$, the normally ordered variances for the quadrature components of the two circularly polarized modes of the field propagating in the Kerr medium. The corresponding formulae are more complicated but, as in the one-mode case, they are exact analytical solutions. It has been shown [28, 29, 81] that the same degree of squeezing can be obtained for the two-mode fields as for the one-mode field. The detailed discussion can be found in the papers cited above. Also, similarly to the one-mode case, it is possible to show that during the evolution the two-mode state evolves to macroscopic superpositions of coherent states when the evolution time is chosen appropriately [35, 82]. The two-mode Kerr states are obtained by applying the unitary evolution operator

$$
\begin{aligned}
U_K(\tau) &= \exp(-iHt/\hbar) \\
&= \exp\left[i\frac{\tau}{2}\left(a_+^{\dagger\,2} a_+^2 + a_-^{\dagger\,2} a_-^2 + 4d\, a_+^\dagger a_-^\dagger a_- a_+\right)\right] \\
&= \exp\left[i\frac{\tau}{2}\left(\hat{n}_+(\hat{n}_+ - 1) + \hat{n}_-(\hat{n}_- - 1) + 4d\,\hat{n}_+\hat{n}_-\right)\right]
\end{aligned}
\tag{104}
$$

with the Hamiltonian (21) and $\tau = -\kappa t$, to the initial state $|\psi(0)\rangle$. Assuming that the initial state is a coherent state $|\alpha_+, \alpha_-\rangle$ with α_+ and α_- being the amplitudes for the two circular polarization components, we get for the two-mode Kerr state the formula

$$
\begin{aligned}
|\psi_K(\tau)\rangle = \sum_{n_+, n_-} b_{n_+} b_{n_-} \exp\Big\{ & i(n_+\varphi_+ + n_-\varphi_-) + i\frac{\tau}{2}\left[n_+(n_+ - 1)\right. \\
& \left. + n_-(n_- - 1) + 4d\, n_+ n_-\right]\Big\} |n_+, n_-\rangle ,
\end{aligned}
\tag{105}
$$

where

$$b_{n_\pm} = \exp(-|\alpha_\pm|^2/2)\frac{|\alpha_\pm|^{n_\pm}}{\sqrt{n_\pm!}}, \quad \alpha_\pm = |\alpha_\pm| \exp(i\varphi_\pm). \tag{106}$$

The two-mode Kerr states (105) cannot be factorized into the one-mode Kerr states (64) because of the presence of the term $4d\, n_+ n_-$ in the exponential, which describes coupling between the two circularly polarized modes. The coupling is defined by the asymmetry parameter d, given by (23), which accounts for the nonlinear properties of the medium. The two-mode Kerr states are, generally, *entangled states*. When $2d$ is an integer number and the evolution time τ is chosen properly as a fraction M/N of the period 2π, the states (105) become discrete superpositions of coherent states [35, 82]. For example, if $\tau = \pi$ and $d = 1/2$ the state becomes

$$|\psi_K(\tau = \pi)\rangle_{d=1/2} = \frac{1}{\sqrt{2}}\left(\mathrm{e}^{-i\pi/4}|i\alpha_+, i\alpha_-\rangle + \mathrm{e}^{i\pi/4}| - i\alpha_+, -i\alpha_-\rangle\right).$$

Similarly to the one-mode case, a great variety of superposition states is possible, but the coupling between the modes can cause a suppression of the number of states in the superposition [82]. From the point of view of squeezing and superpositions of coherent states, the two-mode Kerr states lead to results very similar to the results known from the one-mode Kerr states. The situation becomes different when it comes to photon statistics and polarization of the field.

5.2 Photon statistics

As we have already discussed in section 3, the number operators for the two circular components of the field commute with the Hamiltonian (21), thus the photon statistics of the circular components are constants of motion. However, this is not the case if the Cartesian basis is used — in this case the number operators do not commute with the Hamiltonian (20). Now, we encounter a very interesting situation: let us assume that the incoming field is polarized along the x axis, and we measure the orthogonal y component of the outgoing field. Classically, in isotropic medium, the polarization x of the field should be preserved, and we should not observe any field with y polarization. For quantum fields, however, there are quantum fluctuations, and photons with polarization y can appear despite the fact that the medium is isotropic. Only circular polarization is preserved in an isotropic medium!

In section 4.2 we discussed a possibility of narrowing the photon number distribution, showing that a sub-Poissonian photon statistics is obtained in the Mach–Zehnder interferometer with a Kerr medium in one arm. The state produced in the Kerr medium was the one-mode Kerr state, but it was displaced by the reference beam propagating along the other arm, and this displacement was crucial for getting the sub-Poissonian statistics. Now, we shall show that in the two-mode case, sub-Poissonian photon statistics is possible without the displacement of the Kerr state. It is sufficient to measure the Cartesian component (x or y) of the outgoing field. For initially coherent state $|\alpha_+, \alpha_-\rangle$, from (26) and (27), we get

$$\langle a_x^\dagger(\tau)a_x(\tau)\rangle \quad = \quad \frac{1}{2}\langle [a_+^\dagger(\tau) + a_-^\dagger(\tau)][a_+(\tau) + a_-(\tau)]\rangle$$

$$= \frac{1}{2}\left(|\alpha_+|^2 + |\alpha_-|^2\right) + |\alpha_+\alpha_-|\,\mathrm{Re}\Big\{\exp\Big[i(\varphi_+ - \varphi_-) + \left(\mathrm{e}^{i\tau'} - 1\right)|\alpha_+|^2 + \left(\mathrm{e}^{-i\tau'} - 1\right)|\alpha_-|^2\Big]\Big\}, \tag{107}$$

where $\tau' = \tau(1 - 2d)$. It is worth emphasizing here that the classical expression can be obtained from (107) by the replacement $\mathrm{e}^{i\tau'} - 1 \to i\tau'$ which means the linear approximation to the quantum exponential.

In the same manner we can derive the expression for the second-order correlation function

$$\begin{aligned}\langle a_x^{\dagger 2}(\tau)a_x^2(\tau)\rangle &= \frac{1}{4}\left(|\alpha_+|^4 + |\alpha_-|^4 + 4|\alpha_+\alpha_-|^2\right)\\ &+\frac{1}{2}\,\mathrm{Re}\Big\{|\alpha_+\alpha_-|^2\exp\Big[2i(\varphi_+ - \varphi_-) + \left(\mathrm{e}^{2i\tau'} - 1\right)|\alpha_+|^2 + \left(\mathrm{e}^{-2i\tau'} - 1\right)|\alpha_-|^2\Big]\\ &+ 2|\alpha_+^3\alpha_-|\exp\Big[i(\varphi_+ - \varphi_-) + i\tau' + \left(\mathrm{e}^{i\tau'} - 1\right)|\alpha_+|^2 + \left(\mathrm{e}^{-i\tau'} - 1\right)|\alpha_-|^2\Big]\\ &+ 2|\alpha_+\alpha_-^3|\exp\Big[i(\varphi_- - \varphi_+) + i\tau' + \left(\mathrm{e}^{-i\tau'} - 1\right)|\alpha_+|^2 + \left(\mathrm{e}^{i\tau'} - 1\right)|\alpha_-|^2\Big]\Big\}.\end{aligned} \tag{108}$$

For an elliptically polarized field coming into the Kerr medium there are certain phase relations between the two amplitudes α_+ and α_-. Calculating the expectation values of the Stokes operators (28) in the coherent state $|\alpha_+, \alpha_-\rangle$ and comparing the results to the classical Stokes parameters (14), we get

$$\begin{aligned} s_0 &= \langle S_0\rangle = |\alpha_+|^2 + |\alpha_-|^2 = |\alpha_0|^2\,,\\ s_1 &= \langle S_1\rangle = 2\,\mathrm{Re}(\alpha_+^*\alpha_-) = s_0\cos 2\eta\cos 2\theta\,,\\ s_2 &= \langle S_2\rangle = 2\,\mathrm{Im}(\alpha_+^*\alpha_-) = s_0\cos 2\eta\sin 2\theta\,,\\ s_3 &= \langle S_3\rangle = s_0\sin 2\eta\,,\end{aligned} \tag{109}$$

where $\theta = -(\varphi_+ - \varphi_-)/2$ is the azimuth of the polarization ellipse and η is its ellipticity. With this notation, the amplitudes α_+ and α_- of the two circular components of the elliptically polarized field, with the amplitude α_0, take the following form

$$\alpha_\pm = \frac{\alpha_0}{\sqrt{2}}(\cos\eta \pm \sin\eta)\,\mathrm{e}^{\mp i\theta}\,. \tag{110}$$

Let us assume, for a moment, that the field is linearly polarized ($\eta = 0$) with the azimuth $\theta = \pi/2$, that is, perpendicularly to the observed polarization component. Then, from (107), we obtain

$$\langle a_x^\dagger(\tau)a_x(\tau)\rangle = \frac{|\alpha_0|^2}{2}\left[1 - \exp\left(-2|\alpha_0|^2\sin^2\frac{\tau'}{2}\right)\right], \tag{111}$$

which means that, due to the presence of the exponential, photons with polarization perpendicular to the polarization of the incoming field appear. Classically, no field can be observed in the perpendicular component. In the classical formula the exponential must be replaced by unity and the result is zero.

To check whether it is possible to obtain sub-Poissonian photon statistics i the x component of the outgoing beam, we can calculate the Mandel q paramete defined by (61), which can be expressed by (107) and (108) in the following way

$$q = \left\{ \frac{\langle a_x^{\dagger 2}(\tau) a_x^2(\tau) \rangle}{\langle a_x^\dagger(\tau) a_x(\tau) \rangle^2} - 1 \right\} \langle a_x^\dagger(\tau) a_x(\tau) \rangle . \quad (112$$

On inserting (107) and (108) into (112), assuming that $\tau' \ll 1$ and keeping onl the lowest-order terms, we get the following simple expression for the Mandel parameter:

$$q = \frac{|\alpha_0|^2 \tau' \sin 4\eta \sin \vartheta + \left[|\alpha_0|^2 \tau' \cos 2\eta \sin \vartheta \right]^2}{2(1 + \cos 2\eta \cos \vartheta)} , \quad (113$$

where

$$\vartheta = 2\theta - |\alpha_0|^2 \tau' \sin 2\eta . \quad (114$$

From (113) it is evident that the q parameter is equal to zero when the field i

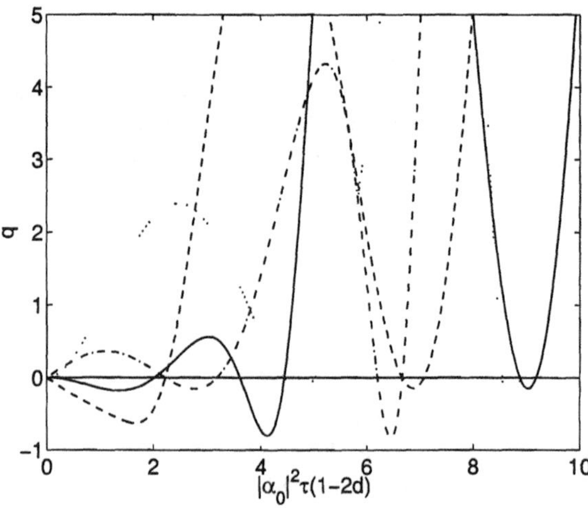

Figure 11: The Mandel q parameter for the x component of the outgoing field versu $|\alpha_0|^2\tau'$. The parameters are: $\eta = \pi/8$ and $\theta = 0$ (solid), $-\pi/4$ (dashed), $\pi/4$ (dashed dotted), $\pi/2$ (dotted).

circularly polarized ($\eta = \pm\pi/2$). This confirms our earlier findings that for cir cularly polarized light, the photon statistics does not change. It is also clear tha only the first term in the numerator can take negative values, and the optimal po larization is elliptical with $\eta = \pi/8$ ($\sin 4\eta = 1$). In Figure 11 we plot the Mande q parameter calculated according to (113) for several values of the azimuth θ, as suming the ellipticity parameter $\eta = \pi/8$. It is seen that there are regions wher the parameter q takes negative values, making possible observation of the sub Poissonian photon statistics by measuring a Cartesian component of the outgoin field with the appropriate choice of parameters of the incoming field. Projecting the two-mode Kerr state onto the Cartesian basis is crucial for obtaining the sub-Poissonian statistics — in the circular basis photon statistics remain Poissonian for an initially coherent field.

5.3 Polarization

The polarization of initially elliptically polarized light propagating in a nonlinear Kerr medium is changed due to nonlinear interaction. Classically the polarization of light is defined by the Stokes parameters (14). For quantum fields we can introduce the Stokes operators (28), the expectation values of which, in a given state of the field, give the Stokes parameters of the quantum field and define its polarization. Using this idea we can deduce that the azimuth and the ellipticity of the polarization ellipse are defined by the relations

$$\tan 2\theta = \frac{\langle S_2\rangle}{\langle S_1\rangle} = \frac{s_1}{s_2}, \quad \tan 2\eta = \frac{\langle S_3\rangle}{\sqrt{\langle S_1\rangle^2 + \langle S_2\rangle^2}} = \frac{s_3}{\sqrt{s_1^2 + s_2^2}}, \tag{115}$$

and the quantum degree of polarization has its classical form (15) in which s_1, s_2, and s_3 are understood as corresponding expectation values of the Stokes operators. The quantum evolution of the Stokes parameters of light propagating in a Kerr medium has been studied by Tanaś and Kielich [83]. Taking the expectation value, in the initial coherent state $|\alpha_+, \alpha_-\rangle$, of the operator solutions (26) it is straightforward to derive the formulae for the Stokes parameters of the beam

$$\begin{aligned}
\langle S_0(\tau)\rangle &= \langle a_+^\dagger(\tau)a_+(\tau)\rangle = |\alpha_+|^2 + |\alpha_-|^2 = |\alpha_0|^2, \\
\langle S_1(\tau)\rangle &= 2\,\mathrm{Re}\left\{\alpha_+^*\alpha_- \exp[(\mathrm{e}^{-i\tau'} - 1)|\alpha_+|^2 + (\mathrm{e}^{i\tau'} - 1)|\alpha_-|^2\right\}, \\
\langle S_2(\tau)\rangle &= 2\,\mathrm{Im}\left\{\alpha_+^*\alpha_- \exp[(\mathrm{e}^{-i\tau'} - 1)|\alpha_+|^2 + (\mathrm{e}^{i\tau'} - 1)|\alpha_-|^2\right\}, \\
\langle S_3(\tau)\rangle &= |\alpha_+|^2 - |\alpha_-|^2,
\end{aligned} \tag{116}$$

where α_+ and α_- are given by (110), and $\tau' = \tau(1-2d)$. Since $a_+^\dagger a_+$ and $a_-^\dagger a_-$ are constants of motion, the Stokes parameters $\langle S_0\rangle$ and $\langle S_3\rangle$ do not alter during the propagation, and only the parameters $\langle S_1(\tau)\rangle$ and $\langle S_2(\tau)\rangle$ evolve in time.

Applying the definitions (115) we find for the azimuth and ellipticity

$$\begin{aligned}
\tan 2\theta(\tau) &= \tan\left(2\theta - |\alpha_0|^2 \sin 2\eta\, \sin\tau'\right), \\
\tan 2\eta(\tau) &= \tan 2\eta\, \exp\left(-2|\alpha_0|^2 \sin\frac{\tau'}{2}\right).
\end{aligned} \tag{117}$$

The classical solutions are obtained from (117) by replacing $\sin\tau' \to \tau'$ and dropping the exponential. The ellipticity of the beam does not change in the classical description, while the azimuth of the polarization ellipse does change if the ellipticity of the beam is nonzero. This is the well-known classical effect of rotation of the polarization ellipse [1]. However, for quantum fields both the azimuth and the ellipticity evolve in a periodic way. It is interesting to notice that, for a linearly polarized incoming field ($\eta = 0$), both the azimuth and the ellipticity are preserved during the evolution, even in the quantum case. This result seems to be in contradiction to the fact, discussed earlier, that the mean number of photons in the Cartesian component alters in the quantum case, and the linear polarization is not preserved during the evolution. This contradiction can be resolved if we take a look at the degree of polarization (15). The result is the following:

$$\mathcal{P}^2(\tau) = \sin^2 2\eta + \cos^2 2\eta \, \exp\left(-4|\alpha_0|^2 \sin^2 \frac{\tau'}{2}\right) . \tag{118}$$

It is now clear that the degree of polarization becomes smaller than unity due

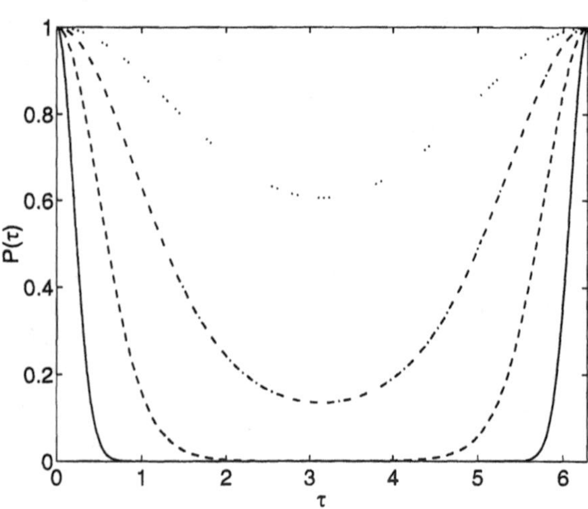

Figure 12: Degree of polarization $P(\tau)$ for $\eta = 0$, $d = 0$ and $\alpha_0 = 5$ (solid), 2 (dashed), 1 (dashed-dotted), 0.5 (dotted).

to the quantum fluctuations of the field. Classically, this quantity remains unity all the time. The polarization of the initially completely polarized light is degraded as a result of quantum fluctuations. The linearly polarized beam preserves the direction of the polarization in that part of the field that remains polarized, whereas part of the incoming intensity becomes unpolarized (isotropic). If there is no damping, the evolution is periodic, and after the initial degradation the polarization is restored to its initial state, and this process is periodically repeated. In Figure 12 we show how fast the degree of polarization is degraded as the intensity of the field increases. In the presence of damping the periodicity of the evolution is lost [84]. Another interesting feature is the fact that, for circularly polarized light $\eta = \pm\pi/4$, the degree of polarization is also unity in the quantum case. The circular polarization is not affected by the quantum fluctuations of the field.

5.4 Quantum Stokes parameters

The degree of polarization discussed in the previous section has been defined by the expectation values of the Stokes operators (28). However, the noncommutability of the Stokes operators, as given by the commutator (29), puts well-known limits on the measurements of the physical quantities represented by the Stokes operators. For example, we have the following Heisenberg uncertainty relation:

$$\langle(\Delta S_1)^2\rangle\langle(\Delta S_2)^2\rangle \geq |\langle S_3\rangle|^2 . \tag{119}$$

Since, according to (30), the square of the "total spin" is conserved, it is easy to calculate the variance of the "total spin". This was done by Tanaś and Kielich [83], and the result is the following:

$$\begin{aligned}\langle \mathbf{S}^2\rangle - \langle \mathbf{S}\rangle^2 &= \langle S_1^2\rangle + \langle S_2^2\rangle + \langle S_3^2\rangle - \left(\langle S_1\rangle^2 + \langle S_2\rangle^2 + \langle S_3\rangle^2\right) \\ &= |\alpha_0|^2 \cos^2 2\eta \left[1 - \exp\left(-4|\alpha_0|^2 \sin^2 \frac{\tau'}{2}\right)\right] + 3|\alpha_0|^2 \\ &= 3|\alpha_0|^2 + |\alpha_0|^2 \left[1 - \mathcal{P}^2(\tau')\right] , \end{aligned} \tag{120}$$

where we have assumed that the initial state is, as before, a coherent state. The "total spin" variance (120) takes its minimum value when the field is fully polarized ($\mathcal{P} = 1$). According to the uncertainty relation (119), the two Stokes observables S_1 and S_2 cannot be measured simultaneously to a high degree of accuracy. The variances of the two observables are given by

$$\begin{aligned}\langle(\Delta S_{1,2})^2\rangle &= |\alpha_0|^2 + \frac{|\alpha_0|^2}{2} \cos^2 2\eta \left\{1 - \exp\left(- 4|\alpha_0|^2 \sin^2 \frac{\tau'}{2}\right)\right. \\ &\pm\Big[\exp\left(-2|\alpha_0|^2 \sin^2 \tau'\right) \cos\left(4\theta - |\alpha_0|^2 \sin 2\eta \sin 2\tau'\right) \\ &\left. - \exp\left(- 4|\alpha_0|^2 \sin^2 \frac{\tau'}{2}\right) \cos\left(4\theta - 2|\alpha_0|^2 \sin 2\eta \sin \tau'\right)\Big]\right\} .\end{aligned}$$

Both variances are equal to $|\alpha_0|^2$ when $\eta = \pm\pi/4$ (circularly polarized light) and do not change during the propagation. The quantum fluctuations in the Stokes parameters are intensity-dependent and periodic in τ'. However, they never become lower than $|\alpha_0|^2$ for the initially coherent state. Quantum mechanics imposes limits on the accuracy of measurements of the Stokes parameters.

5.5 Quantum phase properties

The azimuth $\theta = -(\varphi_+ - \varphi_-)/2$ of the polarization ellipse is classically related to the phase difference of the two circular field components. Thus the rotation of the polarization ellipse can be interpreted as a change of this phase difference of the resulting field. Looking at (115), it is tempting to write

$$2\theta = \tan^{-1}\left(\langle S_1\rangle/\langle S_2\rangle\right) , \tag{121}$$

however, this is not the mean value of any phase-difference operator. Another possibility is to take the normalized Stokes parameter s_1 as a measure of the phase difference cosine defined as

$$\cos 2\theta = \frac{\langle S_1\rangle}{\sqrt{\langle S_1\rangle^2 + \langle S_2\rangle^2}} . \tag{122}$$

Also in this case we cannot say that (122) is the expectation value of the cosine function of the phase-difference operator. In section 4.4 we introduced the Hermitian phase formalism to describe phase properties of the one-mode field. This formalism can be applied to two-mode fields, and the phase-difference operator is defined as the difference of the phase operators of the two modes. Phase properties of the elliptically polarized light propagating in Kerr media have been discussed in detail in [69, 85, 86].

Here, we want to present another definition of the phase-difference operator introduced by Luis and Sánchez-Soto [87] and based on the polar decomposition of the Stokes operators. If we denote $s_{\pm} = (s_1 \pm s_2)/2$, then the classical phase difference between two modes is unambiguously obtained as

$$s_{+} = \mathrm{e}^{-i\varphi}\sqrt{s_{-}s_{+}}\,, \tag{123}$$

where $\varphi = \varphi_{+} - \varphi_{-}$. Following this classical suggestion one can try to decompose the quantum Stokes operators $S_{\pm} = (S_1 \pm S_2)/2$ in the form

$$S_{-} = E\sqrt{S_{+}S_{-}}\,, \quad S_{+} = E^{\dagger}\sqrt{S_{-}S_{+}}\,. \tag{124}$$

When E is unitary, it defines a Hermitian phase-difference operator by $E = \mathrm{e}^{i\Phi}$. Since the matrix elements $\langle n_{+}, 0|E|0, n_{-}\rangle$ are undefined, E cannot be uniquely determined by the unitarity requirement. We thus must impose some additional conditions. The unique condition compatible with the decomposition (124) and unitarity is the commutation relation

$$[E, S_0] = 0\,, \tag{125}$$

which reflects the fact that in the classical domain the form of the polarization ellipse is independent of the intensity. Since the operator E commutes with the total number of photons $a_{+}^{\dagger}a_{+} + a_{-}^{\dagger}a_{-}$, we can study its restriction to the subspace $\mathcal{H}_n$ containing n quanta. Calling this restriction $E^{(n)}$, equation (124) can be solved giving the unitary operator [87]

$$E^{(n)} = \sum_{m=1}^{n} |m-1, n-m+1\rangle\langle m, n-m| + \exp\left[i(n+1)\varphi_0^{(n)}\right]|n, 0\rangle\langle 0, n|\,, \tag{126}$$

where $\varphi_0^{(n)}$ is an arbitrary phase. Since it seems reasonable that the phase difference between two states is independent of any phase reference, we must impose further conditions on E. To this end, let us introduce the exchange operator Π,

$$\Pi|n, m\rangle = |m, n\rangle\,. \tag{127}$$

It is easy to see that $\Pi S_{-}\Pi^{\dagger} = S_{+}$ and therefore we must have

$$\Pi E \Pi^{\dagger} = E^{\dagger}\,, \tag{128}$$

which is the expected action of the exchange operator on the exponential of the phase. Imposing now condition (128) on (126), we easily get that the allowed values of $\varphi_0^{(n)}$ are

$$\varphi_0^{(n)} = \frac{l\pi}{n+1}\,, \tag{129}$$

l being an integer. In each subspace $\mathcal{H}_n$ there are $n+1$ orthonormal states verifying that

$$E^{(n)}|\varphi_r^{(n)}\rangle = \exp\left[i\varphi_r^{(n)}\right]|\varphi_r^{(n)}\rangle \tag{130}$$

with $r = 0, \ldots, n$. These states can be expressed in the number basis as

$$|\varphi_r^{(n)}\rangle = \frac{1}{\sqrt{n+1}} \sum_{n_+=0}^{n} \exp\left[in_+\varphi_r^{(n)}\right]|n_+, n-n_+\rangle\,, \tag{131}$$

where $\varphi_r^{(n)} = \varphi_0^{(n)} + 2\pi r/(n+1)$.

The expression for E on the whole space is

$$E = \sum_{n=0}^{\infty} E^{(n)} = \sum_{n=0}^{\infty}\sum_{r=0}^{n} |\varphi_r^{(n)}\rangle \exp\left[i\varphi_r^{(n)}\right]\langle\varphi_r^{(n)}|\,, \tag{132}$$

which is the operator introduced by Luis and Sánchez-Soto [87]. Since E is unitary, it defines a Hermitian phase-difference operator

$$\Phi = \sum_{n=0}^{\infty}\sum_{r=0}^{n} |\varphi_r^{(n)}\rangle\langle\varphi_r^{(n)}|\,. \tag{133}$$

In order to have the eigenvalues for Φ in the interval $(-\pi, \pi)$ we must take

$$\varphi_0^{(n)} = -\frac{n\pi}{n+1}\,. \tag{134}$$

The Luis and Sánchez-Soto (LSS) phase-difference operator is essentially different from the Pegg–Barnett (PB) phase-difference operator, which is simply the difference between the phase operators of the individual modes. The LSS operator takes into account, for example, the entanglement of the quantum states of the two-mode field. The expectation values of the phase-difference operator (and/or its functions) in the two-mode Kerr state can be calculated according to the rule

$$\langle\Phi(\tau)\rangle = \sum_{n=0}^{\infty}\sum_{r=0}^{n} \varphi_r^{(n)} |\langle\varphi_r^{(n)}|\psi_K(\tau)\rangle|^2\,. \tag{135}$$

The comparison of the phase properties for the two-mode Kerr states has been made in [88]. For large n both PB and LSS approaches give the same results, but for a not very large mean number of photons the discrete character of the LSS operator is evident. The two-mode Kerr states make a good testing ground to see such differences.

5.6 Dissipation

As we have seen in section 4.5, in the case of one-mode Kerr states, dissipation plays an important role in washing out the quantum periodicity of the field evolution. The same, of course, is also true for two-mode Kerr states. Similarly to the one-mode case, one can write the master equation describing the field evolution i the presence of damping (in the interaction picture)

$$\frac{\partial \rho}{\partial t} = \frac{1}{i\hbar}[H,\rho] + \sum_{i=+,-}\left\{\frac{\gamma_i}{2}([a_i\rho, a_i^\dagger] + [a_i, \rho a_i^\dagger]) + \gamma_i \bar{n}_i[[a_i,\rho], a_i^\dagger]\right\}, \quad (136)$$

where γ_i are the damping rates and $\bar{n}_i$ are the mean numbers of thermal photon. The exact solution of the master equation (136) is possible and has been foun by Chaturvedi and Srinivasan [89]. Restricting our considerations to the case of zero-temperature reservoir ($\bar{n}_i = 0$) and an initially coherent state of the field, th solution of the master equation (136) has the form [86]

$$\begin{aligned}
\rho_{m_+,m_-;n_+,n_-}(\tau) &= \langle m_+, m_-|\rho(\tau)|n_+, n_-\rangle = b_{m_+} b_{n_+} b_{m_-} b_{n_-} \\
&\times \exp\left\{i\left[\left(\varphi_+ + \frac{\tau}{2}\right)(m_+ - n_+) + \left(\varphi_- + \frac{\tau}{2}\right)(m_- - n_-)\right]\right\} \\
&\times f_{m_+-n_+;m_--n_-}^{(m_++n_+)/2}(\tau)\, f_{m_--n_-;m_+-n_+}^{(m_-+n_-)/2}(\tau) \\
&\times \exp\left\{|\alpha_+|^2 \lambda \frac{1 - f_{m_+-n_+;m_--n_-}(\tau)}{\lambda + i[m_+ - n_+ + 2d(m_- - n_-)]}\right\} \\
&\times \exp\left\{|\alpha_-|^2 \lambda \frac{1 - f_{m_--n_-;m_+-n_+}(\tau)}{\lambda + i[m_- - n_- + 2d(m_+ - n_+)]}\right\}, \quad (137)
\end{aligned}$$

where we have used the notation

$$f_{m;n}(\tau) = \exp\{-\tau[\lambda + i(m + 2dn)]\},$$
$$\lambda = \gamma_+/|\kappa| = \gamma_-/|\kappa| > 0, \quad \tau = -\kappa t > 0,$$

and other quantities are defined in (106).

The solution (137) allows for calculations of any expectation values of th field operators for the field propagating in a Kerr medium with dissipation. Fo example, the quantum effects on the polarization of light have been studied in [84 and quantum phase properties of elliptically polarized light propagating in a Ke medium in [86]. In the presence of dissipation the Stokes parameters take th following form [90]

$$\begin{aligned}
\langle S_0\rangle &= |\alpha_0|^2 e^{-\lambda\tau}, \\
\langle S_1\rangle &= |\alpha_0|^2 \cos 2\eta \exp\left[-\lambda\tau + \Gamma(\tau) + |\alpha_0|^2\left(e^{-\lambda\tau}\cos\tau' - 1\right)\right] \\
&\quad \times \cos\left[2\theta + \Lambda(\tau) - |\alpha_0|^2 \sin 2\eta\, e^{-\lambda\tau}\sin\tau'\right], \\
\langle S_2\rangle &= |\alpha_0|^2 \cos 2\eta \exp\left[-\lambda\tau + \Gamma(\tau) + |\alpha_0|^2\left(e^{-\lambda\tau}\cos\tau' - 1\right)\right] \\
&\quad \times \sin\left[2\theta + \Lambda(\tau) - |\alpha_0|^2 \sin 2\eta\, e^{-\lambda\tau}\sin\tau'\right], \\
\langle S_3\rangle &= |\alpha_0|^2 \sin 2\eta\, e^{-\lambda\tau}, \quad (138)
\end{aligned}$$

where

$$\Gamma(\tau) = \frac{|\alpha_0|^2}{(\lambda\tau)^2 + (\tau')^2}\left[(\lambda\tau)^2(1 - e^{-\lambda\tau}\cos\tau') + (\lambda\tau)\tau' e^{-\lambda\tau}\sin\tau'\right],$$

$$\Lambda(\tau) = \frac{|\alpha_0|^2 \sin 2\eta}{(\lambda\tau)^2 + (\tau')^2} \left[(\lambda\tau)^2 \mathrm{e}^{-\lambda\tau} \sin\tau' - (\lambda\tau)\tau'(1 - \mathrm{e}^{-\lambda\tau} \cos\tau') \right] .$$

From (115) and (138) one obtains for the azimuth and ellipticity of the polarization ellipse the formulae

$$\begin{aligned} \tan 2\theta(\tau) &= \tan\left[2\theta + \Lambda(\tau) - |\alpha_0|^2 \sin 2\eta\, \mathrm{e}^{-\lambda\tau} \sin\tau'\right] , \\ \tan 2\eta(\tau) &= \exp\left[-\Gamma(\tau) - |\alpha_0|^2 \left(\mathrm{e}^{-\lambda\tau} \cos\tau' - 1\right)\right] \tan 2\eta , \end{aligned}$$

and for the degree of polarization

$$\mathcal{P}^2(\tau) = \sin^2 2\eta + \cos^2 2\eta \exp\left[2|\alpha_0|^2 \left(\mathrm{e}^{-\lambda\tau} \cos\tau' - 1\right) + 2\Gamma(\tau)\right] .$$

For $\lambda = 0$, i.e., if there is no dissipation in the system, one easily regains formulae (117) and (118). Again, it is apparent from the above formulae that the dissipation removes the periodicity of the quantum evolution. The two-mode case is more complex than the one-mode case, but even in this case it is possible to obtain exact analytical solutions to the problem of propagation of quantum fields in Kerr media.

6 Conclusion

As we have shown in this chapter, quantum properties of light propagating through the nonlinear, isotropic Kerr medium manifest themselves in a rich variety of nonclassical phenomena. The nonclassical states of the optical field that appear during the propagation — the *Kerr states* — have many interesting features and only some of them have been reviewed here. There are, of course, many other quantum effects associated with the propagation of light in Kerr media that have not been discussed here, for example, a possibility of generating Fock states [91–95], quantum theory of nonlinear loop mirrors [96], generation of number-phase squeezed states [97], coherent states in finite dimensional Hilbert states [98,99], generation of amplitude-squeezed states from the quadrature-squeezed state propagating in a Kerr medium [100,101], and others [102,103], to list some of them.

A nice feature of the simplified model of propagation of the quantum field in Kerr media presented in this chapter is the fact that it can be solved analytically for both the one- and two-mode cases including dissipation. This is a rather rare feature when it comes to the quantum nonlinear problem. The analytical solutions allow for deeper insight into the physics hidden in the equations, and the model discussed here is a good testing ground for studying quantum properties of optical fields. The fields, no matter how strong they are, exhibit nonclassical properties that cannot be explained without taking explicitly into account the quantum character of the field. Such uniquely quantum features of optical fields have been the subject of this review.

Acknowledgements

I would like to thank my colleagues, in particular Ts. Gantsog, A. Miranowicz, and also L. Sánchez-Soto and A. Luis for their contributions to the results presented here. I thank Poznań Supercomputing and Networking Center for access to the computing facilities.

Bibliography

[1] Maker, P.D., Terhune, R.W., and Savage, C.M., Intensity–dependent changes in refractive index of liquids. *Phys. Rev. Lett.* (1964) **12** 507–509.

[2] Tanaš, R., Squeezed states of an anharmonic oscillator. In: *Coherence and Quantum Optics V* (Proceedings of the conference held in Rochester in 1983) (L. Mandel and E. Wolf, eds.), pp. 645–648. Plenum, New York, 1984.

[3] Milburn, G.J., Quantum and classical Liouville dynamics of the anharmonic oscillator. *Phys. Rev.* A (1986) **33** 674–685.

[4] Milburn, G.J. and Holmes, C.A., Dissipative quantum and classical Liouville mechanics of the anharmonic oscillator. *Phys. Rev. Lett.* (1986) **56** 2237–2240.

[5] Yurke, B. and Stoler, D., Generating quantum mechanical superpositions of macroscopically distinguishable states via amplitude dispersion. *Phys. Rev. Lett.* (1986) **57** 13–16.

[6] Kitagawa, M. and Yamamoto, Y., Number–phase minimum uncertainty state with reduced number uncertainty in a Kerr nonlinear interferometer. *Phys. Rev.* A (1986) **34** 3974–3988.

[7] Gerry, C.C., Application of $su(1,1)$ coherent states to the interaction of squeezed light in an anharmonic oscillator. *Phys. Rev.* A (1987) **35** 2146–2149.

[8] Tombesi, P. and Mecozzi, A., Generation of macroscopically distinguishable quantum states and detection by the squeezed–vacuum technique. *J. Opt. Soc. Am.* B (1987) **4** 1700–1709.

[9] Gerry, C.C. and Rodrigues, S., Time evolution of squeezing and antibunching in an optically bistable two–photon medium. *Phys. Rev.* A (1987) **36** 5444–5447.

[10] Agarwal, G.S., Squeezing in two photon absorption from a strong coherent beam. *Opt. Commun.* (1987) **62** 190–192.

[11] Gerry, C.C. and Vrscay, E.R., Squeezed vacuum states in optically bistable two–photon media. *Phys. Rev.* A (1988) **37** 4265–4268.

[12] Lynch, R., Phase fluctuations in the coherent–light anharmonic–oscillator system via measured phase operators. *Opt. Commun.* (1988) **67** 67–70.

[13] Peřinová, V., and Lukš, A., Third–order nonlinear dissipative oscillator with an initial squeezed state. *J. Mod. Opt.* (1988) **35** 1513–1531.

[14] Tanaś, R., Higher–order squeezing from an anharmonic–oscillator. Comment. *Phys. Rev.* A (1988) **38** 1091–1093.

[15] Milburn, G.J., Mecozzi, A., and Tombesi, P., Squeezed–state superpositions in a damped nonlinear oscillator. *J. Mod. Opt.* (1989) **36** 1607–1614.

[16] Daniel, D.J. and Milburn, G.J., Destruction of quantum coherence in a nonlinear oscillator via attenuation and amplification. *Phys. Rev.* A (1989) **39** 4628–4640.

[17] Tanaś, R., Squeezing from an anharmonic–oscillator model: $(a^\dagger)^2a^2$ versus $(a^\dagger a)^2$ interaction hamiltonians. *Phys. Lett.* A (1989) **141** 217–220.

[18] Bužek, V. and Jex, I., Squeezing properties of coupled nonlinear oscillators. *Int. J. Mod. Phys.* B (1990) **4** 659–676.

[19] Gerry, C.C., Quantum and classical dynamics of an anharmonic oscillator. *Phys. Lett.* A (1990) **146** 363–368.

[20] Miranowicz, A., Tanaś, R., and Kielich, S., Generation of discrete superpositions of coherent states in the anharmonic oscillator model, *Quant. Opt.* (1990) **2** 253–265.

[21] Peřinová, V. and Lukš, A., Exact quantum statistics of a nonlinear dissipative oscillator evolving from an arbitrary state. *Phys. Rev.* A (1990) **41** 414–420.

[22] Gerry, C.C., Quantum dynamics of a nonlinear birefringent model with anticorrelated field states. *J. Mod. Opt.* (1991) **38** 1773–1786.

[23] Joshi, A. and Puri, R.R., Quantum effects in the interaction of an anharmonic oscillator with binomial and squeezed coherent fields. *J. Mod. opt.* (1991) **38** 473–480.

[24] Tanaś, R., Miranowicz, A., and Kielich, S., Squeezing and its graphical representations in the anharmonic oscillator model. *Phys. Rev.* A (1991) **43** 4014–4021.

[25] Ritze, H.H. and Bandilla, A., Quantum effects of a nonlinear interferometer with a kerr cell. *Opt. Commun.* (1979) **29** 126–130.

[26] Tanaś, R. and Kielich, S., Polarization dependence of photon antibunching phenomena involving light propagation in isotropic media. *Opt. Commun.* (1979) **30** 443–446.

[27] Ritze, H.H., Photon statistics and self–induced gyrotropic birefringence. *Z. Phys.* B (1980) **39** 353–360.

[28] Tanaś, R. and Kielich, S., Self–squeezing of light propagating through nonlinear optically isotropic media. *Opt. Commun.* (1983) **45** 351–356.

[29] Tanaś, R. and Kielich, S., On the possibility of almost complete self–squeezing of strong electromagnetic fields. *Opt. Acta* (1984) **31** 81–95.

[30] Imoto, N., Haus, H.A., and Yamamoto, Y., Quantum nondemolition measurement of the photon number via the optical Kerr effect. *Phys. Rev.* A (1985) **32** 2287–2292.

[31] Shelby, R.M., Levenson, M.D., Walls, D.F., Aspect, A., and Milburn, G.J., Generation of squeezed states of light with a fiberoptic ring interferometer. *Phys. Rev.* A (1986) **33** 4008–4025.

[32] Kielich, S., Tanaś, R., and Zawodny, R., Intensity–dependent Faraday effect as a tool for controlling the process of light self–squeezing. *Phys. Rev.* A (1987) **36** 5670–5676.

[33] Kennedy, T.A.B. and Drummond, P.D., Quantum field superpositions via self–phase modulation of coherent wave packets. *Phys. Rev.* A (1988) **38** 1319–1326.

[34] Agarwal, G.S., Visibility index as a simple measure of the quantum effects in light propagation through a nonlinear dispersive element. *Opt. Commun.* (1989) **72** 253–255.

[35] Agarwal, G.S. and Puri, R.R., Quantum theory of propagation of elliptically polarized light through a Kerr medium. *Phys. Rev.* A (1989) **40** 5179–5186.

[36] Horák, R., Quantum light polarization effects in low–birefringent Kerr media. *Opt. Commun.* (1989) **72** 239–243.

[37] Tanaś, R. and Kielich, S., Role of the higher optical Kerr nonlinearities in self–squeezing of light, *Quant. Opt.* (1990) **2** 23–33.

[38] Wright, E.M., Quantum theory of self–phase modulation. *J. Opt. Soc. Am.* B (1990) **7** 1142–1146.

[39] Wilson-Gordon, A.D., Bužek, V., and Knight, P.L., Statistical and phase properties of displaced Kerr states. *Phys. Rev.* A (1991) **44** 7647–7656.

[40] Wright, E.M., Quantum theory of soliton propagation in an optical fiber using the Hartree approximation. *Phys. Rev.* A (1991) **43** 3836–3844.

[41] Brosseau, C., Barakat, R., and Rockower, E., Statistics of the Stokes parameters for Gaussian distributed fields. *Opt. Commun.* (1991) **82** 204–208.

[42] Mihalache, D. and Baboiu, D., Self–squeezing of light in saturable nonlinear media. *Phys. Lett.* A (1991) **159** 303–306.

[43] Werner, M.J. and Risken, H., Quasiprobability distributions for the cavity–damped Jaynes–Cummings model with an additional Kerr medium. *Phys. Rev.* A (1991) **44**

4623–4632.

[44] Sanders, B.C. and Milburn, G.J., Quantum limits to all optical phase shifts in a Kerr nonlinear medium. *Phys. Rev.* A (1992) **45** 1919–1923.

[45] Kielich, S., *Nonlinear Molecular Optics*. Nauka, Moscow, 1981.

[46] Shen, Y.R., *The Principles of Nonlinear Optics*. Wiley, New York, 1985.

[47] Kelley, P.L., Self–focusing of optical beams. *Phys. Rev. Lett.* (1965) **15** 1005–1008.

[48] Born, M. and Wolf, E., *Principles of Optics*, Pergamon, Oxford, 1980.

[49] Hillery, M. and Mlodinov, L.D., Quantization of electrodynamics in nonlinear dielectric media. *Phys. Rev.* A (1984) **30** 1860–1865.

[50] Kennedy, T.A.B. and Wright, E.M., Quantization and phase–space methods for slowly varying optical fields in a dispersive nonlinear medium. *Phys. Rev.* A (1988) **38** 212–221.

[51] Drummond, P.D., Electromagnetic quantization in dispersive inhomogeneous nonlinear dielectrics. *Phys. Rev.* A (1990) **42** 6845–6857.

[52] Abram, I. and Cohen, E., Quantum theory for light propagation in a nonlinear effective medium. *Phys. Rev.* A (1991) **44** 500–517.

[53] Drummond, P.D., Quantum theory of nonlinear dispersive media, In: *Modern Nonlinear Optics, Part 3*, vol. LXXXV of *Advances in Chemical Physics* (M. Evans and S. Kielich, eds.), pp. 379–412. Wiley, New York, 1994.

[54] Shen, Y.R., Quantum statistics of nonlinear optics. *Phys. Rev.* (1967) **155** 921–931.

[55] Drummond, P.D. and Walls, D.F., Quantum theory of optical bistability. 1. Nonlinear polarizability model. *J. Phys.* A (1980) **13** 725–741.

[56] Jauch, J. M. and Rohrlich, F., *The Theory of Photons and Electrons*, Addison–Wesley, Reading, 1959.

[57] Loudon, R. and Knight, P.L., Squeezed light. *J. Mod. Opt.* (1987) **34** 709–759.

[58] Mandel, L., Squeezing and photon antibunching in harmonic generation. *Opt. Commun.* (1982) **42** 437–439.

[59] Loudon, R., Graphical representation of squeezed–state variances. *Opt. Commun.* (1989) **70** 109–114.

[60] Lukš, A., Peřinová, V., and Peřina, J., Principal squeezing of vacuum fluctuations. *Opt. Commun.* (1988) **67** 149–151.

[61] Peřinová, V., Vrana, V., Lukš, A., and Křepelka, J., Quantum statistics of displaced Kerr states. *Phys. Rev.* A (1995) **51** 2499–2515.

[62] Titulaer, U. and Glauber, R.J., Density operators for coherent fields. *Phys. Rev.* (1965) **145** 1041–1050.

[63] Stoler, D., Generalized coherent states. *Phys. Rev. D* (1971) **4** 2309–2312.

[64] Bialynicka–Birula, Z., Properties of the generalized coherent state. *Phys. Rev.* (1968) **173** 1207–1209.

[65] Dodonov, V.V., Malkin, I.A., and Man'ko, V.I., Even and odd coherent states and excitations of a singular oscillator. *Physica* (1974) **72** 597–615.

[66] Dodonov, V.V., Kalmykov, S.Y., and Man'ko, V.I., Statistical properties of Schrödinger real and imaginary cat states. *Phys. Lett.* A (1995) **199** 123–130.

[67] Pegg, D.T. and Barnett, S.M., Unitary phase operator in quantum mechanics. *Europhys. Lett.* (1988) **6** 483–487.

[68] Pegg, D.T. and Barnett, S.M., Phase properties of the quantized single–mode electromagnetic field. *Phys. Rev.* A (1989) **39** 1665–1675.

[69] Tanaś, R., Miranowicz, A., and Gantsog, T., Quantum phase properties of nonlinear optical phenomena. In: *Progress in Optics, vol. XXXV* (E. Wolf, ed.), pp. 355–446. North Holland, Amsterdam, 1996.

[70] Peřinová, V., Lukš, A., and Peřina, J., *Phase in Optics*. World Scientific, Singapore,

1998.

[71] Susskind, L. and Glogower, J., Quantum mechanical phase and time operator. *Physics* (1964) **1** 49–62.

[72] Carruthers, P. and Nieto, M.M., The phase–angle variables in quantum mechanics. *Rev. Mod. Phys.* (1968) **40** 411–440.

[73] Gantsog, T. and Tanaś, R., Phase properties of self–squeezed states generated by the anharmonic oscillator. *J. Mod. Opt.* (1991) **38** 1021–1034.

[74] Vaccaro, J. A. and Orłowski, A., Phase properties of Kerr media via variance and entropy as measures of uncertainty. *Phys. Rev.* A (1995) **51** 4172–4180.

[75] Milburn, G.J. and Holmes, C.A., Quantum coherence and classical chaos in a pulsed parametric oscillator with a Kerr nonlinearity. *Phys. Rev.* A (1991) **44** 4704–4711.

[76] Gantsog, T. and Tanaś, R., Phase properties of a damped anharmonic oscillator. *Phys. Rev.* A (1991) **44** 2086–2093.

[77] Brisudová, M., Nonlinear dissipative oscillator with displaced number states. *J. Mod. Opt.* (1991) **38** 2505–2519.

[78] Brisudová, M.M., Creating multiphase coherent states in a damped anharmonic oscillator. *Phys. Rev.* A (1992) **46** 1696–1699.

[79] Peřinová, V. and Křepelka, J., Free and dissipative evolution of squeezed and displaced number states in the third–order nonlinear oscillator. *Phys. Rev.* A (1993) **48** 3881–3889.

[80] Mancini, S. and Tombesi, P., Quantum dynamics of a dissipative Kerr medium with time–dependent parameters. *Phys. Rev.* A (1995) **52** 2475–2478.

[81] Kielich, S., Tanaś, R., and Zawodny, R., Self–squeezing of elliptically polarized light propagating in a Kerr–like optically active medium. In: *Modern Nonlinear Optics, Part 1*, vol. LXXXV of *Advances in Chemical Physics* (M. Evans and S. Kielich, eds.), pp. 541–594. Wiley, New York, 1993.

[82] Gantsog, T. and Tanaś, R., Discrete superpositions of coherent states and phase properties of elliptically polarized light propagating in a Kerr medium. *Quant. Opt.* (1991) **3** 33–48.

[83] Tanaś, R. and Kielich, S., Quantum fluctuations in the Stokes parameters of light propagating in a Kerr medium. *J. Mod. Opt.* (1990) **37** 1935–1945.

[84] Tanaś, R. and Gantsog, T., Quantum effects on the polarization of light propagating in a Kerr medium. *Opt. Commun.* (1992) **87** 369–377.

[85] Gantsog, T. and Tanaś, R., Phase properties of elliptically polarized light propagating in a Kerr medium. *J. Mod. Opt.* (1991) **38** 1537–1558.

[86] Tanaś, R. and Gantsog, T., Phase properties of elliptically polarized light propagating in a Kerr medium with dissipation. *J. Opt. Soc. Am.* B (1991) **8** 2505–2512.

[87] Luis, A. and Sánchez–Soto, L.L., Phase–difference operator. *Phys. Rev.* A (1993) **48** 4702–4708.

[88] Luis, A., Sánchez–Soto, L.L., and Tanaś, R., Phase properties of light propagating in a Kerr medium: Stokes parameters versus Pegg–Barnett predictions. *Phys. Rev.* A (1995) **51** 1634–1643.

[89] Chaturvedi, S. and Srinivasan, V., Class of exactly solvable master–equations describing coupled nonlinear oscillators. *Phys. Rev.* A (1991) **43** 4054–4057.

[90] Tanaś, R. and Gantsog, T., Quantum fluctuations in the Stokes parameters of light propagating in a Kerr medium with dissipation. *J. Mod. Opt.* (1992) **39** 749–760.

[91] Kuklinski, J.R., Generation of Fock states of the electromagnetic field in a high–q cavity through the Anderson localization. *Phys. Rev. Lett.* (1990) **64** 2507–2510.

[92] Leoński, W. and Tanaś, R., Possibility of producing the one–photon state in a kicked cavity with a nonlinear Kerr medium. *Phys. Rev.* A (1994) **49** R20–R23.

[93] Leoński, W., Fock states in a Kerr medium with parametric pumping. *Phys. Rev.* A (1996) **54** 3369–3372.

[94] Leoński, W., Dyrting, S., and Tanaś, R., Fock states generation in a kicked cavity with a nonlinear medium. *J. Mod. Opt.* (1997) **44** 2105–2123.

[95] Leoński, W., Miranowicz, A., and Tanaś, R., Quasi-periodic and periodic evolution of cavity field in a nonlinear medium. *Laser Physics* (1997) **7** 126–130.

[96] Blow, K.J., Loudon, R., and Phoenix, S.J.D., Quantum theory of nonlinear loop mirrors. *Phys. Rev.* A (1992) **45** 8064–8073.

[97] Collett, M.J., Generation of number–phase squeezed states. *Phys. Rev. Lett.* (1993) **70** 3400–3403.

[98] Miranowicz, A., Piątek, K., and Tanaś, R., Coherent states in a finite–dimensional Hilbert–space. *Phys. Rev.* A (1994) **50** 3423–3426.

[99] Miranowicz, A., Leoński, W., Dyrting, S., and Tanaś, R., Quantum state engineering in finite–dimensional Hilbert space. *Acta Phys. Slovaca* (1996) **46** 451–456.

[100] Sundar, K., Highly amplitude–squeezed states of the radiation field. *Phys. Rev. Lett.* (1995) **75** 2116–2119.

[101] Sundar, K., Amplitude–squeezed quantum states produced by the evolution of a quadrature–squeezed coherent state in a Kerr medium. *Phys. Rev.* A (1996) **5**[illegible] 1096–1111.

[102] Rivera, A.L., Atakishiyev, N.M., Chumakov, S.M., and Wolf, K.B., Evolution under polynomial Hamiltonians in quantum and optical phase spaces. *Phys. Rev.* A (1997) **55** 876–889.

[103] Chumakov, S.M., Frank, A., and Wolf, K.B., Finite Kerr medium: Macroscopic quantum superposition states and Wigner functions on the sphere. *Phys. Rev.* A (1999) **60** 1817–1823.

Chapter 7

From the Jaynes–Cummings model to collective interactions

S.M. Chumakov, A.B. Klimov, M. Kozierowski

1 Introduction

In quantum optics, three famous models play distinguished roles. The Dicke model (DM) [1] describes the interaction of a quantized radiation field with a small sample of A two-level atoms located within a distance much smaller than the wavelength of the radiation. The simplest case of $A = 1$ is known under the name of the Jaynes–Cummings model (JCM) [2] (see also [3]). The interaction of a group of two-level atoms with a single-mode cavity field has also been considered by Tavis and Cummings [4] and this particular type of DM is also termed the Tavis–Cummings model (TCM). From the mathematical point of view, the Dicke model is equivalent to three-wave mixing and second harmonics generation. Different aspects of these models and their applications have been considered in several reviews [5–11] and textbooks [12–17]. Our aim is to show how the use of algebraic methods [18] allows us to understand the dynamics of the coupled system '*atom(s) + quantized electromagnetic field*'.

This chapter is organized as follows. In the next section we recall the basic Hamiltonian describing the interaction between the atom and the field mode. The JCM under the rotating wave approximation will be considered in sections 3–10. We will discuss the JCM evolution operator, the dynamical features (among others, the well-known collapses and revivals of the atomic inversion oscillations), trapping states, the factorization of the wave function and the representation of the dynamics in the field phase space. We stress the fact that the JCM dynamics in the case of the strong field can be completely characterized in terms of the factorized wave functions. In turn, in section 11 we will show how to include corrections due to the counter-rotating terms. In sections 12–14 we shall briefly discuss collective atomic systems. The approximate analytic approach and different dynamical regimes for the Dicke model are discussed in section 12. In section 13 we consider in more detail the strong field case (when the number of photons is much larger than the number of atomic excitations). We will show how the dynamical features of the JCM (the structure of the evolution operator, factorized states, collapses and revivals of the atomic inversion) are generalized to the case of collective atomic systems interacting with a strong quantum field. In section 14 we will introduce the field dissipation (considering real cavities at zero temperature) and show that the factorized states are equally important in this case.

2 The interaction Hamiltonian

The operator of a free electromagnetic field in a perfect cavity has the form (in the Coulomb gauge) [12–17, 19–24]

$$\vec{E} = i \sum_k \vec{\epsilon}_k \sqrt{2\pi\hbar\omega_k/V} \left(a_k e^{i\vec{k}\vec{r}} - a_k^\dagger e^{-i\vec{k}\vec{r}} \right),$$

where $a_k, a_k^\dagger$ are the annihilation and creation operators of photons with the wave vector $\vec{k}$, frequency ω_k and polarization $\vec{\epsilon}_k$, satisfying the common boson commutation relations $[a_k, a_l^\dagger] = \delta_{kl}$; V stands for the cavity volume. The free field Hamiltonian has the form

$$H_f = \sum_k \hbar\omega_k (a_k^\dagger a_k + \tfrac{1}{2})$$

Consider a collection of identical two-level atoms with the upper energy level $|1\rangle$ and the lower energy level $|0\rangle$. The free atomic Hamiltonian reads

$$H_a = \tfrac{1}{2}\omega_a \hbar \sum_j s_{z_j},$$

where $s_{z_j} = |1_j\rangle\langle 1_j| - |0_j\rangle\langle 0_j|$ and $\hbar\omega_a$ is the energy separation between two levels. The electric dipole interaction of the cavity field with the atomic system is described by the Hamiltonian

$$V_{fa} = -\sum_j \left(\vec{d}_j \cdot \vec{E}_j \right),$$

where $\vec{E}_j$ is the electric field at the point $\vec{r}_j$ which corresponds to the position of the j-th atom and $\vec{d}_j$ is the electric dipole operator of this atom given by

$$\vec{d}_j = \vec{d}_j s_{+_j} + \vec{d}_j^* s_{-j}.$$

Here $s_{+_j} = |1_j\rangle\langle 0_j|$ and $s_{-_j} = |0_j\rangle\langle 1_j|$ are the atomic transition operators and $\vec{d}_j = e\langle 1_j|\vec{r}|0_j\rangle$ is the matrix element of the electric dipole momentum (which are assumed to be the same for all the atoms). Note that the operators s_{z_j}, $s_{\pm_j}$ obey the algebraic relations for the common Pauli matrices.

If the dimensions of the atomic system are small in comparison with the wavelength of the field (but the wave functions of different atoms do not overlap, to avoid the dipole–dipole interaction) one can apply the so-called dipole approximation: $e^{i\vec{k}\vec{r}_j} \approx 1$. Then the interaction Hamiltonian can be rewritten in the form

$$V_{fa} = \sum_k (g_k S_+ a_k + g_k^* S_- a_k^\dagger - g_k S_+ a_k^\dagger - g_k^* S_- a_k). \tag{1}$$

Here the interaction constant is given by

$$g_k = -id_k \sqrt{\pi\hbar\omega_k/(2V)}\,,$$

where $d_k = (\vec{d}\cdot\vec{\epsilon}_k)$ is a projection of the atomic dipole on the polarization direction of the k-th field mode. The atomic dipole matrix element can be chosen as purely imaginary without loss of generality, which corresponds to the positive coupling constants g_k. In equation (1) we have introduced collective atomic operators

$$S_z = \tfrac{1}{2}\sum_j s_{zj}, \quad S_\pm = \sum_j s_{\pm j}. \tag{2}$$

Let us assume now that only one cavity mode of frequency ω is important and all of the other field modes, which can exist inside this cavity, are far from resonance with the atomic transition under study. We arrive at a single mode Hamiltonian:

$$H = H_0 + g\left(S_+ a + S_- a^\dagger - S_+ a^\dagger - S_- a\right), \tag{3}$$

$$H_0 = \hbar\omega_a S_z + \hbar\omega(a^\dagger a + \tfrac{1}{2}). \tag{4}$$

The interaction Hamiltonian (3) can be naturally divided into two parts:

$$V_r = g\left(S_- a^\dagger + aS_+\right), \qquad V_a = g\left(a^\dagger S_+ + aS_-\right). \tag{5}$$

The term $a^\dagger S_-$ corresponds to the atomic transition from the excited to the ground state with the emission of a photon; the term aS_+ corresponds to the excitation of the atom with the absorption of a photon. Hence, the part V_r of the interaction Hamiltonian conserves the number of excitations in the system. Correspondingly, the excitation number operator

$$\hat{N} = S_z + a^\dagger a \tag{6}$$

commutes with V_r: $[V_r, \hat{N}] = 0$. On the other hand, the part V_a of the interaction Hamiltonian does not conserve the excitation number; the term $a^\dagger S_+$ describes the atomic excitation with the simultaneous creation of a photon while the term aS_- corresponds to the inverse process. In the interaction picture, the Hamiltonian (3) has the form

$$\begin{aligned} H_I &= e^{\frac{i}{\hbar}H_0 t} H e^{-\frac{i}{\hbar}H_0 t} = g\left(a^\dagger S_- e^{i(\omega-\omega_a)t} + aS_+ e^{-i(\omega-\omega_a)t}\right) \\ &\quad - g\left(a^\dagger S_+ e^{i(\omega+\omega_a)t} + aS_- e^{-i(\omega+\omega_a)t}\right). \end{aligned}$$

In the case when $\omega \sim \omega_a$ the first part of the above expression is almost independent of time, while the second part rapidly oscillates and cannot significantly affect the slow evolution. We thus may neglect the rapidly oscillating terms, making the so-called "rotating wave approximation". In the Shrödinger picture, we neglect the term V_a (the "anti-rotating part") and the Hamiltonian takes the form

$$H = H_0 + V_r. \tag{7}$$

This explains the subscript in the notation V_r. It is clear that under the rotating wave approximation the excitation number becomes the integral of motion

$$i\hbar\dot{\hat{N}} = [\hat{N}, H] = [\hat{N}, V] = 0.$$

In the case of a single atom, Hamiltonian (7) describes the Jaynes–Cummings Model (JCM) [2,3,5,11]. The case of many atoms corresponds to the Dicke Model (DM) [1]. The term V_a plays an important role in the off-resonance case, when the atomic transition frequency is different from the field frequency. For instance, this term leads to multiphoton processes when the multiphoton resonance conditions are satisfied.

3 The spectrum and wave functions

Here we begin the discussion of the dynamics governed by the Hamiltonian (7). Using the units when $\hbar = 1$, it can be rewritten in the form,

$$H = \omega(\hat{N} + \tfrac{1}{2}) + \frac{\Delta}{2}\sigma_z + g\left(a^\dagger\sigma_- + a\sigma_+\right), \tag{8}$$

where $\Delta = \omega_a - \omega$ stands for detuning, and the excitation number operator in a single atom case is $\hat{N} = a^\dagger a + \sigma_z/2$. This Hamiltonian acts in the tensor product of the two-dimensional atomic Hilbert space and the infinite-dimensional field space with the basis

$$|k\rangle_a \otimes |n\rangle_f, \quad n = 0, 1, 2\ldots, \quad k = 0, 1, \tag{9}$$

where $|n\rangle_f$ is the Fock state of the field, $a^\dagger a|n\rangle_f = n|n\rangle_f$, and $|k\rangle_a$ is the eigenstate of the operator σ_z, $\sigma_z|k\rangle_a = (2k-1)|k\rangle_a$.

If the initial state belongs to the subspace with a given excitation number N, then the evolution is restricted to this subspace (as follows from the conservation of the excitation number). We may introduce a different notation for the basis vectors (9), stressing the fact that they are eigenvectors of the excitation operator,

$$|N, k\rangle = |k\rangle_a \otimes |N-k\rangle_f, \quad \hat{N}|N, k\rangle = (N - \tfrac{1}{2})|N, k\rangle, \quad k = 0, 1. \tag{10}$$

Subspaces with N excitations are two-dimensional (except for the case $N = 0$). One may use the vector notation,

$$|N, 0\rangle = \begin{bmatrix} 0 \\ 1 \end{bmatrix}, \quad |N, 1\rangle = \begin{bmatrix} 1 \\ 0 \end{bmatrix}.$$

Inside a single subspace, the Hamiltonian takes the matrix form,

$$H = \omega N \begin{bmatrix} 1 & 0 \\ 0 & 1 \end{bmatrix} + \begin{bmatrix} \Delta/2 & g\sqrt{N} \\ g\sqrt{N} & -\Delta/2 \end{bmatrix}. \tag{11}$$

The eigenvalues of the Hamiltonian read [2,3]

$$\Lambda_{N,0} = \omega N + \Omega, \quad \Lambda_{N,1} = \omega N - \Omega, \quad \Omega = \sqrt{g^2 N + \Delta^2/4}$$

and the normalized eigenvectors are

$$\underline{|N,0\rangle} = \begin{bmatrix} x_N \\ y_N \end{bmatrix}, \quad \underline{|N,1\rangle} = \begin{bmatrix} -y_N \\ x_N \end{bmatrix}, \tag{12}$$

where

$$x_N = \frac{g\sqrt{N}}{\sqrt{(\Omega - \Delta/2)^2 + g^2 N}}, \qquad y_N = \frac{\Omega - \Delta/2}{\sqrt{(\Omega - \Delta/2)^2 + g^2 N}}.$$

To determine the evolution of an arbitrary initial state, $|\zeta\rangle_a \otimes |\phi\rangle_f$ we need to expand it in the dressed basis $\underline{|N,k\rangle}$. For instance, for the initial state

$$|\psi(0)\rangle = |0\rangle_a \otimes |\phi\rangle_f, \quad |\phi\rangle_f = \sum_{n=0}^{\infty} \phi_n |n\rangle_f, \tag{13}$$

we have $|\psi(0)\rangle = \sum_{N=0}^{\infty} \phi_N |N,0\rangle$, with $|N,0\rangle = y_N \underline{|N,0\rangle} + x_N \underline{|N,1\rangle}$ due to equation (12). Therefore, the evolution of the initial state

$$|\psi(0)\rangle = \sum_{N=0}^{\infty} \phi_N \left(y_N \underline{|N,0\rangle} + x_N \underline{|N,1\rangle} \right)$$

is given as

$$|\psi(t)\rangle = \sum_{N=0}^{\infty} \phi_N \left[y_N e^{-i\Lambda_{N,0} t} \underline{|N,0\rangle} + x_N e^{-i\Lambda_{N,1} t} \underline{|N,1\rangle} \right].$$

Note that in the case of the exact resonance ($\Delta = 0$), the states $\underline{|N,k\rangle}$ do not depend on number of the subspace N:

$$\underline{|N,0\rangle} = \frac{1}{\sqrt{2}} \begin{bmatrix} 1 \\ 1 \end{bmatrix}, \qquad \underline{|N,1\rangle} = \frac{1}{\sqrt{2}} \begin{bmatrix} -1 \\ 1 \end{bmatrix}.$$

4 Evolution operator

In this section we will present the evolution operator for the JCM. Since the excitation number operator commutes with the Hamiltonian, $[\hat{H}, \hat{N}] = 0$, the evolution operator takes the form

$$U(t) = e^{-iHt} = e^{-i(\hat{N}+1/2)\omega t} U_I, \quad U_I = e^{-iVt},$$

where the excitation number operator and interaction Hamiltonian can be written as 2×2 matrices with respect to the atomic variables (which should not be confused with matrix notations inside a subspace with a given N introduced in the previous section)

$$\hat{N} = \begin{bmatrix} \hat{n} + 1/2 & 0 \\ 0 & \hat{n} - 1/2 \end{bmatrix}, \qquad \hat{n} = a^{\dagger}a, \qquad V = \begin{bmatrix} \Delta/2 & ga \\ ga^{\dagger} & -\Delta/2 \end{bmatrix}.$$

Using the Taylor expansion of the exponential one may find an explicit form of the evolution operator (see, e.g., [5]):

$$U_I(t) = \begin{bmatrix} \cos\hat{\Omega}_{n+1}t - i\dfrac{\Delta}{2\hat{\Omega}_{n+1}}\sin\hat{\Omega}_{n+1}t & -ig\dfrac{\sin\hat{\Omega}_{n+1}t}{\hat{\Omega}_{n+1}}a \\ -ig\dfrac{\sin\hat{\Omega}_{n}t}{\hat{\Omega}_{n}}a^{\dagger} & \cos\hat{\Omega}_{n}t + i\dfrac{\Delta}{2\hat{\Omega}_{n}}\sin\hat{\Omega}_{n}t \end{bmatrix}, \tag{14}$$

where

$$\hat{\Omega}_n = \sqrt{g^2\hat{n} + \Delta^2/4}. \tag{15}$$

We would like to mention here the convenient commutation relations

$$f(\hat{n})a = af(\hat{n}-1), \quad f(\hat{n})a^{\dagger} = a^{\dagger}f(\hat{n}+1), \tag{16}$$

where $f(\hat{n})$ is an arbitrary function of the photon number operator.

In the case of exact resonance, equation (14) simplifies as

$$U_I(t) = \begin{bmatrix} \cos gt\sqrt{\hat{n}+1} & -i\dfrac{\sin gt\sqrt{\hat{n}+1}}{\sqrt{\hat{n}+1}}a \\ -i\dfrac{\sin gt\sqrt{\hat{n}}}{\sqrt{\hat{n}}}a^{\dagger} & \cos gt\sqrt{\hat{n}} \end{bmatrix}.$$

We can rewrite this formula in a different way using the phase operator for the field, (see, e.g., [21])

$$e^{i\hat{\phi}} = \frac{1}{\sqrt{\hat{n}+1}}a, \quad e^{-i\hat{\phi}} = a^{\dagger}\frac{1}{\sqrt{\hat{n}+1}}, \quad \left[e^{i\hat{\phi}}, e^{-i\hat{\phi}}\right] = |0\rangle_f\,\langle 0|_f, \tag{17}$$

$$e^{-i\hat{\phi}}|n\rangle_f = |n+1\rangle_f, \quad e^{i\hat{\phi}}|n\rangle_f = |n-1\rangle_f, \quad n>0, \quad e^{i\hat{\phi}}|0\rangle_f = 0.$$

The on-resonance evolution operator takes the form,

$$U_I(t) = \begin{bmatrix} \cos gt\sqrt{\hat{n}+1} & -i\sin gt\sqrt{\hat{n}+1}\,e^{i\hat{\phi}} \\ -i\sin gt\sqrt{\hat{n}}\,e^{-i\hat{\phi}} & \cos gt\sqrt{\hat{n}} \end{bmatrix}. \tag{18}$$

This form coincides with the atomic evolution operator in the external classical field, if $e^{i\hat{\phi}}$ and $\hat{n} \approx \hat{n}+1$ are replaced by the classical field phase and amplitude.

Acting with this evolution operator on the initial state we obtain the time-dependent wave function,

$$|\psi(t)\rangle = U_I(t)\,|\psi(0)\rangle.$$

For the initial state $|\psi(0)\rangle = |\phi\rangle_f |1\rangle_a$, where $|\phi\rangle_f = \sum_n \phi_n |n\rangle_f$ is an arbitrary field state, taking into account that

$$\cos gt\sqrt{\hat{n}+1}|\phi\rangle_f = \sum_n \phi_n \cos gt\sqrt{n+1}|n\rangle_f,$$

$$\left(\sin(gt\sqrt{\hat{n}})/\sqrt{\hat{n}}\right) a^\dagger|\phi\rangle_f = \sum_n \phi_n \sin gt\sqrt{n+1}|n+1\rangle_f,$$

we obtain

$$U_I(t)|\phi\rangle_f|1\rangle_a = \sum_n \phi_n \cos gt\sqrt{n+1}|n\rangle_f|1\rangle_a - i\sum_n \phi_{n-1} \sin gt\sqrt{n}|n\rangle_{|}0\rangle_a.$$

The evolution operator enables us to find an arbitrary operator in the Heisenberg representation. For instance, for the atomic inversion operator σ_z under exact resonance we have, using commutation relations (16),

$$\sigma_z(t) = U^\dagger \sigma_z U = \begin{bmatrix} \cos 2gt\sqrt{\hat{n}+1} & -i\dfrac{\sin 2gt\sqrt{\hat{n}+1}}{\sqrt{\hat{n}+1}}a \\ i\dfrac{\sin 2gt\sqrt{\hat{n}}}{\sqrt{\hat{n}}}a^\dagger & -\cos 2gt\sqrt{\hat{n}} \end{bmatrix}.$$

To find the atomic inversion evolution, one needs to calculate the average of the above operator for a given initial state. For instance, if we take the atom initially in its ground state, we obtain

$${}_a\langle 0|\sigma_z(t)|0\rangle_a = -\cos 2gt\sqrt{a^\dagger a},$$

which is still the field operator. For an arbitrary initial field state, we have $|\phi\rangle_f = \sum_{n=0}^{\infty} \phi_n |n\rangle_f$, and we arrive at the result

$${}_f\langle\phi|\,{}_a\langle 0|\sigma_z(t)|0\rangle_a\,|\phi\rangle_f = -\sum_n \phi_n \cos 2gt\sqrt{n}.$$

Let us end this section with a discussion of the case of a large detuning (which is also called the dispersive limit), $\Delta \gg g\sqrt{\bar{n}}$, where $\bar{n}$ is the average number of photons in the field. Expanding the operator $\hat{\Omega}_n$ (15) in series with a small parameter $g\sqrt{n}/\Delta$,

$$\hat{\Omega}_n \approx \Delta/2 + g^2 n/\Delta + \ldots, \tag{19}$$

we obtain the following approximation for the evolution operator (14):

$$U_I(t) \approx e^{-i\xi t/2} \begin{bmatrix} e^{-i[\Delta/2+\xi(\hat{n}+1/2)]t} & 0 \\ 0 & e^{i[\Delta/2+\xi(\hat{n}+1/2)]t} \end{bmatrix},$$

where $\xi = g^2/\Delta$. Since the evolution operator in the dispersive limit takes the diagonal form, the effect of the field on the atom reduces to the phase shifts, different for the excited and ground atomic states. The effective interaction Hamiltonian for the JCM in the dispersive limit has the form

$$V \simeq [\Delta/2 + \xi(\hat{n} + \tfrac{1}{2})]\sigma_z + \xi/2.$$

5 The classical field limit

Let us introduce a new operator $\hat{Q}$ [25] in the bare atomic basis

$$\hat{Q} = \begin{bmatrix} e^{i\hat{\phi}} & 0 \\ 0 & 1 \end{bmatrix}, \tag{20}$$

where $e^{i\hat{\phi}}$ is the phase operator (17). It is well known that the phase operators defined in this way are not unitary and therefore, the operator $\hat{Q}$ is not unitary:

$$\left[\hat{Q}, \hat{Q}^{\dagger}\right] = |0\rangle_{f\,f}\langle 0|\, \sigma_{+}\sigma_{-}.$$

However, operator $\hat{Q}$ is unitary on the field states $|n\rangle_f$ if $n \geq 1$, i.e. if one can neglect the contribution of the vacuum state. Another representation is

$$\hat{Q} = \exp\left[\tfrac{1}{2} i\hat{\phi}\left(\sigma_z + 1\right)\right] = \sum_{k=0,1} e^{i\hat{\phi}k} |k\rangle_{a\,a}\langle k|.$$

Assuming that the Fock states with small photon numbers do not contribute significantly to the initial field state, we apply the $\hat{Q}$ transformation to the JCM Hamiltonian in the exact resonance case:

$$\hat{Q}^{\dagger}\hat{n}\hat{Q} = \hat{n} - \frac{\sigma_z + 1}{2}, \quad \hat{Q}^{\dagger}\sigma_z\hat{Q} = \sigma_z, \quad \widetilde{H}_0 = \hat{Q}^{\dagger}H_0\hat{Q} = \omega\hat{n}.$$

On the other hand,

$$\hat{Q}^{\dagger}\sigma_{\pm}\hat{Q} = \sigma_{\pm}e^{\mp\hat{\phi}}$$

and the transformed interaction Hamiltonian takes the following form:

$$\widetilde{V} = \hat{Q}^{\dagger}V\hat{Q} = g\begin{bmatrix} 0 & e^{-i\hat{\phi}}a \\ a^{\dagger}e^{i\hat{\phi}} & 0 \end{bmatrix} = 2g\sqrt{\hat{n}}\,\sigma_x.$$

The transformed Hamiltonian is a diagonal operator in the field space. The $\hat{n}$ operator commutes with all the other components of the transformed Hamiltonian and may be treated as a *c*-number when calculating the transformed evolution operator $\widetilde{U}(t) = \exp\left(-i\widetilde{H}t\right)$

$$\widetilde{H} = \widetilde{H}_0 + \widetilde{V} = \omega\hat{n} + 2g\sqrt{\hat{n}}\,\sigma_x,$$

$$\widetilde{U}(t) = \exp(-i\omega\hat{n})\exp(-2igt\sqrt{\hat{n}}\,\sigma_x),$$

or, in the matrix form

$$\widetilde{U}(t) = \exp\left(-i\omega\hat{n}\right)\begin{bmatrix} \cos gt\sqrt{\hat{n}} & -i\sin gt\sqrt{\hat{n}} \\ -i\sin gt\sqrt{\hat{n}} & \cos gt\sqrt{\hat{n}} \end{bmatrix}.$$

This coincides with the atomic evolution operator in the classical field of intensity $\hat{n}$ and phase equal to zero. Finally, one can recover the quantum field evolution operator (18) making the inverse transformation,

$$U(t) = \hat{Q}\widetilde{U}(t)\hat{Q}^{\dagger}.$$

We have thus reconstructed the exact evolution operator for a two-level atom interacting with a quantum field mode, knowing the functional form of the atomic evolution operator in an external classical field. Surprisingly, this procedure can be generalized to the case of an arbitrary atomic system interacting with a quantum cavity field, provided that the field energy is much larger than the energy of the atomic system. In this general case the quantum field evolution operator can be approximately reconstructed, if one knows the functional form of the corresponding atomic evolution operator in an external classical field [25]. One may speak in this sense about the "semiclassical quantization" of the atomic evolution operator. This procedure will be described in section 12.

6 Collapses and revivals

Collapses and revivals of oscillations of the atomic inversion $\langle\sigma_z(t)\rangle$ represent one of the most important features of the JCM which, in fact, motivated great attention to this model in the 1980s [26]. Consider the atom initially in one of its bare states ($|k\rangle_a$, $k = 1$ or 0) and the field in a coherent state (CS) with a large average photon number $\bar{n} \gg 1$ (for instance, $\bar{n} = 100$ may be considered as a very "high" photon number in this case). Then the field phase is well defined: $\overline{\Delta\phi} \sim 1/2\sqrt{\bar{n}}$. Fluctuations of the photon number grow with $\bar{n}$, $\overline{\Delta n} = \sqrt{\langle n^2\rangle - \bar{n}^2} \sim \sqrt{\bar{n}}$; however, the relative fluctuations diminish, $\overline{\Delta n}/\bar{n} \sim 1/\sqrt{\bar{n}}$. Thus, one can say that the amplitude of the field is relatively well defined. Therefore, one can expect that interaction with a strong coherent field leads to oscillations of the atomic inversion with the frequency $2\bar{\Omega} = 2g\sqrt{\bar{n}}$, as happens in the case of the external classical field (Rabi oscillations). However, the behaviour of the atomic inversion is much more complicated. $\langle\sigma_z(t)\rangle$ oscillates with the Rabi frequency only for very short times; at times $gt \sim 1$ oscillations collapse and the atomic inversion remains constant for a large period of time; however, the oscillations revive at time $gT_r \sim 2\pi\sqrt{\bar{n}}$. This process is repeated for several times but successive revivals are less and less pronounced, and finally they are changed into a "quasi-chaotic behaviour". Therefore, this simple model reveals in a very clear form the difference between classic and quantum field dynamics.

Let us now present a qualitative discussion of this phenomenon [26]. Consider the atom initially in its ground state and the field in a coherent state $|\alpha\rangle_f$ with $\bar{n} = |\alpha|^2 \gg 1$. The atomic inversion is given by the formula

$$\langle\sigma_z(t)\rangle = \langle\psi(t)|\sigma_z(0)|\psi(t)\rangle, \quad |\psi(t)\rangle = U(t)\,|\alpha\rangle_f \otimes |0\rangle_a. \tag{21}$$

From (14) we obtain

$$\langle\sigma_z(t)\rangle = -\sum_{n=0}^{\infty} P_n \left[\frac{\Delta^2}{4\Omega_n^2} + \left(1 - \frac{\Delta^2}{4\Omega_n^2}\right)\cos 2\Omega_n t\right], \tag{22}$$

where

$$P_n = e^{-\bar{n}}\bar{n}^n/n!$$

is the Poissonian distribution and

$$\Omega_n = \sqrt{\Delta^2/4 + g^2 n}. \tag{23}$$

The frequencies Ω_n are often called the quantum Rabi frequencies. Recall that the

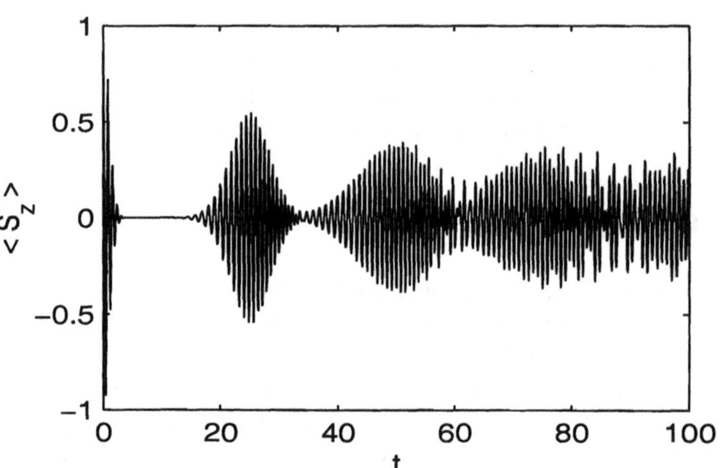

Figure 1: Collapses and revivals of the atomic inversion, $\langle \sigma_z(t) \rangle$, in the JCM. The atom is initially excited and the field is taken in the initial coherent state with $\bar{n} = 16$.

Poissonian distribution has a relatively sharp peak with spread $\overline{\Delta n} \sim \sqrt{\bar{n}}$ in the vicinity of the point $n \sim \bar{n}$. Therefore the frequencies which essentially contribute to the sum (22) correspond to the values of n from the interval $|n - \bar{n}| < \overline{\Delta n}$, where we may use the expansion of quantum Rabi frequencies around the value of the classical Rabi frequency $\bar{\Omega} = \Omega(\bar{n}) = \sqrt{\Delta^2/4 + g^2 \bar{n}}$:

$$\Omega_n = \sqrt{\Delta^2/4 + g^2 n} = \bar{\Omega} + \frac{g^2 \Delta n}{2\bar{\Omega}} - \frac{(g^2 \Delta n)^2}{8\bar{\Omega}^3} + \dots . \tag{24}$$

Hence, the spread of the frequencies is of the order

$$\Delta\Omega_n \sim g^2 \overline{\Delta n}/\bar{\Omega} \sim \frac{g}{\sqrt{1 + \Delta^2/4g^2\bar{n}}},$$

This means that the cosine functions in the sum (22) become out of phase at time $T_c \sim 1/\Delta\Omega$ and oscillations collapse. The collapse time is given as

$$T_c \sim (\Delta\Omega)^{-1} = g^{-1}\sqrt{1 + \Delta^2/(4g^2\bar{n})}.$$

On the other hand, oscillations with neighbouring frequencies again become rephased at the time instants when

$$2\pi k = 2t\left[\Omega(\bar{n}+1) - \Omega(\bar{n})\right] \approx g^2 t/\bar{\Omega}, \quad k = 1, 2, \dots$$

e., the system revives at the following times:

$$T_r \sim 2\pi k\bar{\Omega}/g^2. \tag{25}$$

Therefore, there are three different time scales in the JCM: the period of a single Rabi oscillation $T_{Rabi} \sim 1/2\bar{\Omega}$ (which is very short if $\bar{n} \gg 1$); the collapse time $T_c \sim \sqrt{1+\Delta^2/(4g^2\bar{n})}$; and the revival time $T_r \sim 2\pi\bar{\Omega}/g^2$. It is clear that $T_{Rabi} \ll T_c \ll T_r$. These time scales are seen well in Figure 1 where the typical time evolution of the atomic inversion, $\langle \sigma_z(t) \rangle$, is shown.

Let us turn now to a quantitative description of the problem, considering separately the two different limits: of a large detuning $\Delta \gg g\sqrt{\bar{n}}$ (the dispersive limit) and of a zero detuning $\Delta = 0$.

5.1 The dispersive limit

In the case of large detuning, $\Delta \gg g\sqrt{\bar{n}}$, one can expand the quantum Rabi frequencies, Ω_n, in a power series with a small parameter $g\sqrt{\bar{n}}/\Delta$ (in the same manner as it was done in equation (19)). In this case we restrict ourselves to the linear approximation for the frequencies:

$$\Omega_n \approx \Delta/2 + \xi n + ..., \quad \xi = g^2/\Delta.$$

Then the sum in equation (22) is transformed to the form

$$\langle \sigma_z(t) \rangle = -1 + (4\xi/\Delta)\left[\bar{n} - \sum_{n=0}^{\infty} P_n n \cos\left(\Delta + 2\xi n\right) t\right],$$

where the terms of the order $O\left((\xi\bar{n}/\Delta)^2\right)$ have been neglected. This sum contains harmonic frequencies and can be easily calculated. One finds

$$\begin{aligned} \sum_n P_n n e^{i2\xi n t} &= \bar{n} e^{-\bar{n}} \frac{\partial}{\partial \bar{n}} \sum_n \frac{[\bar{n}\exp(i2\xi t)]^n}{n!} \\ &= \bar{n} \exp\left[\bar{n}(\cos 2\xi t - 1) + i(\bar{n}\sin 2\xi t + 2\xi t)\right]. \end{aligned}$$

The envelope of the Rabi oscillations, $\exp\left(-2\bar{n}\sin^2 \xi t\right)$, diminishes when

$$2\bar{n}\sin^2 \xi t \sim 2\bar{n}(\xi t)^2 \sim 1,$$

which determines the collapse time,

$$T_c \sim \frac{1}{\sqrt{2\bar{n}}\xi} = \frac{\Delta}{g^2\sqrt{2\bar{n}}}.$$

The function $\langle \sigma_z(t) \rangle$ does not oscillate when $\bar{n}\sin^2(\xi t) \gg 1$, and then oscillations revive again at the time instants $\xi t \sim \pi k$, $k = 1, 2, \ldots$, which gives the revival times:

$$kT_r = k\pi\Delta/g^2.$$

The phase of the oscillations $\approx \Delta + \bar{n}\sin 2\xi t$, in the vicinities of the revival time $kT_r - T_c < t < kT_r + T_c$, can be approximated as $2\bar{\Omega}t - 2\pi k\bar{n}$ which leads us to the expression

$$\langle \sigma_z(t) \rangle \approx -1 + \frac{4\xi\bar{n}}{\Delta}\left\{1 - e^{-2\bar{n}\sin^2 \xi t}\cos\left[(\Delta + 2\xi\bar{n})\,t - 2\pi k\bar{n}\right]\right\}.$$

In this case the behaviour of the atomic inversion is regular. Rabi oscillations (of a very small amplitude) revive periodically and subsequent revivals are identical to each other. It is a general property of the harmonic approximation which works well in the case under study. Note that the revival times do not depend on the field intensity in this limit. For other studies of the dispersive regime see, e.g., [27].

6.2 Exact resonance

Replacing $\Delta = 0$ in equation (22) we obtain

$$\langle \sigma_z(t) \rangle = -\sum_{n=0}^{\infty} P_n \cos 2gt\sqrt{n}. \tag{26}$$

This sum is essentially anharmonic, and it is necessary to take into account the quadratic terms in the expansion for the Rabi frequencies (24). Different analytical methods were applied to find a closed expression for this sum in the strong field case, $\bar{n} \gg 1$ [26, 28, 29]. All these methods lead to similar (though different in small details) expressions for the atomic inversion as a sum of oscillating packets with Gaussian envelopes. Each packet corresponds to a single revival of the Rabi oscillations with growing spreads and decreasing maxima of subsequent revivals. Here we will show that the sum (26) can be approximately written in terms of the Jacobi function Θ_3 (see, e.g., [30]). After that the analytic expression for the collapses and revivals arises naturally, by using the well-known properties of the Θ_3 function. In fact, this is a general method of treating the sums with quadratic frequencies.

The sum (26) can be rewritten in the form

$$\langle \sigma_z(t) \rangle = -\mathrm{Re}\, W_+ , \quad W_+ = \sum_{n=0} P_n e^{i2gt\sqrt{n}} .$$

Let us start with the case of an integer average photon number $\bar{n}$. Expanding the square root in a Taylor series up to the second-order terms, we have

$$\sqrt{n} \approx \sqrt{\bar{n}} + \frac{m}{2\sqrt{\bar{n}}} - \frac{m^2}{8\bar{n}^{3/2}}, \quad m = n - \bar{n}, \tag{27}$$

where m is integer. For strong fields ($\bar{n} \gg 1$) one may approximate with a high accuracy the Poissonian distribution by the Gaussian one,

$$P_n \approx (2\pi\bar{n})^{-1/2} \exp\left(-m^2/2\bar{n}\right).$$

Then the sum W_+ takes the form of the Jacobi Θ_3 function:

$$W_+ = \frac{\Theta_3(z|\mu)}{\sqrt{2\pi\bar{n}}}\, e^{igt2\sqrt{\bar{n}}},$$

where, by definition,

$$\Theta_3(z|\mu) = \sum_{m=-\infty}^{\infty} e^{izm-\mu m^2},$$

and

$$z = \frac{gt}{\sqrt{\bar{n}}}, \qquad \mu = \frac{1}{2\bar{n}}\left(1 + \frac{igt}{2\sqrt{\bar{n}}}\right), \qquad \sigma = 2\bar{n}\mu = 1 + \frac{igt}{2\sqrt{\bar{n}}}.$$

Using the Poisson summation formula

$$\sum_{n=-\infty}^{\infty} f(n) = \sum_{m=-\infty}^{\infty} \int_{\infty}^{\infty} dx\, f(x) e^{2\pi imx},$$

we obtain another representation for the Jacobi function,

$$\Theta_3(z|\mu) = \sqrt{\frac{\pi}{\mu}} \sum_k \exp\left[-\frac{(z-2\pi m)^2}{4\mu}\right].$$

The sum W_+ can thus be approximated as

$$W_+ = \frac{e^{igt2\sqrt{\bar{n}}}}{\sqrt{\sigma}} \sum_k \exp\left[-\frac{g^2(t-T_r k)^2}{2\sigma}\right], \qquad T_r = 2\pi\sqrt{\bar{n}}/g.$$

Taking into account that

$$\frac{1}{\sigma} = \frac{e^{-i\phi}}{|\sigma|}, \qquad \phi = \arctan\frac{gt}{2\sqrt{\bar{n}}}, \qquad |\sigma|^2 = \left[1 + \frac{g^2t^2}{4\bar{n}}\right],$$

we obtain

$$\langle\sigma_z(t)\rangle = -\mathrm{Re}\, W_+ = |\sigma|^{-1/2} \sum_k e^{-\tau_k^2/2|\sigma|^2} \cos\left[2g\sqrt{\bar{n}}t + \frac{\tau_k^2 gt}{4\sqrt{\bar{n}}|\sigma|^2} - \frac{\phi}{2}\right].$$

Here we have introduced the local time $\tau_k = g(t - T_r k)$ in the vicinity of the k-th revival. The spread of each revival, $\overline{\Delta\tau_k}/g$, has the same order as the collapse time T_c. Remembering that the oscillations disappear between different revivals, we can use the values of parameters $\sigma(t)$ and $\phi(t)$ at revival times $t = T_r k$ (indeed, these are slowly varying functions, which do not change significantly in the time range $kT_r - T_c < \tau_k < kT_r + T_c$),

$$|\sigma(t = T_r k)|^2 \to 1 + \pi^2 k^2, \qquad \phi(t = T_r k) \to \phi_k = \arctan \pi k.$$

Finally, we obtain

$$\langle\sigma_z(t)\rangle = -\sum_k \exp\left\{-\frac{\tau_k^2}{2(1+\pi^2k^2)}\right\}\frac{\cos\left[\psi_k(t)\right]}{(1+\pi^2k^2)^{1/4}}, \tag{28}$$

$$\psi_k(t) = 2g\sqrt{\bar{n}}t + \frac{\tau_k^2\pi k}{2(1+\pi^2k^2)} - \frac{\phi_k}{2}. \tag{29}$$

The summation index k in (28) corresponds to the revival number, for the k-th term is essentially different from zero only in the vicinity of the revival time $T_r k$. For instance, if $\tau_0 = gt \sim 1 \ll 2\pi\sqrt{\bar{n}}$ (within the interval of the initial collapse) only the term $k = 0$ is different from zero in the sum (28). In turn, the first revival appears when $t \sim T_r$ and $\tau_1 \sim 1 \ll 2\pi\sqrt{\bar{n}}$. Revivals have Gaussian envelopes with the spread $|\sigma(kT_r)|^2 = 1 + \pi^2k^2$ and the maxima decreasing as $(1+\pi^2k^2)^{-1/4}$, which means that every subsequent revival is wider and lower than the preceding one. The Rabi oscillations of the frequency $2g\sqrt{\bar{n}}$ inside every revival are modulated by the phase shift $\tau_k^2\pi k/2(1+\pi^2k^2) - \phi_k/2$.

For non-integer values of the average photon number $\bar{n}$ the above formulae must be modified in such a way that the values of the index m in equation (27) remain integer. To this end, let us introduce

$$\bar{n} = [\bar{n}] + \delta, \qquad n = [\bar{n}] + m = \bar{n} - \delta + m,$$

where $[\bar{n}]$ is the integer part of $\bar{n}$. Thus we have

$$P_m \approx (2\pi\bar{n})^{-1/2}\exp\left[-(m-\delta)^2/2\bar{n}\right],$$

$$\sqrt{n} = \sqrt{[\bar{n}]} + \frac{m}{2\sqrt{\bar{n}}} - \frac{m^2}{8\bar{n}^{3/2}} + O(1/\bar{n}).$$

Correspondingly, $\bar{n}$ in the phase $\psi_k(t)$ (29) can be replaced by $[\bar{n}]$.

7 The JCM with an initial thermal field

Let us consider now the JCM with the initial field in a thermal state. If the atom is taken initially in its ground state, the atomic inversion is given by the same equation (26), and the photon number distribution has a form

$$P_n = (1-q)\,q^n, \qquad q = \frac{\bar{n}}{\bar{n}+1}, \tag{30}$$

where $\bar{n}$ is the average photon number. In this case the collapse–revival structure disappears [31] (see, also, [32–35]). The initial atomic inversion collapses to the mid-level and does not oscillate for some period of time, which is followed by a quasichaotic behaviour (see Figure 2). To evaluate the series (26) we use the Abel–Plana summation formula (see [36], Chapter VII, Miscellaneous examples, example 7):

$$\sum_{n=0}^{\infty} f(n) = \frac{1}{2}f(0) + \int_0^{\infty} dx\, f(x) + i\int_0^{\infty} dx\,\frac{f(ix)-f(-ix)}{\exp(2\pi x)-1}.$$

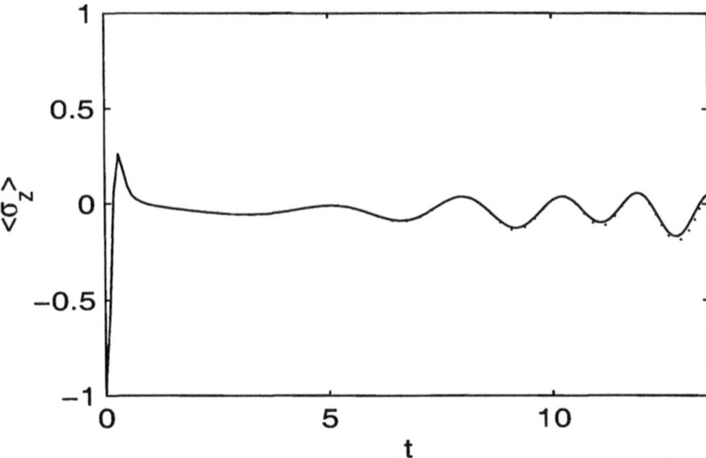

Figure 2: Dynamics of the atomic inversion in the JCM with the initial thermal field state, $\bar{n} = 25$, and the atom initially in its ground state. The solid line corresponds to the numerical calculation and the dotted line to the analytic approximation.

For the atomic inversion we find

$$\langle \sigma_z(t) \rangle = -(1-q)\left[\tfrac{1}{2} + I_0(t) + I_1(t)\right], \tag{31}$$

$$\begin{aligned} I_0(t) &= \mathrm{Re}\int_0^\infty dx \, \exp\left[x \ln q + 2igt\sqrt{x}\right] \\ &= \frac{1}{|\ln q|}\left[1 + \sqrt{\pi} z e^{-z^2} \,\mathrm{Im}\,(1 - \mathrm{erf}(-iz))\right], \end{aligned} \tag{32}$$

$$I_1(t) = \int_0^\infty dx \, \frac{[\chi(t,x) + \chi(-t,x)]}{\exp(2\pi x) - 1}, \quad z = \frac{2gt}{\sqrt{|\ln q|}},$$

$$\chi(t,x) = \exp\left(gt\sqrt{2x}\right)\sin\left(gt\sqrt{2x} - x\ln q\right).$$

The integral I_0 in (31) corresponds to the semiclassical limit (when the series is replaced by the integral) and describes the initial part of the curve (initial collapse) [31]. In the strong field case, $q = 1 - \epsilon$ and $-\ln q \approx \epsilon \sim 1/\bar{n}$. Then, for short times, $gt \sim \sqrt{\epsilon}$, the integral $I_0 \sim 1/\epsilon$, while it rapidly tends to zero as $\sim (gt)^{-2}$ for $gt > \sqrt{\epsilon}$. Note that the average photon number $\bar{n}$ appears only in the scale factors in (32), which determine the scales of time and of I_0 itself.

The integral I_1 gives quantum corrections which are responsible for the quasichaotic behaviour. The principal contribution to the integral is given by the vicinity of the point $x_0 = (gt)^2/(8\pi^2)$ (of width $\Delta x \sim gt\pi^{-3/2}/2$). If $x_0\epsilon \ll 1$, which means that $gt \ll \pi\sqrt{8\bar{n}}$, the integral I_1 is almost independent of the average photon number and is approximated as

$$I_1(t) = 2\int_0^{\infty} dx\, \frac{\sinh\left(gt\sqrt{2x}\right)\sin\left(gt\sqrt{2x}\right)}{\exp(2\pi x)-1} + O(\epsilon). \tag{33}$$

In Figure 2 we show the results of both the exact (numerical) and analytical calculations carried out with the help of formulae (31)–(33).

To conclude, the quantum corrections to the semiclassical dynamics of the atomic inversion in the strong thermal field do not depend on the intensity of the field in the large time range $gt \ll \pi\sqrt{8\bar{n}}$. This range corresponds to the beginning of the "quasichaotic" behaviour.

8 Trapping states

The behaviour of the atom interacting with a quantum field is different from that in a classical field. However, there exist special initial states of the atom for which the quantum nature of the field almost does not affect the atomic dynamics [37].

In the classical field limit the photon creation and annihilation operators in the JCM Hamiltonian should be replaced by the complex numbers (field amplitudes)

$$\hat{a} \to \alpha \equiv \sqrt{\bar{n}}e^{i\varphi}, \qquad \hat{a}^{\dagger} \to \bar{\alpha} \equiv \sqrt{\bar{n}}e^{-i\varphi}, \tag{34}$$

where $\bar{n}$ is the classical field intensity in units of the photon number and φ is the classical field phase. The resulting Hamiltonian describes the atom in an external classical field of amplitude $\sqrt{\bar{n}}$ and phase φ:

$$H_{cl} = g\sqrt{\bar{n}}\left(e^{-i\varphi}\sigma_- + e^{i\varphi}\sigma_+\right) = g\begin{bmatrix} 0 & \alpha \\ \bar{\alpha} & 0 \end{bmatrix}. \tag{35}$$

The atomic evolution operator in the external classical field takes the form

$$U_{cl}(t) = \begin{bmatrix} \cos(\Omega t) & -i\sin(\Omega t)\,e^{i\varphi} \\ -i\sin(\Omega t)\,e^{-i\varphi} & \cos(\Omega t) \end{bmatrix},$$

where $\Omega = g\sqrt{\bar{n}}$ is the classical Rabi frequency. Let us denote the eigenvalues and eigenstates of H_{cl} as $\lambda_p\sqrt{\bar{n}}$ and $|\underline{p,\varphi}\rangle_a$:

$$H_{cl}|\underline{p,\varphi}\rangle_a = \lambda_p\sqrt{\bar{n}}|\underline{p,\varphi}\rangle_a, \qquad p = 0,1, \tag{36}$$

where the argument φ indicates the dependence of the eigenvectors of the semiclassical Hamiltonian of the phase of this classical field. H_{cl} can be rewritten in the form

$$H_{cl} = \sqrt{\bar{n}}e^{i\frac{1}{2}\varphi\sigma_z}\left(\sigma_- + \sigma_+\right)e^{-i\frac{1}{2}\varphi\sigma_z},$$

and hence

$$|\underline{p,\varphi}\rangle_a = e^{i\frac{1}{2}\varphi\sigma_z}|\underline{p}\rangle_a, \tag{37}$$

where $|\underline{p}\rangle_a = |\underline{p,0}\rangle_a$ are eigenvectors of the operator $2\sigma_x = \sigma_- + \sigma_+$,

$$2\sigma_x|\underline{p}\rangle_a = \lambda_p|\underline{p}\rangle_a, \qquad \lambda_p = 2p-1, \qquad p = 0,1. \tag{38}$$

n the standard representation the normalized eigenvectors $|\underline{p}\rangle_a$ are:

$$|\underline{0}\rangle_a = \frac{1}{\sqrt{2}}\begin{bmatrix} 1 \\ -1 \end{bmatrix}, \qquad |\underline{1}\rangle_a = \frac{1}{\sqrt{2}}\begin{bmatrix} 1 \\ 1 \end{bmatrix}.$$

nd the vectors $|\underline{p,\varphi}\rangle_a$ have the form

$$|\underline{0,\varphi}\rangle_a = \frac{1}{\sqrt{2}}\begin{bmatrix} e^{i\varphi/2} \\ -e^{-i\varphi/2} \end{bmatrix}, \qquad |\underline{1,\varphi}\rangle_a = \frac{1}{\sqrt{2}}\begin{bmatrix} e^{i\varphi/2} \\ e^{-i\varphi/2} \end{bmatrix}.$$

Ve will call the vectors $|\underline{p,\varphi}\rangle_a$ semiclassical states. Their evolution under the lassical field Hamiltonian is reduced to the global phase factor

$$\exp\left(-itH_{cl}\right)|\underline{p,\varphi}\rangle_a = e^{-i\lambda_p\sqrt{\bar{n}}t}|\underline{p,\varphi}\rangle_a.$$

This means that the atomic inversion will always be zero in these states, since the perator σ_z has off-diagonal form in the basis of eigenvectors of the operator σ_x: $\langle\underline{p}|\sigma_z|\underline{p}\rangle = 0$.

Let us return now to the quantum field case and calculate the average value of he atomic inversion

$$\langle\sigma_z(t)\rangle = \langle\psi(t)|\sigma_z(0)|\psi(t)\rangle,$$

vhen the initial field is in a coherent state $|\alpha\rangle_f$, $\alpha = \sqrt{\bar{n}}e^{i\phi}$ with high photon umber $\bar{n} \gg 1$, and the atom is initially taken to be in its semiclassical state, $|\underline{p,\varphi}\rangle_a$, for instance $|\underline{1,\varphi}\rangle_a$:

$$|in\rangle = |\alpha\rangle_f \otimes \frac{1}{\sqrt{2}}\begin{bmatrix} e^{i\varphi/2} \\ e^{-i\varphi/2} \end{bmatrix}, \tag{39}$$

The evolution of this state for the exact resonance case can be found applying the volution operator (18) to the state (39):

$$|\psi(t)\rangle = \frac{1}{\sqrt{2}}\begin{bmatrix} \cos gt\sqrt{\hat{n}+1}e^{i\frac{1}{2}\varphi} - i\sin gt\sqrt{\hat{n}+1}e^{i\left(\hat{\phi}-\frac{1}{2}\varphi\right)} \\ \cos gt\sqrt{\hat{n}}e^{-i\frac{1}{2}\varphi} - i\sin gt\sqrt{\hat{n}}e^{-i\left(\hat{\phi}-\frac{1}{2}\varphi\right)} \end{bmatrix}|\alpha\rangle.$$

Remember that $e^{\pm i\hat{\phi}}$ are the field phase operators. In the coherent state, the phase fluctuations are proportional to $(2|\alpha|)^{-1}$ and the phase is well defined when $\bar{n} = |\alpha|^2 \gg 1$. In other words,

$$e^{\pm i\frac{1}{2}\hat{\phi}}|\alpha\rangle = e^{\pm i\frac{1}{2}\varphi}|\alpha\rangle + O(\bar{n}^{-1/2}).$$

Hence we have

$$|\psi(t)\rangle \approx \frac{1}{\sqrt{2}}\begin{bmatrix} e^{-igt\sqrt{\hat{n}+1}}e^{i\frac{1}{2}\varphi} \\ e^{-igt\sqrt{\hat{n}}}e^{-i\frac{1}{2}\varphi} \end{bmatrix}|\alpha\rangle + O(\bar{n}^{-1/2}), \tag{40}$$

and the atomic inversion evolves according to the formula

$$\langle\sigma_z(t)\rangle = \langle\psi(t)|\sigma_z(0)|\psi(t)\rangle = O\left(\bar{n}^{-1/2}\right) \to 0, \qquad \bar{n} \to \infty.$$

This means that the states $|\underline{p},\varphi\rangle_a$ are "trapping states"; there is no energy exchange between the atom and the field within an accuracy of $O\left(1/\sqrt{\bar{n}}\right)$. In this sense the initial states (39) evolve as in the case of an external classical field. That is why we will call the states

$$|\underline{p,\varphi}\rangle_a|\alpha\rangle_f = e^{i\frac{1}{2}\varphi\sigma_z}|\underline{p}\rangle_a|\alpha\rangle_f \tag{41}$$

"semiclassical states of the total system".

9 Factorization of the wave function

The semiclassical states are, in fact, more than simply trapping states. To see this let us further transform the wave function (40). Expanding the square root $\sqrt{\hat{n}+1}$ in a power series with a small parameter $1/\bar{n} \ll 1$,

$$\sqrt{\hat{n}+1} \approx \sqrt{\hat{n}} + \frac{1}{2\sqrt{\hat{n}}} + O(\bar{n}^{-3/2}),$$

we rewrite equation (40) in the form

$$|\psi(t)\rangle \approx \frac{1}{\sqrt{2}} e^{-igt\sqrt{\hat{n}}} \begin{bmatrix} e^{-igt/(2\sqrt{\bar{n}})} e^{i\frac{1}{2}\varphi} \\ e^{-i\frac{1}{2}\varphi} \end{bmatrix} |\alpha\rangle,$$

where we have substituted $1/\sqrt{\hat{n}} \to 1/\sqrt{\bar{n}}$. This leads to an error in the frequencies of the order $O(1/\bar{n})$. Indeed,

$$\frac{1}{\sqrt{\hat{n}}} = \frac{1}{\sqrt{\bar{n}+\Delta\hat{n}}} \approx \frac{1}{\sqrt{\bar{n}}} - \frac{\Delta\hat{n}}{2\bar{n}^{3/2}}, \qquad \Delta\hat{n} = \hat{n} - \bar{n} \sim \sqrt{\bar{n}}.$$

Therefore, our approximation is valid if $gt < \bar{n}$. Taking into account that

$$\exp\left[-i\frac{\omega_1 t}{2}(\sigma_z+1)\right]|\underline{1}\rangle_a = \frac{1}{\sqrt{2}}\begin{bmatrix} e^{-i\omega_1 t} \\ 1 \end{bmatrix}, \qquad \omega_1 = \frac{g}{2\sqrt{\bar{n}}},$$

we conclude that the wave function of the system for the initial state (39) is approximately written in the following form [38]:

$$|\psi(t)\rangle \approx |A(t)\rangle_a|\Phi(t)\rangle_f = \frac{1}{\sqrt{2}}\begin{bmatrix} e^{-i\omega_1 t} \\ 1 \end{bmatrix} \otimes e^{-igt\sqrt{\hat{n}}}|\alpha\rangle.$$

In general, for the initial atomic state $|\underline{p,\varphi}\rangle$, $(p=0,1)$, the result reads

$$\begin{aligned} |\psi(t)\rangle &\approx |A_p(t)\rangle_a|\Phi_p(t)\rangle_f, &(42)\\ |A_p(t)\rangle_a &= \exp\left[-i\omega_p t\,(\sigma_z+1)/2\right]|\underline{p,\varphi}\rangle_a, &(43)\\ |\Phi(t)\rangle_f &= e^{-i\lambda_p\, gt\sqrt{\hat{n}}}|\alpha\rangle, &(44)\end{aligned}$$

where

$$\omega_p = \lambda_p g/(2\sqrt{\bar{n}}), \tag{45}$$

and λ_p is an eigenvalue of the semiclassical Hamiltonian (36), (38).

Equation (42) means that if the system is initially prepared in its semiclassical state (41), then the total wave function remains approximately factorized into the product of the atomic and field wave functions [38]. Usually, the factorization means that the subsystems do not interact. This is not our case, however, since the evolution of the atomic part of the wave function depends on the initial field intensity $\bar{n}$ and the field part depends on the atomic initial parameter λ_p. Moreover, the evolution of every subsystem (the field or atom) alone is not unitary: the states $|\underline{1,\varphi}\rangle_a$, $|\underline{0,\varphi}\rangle_a$ evolve in such a manner that they coincide (up to the phase factor) at a special time instant $t_0 = \pi/\omega_1 = \pi\sqrt{\bar{n}}/g = T_r/2$:

$$|A_1(t_0)\rangle_a = \frac{1}{\sqrt{2}}\begin{bmatrix} -ie^{i\frac{1}{2}\varphi} \\ e^{-i\frac{1}{2}\varphi} \end{bmatrix} = -|A_0(t_0)\rangle_a$$

But the states $|\underline{0,\varphi}\rangle_a$ and $|\underline{1,\varphi}\rangle_a$ are orthogonal, and thus must remain orthogonal in the course of a unitary evolution.

Let us study the evolution of the field part of the factorized wave function (44). This evolution is nonlinear and is governed by the effective field Hamiltonian,

$$H_{eff} = \lambda_p g\sqrt{\hat{n}}.$$

To better understand it, one may expand the square root $\sqrt{\hat{n}} = \sqrt{\bar{n} + \Delta\hat{n}}$ in powers of the operator $\Delta\hat{n}/\bar{n} \sim 1/\sqrt{\bar{n}}$:

$$H_{eff} = \lambda_p g\sqrt{\hat{n}} \approx \lambda_p g\left(\sqrt{\bar{n}} + \frac{\Delta\hat{n}}{2\sqrt{\bar{n}}} - \frac{(\Delta\hat{n})^2}{8\bar{n}^{3/2}}\right). \tag{46}$$

Applying the linearized operator to the initial coherent state and neglecting the quadratic term $\sim 1/\sqrt{\bar{n}}$ for times $gt \ll \sqrt{\bar{n}}$ we have

$$|\Phi(t)\rangle_f = e^{-i\lambda_p gt\sqrt{\hat{n}}}|\alpha\rangle \approx e^{-i\lambda_p gt\sqrt{\bar{n}}/2}|e^{-i\omega_p t}\alpha\rangle. \tag{47}$$

Here we have used the fact that $e^{-i\omega_p t\hat{n}}|\alpha\rangle = |e^{-i\omega_p t}\alpha\rangle$. This means that for short times (before the first revival), the initial coherent state simply rotates with the frequency ω_p in the phase plane of the field. For larger times one has to take into account anharmonic terms in equation (46) which deform the shape of the initial coherent state.

It is important that the evolution of an arbitrary initial atomic state can be described in terms of the factorized states. To see this, we recall that the states $|\underline{p,\varphi}\rangle_a$ form a complete basis in the atomic Hilbert space, and arbitrary initial atomic state can be written in this basis as

$$|\psi(0)\rangle_a = \sum_{p=0}^{1} C_p|\underline{p,\varphi}\rangle_a\,, \qquad C_p = {}_a\langle\underline{p,\varphi}|\psi(0)\rangle_a\,.$$

Thus, the evolution of this state in the presence of the quantum initially coherent field is given as a superposition of the factorized states,

$$|\psi(t)\rangle = \sum_{p=0}^{1} C_p|A_p(t)\rangle_a|\Phi_p(t)\rangle_f. \tag{48}$$

Let us take, for example, the atom in the initial ground state, $|\psi(0)\rangle_a = |0\rangle_a$, and the field in the CS with zero phase, $\alpha^2 = \bar{n}$, and calculate the atomic inversion evolution using the factorized wave functions (42). We have

$$\langle\psi(t)|\sigma_z|\psi(t)\rangle \approx \sum_{p,q=0}^{1} C_p \bar{C}_q \, {}_a\langle A_q(t)|\sigma_z|A_p(t)\rangle_a \, {}_f\langle\Phi_q(t)|\Phi_p(t)\rangle_f , \quad (49)$$

where $C_1 = -C_0 = 1/\sqrt{2}$. Using the definition of the atomic states (43), one can easily find the atomic matrix element

$${}_a\langle A_q(t)|\sigma_z|A_p(t)\rangle_a = {}_a\langle \underline{q}|\sigma_z \exp\left[-i\frac{\omega_p - \omega_q}{2}(\sigma_z + 1)\,t\right] |\underline{p}\rangle_a . \quad (50)$$

In turn, the overlap integral between the field states

$${}_f\langle\Phi_q(t)|\Phi_p(t)\rangle_f = \sum_n P_n e^{-i(\lambda_p - \lambda_q)\,gt\sqrt{n}} \quad (51)$$

has a form of an anharmonic series. To get a qualitative image, one may use the liner approximation for short times (47)

$${}_f\langle\Phi_q(t)|\Phi_p(t)\rangle_f \approx \exp\left[-i(\lambda_p - \lambda_q)\,gt\sqrt{\bar{n}}/2\right] \langle e^{-i\omega_q t}\alpha|e^{-i\omega_p t}\alpha\rangle . \quad (52)$$

The rapidly oscillating factor $\exp\left[-i(\lambda_p - \lambda_q)\,gt\sqrt{\bar{n}}/2\right]$ represents here the Rabi oscillations. Their slowly varying amplitude takes the form,

$$|\langle e^{-i\omega_q t}\alpha|e^{-i\omega_p t}\alpha\rangle| = \exp\left[-\sqrt{\bar{n}}|e^{-i\omega_q t} - e^{-i\omega_p t}|^2/2\right] . \quad (53)$$

The amplitude has its maximum at $t = 0$, and the oscillations collapse when

$$\sqrt{\bar{n}}|e^{-i\omega_q t} - e^{-i\omega_p t}| \sim 1 .$$

Since $\omega_p \sim 1/\sqrt{\bar{n}}$ (we are interested here in short times), we can expand the exponents in series and obtain the collapse time

$$T_c^{-1} = (\omega_p - \omega_q)\sqrt{\bar{n}} .$$

On the other hand, it is seen from equation (53) that the amplitude of oscillations becomes close to 1 when

$$(\omega_p - \omega_q)\,T_r = 2\pi k, \qquad k = 1, 2, \ldots,$$

which gives us the times of possible revivals. For these time instants the atomic matrix element (50) is almost independent of time, since

$$\exp\left[-i(\omega_p - \omega_q)(\sigma_z + 1)\,T_r/2\right] = \exp\left[-i\pi\,(\sigma_z + 1)\right] = 1,$$

and therefore,

$${}_{at}\langle A_q(t)|\sigma_z|A_p(t)\rangle_{at} \approx {}_{at}\langle \underline{q}|\sigma_z|\underline{p}\rangle_{at} = \delta_{p,\,q+1} + \delta_{p,\,q-1} . \quad (54)$$

Finally, from (49) using (51) and (54) we find

$$\langle\psi(t)|\sigma_z|\psi(t)\rangle = -\sum_n P_n \cos(2\,gt\sqrt{n}) .$$

We thus recover the former expression for the atomic inversion.

10 Evolution in the field phase space

The factorized wave functions (42) lead to a very transparent dynamical picture in a field phase plane. To visualize the evolution of the field one may use, for example, the Q-function, i.e. the matrix element of the field density matrix between the coherent states:

$$Q(\beta) = \langle\beta|\rho_f|\beta\rangle. \tag{55}$$

The coherent state parameter β takes arbitrary complex values and the complex β-plane represents a classical phase space of the field. The field density matrix is obtained from the total density matrix taking the trace over the atomic variables, $\rho_f = \mathrm{Tr}_a\left(|\psi(t)\rangle\langle\psi(t)|\right)$. For a semiclassical initial state which leads to a factorized wave function, we have $\rho_f \approx |\Phi_p(t)\rangle_f\, {}_f\langle\Phi_p(t)|$. Hence the Q-function for a single semiclassical state is given as

$$Q(\beta) = |\langle\beta|\Phi_p(t)\rangle_f|^2.$$

For short times $t < T_r$, when the linear approximation is valid, the Gaussian hump corresponding to the initial coherent state rotates around the origin of phase plane along the circle of the radius $\sqrt{\bar{n}}$ with the frequency ω_p. For larger times one cannot neglect the anharmonic terms in equation (46). They deform the shape of the initial coherent state. The hump that corresponds to the state $|\Phi(t)\rangle_f$ acquires a typical crescent shape. The Q-function evolution at this stage can be described analytically taking into account the quadratic term in the expansion (46) (see [39], [40] and section 13). The crescent shape leads to squeezing of the field quadrature components [40,41]. Further on, the phase spread exceeds 2π which leads to self-interference in the phase plane. This, in turn, leads to a kind of standing wave at some time instants, which are called fractional revivals [42]. These phenomena for the JCM can also be explained at the level of quadratic approximation to the frequencies. Note that the field Hamiltonian

$$H = \omega\hat{n} + \chi\hat{n}^2 \tag{56}$$

has an independent physical sense and describes the optical Kerr medium [43,44] (see Chapter 6). This is the simplest model that reveals the phase spread due to the nonlinear evolution. It demonstrates, to some extent, features similar to the JCM dynamical features [39]. For instance, the standing waves in the phase space are very well pronounced for a Kerr medium (it is the consequence of the integer spectrum of the nonlinear term $\hat{n}^2$). They are usually called the "Schrödinger cat states" [45,46].

For an arbitrary initial atomic state the situation is more complicated. The initial coherent state is split into two humps, which correspond to the two different semiclassical states in the decomposition (48). These humps rotate with the frequencies $\omega_0 = -\omega_1 = g/2\sqrt{\bar{n}}$ (i.e. clock-wise and counter clock-wise) along the circle of the radius $\sqrt{\bar{n}}$. When the humps are away from each other there are no Rabi oscillations, since the overlap integral between the semiclassical states is zero, in accordance with equations (49) and (52); this corresponds to the collapse region. From this picture one can estimate once again the collapse time.

Indeed, the Q-function of a single semiclassical state is represented by a round hump of radius ~ 1. The humps start their motion from the same point in opposite directions with linear velocities $|\omega_0|\sqrt{\bar{n}}$ and become well separated after time $t \sim 2\omega_0\sqrt{\bar{n}} = 1/g = T_c$. The Rabi oscillations revive when the humps meet each other in the phase plane. This happens at times $t = k\pi/\omega_0 = kT_r$. It is just the interference of different semiclassical states that produces revivals of the Rabi oscillations.

11 The JCM without RWA

In this section we consider the JCM without the rotating wave approximation. Let us recall that the Hamiltonian has the form:

$$\begin{aligned} H &= H_0 + H_1 + H_2, \qquad & H_0 &= \frac{1}{2}\omega_a\sigma_z + \omega_f a^\dagger a, \\ H_1 &= g\left(a\sigma_+ + a^\dagger\sigma_-\right), \qquad & H_2 &= g\left(a^\dagger\sigma_+ + a\sigma_-\right), \end{aligned} \tag{57}$$

where $H_0 + H_1$ is the JCM Hamiltonian under the rotating wave approximation and H_2 represents the counter-rotating terms. (Note that the term H_2 in Eq. (3) has different sign; it can be changed by the appropriate unitary transformation.) In the classical field limit these terms cause a small shift of the resonance frequency [48–51], $\delta_{cl} = g^2\bar{n}/(\omega_f + \omega_a)$, where $\bar{n}$ is the field intensity in the units of the photon number.

We will show that the perturbation theory used by Shirley [49] can be directly generalized to the quantum field case, leading to a simple effective Hamiltonian that takes into account the influence of counter-rotating terms. In the quantum domain the Bloch–Siegert shift depends on the photon number; the corresponding term in the effective Hamiltonian appears to be

$$(g^2/2\underline{\omega})\,\hat{n}\sigma_z\,, \qquad \hat{n} \equiv a^\dagger a\,, \qquad \underline{\omega} \equiv (\omega_a + \omega_f)/2\,.$$

Terms of such a kind are usually related to the "dynamical Stark shift" [52–55]. The eigenvectors and eigenvalues of the effective Hamiltonian can be easily found. We finally apply this approach to find the atomic inversion evolution. It leads to a modification of the shape of the famous JCM collapses and revivals in the RWA.

Diagonalization of the Hamiltonian

We will approximately transform the Hamiltonian (57) to the one which commutes with the excitation number operator $\hat{N}$, applying a sequence of "small unitary" transformations (rotations). At the first step we apply to the Hamiltonian (57) the transformation

$$V = \exp\left[\epsilon A\right], \qquad A = a^\dagger\sigma_+ - a\sigma_-\,.$$

Assuming ϵ to be small we neglect the terms of order ϵ^k, $k \geq 3$, writing

$$VHV^\dagger \approx H + \epsilon[A, H] + \frac{\epsilon^2}{2}\,[A, [A, H]]\,.$$

'he parts of the Hamiltonian are transformed in the following manner

$$
\begin{aligned}
VH_0V^\dagger &\approx H_0 - 2\epsilon\underline{\omega}\left(a^\dagger\sigma_+ + a\sigma_-\right) - \epsilon^2\underline{\omega}\left[\sigma_z(2\hat{n}+1) - 1\right], \quad (58)\\
VH_1V^\dagger &\approx H_1 + \epsilon g\left(a^{\dagger 2} + a^2\right)\sigma_z - \epsilon^2 g\left[\left(\sigma_+ a^{\dagger 3} + \sigma_- a^3\right)\right.\\
&\quad \left. + \left(\sigma_+ a\hat{n} + \hat{n}a^\dagger\sigma_-\right)\right], \quad (59)\\
VH_2V^\dagger &\approx H_2 + \epsilon g\left[\sigma_z(2\hat{n}+1) - 1\right] - 2\epsilon^2 g\left(\sigma_- a\hat{n} + \hat{n}a^\dagger\sigma_+\right). \quad (60)
\end{aligned}
$$

'he second term in (58) cancels the anti-rotating term H_2 if we choose

$$\epsilon = g/2\underline{\omega}.$$

Now, the zeroth-order transformed Hamiltonian commutes with $\hat{N}$ and conserves he number of excitations. Let us note that the parameter ϵ is indeed small for a standard experimental setup. The third term in (58) and the second one in (60) are of the first order and give the quantum Bloch–Siegert shift.

In order to diagonalize the Hamiltonian up to the second order in ϵ in the basis of eigenstates of $\hat{N}$ we apply the second "small" rotation

$$V_1 = \exp\left[\phi B\right], \qquad B = \left(a^{\dagger 2} - a^2\right)\sigma_z, \qquad \phi = \epsilon g/2\omega_f \sim \epsilon^2.$$

Then the second term in the transformed Hamiltonian H_0,

$$V_1 H_0 V_1^\dagger \approx H_0 - \phi\, 2\omega(a^{\dagger 2} + a^2)\sigma_z - 8\phi^2\,\omega(\hat{n} + 1/2),$$

cancels the second term in (59), whereas the last term in the above equation is of an order of ϵ^3. The whole Hamiltonian after the transformations V and V_1 takes the following form

$$
\begin{aligned}
V_1VHV^\dagger V_1^\dagger &\approx H_0 + \tilde{H}_1 + \epsilon g\left[\sigma_z\left(\hat{n} + \tfrac{1}{2}\right) - \tfrac{1}{2}\right] - 2\epsilon^2\frac{\underline{\omega}}{\omega_f}g\left[a^3\sigma_+ + a^{\dagger 3}\sigma_-\right]\\
&\quad + \epsilon^2 g\left[\frac{\omega_a}{\omega_f}\left(a^{\dagger 3}\sigma_+ + a^3\sigma_-\right) + \frac{\omega_a - \omega_f}{\omega_f}\left(\sigma_+\hat{n}a^\dagger + a\hat{n}\sigma_-\right)\right], \quad (61)
\end{aligned}
$$

where $\tilde{H}_1$ is the RWA interaction Hamiltonian with intensity-dependent coupling constant

$$\tilde{H}_1 = g(\hat{n})a^\dagger\sigma_- + \sigma_+ a g(\hat{n}), \qquad g(\hat{n}) = g(1 - \epsilon^2\hat{n}).$$

We notice that the two last terms in (61) can be easily eliminated by applying corresponding transformations with rotation parameters proportional to ϵ^3 which, of course, does not affect the terms of the order of ϵ^2. To get rid of the term $\sim \epsilon^2 g(a^3\sigma_+ + a^{\dagger 3}\sigma_-)$ describing the three photon transitions one should apply the transformation

$$V_3 = \exp\left[\gamma\left(a^3\sigma_+ - a^{\dagger 3}\sigma_-\right)\right], \quad (62)$$

with

$$\gamma = \frac{2\epsilon^2\underline{\omega}g}{\omega_f(\omega_a - 3\omega_f)}. \quad (63)$$

If the atomic transition is far from the three-photon resonance, we get $\gamma \sim \epsilon$ and the term $\epsilon^2 g(a^3\sigma_+ + a^{\dagger 3}\sigma_-)$ can be eliminated. But in the resonant cas when $\omega_a \simeq 3\omega_f$, the parameter γ is great and the transformation (62) can n longer be applied. From equation (63) it is easy to find the width of the thre photon resonance, $\omega_a - 3\omega_f \sim g\epsilon^2$, which is much less than the width of the fir resonance, $\omega_a - \omega_f \sim g$. Let us note that in the three-photon resonance case th term describing the one-photon transitions, $g(a\sigma_+ + a^\dagger\sigma_-)$, can be removed b the appropriate small transformation.

To conclude, the effective Hamiltonian in the region close to the one-photo resonance takes the form

$$H_{eff} = \tfrac{1}{2}\omega_a\,\sigma_z + \omega_f a^\dagger a + \frac{g^2}{2\underline{\omega}}\left[\sigma_z\left(\hat{n}+\tfrac{1}{2}\right)-\tfrac{1}{2}\right] + g(\hat{n})a^\dagger\sigma_- + \sigma_+ a g(\hat{n}), \quad (6$$

where $g(\hat{n}) = g(1-\epsilon^2\hat{n})$. It is clear that $[H_{eff}, \hat{N}] = 0$ and the dynamics wi H_{eff} can be found exactly, as for the common JCM. Corrections produced by th counter-rotating terms appear in the first order as the intensity-dependent detu ing ("dynamical Stark shift") and in the second order as the intensity-depende coupling. One can rewrite the Stark shift term in equation (64) as

$$\epsilon g\left[\sigma_z\left(\hat{n}+\tfrac{1}{2}\right)-\tfrac{1}{2}\right] = \epsilon g\sigma_z\left(\hat{N}+\tfrac{1}{2}\right) - \epsilon g.$$

This means that the atomic transition frequency acquires a different shift in eac invariant subspace.

Let us denote the bare basis as $|N,k\rangle = |N-k\rangle_f|k\rangle_a$ (where $k = 0,1$ is th atomic excitation number) and the dressed basis as $|\underline{N,k}\rangle$, so that the eigenvecto and eigenvalues of the initial Hamiltonian are

$$H|\underline{N,k}\rangle = \lambda_{N,k}|\underline{N,k}\rangle\,.$$

The exact eigenvalues are approximated by those of the effective Hamiltonia (64), which are given by

$$\lambda_{N,k=0,1} = \omega N - \epsilon g + (-1)^k\Omega_N, \qquad \Omega_N = \sqrt{\Delta_N^2/4 + g_N^2}, \qquad (6$$

$$\Delta_N = \Delta + 2N\epsilon g, \qquad g_N = g\sqrt{N}(1-\epsilon^2 N), \qquad \epsilon = g/(2\underline{\omega}).$$

Here the term $2N\epsilon g$ represents the dynamical Stark shift corresponding to th semiclassical Bloch–Siegert shift. The eigenvectors (dressed vectors) in the fir order in ϵ take the following form:

$$\begin{aligned}|\underline{N,k}\rangle &= a_N^k|N,0\rangle + b_N^k|N,1\rangle \\ &\quad + \epsilon\left[\sqrt{N-1}\,b_N^k|N-2,0\rangle - \sqrt{N+1}\,a_N^k|N+2,1\rangle\right], \qquad (6\end{aligned}$$

$$a_N^0 = b_N^1 = \sqrt{\frac{1}{2} - \frac{\Delta_N}{4\Omega_N}}, \qquad a_N^1 = -b_N^0 = -\sqrt{\frac{1}{2} + \frac{\Delta_N}{4\Omega_N}}.$$

In the zeroth order approximation, the two eigenvectors $|\underline{N,k}\rangle$, $k = 0,1$ are loca ized in the subspace with N excitations, and in our notations $k = 0$ corresponds t the higher eigenvalue (for a given N). Small contributions of the subspaces $N \pm$ appear in the first order.

Atomic inversion

Let us assume that initially the field is taken in the coherent state with zero phase and the atom is prepared in its ground state,

$$|\mathrm{in}\rangle = \sum P_n |n\rangle_f |0\rangle_a, \qquad P_n = \sqrt{e^{-\bar{n}}\bar{n}^n/n!}.$$

The wave function evolution is given by

$$|\Psi(t)\rangle = \sum_{N,k} \exp(-it\lambda_{Nk}) C_{Nk} |\underline{N,k}\rangle,$$

where

$$C_{Nk} = \langle \underline{N,k}|\mathrm{in}\rangle = P_N\, a_N^k + \epsilon\sqrt{N-1} P_{N-2}\, b_N^k.$$

The atomic inversion $\langle\Psi(t)|\sigma_z|\Psi(t)\rangle$ versus time is shown in Figure 3 for the values $\omega_a = \omega_f = 50$, $\bar{n} = 30$, and $g = 1$. This picture shows the results of exact numerical diagonalization of the Hamiltonian (1). The graph differs significantly from the well-known atomic inversion picture for the JCM under the RWA [26]. For instance, the rapid oscillations with the frequency $\sim \omega$ are still present when the Rabi oscillations (frequency $\sim \Omega_{\bar{n}}$) have collapsed.

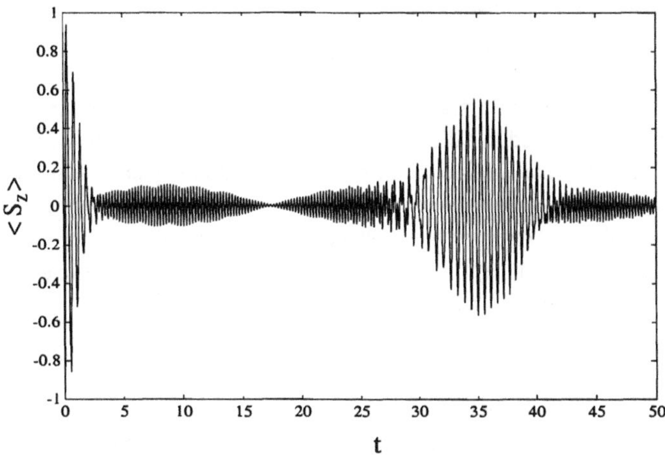

Figure 3: Beyond the rotating wave approximation: atomic inversion for the initial coherent state of the field and the ground state of the atom; $\omega = \omega_a = 50$, $g = 1$, $\bar{n} = 30$.

In the first-order approximation we have

$$|\Psi(t)\rangle \approx |\Psi(t)\rangle^0 + \epsilon|\Psi(t)\rangle^1, \tag{67}$$

$$\begin{aligned} |\Psi(t)\rangle^0 &= \sum_N e^{-i\omega N t} P_N \left\{ \left(\cos\Omega_N t + \frac{i\Delta_N}{2\Omega_N} \sin\Omega_N t \right) |N,0\rangle \right. \\ &\quad \left. - \frac{ig_N}{\Omega_N} \sin\Omega_N t |N,1\rangle \right\}, \end{aligned}$$

$$
\begin{aligned}
|\Psi(t)\rangle^1 &= \sum_N e^{-i\omega N t}\left\{\left[-ie^{-i2\omega t}\sin\Omega_{N+2}\sqrt{N+1}P_{N+2}\frac{g_{N+2}}{\Omega_{N+2}}\right.\right.\\
&\left.-i\sin\Omega_N t\sqrt{N-1}P_{N-2}\frac{g_N}{\Omega_N}\right]|N,0\rangle\\
&+\sqrt{N-1}P_{N-2}\left[e^{i2\omega t}\left(-\cos\Omega_{N-2}t-\frac{i\Delta_{N-2}}{2\Omega_{N-2}}\sin\Omega_{N-2}t\right)\right.\\
&\left.\left.+\cos\Omega_N t-\frac{i\Delta_N}{2\Omega_N}\sin\Omega_N t\right]|N,1\rangle\right\}.
\end{aligned}
$$

The calculation with the wave function (67) is in excellent agreement with the exact results (shown in Figure 3) for the small parameter $\epsilon = g/(\omega_a+\omega_f) = 0.01$.

Classical field limit

In the rotating frame (making the transformation $\exp(i\omega_f\hat{n}t)$) Hamiltonian (57) takes the form

$$H(t) = \tfrac{1}{2}\,\omega_a\sigma_z + g\sigma_x\left(ae^{-i\omega_f t} + a^\dagger e^{i\omega_f t}\right). \tag{68}$$

Usually, the classical field limit is obtained by replacing the field operators by c-numbers: $a \to \alpha$. Then

$$H_{cl}(t) = \tfrac{1}{2}\,\omega_a\sigma_z + g\sigma_x\left(\alpha e^{-i\omega_f t} + \alpha^* e^{i\omega_f t}\right),$$

which is just the Hamiltonian studied by Shirley [49]. However, the quasi-classical limit can be obtained in a different way. Let us take the field initially in a coherent state. Then the relative fluctuations of the field intensity are small, $\Delta n \ll \bar{n}$, and one may substitute $a \to \sqrt{\bar{n}}\exp(i\hat{\phi})$ in the initial Hamiltonian (57), where $\bar{n}$ is the field intensity and $\exp(i\hat{\phi})$ is the phase operator ($\exp i\hat{\phi}|n\rangle = |n-1\rangle$). (Note that $[\exp(i\hat{\phi}), \exp(-i\hat{\phi})] = 0$, if the contribution of the vacuum state is neglected.) Then Hamiltonian (57) takes the form

$$H_{cl} = \omega_f\hat{n} + \tfrac{1}{2}\,\omega_a\sigma_z + g\sqrt{\bar{n}}\sigma_x\left(e^{i\hat{\phi}} + e^{-i\hat{\phi}}\right) + O(1). \tag{69}$$

In the Fock basis this operator has the matrix form which exactly coincides with the Hamiltonian that arises from the Fourier expansion of the Floquet solution discussed by Shirley [49]. Correspondingly, its approximate eigenvalues and eigenvectors are given by equations (65) and (66), where N is replaced by $\bar{n}$. In particular, the dynamical Stark shift $2N\epsilon g$ is transformed into the Bloch–Siegert shift.

In the rotating frame Hamiltonian (69) can be rewritten as

$$H_{cl}(t) = \tfrac{1}{2}\,\omega_a\sigma_z + g\sqrt{\bar{n}}\sigma_x\left(e^{i(\hat{\phi}-\omega_f t)} + e^{-i(\hat{\phi}-\omega_f t)}\right) + O(1).$$

It is clear now that the field operator $\exp(\pm i\hat{\phi})$ commutes with the above Hamiltonian. In this approximation the field variables are still considered as operators; however, the atomic dynamics does not affect the field and the total wave function is factorized into the atomic and field parts. The semiclassical Hamiltonian (69) well describes only the short-time region $gt < 1$ which covers times before the first collapse. For other studies of the models without RWA see, e.g., [56–60].

12 Collective interactions

In this section we show how to take into account collective phenomena. Let us consider a system of A identical two-level atoms interacting with a quantized electromagnetic field in a perfect cavity. The Hamiltonian under the rotating wave approximation has the form $H = H_0 + V$, where

$$H_0 = \omega a^\dagger a + \omega_a S_z , \tag{70}$$
$$V = g\left[a^\dagger S_- + aS_+\right] . \tag{71}$$

We assume that the linear dimensions of the atomic system are small compared with the wavelength of the field ("a small volume approximation") and thus the coupling constant g is the same for all atoms. However, we assume that the wave functions of different atoms do not overlap, which means that we neglect the dipole–dipole interactions between the atoms. Such a system was considered for the first time in Dicke's paper in 1954 [1], and we will call it the *Dicke model* [61] (though the term "the Tavis–Cummings model" is also used [4]). For simplicity, we will restrict ourselves to the exact resonance case $\omega_a = \omega$. Then the excitation number operator $\hat{N} = \hat{n} + S_z$ is proportional to the bare Hamiltonian H_0, and $[\hat{N}, V] = 0$.

Symmetry properties of the initial state under permutations of atoms are important for this system [1, 18]. Here we consider the fully symmetrical case when the atomic collective operators (2) generate the $A + 1$ dimensional representation of the algebra $su(2)$. The $su(2)$ commutation relations are

$$[S_z, S_\pm] = \pm S_\pm, \qquad [S_+, S_-] = 2S_z.$$

Let us denote the basis in this representation as

$$|A, k\rangle_a, \qquad k = 0, 1, \ldots A,$$

where k is the number of symmetric atomic excitations. The action of the atomic operators in this basis is given as follows:

$$\begin{aligned} S_z|A,k\rangle_a &= (k - A/2)\,|A,k\rangle_a, \\ S_+|A,k\rangle_a &= \sqrt{(k+1)(A-k)}\,|A,k+1\rangle_a, \\ S_-|A,k\rangle_a &= \sqrt{k(A-k+1)}\,|A,k-1\rangle_a. \end{aligned} \tag{72}$$

The explicit form of the *Dicke states* $|A, k\rangle_a$ can be produced by the operator S_+ acting on the vacuum state $|A, 0\rangle_a = |0_1\rangle_a|0_2\rangle_a \ldots |0_A\rangle_a$.

It is convenient to use the Schwinger realization for the atomic operators in terms of two fictitious boson modes, $[b_i, b_j^\dagger] = \delta_{ij},\ i, j = 1, 2,$

$$S_z = (b_2^\dagger b_2 - b_1^\dagger b_1)/2, \qquad S_+ = b_1 b_2^\dagger, \qquad S_- = b_1^\dagger b_2.$$

Then the Dicke state is written in terms of Fock states of the modes $|k\rangle_{1,2}$ as

$$|k\rangle_a = |A - k\rangle_1 |k\rangle_2 .$$

The interaction Hamiltonian (71) takes a trilinear form,

$$V = a\, b_1\, b_2^\dagger + a^\dagger\, b_1^\dagger\, b_2. \tag{73}$$

The number of atoms A and the excitation number N are written as

$$A = b_1^\dagger\, b_1 + b_2^\dagger\, b_2, \qquad N = a^\dagger\, a + b_2^\dagger\, b_2\,,$$

and they are integrals of motion for the three-photon Hamiltonian (73). This Hamiltonian is interesting in itself, since it describes such optical processes as frequency conversion, Raman and Brillouin scattering [62, 63]. The operators a, b_1 and b_2 are usually called "signal", "idler" and "pump" modes. From the mathematical point of view, the Hamiltonians (71) and (73) are fully equivalent.

12.1 The zeroth-order approximation

The weak field case

The two cases $A > N$ and $A < N$ are essentially different and will be treated separately. Firstly, let us assume that the number of photons in the cavity is small compared with the number of atoms, and the atoms are initially in their ground states; thus $A > N$. The initial excitation number is conserved, $N = n$, and only the states

$$|N, n\rangle \equiv |A, N-n\rangle_{\rm a}|n\rangle_{\rm f} = |A-N+n\rangle_1|N-n\rangle_2|n\rangle_{\rm f}, \quad 0 \le n \le N < A, \tag{74}$$

participate in the dynamics. When $A > N$, there are many photons in the "idler" mode $|k\rangle_1$. In the $(N+1)$-dimensional subspace (74) the operators b_1 and $b_1^\dagger$ have matrix elements $\sqrt{k}$, where $A-N+1 \le k \le A$. We may replace these operators by a fictitious classical field with the intensity equal to the "average" photon number in the above interval,

$$\sqrt{A - N/2 + 1/2} = \Omega_N^{(d)}/g\,, \qquad N < A\,. \tag{75}$$

Ω_N will play the role of the Rabi frequency. The zeroth-order approximation for the interaction Hamiltonian may be written in the form

$$V_0 = \Omega_N(a\, b_2^\dagger + a^\dagger\, b_2)\,. \tag{76}$$

Another justification for this choice of the zeroth-order approximation will be given below. In the subspace (74) the operators

$$\begin{aligned} L_+ &= a\, b_2^\dagger = L_-^\dagger\,, & L_z &= (b_2^\dagger\, b_2 - a^\dagger\, a)/2\,, \\ L_x &= (L_+ + L_-)/2\,, & L_y &= (L_+ - L_-)/2i\,, \end{aligned} \tag{77}$$

determine the $(N+1)$-dimensional representation of a new "dynamical" algebra $su(2)$. The zeroth-order Hamiltonian (76) can also be written as

$$V_0 = 2\Omega_N L_x\,.$$

In other words, the zeroth-order dynamics is connected with the rotation of the fictitious spin $N/2$ around the x-axis with the frequency $2\Omega_N$. Now, in every subspace with a given N we may construct a perturbation theory, treating the operator $V - V_0$ as a perturbation.

'he strong field case

f the number of photons is larger than the number of atoms, $A<N$, then only the tates

$$|N,n\rangle = |A, N-n\rangle_a |n\rangle_f\,, \qquad N-A \le n \le N, \tag{78}$$

re involved in the evolution, and the dimension of the subspace with N excitaions is $A+1$. If $A \ll N$, the number of real photons is large and the zeroth-order pproximation corresponds to a classical field with intensity equal to the "average" umber of photons in the interval $[N-A+1, N]$,

$$\sqrt{N-A/2+1/2} = \Omega_N/g\,, \qquad A<N, \tag{79}$$

vhich gives us the carrying Rabi frequency in this case. Note that the strength of he effective classical field depends on the index of the subspace N. The dynamcal algebra operators $L_{\pm,z}$ in this case coincide with the atomic operators (72). The frequencies (75) and (79) coincide when $A=N$.

It is easy to find the eigenvectors and eigenvalues in the zeroth-order approxination:

$$V_0|\underline{N,p}\rangle = \Lambda^{(0)}_{Np}|\underline{N,p}\rangle.$$

Iamiltonian V_0 in the weak (strong) field limit is proportional to $2L_x$ from the $N+1$)-dimensional representation of the $su(2)$ algebra. It can be diagonalized y the rotation around the y axis by the angle $\pi/2$:

$$\exp\left(i\frac{\pi}{2}L_y\right) L_x \exp\left(-i\frac{\pi}{2}L_y\right) = L_z\,.$$

Operators L_x and L_z have identical equidistant spectra, $N/2-p$, $p=0,1,\ldots,N$. (Here we write the formulae for the weak field case; the strong field case can be obtained by replacing N and A.) The zeroth-order eigenvalues of the Hamiltonian are

$$\Lambda^{(0)}_{Np} = \Omega_N(N-2p), \qquad p=0,1,\ldots,N\,,$$

where the collective Rabi frequency is defined as follows,

$$\Omega_N = \begin{cases} g\sqrt{N-A/2+1/2}\,, & A \le N, \\ g\sqrt{A-N/2+1/2}\,, & A \ge N, \end{cases} \tag{80}$$

The eigenvectors in the zeroth-order approximation are

$$|\underline{N,p}\rangle = \exp\left(i\frac{\pi}{2}L_y\right)|N,p\rangle, \tag{81}$$

and the components in the bare basis are given by the Wigner d-functions of the argument $\pi/2$ (see, e.g., [64]):

$$\alpha^N_{np} = \langle N,n|\underline{N,p}\rangle = \left(\frac{n!p!}{2^N(N-n)!(N-p)!}\right)^{1/2} \sum_{j=0}^{\min(p,n)} \frac{(-2)^j(N-j)!}{j!(n-j)!(p-j)!}. \tag{82}$$

Note that α^N_{np} does not depend on the atomic number A. Formula (82) is written for the weak field case. In the strong field case it is necessary to interchange N and A.

12.2 Perturbation theory

We describe here how to find corrections to the equidistant spectrum of the zerot order approximation [65–67]. The interaction Hamiltonian (71) in the basis (7 or (78) is reduced to the three-diagonal matrix with non-vanishing matrix elemen

$$\langle N,n-1|V|N,n\rangle = \langle N,n|V|N,n-1\rangle = \sqrt{n\,(N-n+1)(A-N+n)}\,.$$

When $A \gg N$ and $0 \le n \le N$, the factor $A-N+n$ is large and the square ro can be expanded in series over the small parameter

$$\epsilon = (A - N/2 + 1/2)^{-1} = g^2/\Omega_N^2, \qquad A > N. \tag{83}$$

In turn, in the strong field case, $N \gg A$ and $N-A \le n \le N$, the factor n is larg and one may expand the square root using the small parameter

$$\epsilon = (N - A/2 + 1/2)^{-1} = g^2/\Omega_N^2, \qquad N > A. \tag{84}$$

The interaction Hamiltonian is then written as

$$V = 2\Omega_N(V_0 + \epsilon V_1 + \epsilon^2 V_2 + \ldots). \tag{85}$$

For instance, in the weak field case, $N \le A$, the non-vanishing matrix elements the operators in equation (85) are

$$\begin{aligned}
\langle N,n|V_0|N,n-1\rangle &= \sqrt{n\,(N-n+1)}\,,\\
\langle N,n|V_1|N,n-1\rangle &= \frac{1}{2}(n - N/2 - 1/2)\sqrt{n\,(N-n+1)}\,,\\
\langle N,n|V_2|N,n-1\rangle &= -\frac{1}{8}(n - N/2 - 1/2)^2\sqrt{n\,(N-n+1)}\,.
\end{aligned}$$

The eigenvectors and eigenvalues of the Hamiltonian are defined as follows:

$$V|\Psi_{N,p}\rangle = \Lambda_{N,p}|\Psi_{N,p}\rangle.$$

A usual perturbation theory yields the second-order eigenvalues $(0 \le p \le N)$

$$\Lambda^{(2)}_{N,p} = \Omega_N(N-2p)\left\{1 - \frac{\epsilon^2}{32}\left[10p(N-p) - (N-1)(N-2)\right]\right\}, \tag{86}$$

where Ω_N and ϵ are given by (80) and (83) or (84). As seen, the first-order co rections to the eigenvalues vanish. This is a consequence of our choice of Ω_N equation (76) and of the small parameter (83). For the cases $N=1$ and $N=2$ th second-order corrections vanish (as well as the corrections of all higher order and the spectrum is equidistant for any A. However, for $N \ge 3$ the spectrum not equidistant, due to the nonlinear dependence on the index p. The first-ord eigenvector components are given as

$$\begin{aligned}
A^{(1)}_{np} = \alpha^N_{np} + \frac{\epsilon}{8}\Big\{&(N - 2p + 1)\sqrt{p(N - p + 1)}\alpha^N_{np-1}\\
&- (N - 2p - 1)\sqrt{(p + 1)(N - p)}\alpha^N_{np+1}\Big\},
\end{aligned} \tag{87}$$

where the zeroth-order components α_{np}^N are given by equation (82).

The strong field case can be obtained from formulae (86) and (87), by interchanging A and N. For instance, the second-order eigenvalues take the form $(0 \leq p \leq A)$

$$\Lambda_{N,p}^{(2)} = \Omega_N(A-2p)\left\{1 - \frac{\epsilon^2}{32}\left[10\,p(A-p) - (A-1)(A-2)\right]\right\}. \qquad (88)$$

The zeroth-order approximation reproduces the exact solution for the JCM $(A=1)$ and the exact spectrum for two atoms, $A=2$. The eigenvectors in the case $A=2$ differ from the exact ones by terms of order of $1/N$.

12.3 Revivals of the first order

We would like to give here some general remarks about revivals, not only related to the Dicke model (see also, e.g., [29,42,68]). Let the evolution of some observable oscillating with frequencies $2\Omega_n$ be given by an anharmonic series. We include the factor 2 to keep the correspondence with the JCM notations. Relative contributions ("the distribution") of these frequencies are determined by the initial conditions. Assume that this distribution has its maximum at the point $n = \bar{n}$. If Ω_n is a smooth function of n, we may expand it around the point $\bar{n}$:

$$\Omega_n = \Omega_{\bar{n}} + (n-\bar{n})\Omega_{\bar{n}}^{(1)} + (n-\bar{n})^2\Omega_{\bar{n}}^{(2)} + \ldots, \quad \Omega_{\bar{n}}^{(r)} = \frac{1}{r!}\frac{d^r\Omega_n}{dn^r}\bigg|_{n=\bar{n}}. \qquad (89)$$

The first term here leads to rapid oscillations, while the remaining terms determine their envelope. If $|\Omega_{\bar{n}}^{(1)}| \gg |\Omega_{\bar{n}}^{(k)}|$, $k = 2, 3, \ldots$, then the linear term is responsible for collapses and revivals of the oscillations. We can find the revival time by the same arguments as for the JCM with the coherent field. At the revival time the most heavily weighed oscillations are in phase

$$2|\Omega_{\bar{n}+1} - \Omega_{\bar{n}}|T_R \approx 2|\Omega_{\bar{n}}^1|T_R = 2\pi. \qquad (90)$$

The terms with frequencies $2\Omega_{\bar{n}+k}$ and $2\Omega_{\bar{n}}$ are also in phase at that time instant, provided that $k \ll \overline{\Delta n}$, and this leads to the amplification of the oscillations. Therefore, the revival time is determined by the first derivative of the frequency Ω_n (which plays the role of a dispersion curve) at the point $\bar{n}$. For this reason, we will refer to this process as revivals of the first order [69]. JCM revivals are of the first order, as follows from section 6. In the linear (harmonic) approximation revivals are perfectly periodic. The higher-order terms in equation (89) lead to the growing spread and decreasing amplitudes of subsequent revivals. For some special initial distributions, the higher-order terms can significantly modify the shape of revivals, as for ringing revivals in the JCM with the initial field in a strongly squeezed state [70].

Let us now apply these arguments to the dynamics of the Dicke model in the weak field case, $A \gg N$, so that we may restrict ourselves to the zeroth-order approximation. Assume that all atoms were initially in their ground states. Then

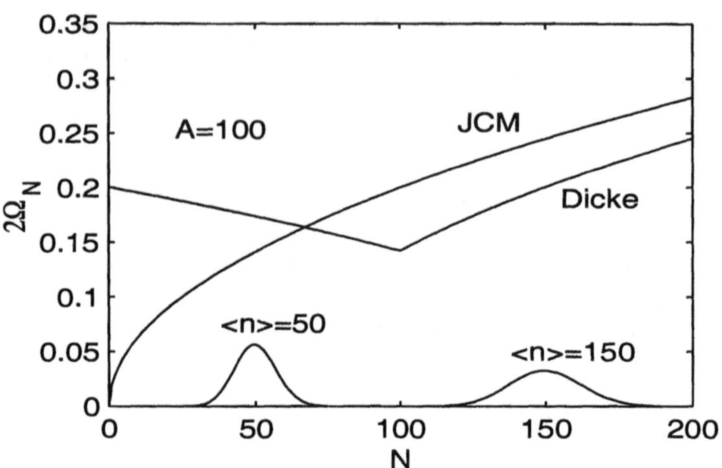

Figure 4: Rabi frequency, $2\Omega_N$, as a function of the excitation number, N, for the case of $A = 100$ atoms ($g = 0.01$); Rabi frequency for the JCM ($A = 1$); Poissonian distribution for $\langle n \rangle = 50$ and $\langle n \rangle = 150$.

the initial photon number distribution $P(n)$ is transformed to the excitation number distribution $P(N)$ which determines the weights of different Rabi frequencies which contribute to the evolution. In Figure 4 we show the zeroth-order Rabi frequencies $2\Omega_n$ (80) for the case $A = 100$ in the interval $0 \le n \le 200$, which cover different dynamical regions. Interference of different frequencies $2\Omega_n$ may lead to collapses and revivals of oscillations in this system. To find the conditions when it happens, we rewrite the linear terms in the expansion (89) for the Rabi frequency (80) in a slightly different form:

$$\Omega_N \approx \Omega_{\bar{n}} + (n - \bar{n})\, \Omega_{\bar{n}}^{(1)} \approx \Omega_0 - n\, \Omega_{\bar{n}}^{(1)}, \tag{91}$$

where now,

$$\Omega_0 = \Omega_{n=0} = g\sqrt{A + 1/2}, \quad \Omega_{\bar{n}}^{(1)} = \left| \frac{d\Omega}{dn} \right|_{n=\bar{n}} = \frac{g}{4\sqrt{A - \bar{n}/2 + 1/2}}.$$

The revival time depends only on the number of atoms:

$$T_R^{(w)} = \pi / \Omega_{\bar{n}}^{(1)} \approx 4\pi\sqrt{A}/g\,.$$

We show the Poissonian distribution with $\bar{n} = 50$ in Figure 4 as an example of the weak field limit. Note that in this limit of the Dicke model, $\bar{n} \ll A$, collapses and revivals also appear for an initial thermal state of the field (in contrast to the JCM, see section 7).

In the strong field limit, $\bar{n} \gg A$, we have from equations (80) and (90) the revival time

$$T_R^{(s)} = \pi / \Omega_{\bar{n}}^{(1)} \approx 2\pi\sqrt{\bar{n}}/g.$$

The Poissonian distribution with $\bar{n}=150$ is shown in Figure 4 as an example of the strong field regime.

For a single atom case, $A=1$, we obtain from equation (80) the JCM Rabi frequency, $2\Omega_n=2g\sqrt{n}$, which is also included in figure 4. The JCM is a particular case of the Dicke model in the strong field regime. It is well known that for the on-resonant JCM with an initially coherent field, collapses and revivals disappear for small photon numbers. The reason is that in this case the sum over the photon numbers includes the region of the greatest nonlinearity of the Rabi frequency. In fact, the derivative of Ω_n at the point $n=0$ is singular and the linear approximation fails. This is also the reason for the absence of collapses and revivals in the JCM with initial thermal field [31] and squeezed vacuum field [71] for any value of $\bar{n}$. In turn, collapses and revivals occur in the Dicke model with initial thermal field if it belongs to the weak field region [72].

It is important to stress that formula (90) is valid for the states to which both odd and even photon numbers contribute. For an initial squeezed vacuum field (to which only even photon number states contribute), the revival time in the Dicke model is half of the time (90) [73]. In this case the nearest-neighbour term to $\bar{n}$ is not $\bar{n}+1$ but $\bar{n}+2$ and the equation (90) must be modified, respectively.

The situation is completely different if the first derivative of the frequency Ω_n vanishes: $\Omega_1^{(\bar{n})}=0$. In general, one can speak of revivals of the k-th order if all of the derivatives $\Omega_r^{(\bar{n})}$, $r=1,2,\ldots k-1$ are zeros, and the first non-vanishing derivative is $\Omega_k^{(\bar{n})}$. In particular, if $\Omega_2^{(\bar{n})}\neq 0$ (and $\Omega_1^{(\bar{n})}=0$) we are faced with revivals of the second order. The shape of these revivals is different and the formula (90) is no longer valid. Let us note that the Rabi frequency is not a smooth function at the point $N=A$ and the vicinity of this point is beyond the scope of the perturbation method under discussion. In this case it is necessary to use other approaches (see, for instance, [74–77]).

12.4 Revivals of the second order

As an example of second-order revivals, we consider the case when the field is taken initially in a Fock state $|n\rangle_f$, and the atomic system is prepared in its ground state $|A,0\rangle$. We consider the weak field case, $n\ll A$. Thus, the initial state of the total system is $|\text{in}\rangle=|n,n\rangle$ and $N=n$. The evolution of the average photon number is given in terms of eigenvalues (86) and eigenvectors (87) as follows,

$$\langle\hat{n}(t)\rangle=\sum_{p,q=0}^{n}\left(\sum_{m=0}^{n} mA_{mp}A_{mq}\right)A_{np}A_{nq}e^{i(\Lambda_{n,q}-\Lambda_{n,p})t}.$$

We keep second-order accuracy in eigenvalues, restricting ourselves to first-order accuracy in eigenvectors. From equations (86) and (87) we have

$$\langle n(t)\rangle=\frac{n}{2}-\frac{\epsilon n(n-1)}{16}+\sum_{p=1}^{n}C_{np}\cos 2\Omega^{(p)}t$$

$$+\frac{\epsilon}{8}\sum_{p=1}^{n}C_{np}\left\{\left[(n-2p)^2-2p+1\right]\cos 2\Omega^{(p)}t+2(p-1)\cos 2\tilde{\Omega}^{(p)}t\right\}. \quad (92)$$

Here the principal term (of the zeroth-order in eigenvectors) contains the second-order Dicke frequencies

$$\Omega^{(p)} = \Lambda^{(2)}_{n,\,p-1} - \Lambda^{(2)}_{n,\,p} = \Omega_n \left\{ 1 + \frac{3\epsilon^2}{16}[5(p-1)(p-n) + (n-1)(n-2)] \right\} \quad (93)$$

weighted with the coefficients

$$C_{np} = \frac{n!}{2^n (n-1)!(p-1)!},$$

which form the binomial distribution multiplied by the factor $n/2^n$. In the first-order approximation in eigenvectors new frequencies appear

$$\tilde{\Omega}^{(p)} = \Lambda^{(2)}_{n,\,p-2} - \Lambda^{(2)}_{n,\,p}. \quad (94)$$

The oscillation amplitudes depend on the atomic number only through the small parameter ϵ.

It is clear from equations (92)–(94) that the most important role in the evolution is played by the terms with $p \sim n/2$. The Dicke frequency (93), considered as a continuous function of p, takes its maximum value at point $p_m = (n+1)/2$. The distribution C_{np} takes its maximum value at the same point (see Figure 5).

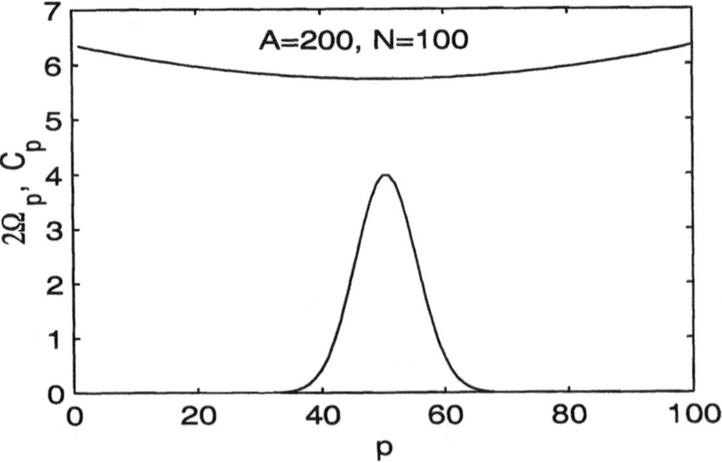

Figure 5: Collective Dicke frequency $2\Omega^{(p)}$ for $A = 200$, $N = 100$, $g = 10^{-2}$, and the binomial distribution of frequencies.

Therefore revivals (see Figure 6) are of the second order. The shape of these revivals is very different from that of the first order. A discussion of their properties can be found in [69,78] (these phenomena were termed "superstructures" in [78]). Let us mention here that the the revival time depends on the parity of the initial

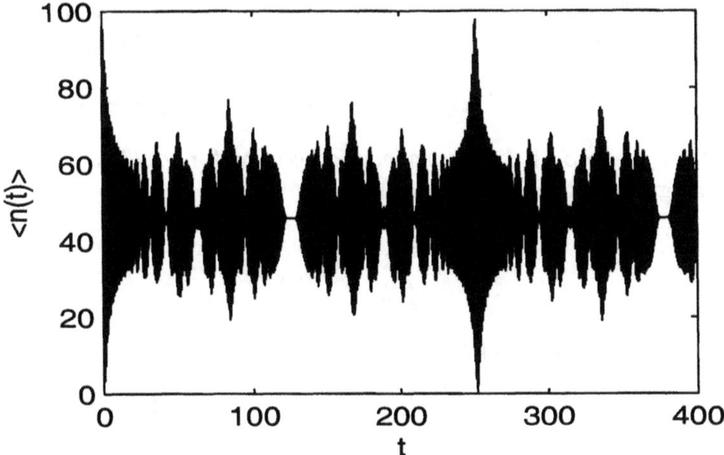

Figure 6: Revivals of the second order in the Dicke model: Photon number evolution for the initial Fock field state with 100 photons and 200 atoms initially taken in their ground states.

photon number n. Namely, the revival time for the same atomic number A and odd n is twice the revival time for the neighbouring even n:

$$T_{R \text{ odd } n} = 2T_{R \text{ even } n} = (16\pi/15g)(A - n/2 + 1/2)^{3/2}. \tag{95}$$

Collapses and revivals recur periodically in the approximation discussed.

The second-order revivals appear also in the JCM with a Kerr medium for special values of the parameters [78, 79]. Then initially coherent field states lead to different revival times depending on whether the mean photon number in the initial coherent state is integer or fractional [80, 81].

13 Collective dynamics in a strong field

We will now study the zeroth-order approximation for the strong field case, introduced in the previous section. We thus consider a collection of A identical two-level atoms interacting with a strong quantum field in a perfect cavity. We will show how the JCM results presented in the previous sections are generalized to the case of a collective atomic system. Namely, we will find the approximate evolution operator and prove the factorization of wave functions for initial semiclassical atomic states.

The evolution operator

We start with the following generalization [25] of operator $\hat{Q}$ (20):

$$\hat{Q} = \exp[i\hat{\phi}\,(S_z + A/2)], \tag{96}$$

where $e^{i\hat{\phi}}$ is the phase operator (17). This operator is unitary only on Fock stat with $n > A > k$. Its action in the bare basis (78) is given as follows,

$$\hat{Q}|N,k\rangle = e^{i\hat{\phi}k}\,|k\rangle_{at}\otimes|n\rangle_f = |k\rangle_a \otimes |n-k\rangle_f = |n,k\rangle.$$

Since the operator $S_z + A/2$ has an integer spectrum (for $-A/2$ is just the grour state energy of the operator S_z), we obtain

$$\hat{Q} = \sum_{k=0}^{A} e^{ik\hat{\phi}}|k\rangle_a\,{}_a\langle k|.$$

One can check the following relations:

$$f(\widehat{n})\hat{Q}^\dagger = \hat{Q}^\dagger f(\widehat{n} + S_z + A/2), \qquad \hat{Q}f(\widehat{n}) = f(\widehat{n} + S_z + A/2)\hat{Q}, \qquad (9$$

$$\hat{Q}^\dagger\,\widehat{n}\,\hat{Q} = \widehat{n} - S_z - A/2, \qquad \hat{Q}^\dagger a\hat{Q} = e^{i\widehat{\phi}}\sqrt{\widehat{n} - S_z - A/2}\,. \qquad (9$$

$$\hat{Q}^\dagger S_\pm \hat{Q} = \exp(\mp i\hat{\phi})S_+, \qquad \widehat{N} = \hat{Q}\,(\widehat{n} - A/2)\,\hat{Q}^\dagger.$$

Now let us apply the $\hat{Q}$ transformation to the Dicke model Hamiltonian

$$H = \omega\widehat{N} + \Delta S_z + V, \qquad \widehat{N} = \widehat{n} + S_z, \qquad V = g\left(aS_+ + a^\dagger S_-\right).$$

Using equations (97) and (98) we obtain $H_a\,(\hat{n}) = \hat{Q}^\dagger H\hat{Q}$, where operator H_a (is diagonal in the field space,

$$H_a\,(\widehat{n}) \equiv \omega\widehat{\nu} + \Delta S_z + g\left(\sqrt{\hat{\nu} - S_z + 1}S_+ + S_-\sqrt{\hat{\nu} - S_z + 1}\right). \qquad (9$$

The field operator $\hat{\nu} = \hat{n} - A/2$, contained in the Hamiltonian H_a, commutes wi all the remaining ingredients and may be treated as a c-number when calculatir the evolution operator $\exp(-itH_a)$. The price which we pay for the eliminatic of the field operators is that the atomic Hamiltonian H_a is nonlinear. If the tot excitation number is much larger than the number of atomic excitations, one ma develop the perturbation theory with the inverse of the field intensity (more pr cisely, with the operator $(\widehat{\nu}+1/2)^{-1}$) as a small parameter. Expanding the squa roots in equation (99) in a series we have

$$H_a = H_{a,0} + (\widehat{\nu} + 1/2)^{-1/2}\,H_{a,1} + O(\widehat{\nu}^{-3/2}), \qquad (10$$

$$H_{a,0} = \omega\widehat{\nu} + \Delta S_z + g\sqrt{\widehat{\nu} + 1/2}\,(S_+ + S_-)\,, \qquad (10$$

$$H_{a,1} = -\frac{g}{4}\,\{S_z,\, S_+ + S_-\}_+\,, \qquad (10$$

where $\{,\}_+$ stands for the anticommutator. One can check that this choice of small parameter leads to vanishing of the first-order corrections to the eigenfr quencies in the resonance case.

The Hamiltonian $H_{a,0}$ has a transparent physical meaning: the zeroth-ord evolution is linear and corresponds to the Bloch vector rotation around the x axi However, the rotation frequency $\Omega(\hat{n}) = g\sqrt{\hat{n} - A/2 + 1/2}$ is a diagonal oper tor. Thus, in fact, one deals with a set of rotations with frequencies which are th eigenvalues of $\Omega(\hat{n})$. Using mathematical terms, one may say that Hamiltonian $H_{a,0}$ belongs to a $(A+1)$-dimensional representation of the $su(2)$ algebra, and the zeroth-order evolution operator

$$U_0(t) = \hat{Q}U_0^{(a)}(t)\hat{Q}^\dagger, \qquad U_0^{(a)}(t) = \exp\left(-itH_{a,0}(\hat{n})\right), \tag{103}$$

performs a finite rotation from the representation of the group $SU(2)$. Since the argument here contains the diagonal operator $\hat{n}$, $U_0^{(a)}(t)$ is not a simple rotation but a set of $SU(2)$ rotations, one for each eigenvalue of $\hat{n}$. The matrix elements of $SU(2)$ rotations are well-known Wigner d-functions (see, e.g., [64]).

In a single-atom case all corrections vanish and the zeroth-order evolution operator U_0 reproduces the exact JCM evolution operator (14). It is also clear that the expansion (100) corresponds to the strong field case of the perturbation theory of the previous section. The advantage of this new form is that we can automatically reconstruct the zeroth-order evolution operator for the quantum field case, $U_0(t)$, knowing the corresponding atomic evolution operator in an external constant classical field $U_0^{(a)}(t)$.

Factorization of the wave function

We now consider the Dicke model dynamics with an initial field in a strong coherent state, $\bar{n} \gg 1$. As for the JCM, special initial atomic states exist, which lead to the approximate factorization of the total wave function into the field and atomic parts [82, 83]. Here we restrict ourselves to the exact resonance case $\Delta = 0$. Consider the initial state

$$|\Psi_p(0)\rangle = |\alpha\rangle_f \otimes |\underline{p,\varphi}\rangle_a, \tag{104}$$

where $\alpha = \sqrt{\bar{n}}e^{i\varphi}$ and $|\underline{p,\varphi}\rangle_a$ is an eigenstate of the Hamiltonian in the classical field limit (compare with Eq. (36)):

$$H_{cl} = e^{i\varphi}S_+ + e^{-i\varphi}S_-, \qquad |\underline{p,\varphi}\rangle_a = e^{i\varphi(S_z+A/2)}|\underline{p}\rangle_a,$$

where $|\underline{p}\rangle_a$ are the eigenvectors of the operator S_x

$$2S_x|\underline{p}\rangle_a = \lambda_p|\underline{p}\rangle_a, \qquad p = 0, 1, .., A, \tag{105}$$

and $\lambda_p = 2p - A$. Coherent states with large photon numbers are approximate eigenstates of the field phase operator: $e^{\pm\hat{\phi}}|\alpha\rangle_f = e^{\pm\phi}|\alpha\rangle_f + O(\bar{n}^{-1/2})$. Thus, one can write

$$\hat{Q}^\dagger|\alpha\rangle_f = e^{-i\hat{\phi}(S_z+A/2)}|\alpha\rangle_f = e^{-i\varphi(S_z+A/2)}|\alpha\rangle_f + O(A\bar{n}^{-1/2}), \tag{106}$$

which leads to an error $O(A\bar{n}^{-1/2})$ in the amplitudes. Applying the evolution operator (103) to the initial state (104) and using (106) we obtain

$$|\Psi(t)\rangle = U_0(t)|\Psi_p(0)\rangle \approx \hat{Q}\exp\left[-2itg\sqrt{\hat{\nu}+1/2}\,S_x\right]|\alpha\rangle_f|\underline{p}\rangle_a.$$

Since $|\underline{p}\rangle_a$ is an eigenstate of operator S_x, we have

$$|\Psi(t)\rangle \approx \hat{Q}\exp\left[-itg\sqrt{\hat{\nu}+1/2}\,\lambda_p\right]|\alpha\rangle_f|\underline{p}\rangle_a.$$

Using the commutation relation

$$\hat{Q}\exp\left[-itg\sqrt{\hat{\nu}+1/2}\,\lambda_p\right] = \exp\left[-itg\sqrt{\hat{\nu}+S_z+A/2+1/2}\,\lambda_p\right]\hat{Q},$$

we get

$$|\Psi(t)\rangle \approx \exp\left[-itg\sqrt{\hat{\nu}+S_z+A/2+1/2}\,\lambda_p\right]|\Psi_p(0)\rangle. \qquad (107)$$

Finally, expanding the square root in a Taylor series

$$\sqrt{\hat{\nu}+S_z+\frac{A+1}{2}} = \sqrt{\hat{\nu}+\frac{1}{2}} + \frac{S_z+A/2}{2\sqrt{\overline{\nu}+1/2}} + O\left(\frac{A^2}{\overline{n}^{3/2}}\right), \qquad (108)$$

we obtain the factorized wave function

$$|\Psi(t)\rangle = |\Phi_p(t)\rangle_f \otimes |A_p(t)\rangle_{at}, \qquad (109)$$

$$|\Phi_p(t)\rangle_f = \exp\left[-it\sqrt{\hat{\nu}+1/2}\,\lambda_p\right]|\alpha\rangle_f, \qquad (110)$$

$$|A_p(t)\rangle_{at} = \exp\left(-it\frac{S_z+A/2}{2\sqrt{\overline{\nu}+1/2}}\lambda_p\right)|\underline{p,\varphi}\rangle_a, \qquad (111)$$

where $\bar{\nu} = \bar{n} - A/2$. In a single-atom case, we recover the corresponding result for the JCM. The factorization (109) is valid for times $gt < \sqrt{\bar{n}}$ (due to the accuracy of approximation (108) in frequencies) with the error in the amplitudes $O(A/\sqrt{\bar{n}})$ (in accordance with the approximation (106)). Note that the photon distribution of the field state remains Poissonian.

As in the JCM case, this factorization does not mean that subsystems evolve without interaction, and the evolution of the field and atomic parts of the factorized wave function is not unitary. From equation (109) it follows that the states $|\Psi_p(0)\rangle = |\alpha\rangle_f \otimes ||\underline{p,\phi}\rangle_a;\, p = 0, 1, ..., A$ are trapping states

$$\langle\Psi_p(t)|S_z|\Psi_p(t)\rangle = \langle\Psi_p(0)|U^\dagger S_z U|\Psi_p(0)\rangle \simeq \langle\Psi(0)|S_z|\Psi(0)\rangle. \qquad (112)$$

Thus, the atomic inversion remains (with a high accuracy) constant for these initial states. In turn, bare initial atomic states lead to collapsed and revived Rabi oscillations, as we will see further on.

Evolution in the field phase space

The evolution of the field part of the factorized wave function is basically the same as for the JCM. The effective Hamiltonian for the field is [82]

$$H_{\text{eff}} = g\lambda_p\sqrt{\hat{n}-A/2+1/2}, \qquad (113)$$

and the dynamics can be analyzed using the same expansion of this Hamiltonian as in the JCM case (see equation (46) and subsection 12.3). The field Q-function

(55) once again has the form $Q_p(\beta)=|\langle\beta|\Phi_p(t)\rangle_f|^2$. For instance, for short times, $gt \ll \sqrt{\bar{n}}$, the initial coherent state rotates around the origin of phase plane with the frequency

$$\omega_p = \frac{\lambda_p}{2\sqrt{\bar{n} - A/2 + 1/2}}, \tag{114}$$

and the Q-function has a Gaussian shape

$$Q(\beta) = |\langle\beta|e^{-i\omega_p t}\alpha\rangle_f|^2 = \exp\left[-|\beta - e^{-i\omega_p t}\alpha|^2\right].$$

Here we take the initial coherent state with zero phase, $\varphi = 0$, for simplicity. For longer times the Gaussian shape is deformed due to the nonlinearity of the effective Hamiltonian (113). Using the expansion analogous to (46) one can find the following analytic form for $Q(\beta)$ ($\beta = r\,e^{i\phi}$):

$$Q_p(\beta) = \frac{\exp\left[-(r-r_0)^2\right]}{\sqrt{rr_0\mu(t)}} \sum_{k=-\infty}^{+\infty} \exp\left[-\frac{(\phi + \phi_0(t) - 2\pi k)^2}{\mu(t)}\right].$$

Each term in this sum has a Gaussian shape centred at the phase

$$\phi_0(t) = \frac{\tan\lambda_p}{2\sqrt{rr_0 - A/2 + 1/2}},$$

and with the variance

$$\mu(t) = \frac{1}{rr_0} + \left(\frac{g\lambda_p t}{2rr_0}\right)^2, \qquad r_0 \equiv \sqrt{\bar{n}},$$

depending on the intensity r^2. This corresponds to the phase spread. At a given time, not all of the terms contribute to the sum. For instance, for short times $gt < 2\pi\sqrt{\bar{n}}/\lambda_p$ only the $k=0$ term is important. Thus, the curves of the constant level for the Q-function are given by the equation

$$(r - r_0)^2 + (\phi - \phi_0(t))^2/\mu(t) = \text{const.}$$

The evolution of an arbitrary initial atomic state $|\psi(0)\rangle_{at}$ can be described in terms of the factorized states. Assuming again the zero phase of the initial field CS, we expand the initial atomic state in the basis of semiclassical states

$$|\psi(0)\rangle_a = \sum_{p=0}^{A} C_p|\underline{p}\rangle_a, \qquad C_p = {}_a\langle\underline{p}|\psi(0)\rangle_a. \tag{115}$$

The state of the total system is given as a superposition of factorized states

$$|\Psi(t)\rangle = \sum_{p=0}^{A} C_p|\Phi_p(t)\rangle_f \otimes |A_p(t)\rangle_a. \tag{116}$$

An arbitrary atomic initial state produces in the field phase plane $A+1$ humps, one for every semiclassical state in the expansion (116). Every hump rotates along the circle of the radius $\sqrt{\bar{n}}$ with the angular velocity ω_p (114). To understand what happens when different humps collide, let us calculate the atomic inversion for an arbitrary initial state (115) using the factorization approximation (116):

$$\langle\Psi(t)|S_z|\Psi(t)\rangle \approx \sum_{p,q=0}^{A} C_p C_q^* \,{}_{at}\langle A_q(t)|S_z|A_p(t)\rangle_{at}\,{}_f\langle\Phi_q(t)|\Phi_p(t)\rangle_f .$$

It is easy to see that

$$_{at}\langle A_q(t)|S_z|A_p(t)\rangle_{at} = \langle\underline{q}|S_z \exp\left[-i(\omega_p-\omega_q)\,(S_z+A/2)\,t\right]|\underline{p}\rangle_a . \qquad (117)$$

The overlap integral for the field states

$$_f\langle\Phi_q(t)|\Phi_p(t)\rangle_f = \sum_n P_n \exp\left[-i\,t(\lambda_p-\lambda_q)\sqrt{n-A/2+1/2}\right]$$

can be approximated for short times [as in the JCM case (51)–(53)] by

$$_f\langle\Phi_q(t)|\Phi_p(t)\rangle_f \approx e^{-it(\lambda_p-\lambda_q)\sqrt{\bar{n}-A/2+1/2}/2}\langle e^{-i\omega_q t}\alpha|e^{-i\omega_p t}\alpha\rangle_f .$$

Here, the rapidly oscillating factor is responsible for the Rabi oscillations. Considering the envelope of these oscillations

$$|\langle e^{-i\omega_q t}\alpha|e^{-i\omega_p t}\alpha\rangle_f| = \exp\left[-\bar{n}|e^{-i\omega_q t}-e^{-i\omega_p t}|^2/2\right],$$

one can see that oscillations collapse when $\sqrt{\bar{n}}|e^{-i\omega_q t}-e^{-i\omega_p t}| \sim 1$. Taking into account that $\omega_p \sim 1/\sqrt{\bar{n}}$, we can expand the exponents in series, which gives for the collapse time

$$T_c^{-1} \sim (\omega_p-\omega_q)\sqrt{\bar{n}} .$$

On the other hand, different humps collide when $(\omega_p-\omega_q)\,T_r = 2\pi$, which gives us possible revival times. At these time instants atomic matrix elements (117) are almost independent of time, for

$$\exp\left[-i(\omega_p-\omega_q)\,(S_z+A/2)\,T_r\right] = \exp\left[-i2\pi\,(S_z+A/2)\right] = 1.$$

Therefore, at the revival time $t \sim T_r$ we have

$$\begin{aligned} &_{at}\langle A_q(t)|S_z|A_p(t)\rangle_{at} \approx {}_{at}\langle\underline{q}|S_z|\underline{p}\rangle_{at} \\ &= \frac{1}{2}\sqrt{(A-p)(p+1)}\,\delta_{p+1,q} + \frac{1}{2}\sqrt{(A-p+1)p}\,\delta_{p-1,q} . \end{aligned} \qquad (118)$$

Thus, not every collision of the humps is accompanied by a revival. Revivals happen only if the hump p collides with the hump $p\pm1$. Finally, the revival time reads

$$T_r = 2\pi g^{-1}\sqrt{\bar{n}-A/2+1/2},$$

while the collapse time has the form

$$T_c = g^{-1}\sqrt{(2\bar{n}-A+1)/(2\bar{n})}.$$

The revival time is almost independent of the number of atoms if $\bar{n} \gg A$.

4 Dissipative dynamics

To show how the dynamical symmetry methods work in the presence of dissipation (due to the absorption of photons by the cavity walls), let us consider the resonant case of the Dicke model in a non-ideal cavity [84,85] (see also [86–90]). We restrict ourselves to the case of zero temperature of a thermal reservoir. Such a system is described by a density matrix ρ obeying the standard master equation

$$\partial_t \rho = -ig[aS_+ + a^\dagger S_-, \rho] + \frac{\gamma}{2}(2a\rho a^\dagger - a^\dagger a\rho - \rho a^\dagger a)\,, \tag{119}$$

where γ is the cavity decay constant. Now, the system loses photons in the course of evolution and therefore the number of excitations is no longer conserved. However, for highly excited initial field states, the system remains in the strong field region for some period of time, and we can use a technique similar to that described in the preceding sections. Namely, let us apply the $\hat{Q}$ transformation (96) to the master equation (119). In the zeroth-order approximation (101) for the transformed density matrix

$$\underline{\rho} = \hat{Q}^\dagger \rho \hat{Q}, \tag{120}$$

we get the following equation:

$$\partial_t \underline{\rho} = -2ig[\sqrt{\hat{\nu}+1/2}\,S_x\,,\,\underline{\rho}] + L_0\underline{\rho} + L_1\underline{\rho}, \tag{121}$$

$$L_0\underline{\rho} = \frac{\gamma}{2}\left(2a\underline{\rho}a^\dagger - a^\dagger a\underline{\rho} - \underline{\rho}a^\dagger a\right),$$

$$L_1\underline{\rho} = \frac{\gamma}{2}(2a\underline{\rho}a^\dagger - 2a\underline{\rho}a^\dagger + \{S_z + A/2, \underline{\rho}\})\,.$$

The transformed equation (121) is diagonal with respect to the field indices, and is convenient for numerical calculations. To find an analytical solution let us note that the contribution of operator L_1 can be neglected in the case of large photon numbers, since

$$L_1\underline{\rho} = \frac{\gamma}{2}\left(-\left\{\frac{S_z + A/2}{(\hat{n}+1)}, a\underline{\rho}a^\dagger\right\} + \left\{S_z + \frac{A}{2}, \underline{\rho}\right\}\right) + O\left(\frac{\gamma}{\hat{n}+1}\right),$$

and $L_1\underline{\rho} \approx O(\gamma/\hat{n})$. We thus assume that $\gamma \ll g\bar{n}$. In the Fock basis equation (121) has the form

$$\partial_t \underline{\rho}_{nm} = -2ig(S_x\sqrt{n_A}\,\underline{\rho}_{nm} - \underline{\rho}_{nm}\sqrt{m_A}\,S_x) + (L_0\underline{\rho})_{nm}, \tag{122}$$

where $n_A = n + 1/2 - A/2$. The matrix elements $\underline{\rho}_{nm}$ are still operators in the atomic space. Equation (122) takes a simple form in the basis in which the atomic operator S_x is diagonal (semiclassical states)

$$S_x|\underline{p}\rangle_a = \lambda_p|\underline{p}\rangle_a\,, \qquad \lambda_p = p - A/2, \qquad p = 0, 1, \ldots, A\,. \tag{123}$$

The bare atomic state is written in the semiclassical basis as follows:

$$|k\rangle_a = \sum_p C_{kp}|\underline{p}\rangle_a, \tag{124}$$

where the matrix elements C_{kp} are given in terms of the Wigner d-functions $d^A_{pk}(\theta) = \langle p|\exp(i\theta S_x)|k\rangle$ (in notation of Ref. [64])

$$C_{kp} = \langle \underline{p}|k\rangle = \langle p|\exp(i\pi/2\, S_y)|k\rangle = i^{k-p}\, d^A_{pk}(\pi/2). \tag{125}$$

The matrix elements of the inverse transformation $|\underline{p}\rangle_a = \sum_p C^*_{pk}|k\rangle_a$ are $C^*_{pk} = \langle k|\underline{p}\rangle = C_{kp}$. Note that coefficients C_{kp} differ only by the phases of each column from the coefficients α^A_{kp} (82). For the density matrix elements of $\underline{\rho}_{nm}$ in the semiclassical atomic basis $x_{qp,nm} = \langle \underline{q}|\underline{\rho}_{nm}|\underline{p}\rangle$ we have

$$\partial_t x_{qp,nm} = -f^{qp}_{nm} x_{qp,nm} + \gamma\sqrt{(n+1)(m+1)}\, x_{qp,n+1m+1}\,, \tag{126}$$

$$f^{qp}_{nm} = 2ig\,(\lambda_q\sqrt{n_A} - \lambda_p\sqrt{m_A}) + \gamma(n+m)/2\,. \tag{127}$$

Making the change of variables (here we omit the atomic indices)

$$x_{nm} = y_{nm}/\left(\gamma^n\sqrt{n!m!}\right),$$

we get

$$\partial_t y_{nm} = -f_{nm}y_{nm} + y_{n+1m+1}\,.$$

Applying the Laplace transformation we arrive at the recursive relation

$$\overline{y}_{n+1m+1}(s) - (f_{nm}+s)\overline{y}_{nm}(s) = -\overline{y}_{nm}(0),$$

where $\overline{y}_{nm}(s)$ stands for the Laplace transform of y_{nm}. The solution of the above equation has the form

$$\overline{y}_{nm}(s) = \sum_{k=0}^{\infty} \overline{y}_{n+k\ m+k}(0)\left[\prod_{j=0}^{k}(f_{n+j\ m+j}+s)\right]^{-1}. \tag{128}$$

and the inverse transformation gives

$$y_{nm}(t) = \sum_{k=0}^{\infty} y_{n+k\ m+k}(0) I^{(q,p)}_{k,\ nm}(t). \tag{129}$$

Therefore, the solution of equation (126) takes the form

$$x_{qp,nm}(t) = \sum_{k=0}^{\infty}\gamma^k\sqrt{\frac{(n+k)!}{n!}\frac{(m+k)!}{m!}}\, x_{qp,n+k\,m+k}(0)\, I^{(qp)}_{k,nm}(t)\,, \tag{130}$$

where

$$I^{(qp)}_{0,nm}(t) = \exp(-f_{qp,nm}t)\,,$$

$$I^{(qp)}_{k,nm}(t) = \sum_{j=0}^{k}\frac{\exp(-f^{qp}_{n+j\,m+j}t)}{\prod_{i=0}^{k}(f^{qp}_{n+i\,m+i} - f^{qp}_{n+j\,m+j})|_{i\neq j}}\,, \qquad k>0. \tag{131}$$

Initial conditions in the semiclassical basis for equation (130) can be written in terms of the initial conditions in the bare basis as follows:

$$x_{qp,nm}(0) = {}_a\langle \underline{q}|\langle n|\hat{Q}^\dagger \rho(0)\hat{Q}\,|m\rangle|\underline{p}\rangle_a = \sum_{k,l} C_{kq}C_{lp}\rho_{kl,n-k\,m-l}(0), \quad (132)$$

where $\rho(0)$ is the initial density matrix of the total system.

The dynamics without cavity losses can be recovered if $\gamma = 0$. In this case only the term $k = 0$ survives in the sum (130). On the other hand, we can reproduce the solution for the quantum field evolution in a dissipative cavity (harmonic oscillator with dissipation) taking $g = 0$ in equation (130). Let us consider particular cases of initial field states.

Initial Fock state of the field

If the initial state of the field is $|N\rangle_f$ (with $N \gg A$) and the atomic system is taken in its ground state, the initial density matrix has the form

$$\rho(0) = |0\rangle_{at\,at}\langle 0| \otimes |N\rangle_{f\;f}\langle N|.$$

Then it follows from equation (132) that

$$x_{qp,nm}(0) = \langle \underline{p}|0\rangle\langle \underline{q}|0\rangle \delta_{nN}\,\delta_{mN}\,. \quad (133)$$

On insertion of (133) into (130) we obtain the elements of the Q-transformed density matrix in the semiclassical basis

$$x_{qp,nm}(t) = \gamma^{N-n}\frac{N!}{n!}I^{(qp)}_{N-n,nn}(t)\,C_{0p}\,C_{0q}\,\delta_{mn}\,, \quad (134)$$

and in the bare atomic basis

$$\underline{\rho}_{kl,nm}(t) = \delta_{mn}\sum_{q,p=0}^{A} C_{kq}\,C_{lp}\,x_{qp,nn}(t). \quad (135)$$

In this case the density matrix is diagonal in the field space. Using equations (134) and (135) we can calculate the atomic inversion

$$\langle S_z\rangle = -\frac{A}{2}\mathrm{Re}\sum_n \gamma^{N-n}\frac{N!}{n!}I^{(p\,p-1)}_{N-n,nn}\,,$$

where $I^{(p\,p-1)}_{N-n,nn}$ is defined by (131) and is independent of the atomic index p.

Initial coherent state of the field

For an initially coherent field state with $\bar{n} \gg A$, the functions (127) vary slowly near the point $\overline{n}$ and one may make the following approximation:

$$f^{qp}_{n+j,m+j} \simeq f^{qp}_{nm} + j\gamma'_{qp,nm}\,,$$

where

$$\gamma'_{qp,nm} = \gamma + ig\left(\frac{\lambda_q}{\sqrt{n_A}} - \frac{\lambda_p}{\sqrt{m_A}}\right). \tag{136}$$

Using equation (136) to simplify the coefficients (131), we obtain

$$I^{(qp)}_{k,nm}(t) = \frac{1}{k!}\left[\frac{1-\exp(-\gamma'_{qp,nm}t)}{\gamma'_{qp,nm}}\right]^k \exp(-f^{qp}_{qp,nm}t) . \tag{137}$$

Let us take the initial coherent state with zero phase and an arbitrary initial atomic state

$$|\text{in}\rangle_a = \sum_p C_p|\underline{p}\rangle_a,$$

that corresponds to the initial density matrix $\rho(0) = |\text{in}\rangle\langle\text{in}| \otimes |\alpha\rangle\langle\alpha|$. The initial condition takes the following form in the semiclassical basis (132):

$$x_{qp,nm}(0) = \sum_{ll'=0}^{A}\sum_{rr'=0}^{A} C_{lq}C_{l'p}C_{lt}C_{l'r'}C_rC_{r'}\,\alpha_{n-l}\alpha_{m-l'}\,,$$

where C_{lp} are defined by equation (125) and $p_n = \sqrt{e^{-\bar{n}}\,\bar{n}^n/n!}$. One may use the form $p_{n+l} = p_n + O(A\bar{n}^{-1/2})$ to simplify the initial condition under study. Neglecting the terms of order $O\left(A/\sqrt{\bar{n}}\right)$ and taking into account that $\sum_l C_{lr}C_{lq} = \delta_{rq}$, we get

$$x_{qp,nm}(0) \approx c_q\, c_p\, p_n p_m. \tag{138}$$

It is worth noting that the above equation is exact when the ground state is chosen as the initial atomic state. In this case $c_p = C_{0p}$ and the $\hat{Q}$ transformation is reduced to the unit operator.

On insertion of (137) and (138) into equation (130) we obtain the density matrix elements in the semiclassical basis

$$x_{qp,nm}(t) = c_q\, c_p\, p_n(t)p_m(t)F_{qp,nm}(t), \tag{139}$$

where $p_n(t)$ is given by the square root of the Poisson distribution with a time-dependent average photon number:

$$p_n(t) = \sqrt{e^{-\bar{n}(t)}\,[\bar{n}(t)]^n/n!}\,, \qquad \bar{n}(t) = \bar{n}e^{-\gamma t}.$$

The functions $F_{qp,nm}(t)$ are given by

$$F_{qp,nm}(t) = \exp\left[-2igt(\lambda_q\sqrt{n_A} - \lambda_p\sqrt{m_A}) - \mu_{qp}(t)\right], \tag{140}$$

$$\mu_{qp}(t) = \bar{n}\left[(1-e^{-\gamma t}) - (\gamma/\gamma')\,(1-e^{-\gamma' t})\right], \tag{141}$$

$$\gamma' = \gamma + \frac{ig(\lambda_q - \lambda_p)}{\sqrt{\bar{n}(t) - A/2 + 1/2}}. \tag{142}$$

In order to obtain this expression we have approximated $\gamma'_{qp,nm}$ (defined by Eq (136)) by its value at the point $n = m = \bar{n}(t)$.

The Q-transformed density matrix (139) in the bare atomic basis can be written in the form

$$\underline{\rho}_{kl,nm}(t) = p_n(t)p_m(t) \sum_{q,p=0}^{A} C_{kq}\, C_{lp}\, c_q\, c_p\, F_{pq,nm}(t). \tag{143}$$

Making the inverse Q transformation, $\rho = Q\underline{\rho}Q^\dagger$, we get the density matrix in the bare atomic basis

$$\rho_{kl,nm}(t) = \underline{\rho}_{kl,n+k,m+l}(t)\,. \tag{144}$$

Equations (139)–(144) describe the dissipative dynamics for the case of a strong initial coherent state. Now it is easy to calculate the expectation value of any observable. In order to illustrate the use of equation (144), we calculate here the atomic inversion for the case when the atomic system is prepared in its ground state. The atomic density matrix in the bare basis reads

$$\rho_{kl}^{(at)} = \sum_n \underline{\rho}_{kl,k+n\, l+n}(t).$$

Therefore

$$\langle S_z(t)\rangle = \sum_{k=0}^{A}(k - A/2)\rho_{kk}^{(at)} = \sum_n \sum_{k=0}^{A}(k - A/2)\, \underline{\rho}_{kk,k+n\, k+n}(t),$$

where $\underline{\rho}_{kk,k+n\, k+n}(t)$ is defined by equation (144) with

$$c_p = C_{0p} = i^{-p} d_{0p}^A(\pi/2) = \sqrt{\frac{A!}{(A-k)!k!}} \sin^k\frac{\theta}{2} \cos^{A-k}\frac{\theta}{2}\,.$$

We get

$$\langle S_z\rangle = -\frac{A}{2}\,\mathrm{Re}\sum_n P_n(t)\, \exp\left[-2igt\sqrt{n_A} - \mu(t)\right], \tag{145}$$

where $\mu(t)$ is given by equation (141) with $\gamma' = ig/\sqrt{\bar{n}_A} + \gamma$.

Factorized dynamics

From the previous sections we know that the evolution of the initial semiclassical atomic states interacting with a strong coherent field leads to the factorization of the wave function. We will show now that an analogue of the wave function factorization also occurs in the dissipative case. Let us consider the initial atomic density matrix $\rho^{(at)}(0) = |\underline{q}\rangle\langle\underline{p}|$ and the initial coherent field: $\rho^{[qp]}(0) = |\underline{q}\rangle\langle\underline{p}| \otimes |\alpha\rangle\langle\alpha|$. From equations (143) and (144) we get the following expression for the atomic density matrix in the bare basis

$$\rho_{kl,nm}^{[qp]}(t) = p_{n+k}(t)p_{m+l}(t)C_{kq}\, C_{lp}\, F_{pq,n+k\, m+l}(t). \tag{146}$$

For not very long times, while $A \ll \sqrt{\bar{n}e^{-\gamma t}}$ still holds, the time-dependent Poissonian distribution in equation (146) can be approximated as follows:

$$p_{n+k}(t) \approx p_n(t) + O(A/\sqrt{\bar{n}(t)})\,. \tag{147}$$

We now expand the square roots in the frequencies contained in the functions $F_{qp,n+k\;m+k}$ as follows:

$$\sqrt{n_A + k} \approx \sqrt{n_A} + \frac{k}{2\sqrt{n_A}} + O(A^2\bar{n}^{-3/2})\,. \tag{148}$$

Substituting expressions (147) and (148) into (146) we obtain

$$\begin{aligned}\rho^{[qp]}_{kl,nm} &\approx \exp(-\mu_{qp})p_n(t)p_m(t)\,C_{kq}C_{lp} \\ &\times \exp\left[-2igt(\lambda_q\sqrt{n_A} - \lambda_p\sqrt{m_A}) - igt\left(\frac{k\lambda_q}{\sqrt{n_A}} - \frac{l\lambda_p}{\sqrt{m_A}}\right)\right]. \end{aligned} \tag{149}$$

In fact, we need one more approximation and write the slow frequencies in the last exponential in equation (149) as

$$k/\sqrt{n_A} \approx k/\sqrt{\bar{n}_A(t)} + O\left(A/\bar{n}(t)\right)\,. \tag{150}$$

In such a way the accuracy is further reduced, but we gain a surprisingly simple and clear physical picture; the total density matrix acquires the form of a product of the field and atomic density matrices, which represent the pure states of the field and atoms:

$$\rho^{[qp]}(t) \approx \exp(-\mu_{qp}(t))|\Phi_q(t)\rangle\langle\Phi_p(t)| \otimes |A_q(t)\rangle\langle A_p(t)|\,, \tag{151}$$

where the field and atomic states are given as

$$\begin{aligned}|\Phi_q(t)\rangle &= \sum_n p_n(t)\exp\left[-2igt\lambda_q\sqrt{n_A}\right]|n\rangle_f\,, \\ |A_q(t)\rangle &= \exp\left[-i\frac{gt\lambda_q}{\sqrt{\bar{n}_A(t)}}\left(S_z + A/2\right)\right]|\underline{q}\rangle\,.\end{aligned}$$

This solution differs in two points from the corresponding solution in the unitary case: (a) the Poisson distribution appears with a time-dependent average photon number, $\bar{n}(t) = \bar{n}e^{-\gamma t}$; (b) density matrices corresponding to different initial conditions $|\underline{p}\rangle\langle\underline{q}|$ are multiplied by factors $\exp(-\mu_{qp}(t))$. This does not affect the diagonal initial conditions in the semiclassical basis, since $\mu_{qq} = 0$. However, the factors $\exp(-\mu_{qp}(t))$ rapidly decrease for $q \neq p$, which results in rapid loss of coherence among different semiclassical trajectories. From the explicit expression (141) for $\mu_{qp}(t)$ we can determine the decoherence time, when different semiclassical states lose coherence among each other. This time is defined by the condition $\mu_{qp}(t_{\text{decoh}}) \sim 1$ and for a sufficiently strong initial field, $\gamma\bar{n}^{3/2} \gg 1$, the decoherence time is almost independent of the initial photon number $\bar{n}$ and is much smaller than the revival time:

$$gt^{(qp)}_{\text{decoh}} \sim \left(\frac{6g}{\gamma\,(q-p)^2}\right)^{1/3} \ll gt_R. \tag{152}$$

We see that semiclassical states lose mutual coherence more rapidly when $|q-p|$ grows. On the other hand, in the case of sufficiently small losses, $\gamma \ll g$, we have $gt_{\mathrm{decoh}}^{(qp)} \gg 1$, i.e., the decoherence time is larger than the collapse time.

For an arbitrary initial atomic state $|\mathrm{in}\rangle = \sum_p c_q |\underline{q}\rangle$ the initial density matrix is $\rho(0) = \sum_{qp} c_q c_p^* \rho^{[qp]}(0)$. Then the solution can be composed in terms of the factorized "elementary solutions" (151) as $\rho(t) = \sum_{qp} c_q c_p^* \rho^{[qp]}(t)$. For times larger than the decoherence time ($t > t_{\mathrm{decoh}}$) the density matrix takes the form of a statistical mixture

$$\rho_f(t) \approx \sum_p |c_p|^2 \rho_p(t), \qquad \rho_p(t) = |\Phi_p(t)\rangle\langle\Phi_p(t)| \,. \tag{153}$$

The factorization (151) leads to a clear dynamical image in the field phase space: every semiclassical state is represented by a hump rotating around the origin along a spiral whose radius decreases slowly with time, $r(t) = \sqrt{\bar{n}} e^{-\gamma t/2}$. In the absence of dissipation, when two neighbouring semiclassical states collide, they produce a revival of the Rabi oscillations of the atomic inversion. In the dissipative case, different semiclassical states rapidly lose their mutual coherence which results in suppression of the revival amplitude. On the other hand, squeezing in the field quadratures is not affected by dissipation either for short or long interaction times [91]. In fact, squeezing for short times is produced due to the interference between different semiclassical states and reaches its maximal value for times of order of the collapse time, $gt_{\mathrm{sq}} \sim 1$, which (for typical values of the model parameters) is less than the decoherence time (152). For long interaction times (around the first revival time) squeezing appears due to a specific deformation of the shape of the field states associated with each initial semiclassical state. This allows us to conclude that squeezing can be destroyed only if the shape of each ρ_p in (153) is essentially changed. This may only happen in the strong dissipation case $\gamma\sqrt{\bar{n}} \geq g$, when the average photon number significantly decreases for times around the first revival moment: $\bar{n}(t_R) \ll \bar{n}$.

To estimate the accuracy of the factorization, we evaluate the purity parameter $P = \operatorname{Tr}\left(\rho^{(at)}\right)^2$ for the initial semiclassical atomic state $|\underline{q}\rangle\langle\underline{q}|$. After some algebra we obtain from equation (149)

$$P \approx \sum_{k,l=0}^{A} C_{qk}^2 C_{ql}^2 \exp\left[-4\bar{n}(t) \sin^2 \frac{gt\,(k-l)}{8\bar{n}(t)^{3/2}}\right],$$

where coefficients C_{kp} are given by equation (125). The purity parameter essentially differs from unity if $gt \sim \bar{n}(t)$. This means that the factorization holds for times $gt \ll \bar{n}(t)$. In the case of weak dissipation, $\gamma\sqrt{\bar{n}} \ll g$, this condition leads to $gt \ll \bar{n}$. However, gt may be still greater than $\sqrt{\bar{n}}$, which means that the factorization survives for several revival times.

For the most recent studies devoted to the subjects considered in this chapter see, e.g., references [92–111].

Bibliography

[1] Dicke, R., Coherence in spontaneous radiation processes, *Phys. Rev.* (1954) **93** 9 110.

[2] Jaynes, E.T. and Cummings, F.W., Comparison of quantum and semiclassical ra ation theories with application to the beam maser. *Proc. IEEE* (1963) **51** 89–108.

[3] Paul, H., Induzierte Emission bei starker Einstrahlung *Ann. Phys. (Leipzig)* (196 **11** 411–412.

[4] Tavis, M., and Cummings, F.W., Exact solution for an N-molecule-radiation-fie Hamiltonian, *Phys. Rev.* (1968) **170** 379–384.

[5] Stenholm, S., Quantum theory of electromagnetic fields interacting with atoms a molecules. *Phys. Rep.* (1973) **6** 1–121.

[6] Haroche, S. and Raimond, J.M., Radiative properties of Rydberg states in resona cavities. *Adv. Atom. Mol. Phys.* (1985) **20** 347–411.

[7] Yoo, H.-I. and Eberly, J.H., Dynamical theory of an atom with two or three leve interacting with quantized cavity fields. *Phys. Rep.* (1985) **118** 239–337.

[8] Milonni, P.W. and Singh, S., Some recent developments in the fundamental theo of light. *Adv. Atom. Mol. Phys.* (1990) **28** 75–142.

[9] Bužek, V., N-level atom interacting with single-mode radiation field: an exac solvable model with multiphoton transitions and intensity-dependent coupling. *Mod. Opt.* (1990) **37** 1033–1053.

[10] Kien, F.L. and Shumovsky, A.S., Interaction of an atom with a cavity radiation fie *Int. J. Mod. Phys.* B (1991) **5** 2287–2322.

[11] Shore, B.W. and Knight, P.L., Topical review: The Jaynes–Cummings model. *Mod. Opt.* (1993) **40** 1195–1238.

[12] Louisell, W.H., *Quantum Statistical Properties of Radiation*. Wiley, New Yo 1973.

[13] Allen, L. and Eberly, J.H., *Optical Resonance and Two-Level Atoms*. Wiley, Ne York, 1975.

[14] Walls, D.F. and Milburn, G.J., *Quantum Optics*. Springer, Berlin, 1994.

[15] Vogel, D. and Welsch, D.-G., *Lectures on Quantum Optics*, Akademie Verla Berlin, 1994.

[16] Scully, M.O. and Zubairy, M.S., *Quantum Optics*, Cambridge University Pre Cambridge, 1997.

[17] Barnett, S.M. and Radmore, P.M., *Methods in Theoretical Quantum Optics*, Clare don Press, Oxford, 1997.

[18] Chumakov, S.M., Dodonov, V.V., and Man'ko, V.I., Soluble models of multilev system interaction with a quantized electromagnetic field. In: *Classical and Qua tum Effects in Electrodynamics. Proc. Lebedev Phys. Inst., vol. 176* (A.A. Kom ed.), pp. 77–126. Nova Science, Commack, 1988.

[19] Heitler, W., *The Quantum Theory of Radiation*, Oxford Univ. Press, New Yo 1954.

[20] J. Peřina, *Coherence of Light*, (Van Nostrand, London, 1972)

[21] Loudon, R., *Quantum Theory of Light*, Oxford Univ. Press, Oxford, 1973.

[22] Peřina, J., *Quantum Statistics of Linear and Nonlinear Optical Phenomena*. Kluw Dordrecht, 1984.

[23] Gardiner, C.W., *Quantum Noise*. Springer, Berlin, 1991.

[24] Meystre, P. and Sargent III, M., *Elements of Quantum Optics*. Springer, Berl 1991.

[25] Klimov, A.B. and Chumakov, S.M., Semiclassical quantization of the evolution o

erator for a class of optical models. *Phys. Lett.* A (1995) **202** 145–154.

[26] Narozhny, N.B., Sanchez-Mondragon, J.J., and Eberly, J.H., Coherence versus incoherence: collapse and revival in a single quantum model, *Phys. Rev. A* (1981) **23** 236–247.

[27] Klimov, A.B., Sanchez-Soto, L.L., and Delgado, J., Mimicking a Kerrlike medium in the dispersive regime of second-harmonic generation. *Opt. Commun.* (2001) **191** 419–426.

[28] Fleischhauer, M. and Schleich, W.P., Revivals made simple: Poisson summation formula as a key to the revivals in the Jaynes–Cummings model. *Phys. Rev.* A (1993) **47** 4258–4269.

[29] Lechte, C., Averbukh, I.S., and Schleih, W.P., Multilevel quantum beats: An analytical approach. *Phys.Rev.* A (1996) **54** 5299–5312.

[30] Abramowitz, M. and Stegun, I.A., *Handbook of Mathematical Functions* (Applied Mathematics Series, vol. 55), National Bureau of Standards, Washington, 1964.

[31] Knight, P.I. and Radmore, P.M., Quantum revivals of a two-level system driven by chaotic radiation, *Phys. Lett.* A (1982) **90** 342–346.

[32] Arancibia-Bulnes, C.A., Chumakov, S.M., and Sanchez-Mondragon, J.J., Atomic dipole dynamics in the thermal quantized cavity field, *J. Mod. Opt.* (1993) **40** 2071–2079.

[33] Chumakov, S.M., Kozierowski, M., and Sanchez-Mondragon, J.J., Analytical approach to the photon statistics in the thermal Jaynes–Cummings model with an initially unexcited atom. *Phys. Rev.* A (1993) **48** 4594–4597.

[34] Kozierowski, M., Poyatos, J.F., and Sanches-Soto, L.L., Generation of sub-Poissonian and squeezed fields in the thermal superposition Jaynes-Cummings model. *Phys. Rev.* A (1995) **51** 2450–2458.

[35] Kozierowski, M., Poyatos, J.F., and Sanches-Soto, L.L., Squeezing in the thermal Jaynes-Cummings model. *J. Mod. Opt.* (1995) **42** 569–578.

[36] Whittaker, E.T. and Watson, G.N., *A Course of Modern Analysis*, Cambridge University Press, London, 1927.

[37] Zaheer, K. and Zubairy, M.S., Phase sensitivity in atom-field interaction via coherent superposition. *Phys. Rev.* A (1989) **39** 2000–2004.

[38] Gea-Banacloche, J., Atom- and field-state evolution in the Jaynes–Cummings model for large initial fields. *Phys. Rev.* A (1991) **44** 5913–5931.

[39] Chumakov, S.M., Klimov, A.B., and Saavedra, C., Competing interactions and quantum nonspreading wave packets. *Phys. Rev.* A (1995) **52** 3153–3156.

[40] Woods, C.W. and Gea-Banacloche, J., Squeezing in the Jaynes–Cummings model for large coherent fields. *J. Mod. Opt.* (1993) **40** 2361–2379.

[41] Kuklinski, J.R. and Madajczyk, J.L., Strong squeezing in the Jaynes–Cummings model. *Phys. Rev.* A (1988) **37** 3175–3178.

[42] Averbukh, I.Sh., Fractional revivals in the Jaynes–Cummings model. *Phys. Rev.* A (1992) **46** R2205–R2208.

[43] Kitagawa, M. and Yamamaoto, Y., Number-phase minimum-uncertainty state with reduced number uncertainty in a Kerr nonlinear interferometer. *Phys. Rev.* A (1986) **34** 3974–3988.

[44] Milburn, G.J., Quantum and classical Liouville dynamics of the anharmonic oscillator. *Phys. Rev.* A (1986) **33** 674–685.

[45] Yurke, B. and Stoler, D., Generating quantum mechanical superposition of macroscopically distinguishable states via amplitude dispersion. *Phys. Rev. Lett.* (1986) **57** 13–16.

[46] Dodonov, V.V., Malkin, I.A., and Man'ko, V.I., Even and odd coherent states and

excitations of a singular oscillator. *Physica* (1974) **72** 597–615.

[47] Chumakov, S.M. and Klimov, A.B., *Algebraic Methods in Quantum Optics*. Universidad de Guadalajara, Guadalajara, Mexico, 1999.

[48] Bloch, F. and Siegert, A., Magnetic resonance for nonrotating fields. *Phys. Re* (1940) **57** 522–527.

[49] Shirley, J.H., Solution of the Schrödinger equation with a Hamiltonian periodic in time. *Phys. Rev.* (1965) **138** B979–B987.

[50] Cohen-Tannoudji, C., Dupont-Roc, J., and Fabre, C., A quantum calculation of th higher order terms in the Bloch–Siegert shift. *J. Phys.* B (1973) **6** L214–L217.

[51] Yabuzaki, T., Nakayama, S., Murakami, Y., and Ogawa, T., Interaction between spin-1/2 atom and a strong rf field. *Phys. Rev.* A (1974) **10** 1955–1963.

[52] Carmichael, H.J. and Walls, D.F., A quantum–mechanical master equation trea ment of the dynamical Stark effect. *J. Phys.* B (1976) **9** 1199–1219.

[53] Puri, R.R. and Bullough, R.K., Quantum electrodynamics of an atom making two photon transitions in an ideal cavity. *J. Opt. Soc. Am.* B (1988) **5** 2021–2028.

[54] Lugiato, L.A., Galatola, P., and Narducci, L.M., Universal normal-form descriptio of squeezing and instabilities in two-photon processes. *Opt. Commun.* (1990) **7** 276–286.

[55] Toor, A.H. and Zubairy, M.S., Validity of the effective Hamiltonian in the two photon atom-field interation. *Phys. Rev.* A (1992) **45** 4951–4959.

[56] Feranchuk, I.D., Komarov, L.I., and Ulyanenkov, A.P., Two-level system in a one mode quantum field: Numerical solution on the basis of the operator method. *J. Phys.* A (1996) **29** 4035–4047.

[57] Ng, K.M., Lo, C.F., and Liu, K.L., Exact eigenstates of the two-photon Jaynes Cummings model with the counter-rotating term. *Eur. Phys. J.* D (1999) **6** 119–12

[58] Ng, K.M., Lo, C.F., and Liu, K.L., Exact eigenstates of the intensity-depende Jaynes–Cummings model with the counter-rotating term. *Physica* A (2000) **27** 463–474.

[59] Tur, E.A., Jaynes–Cummings model: Solutions without rotating-wave approxima tion. *Opt. Spectrosc.* (2000) **89** 574–588.

[60] Debergh, N. and Klimov, A.B., Quasi-exactly solvable approach to the Jaynes Cummings model without rotation wave approximation. *Int. J. Mod. Phys.* A (200 **16** 4057–4068.

[61] Kozierowski, M. and Chumakov, S.M., Exactly solvable Dicke models and radiatio dragging. *J. Russ. Laser Research* (1994) **15** 118–135.

[62] Walls, D.F. and Barakat, R., Quantum-mechanical amplification and frequency con version with a trilinear Hamiltonian. *Phys. Rev.* A (1970) **1** 446–453.

[63] Bandilla, A., Drobný, G., and Jex, I., Nondegenerate parametric interactions an nonclassical effect. *Phys. Rev.* A (1996) **53** 507–516.

[64] Vilenkin, N.Ya. and Klimyk, A.U., *Representations of Lie Groups and Specia Functions*, Vols. 1–3. Kluwer, Dordrecht, 1991.

[65] Kozierowski, M., Mamedov, A.A., and Chumakov, S.M., Spontaneous emission b a system of N two-level atoms in terms of the $SU(2)$-group representations., *Phy Rev.* A (1990) **42** 1762–1766.

[66] Kozierowski, M., Chumakov, S.M., Światłowski, J., and Mamedov, A.A., Collectiv collapses and revivals in spontaneous emission of a partially inverted system of two level atoms: Analytical solution. *Phys. Rev.* A (1992) **46** 7220–7227.

[67] Chumakov, S.M. and Kozierowski, M., Dicke model: quantum nonlinear dynamic and collective phenomena. *Quant. Semiclass. Opt.* (1996) **8** 775–803.

[68] Averbukh, I.S. and Perel'man, N.F., The dynamics of wave packets of highly

excited states of atoms and molecules. *Uspekhi Fiz. Nauk* (1991) **161** 41–81 [*Sov. Phys. - Uspekhi* (1991) **34** 572–591].

[69] Kozierowski, M. and Chumakov, S.M., Photon statistics in spontaneous emission of the Dicke model in a lossless cavity and the generation of the Fock state. *Phys. Rev.* A (1995) **52** 4194–4201.

[70] Satyanarayana, M.V., Rice, P., Vyas, R., and Charmichael, H.J., Ringing revivals in the interaction of a two-level atom with squeezed light. *J. Opt. Soc. Am.* B (1989) **6** 228–237.

[71] Milburn, G.J., Interaction of two-level atom with squeezed light. *Opt. Acta* (1984) **31** 671–679.

[72] Chumakov, S.M. and Sanchez-Mondragon, J.J., Atomic coherence vs field incoherence: Interaction of many atoms with a thermal quantized cavity field. *Opt. Commun.* (1994) **107** 231–234.

[73] Kozierowski, M. and Chumakov, S.M., Squeezed vacuum as an accelerator of revivals in the Dicke model. *Phys. Rev.* A (1995) **52** 4293–4296.

[74] Scharf, G., Time evolution of a quantum mechanical maser model. *Ann. Phys.* (NY) (1974) **83** 71–102.

[75] Karassiov, V.P. and Klimov, A.B., An algebraic appoach to solving evolution problems in some nonlinear quantum models. *Phys. Lett.* A (1994) **189** 43–51.

[76] Karassiov, V.P., $sl(2)$ variational schemes for solving one class of nonlinear quantum models. *Phys. Lett.* A (1998) **238** 19–28.

[77] Karassiov, V.P., Gusev, A.A., and Vinitsky, S.I., Polynomial Lie algebra methods in solving the second-harmonic generation model: some exact and approximate calculations. *Phys. Lett.* A (2002) **295** 247–255.

[78] Góra, P. and Jedrzejek, C., Nonlinear Jaynes–Cummings model. *Phys. Rev.* A (1992) **45** 6816–6828.

[79] Kozierowski, M., Thermal and squeezed vacuum Jaynes–Cummings models with a Kerr medium. *J. Mod. Opt.* (2001) **48** 773–781.

[80] Kozierowski, M. and Chumakov, S.M., On regular dynamics of the off-resonant Jaynes–Cummings model. *Jemná Mechanika a Optika* (1996) **41** 334–337.

[81] Kozierowski, M. and Chumakov, S.M., Revivals in the off-resonant Jaynes-Cummings model with a Kerr medium. *Acta Physica Slovaca* (1997) **47** 307–310.

[82] Chumakov, S.M., Klimov, A.B., and Sanchez-Mondragon, J.J., Collective atomic dynamics in a strong quantum field. *Opt. Commun.* (1995) **118** 529–536.

[83] Chumakov, S.M., Klimov, A.B., and Sanchez-Mondragon, J.J., General properties of quantum optical systems in a strong field limit. *Phys. Rev.* A (1994) **49** 4972–4978.

[84] Saavedra, C., Klimov, A.B., Chumakov, S.M., and Retamal, J.C., Dissipation in collective interaction. *Phys. Rev.* A (1998) **58** 4078–4086.

[85] Gea-Banacloche, J., Jaynes–Cummings model with quasiclassical fields: The effect of dissipation. *Phys. Rev.* A (1993) **47** 2221–2234.

[86] Klimov, A.B., Chumakov, S.M., Retamal, J.C., and Saavedra, C., An algebraic approach to the Jaynes–Cummings model with dissipation. *Phys. Lett.* A (1996) **211** 143–147.

[87] Briegel, H.-J. and Englert, B.-G., Quantum optical master equation: The use of damping basis. *Phys. Rev.* A (1993) **47** 3311–3329.

[88] van Wonderen, A.J., Exact solution of the Jaynes–Cummings model with cavity damping. *Phys. Rev.* A (1997) **56** 3116–3128.

[89] Stefanescu, E., Sandulescu, A., and Scheid, W., The collisional decay of a Fermi system interacting with a many-mode electromagnetic field. *Int. J. Mod. Phys.* E

(2000) **9** 17–50.

[90] Chumakov, S.M., Klimov, A.B., and Saavedra, C., Dispersive atomic evolution in a dissipative-driven cavity. *Phys. Rev.* A (2000) **61** art. 033814.

[91] Retamal, J.C., Saavedra, C., Klimov, A.B., and Chumakov, S.M., Transient squeezing by a collection of two-level atoms. *Phys. Rev.* A (1997) **55** 2413–2425.

[92] Smirnov, M.Z., A modified Jaynes–Cummings model for an atom interacting with a classical multifrequency field. *JETP* (1997) **85** 441–446.

[93] Napoli, A. and Messina, A., Nonclassical features in the dynamics of a new quadratic quantum model of the radiation-matter interaction in a confined space. *Quant. Semiclass. Opt.* (1997) **9** 587–597.

[94] Klimov, A.B. and Chumakov, S.M., Long-time behaviour of atomic inversion for the Jaynes–Cummings model in a strong thermal field. *Phys. Lett.* A (1999) **264** 100–102.

[95] Klimov, A.B., Negro, J., Farias, R., and Chumakov, S.M., Nonlinear dynamics of the two-photon Dicke model. *J. Opt.* B (1999) **1** 562–570.

[96] Iwasawa, H. and Matsuo, K., Exact solutions for two-photon Jaynes–Cummings model with dynamic Stark-shifts. *J. Phys. Soc. Japan* (1999) **68** 3439–3443.

[97] Rybin, A., Miroshnichenko, G., Vadeiko, I., and Timonen, J., Quantum dynamics of the intensity-dependent Tavis–Cummings model. *J. Phys.* A (1999) **32** 8739–8754.

[98] Law, C.K., Chen, T.W., and Leung, P.T., Jaynes–Cummings model in leaky cavities: An exact pure-state approach. *Phys. Rev.* A (2000) **61** 023808.

[99] Klimov, A.B. and Sanchez-Soto, L.L., Method of small rotations and effective Hamiltonians in nonlinear quantum optics. *Phys. Rev.* A (2000) **61** 063802.

[100] Miroshnichenko, G.P. and Smirnov, M.Z., Singularity and squeezing in the coupled system of dressed atom and quantized mode. *Opt. Commun.* (2000) **182** 393–401.

[101] Alscher, A. and Grabert, H., Semiclassical dynamics of the Jaynes–Cummings model. *Eur. Phys. J.* D (2001) **14** 127–136.

[102] Bishop, R.F. and Emary, C., Time evolution of the Rabi Hamiltonian from the unexcited vacuum. *J. Phys.* A (2001) **34** 5635–5651.

[103] Lindner, A., Freese, H., Quehl, G., and Reiss, D., Jaynes–Cummings–model with damping at resonance. *Eur. Phys. J.* D (2001) **17** 99–112.

[104] Ismailov, R.A. and Kazakov, A.Y., Two-photon Jaynes–Cummings systems interacting with a classical field. *JETP* (2001) **93** 1017–1027.

[105] Verlan, E.M. and Razumova, M.A., Correlated coherent states in the Jaynes–Cummings model: I. *Opt. Spectrosc.* (2001) **90** 928–934.

[106] Verlan, E.M. and Razumova, M.A., Excitation of coherent correlated states in the Jaynes–Cummings model by external deterministic forces: II. *Opt. Spectrosc.* (2001) **91** 735–740.

[107] Emary, C. and Bishop, R.F., Bogoliubov transformations and exact isolated solutions for simple nonadiabatic Hamiltonians. *J. Math. Phys.* (2002) **43** 3916–3926.

[108] Gerry, C.C., Conditional state generation in a dispersive atom-cavity field interaction with a continuous external pump field. *Phys. Rev.* A (2002) **65** 063801.

[109] Swain, S. and Ficek, Z., The damped and coherently-driven Jaynes–Cummings model. *J. Opt.* B (2002) **4** S328–S336.

[110] Daoud, M. and Hussin, V., General sets of coherent states and the Jaynes–Cummings model. *J. Phys.* A (2002) **35** 7381–7402.

[111] van Wonderen, A.J. and Lendi, K., Unification of the Jaynes–Cummings model and Planck's radiation law. *J. Math. Phys.* (2002) **43** 4692–4720.

Index

Milton Keynes UK
Ingram Content Group UK Ltd.
UKHW020015071024
449327UK00031B/2798